The Natural World

Scientiam non dedit natura semina scientiae nobis dedit
"Nature has given us not knowledge itself, but the seeds thereof."
Seneca

The Joy of Knowledge Encyclopaedia is affectionately
dedicated to the memory of John Beazley 1932–1977,
Book Designer, Publisher, and Co-Founder of the
publishing house of Mitchell Beazley Limited, by all
his many friends and colleagues in the company.

The Joy of Knowledge Library

General Editor: James Mitchell
With an overall preface by Lord Butler, Master of Trinity College,
University of Cambridge

Science and The Universe	Introduced by Sir Alan Cottrell, Master of Jesus College, University of Cambridge; and Sir Bernard Lovell, Professor of Radio Astronomy, University of Manchester
The Physical Earth	Introduced by Dr William Nierenberg, Director, Scripps Institution of Oceanography, University of California
The Natural World	Introduced by Dr Gwynne Vevers, Assistant Director of Science, the Zoological Society of London
Man and Society	Introduced by Dr Alex Comfort, Professor of Pathology, Irvine Medical School, University of California, Santa Barbara
History and Culture 1 **History and Culture** 2	Introduced by Dr Christopher Hill, Master of Balliol College, University of Oxford
Man and Machines	Introduced by Sir Jack Callard, former Chairman of Imperial Chemical Industries Limited
The Modern World	
Fact Index A–K	
Fact Index L–Z	

The Mitchell Beazley Joy of Knowledge Library

The Natural World

Introduced by Gwynne Vevers, MBE, MA, DPHIL, FLS, F I BIOL,
Assistant Director of Science, the Zoological Society of London

MITCHELL BEAZLEY

The Joy of Knowledge Library

Editorial Director	**Frank Wallis**
Creative Director	**Ed Day**
Project Director	**Harold Bull**

Volume editors
Science and The Universe — John Clark
Lawrence Clarke
The Natural World — Ruth Binney
The Physical Earth — Erik Abranson
Dougal Dixon
Man and Society — Max Monsarrat
History and Culture 1 & 2 — John Tusa
Roger Hearn
Time Chart — Jane Kenrick
Man and Machines — John Clark
Fact Index — Stephen Elliott
Stanley Schindler
John Clark

Art Director	Rod Stribley
Production Editor	Helen Yeomans
Assistant to the Project Director	Graham Darlow
Associate Art Director	Anthony Cobb
Art Buyer	Ted McCausland
Co-editions Manager	Averil Macintyre
Printing Manager	Bob Towell
Information Consultant	Jeremy Weston

Sub-Editors	Don Binney
	Arthur Butterfield
	Charyn Jones
	Jenny Mulherin
	Shiva Naipaul
	David Sharp
	Jack Tresidder
Proof-Readers	Jeff Groman
	Anthony Livesey
Researchers	Peter Furtado
	Malcolm Hart
	Peter Kilkenny
	Ann Kramer
	Lloyd Lindo
	Heather Maisner
	Valerie Nicholson
	Elizabeth Peadon
	John Smallwood
	Jim Somerville

Senior Designer	Sally Smallwood
Designers	Rosamund Briggs
	Mike Brown
	Lynn Cawley
	Nigel Chapman
	Pauline Faulks
	Nicole Fothergill
	Juanita Grout
	Ingrid Jacob
	Carole Johnson
	Chrissie Lloyd
	Aean Pinheiro
	Andrew Sutterby
Senior Picture Researchers	Jenny Golden
	Kate Parish
Picture Researchers	Phyllida Holbeach
	Philippa Lewis
	Caroline Lucas
	Ann Usborne

Assistant to the Editorial Director	Judy Garlick
Assistant to the Section Editors	Sandra Creese
Editorial Assistants	Joyce Evison
	Miranda Grinling
Production Controllers	Jeremy Albutt
	John Olive
	Anthony Bonsels
Production Assistants	Nick Rochez
	John Swan

ISBN 0 85533 105 4

Typesetting by Filmtype Services Limited, England
Photoprint Plates Ltd, Rayleigh, Essex, England

Printed in England by Sackville Smeets

Major contributors and advisers
to The Joy of Knowledge Library

Fabian Acker CEng, MIEE, MIMarE; Professor H.C. Allen MC; Leonard Amey OBE; Neil Ardley BSc; Professor H.R.V. Arnstein DSc, PhD, FIBiol; Russell Ash BA(Dunelm), FRAI; Norman Ashford PhD, CEng, MASCE, MCIT; Professor Robert Ashton; B.W. Atkinson BSc, PhD; Anthony Atmore BA; Professor Philip S. Bagwell BSc(Econ), PhD; Peter Ball MA; Edwin Banks MIOP; Professor Michael Banton; Dulan Barber; Harry Barrett; Professor J.P. Barron MA, DPhil, FSA; Professor W.G. Beasley FBA; Alan Bender PhD, MSc, DIC, ARCS; Lionel Bender BSc; Israel Berkovitch PhD, FRIC, MIChemE; David Berry MA; M.L. Bierbrier PhD; A.T.E. Binsted FBBI (Dipl); David Black; Maurice E.F. Block BA, PhD(Cantab); Richard H. Bomback BSc (London), FRPS; Basil Booth BSc(Hons), PhD, FGS, FRGS; J. Harry Bowen MA(Cantab), PhD(London); Mary Briggs MPS, FLS; John Brodrick BSc (Econ); J.M. Bruce ISO, MA, FRHistS, MRAeS; Professor D.A. Bullough MA, FSA, FRHistS; Tony Buzan BA(Hons) UBC; Dr Alan R. Cane; Dr J.G. de Casparis; Dr Jeremy Catto MA; Denis Chamberlain; E.W. Chanter MA; Professor Colin Cherry DSc(Eng), MIEE; A.H. Christie MA, FRAI, FRAS; Dr Anthony W. Clare MPhil(London), MB, BCh, MRCPI, MRCPsych; Sonia Cole; John R. Collis MA, PhD; Professor Gordon Connell-Smith BA, PhD, FRHistS; Dr A.H. Cook FRS; Professor A.H. Cook FRS; J.A.L. Cooke MA, DPhil; R.W. Cooke BSc, CEng, MICE; B.K. Cooper; Penelope J. Corfield MA; Robin Cormack MA, PhD, FSA; Nona Coxhead; Patricia Crone BA, PhD; Geoffrey P. Crow BSc(Eng), MICE, MIMunE, MInstHE, DIPTE; J.G. Crowther; Professor R.B. Cundall FRIC; Noel Currer-Briggs MA, FSG; Christopher Cviic BA(Zagreb), BSc(Econ, London); Gordon Daniels BSc(Econ, London), DPhil(Oxon); George Darby BA; G.J. Darwin; Dr David Delvin; Robin Denselow BA; Professor Bernard L. Diamond; John Dickson; Paul Dinnage MA; M.L. Dockrill BSc(Econ), MA, PhD; Patricia Dodd BA; James Dowdall; Anne Dowson MA(Cantab); Peter M. Driver BSc, PhD, MIBiol; Rev Professor C.W. Dugmore DD; Herbert L. Edlin BSc, Dip in Forestry; Pamela Egan MA(Oxon); Major S.R. Elliot CD, BComm; Professor H.J. Eysenck PhD, DSc; Dr Peter Fenwick BA, MB, BChir, DPM, MRCPsych; Jim Flegg BSc, PhD, ARCS, MBOU; Andrew M. Fleming MA; Professor Antony Flew MA(Oxon), DLitt(Keele); Wyn K. Ford FRHistS; Paul Freeman DSc(London); G.E. Fussell DLitt, FRHistS; Kenneth W. Gatland FRAS, FBIS; Norman Gelb BA; John Gilbert BA(Hons, London); Professor A.C. Gimson; John Glaves-Smith BA; David Glen; Professor S.J. Goldsack BSc, PhD, FINSTP, FBCS; Richard Gombrich MA, DPhil; A.F. Gomm; Professor A. Goodwin MA; William Gould BA(Wales); Professor J.R. Gray; Christopher Green PhD; Bill Gunston; Professor A. Rupert Hall LittD; Richard Halsey BA(Hons, UEA); Lynette K. Hamblin BSc; Norman Hammond; Professor Thomas G. Harding PhD; Richard Harris; Dr Randall P. Harrison; Cyril Hart MA, PhD, FRICS, FIFor; Anthony P. Harvey; Nigel Hawkes BA(Oxon); F.P. Heath; Peter Hebblethwaite MA(Oxon), LicTheol; Frances Mary Heidensohn BA; Dr Alan Hill MC, FRCP; Robert Hillenbrand MA, DPhil; Professor F.H. Hinsley; Dr Richard Hitchcock; Dorothy Hollingsworth OBE, BSc, FRIC, FIBiol, FIFST, SRD; H.P. Hope BSc (Hons, Agric); Antony Hopkins CBE, FRCM, LRAM, FRSA; Brian Hook; Peter

Howell BPhil, MA(Oxon); Brigadier K. Hunt; Peter Hurst BDS, FDS, LDS, RSCEd, MSc(London); Anthony Hyman MA, PhD; Professor R.S. Illingworth MD, FRCP, DPH, DCH; Oliver Impey MA, DPhil; D.E.G. Irvine PhD; L.M. Irvine BSc; Anne Jamieson cand mag(Copenhagen), MSc(London); Michael A. Janson BSc; Professor P.A. Jewell BSc(Agric), MA, PhD, FIBiol; Hugh Johnson; Commander I.E. Johnston RN; I.P. Jolliffe BSc, MSc, PhD, CompICE, FGS; Dr D.E.H. Jones ARCS, FCS; R.H. Jones PhD, BSc, CEng, MICE, FGS, MASCE; Hugh Kay; Dr Janet Kear; Sam Keen; D.R.C. Kempe BSc, DPhil, FGS; Alan Kendall MA(Cantab); Michael Kenward; John R. King BSc(Eng), DIC, CEng, MIProdE; D.G. King-Hele FRS; Professor J.F. Kirkaldy DSc; Malcolm Kitch; Michael Kitson MA; B.C. Lamb BSc, PhD; Nick Landon; Major J.C. Larminie QDG, Retd; Diana Leat BSc(Econ), PhD; Roger Lewin BSc, PhD; Harold K. Lipset; Norman Longmate MA(Oxon); John Lowry; Kenneth E. Lowther MA; Diana Lucas BA(Hons); Keith Lye BA, FRGS; Dr Peter Lyon; Dr Martin McCauley; Sean McConville BSc; D.F.M. McGregor BSc, PhD(Edin); Jean Macqueen PhD; William Baird MacQuitty MA(Hons), FRGS, FRPS; Jonathan Martin MA; Rev Canon E.L. Mascall DD; Christopher Maynard MSc, DTh; Professor A.J. Meadows; J.S.G. Miller MA, DPhil, BM, BCh; Alaric Millington BSc, DipEd, FIMA; Peter L. Moldon; Patrick Moore OBE; Robin Mowat MA, DPhil; J. Michael Mullin BSc; Alistair Munroe BSc, ARCS; Professor Jacob Needleman; Professor Donald M. Nicol MA, PhD; Gerald Norris; Caroline E. Oakman BA(Hons, Chinese); S. O'Connell MA(Cantab), MInstP; Michael Overman; Di Owen BSc; A.R.D. Pagden MA, FRHistS; Professor E.J. Pagel PhD; Carol Parker BA(Econ), MA(Internat. Aff.); Derek Parker; Julia Parker DFAstrolS; Dr Stanley Parker; Dr Colin Murray Parkes MD, FRC(Psych), DPM; Professor Geoffrey Parrinder MA, PhD, DD(London), DLitt(Lancaster); Moira Paterson; Walter C. Patterson MSc; Sir John H. Peel KCVO, MA, DM, FRCP, FRCS, FRCOG; D.J. Penn; Basil Peters MA. MInstP, FBIS; D.L. Phillips FRCR, MRCOG; B.T. Pickering PhD, DSc; John Picton; Susan Pinkus; Dr C.S. Pitcher MA, DM, FRCPath; Alfred Plaut FRCPsych; A.S. Playfair MRCS, LRCP, DObstRCOG; Dr Antony Polonsky; Joyce Pope BA; B.L. Potter NDA, MRAC, CertEd; Paulette Pratt; Antony Preston; Frank J. Pycroft; Margaret Quass; Dr John Reckless; Trevor Reese BA, PhD, FRHistS; Derek A. Reid BSc, PhD; Clyde Reynolds BSc; John Rivers; Peter Roberts; Colin A. Ronan MSc, FRAS; Professor Richard Rose BA(Johns Hopkins), DPhil(Oxon); Harold Rosenthal; T.G. Rosenthal MA(Cantab); Anne Ross MA, MA(Hons, Celtic Studies), PhD(Archaeol and Celtic Studies, Edin); Georgina Russell MA; Dr Charles Rycroft BA(Cantab), MB(London), FRCPsych; Susan Saunders MSc(Econ); Robert Schell PhD; Anil Seal MA, PhD(Cantab); Michael Sedgwick MA(Oxon); Martin Seymour-Smith BA(Oxon), MA(Oxon); Professor John Shearman; Dr Martin Sherwood; A.C. Simpson BSc; Nigel Sitwell; Dr Alan Sked; Julie and Kenneth Slavin FRGS, FRAI; Alec Xavier Snobel BSc(Econ); Terry Snow BA, ATCL; Rodney Steel; Charles S. Steinger MA, PhD; Geoffrey Stern BSc(Econ); Maryanne Stevens BA(Cantab), MA(London); John Stevenson DPhil, MA; J. Stidworthy MA; D. Michael Stoddart BSc, PhD; Bernard Stonehouse DPhil, MA, BSc, MInstBiol; Anthony Storr FRCP, FRCPsych; Richard Storry; Professor John Taylor; John W.R. Taylor FRHistS, MRAeS, FSLAET; R.B. Taylor BSc(Hons, Microbiol); J. David Thomas MA, PhD; Harvey Tilker PhD; Don Tills PhD, MPhil, MIBiol, FIMLS; Jon Tinker; M. Tregear MA; R.W. Trender; David

Trump MA, PhD, FSA; M.F. Tuke PhD; Christopher Tunney MA; Laurence Urdang Associates (authentication and fact check); Sally Walters BSc; Christopher Wardle; Dr D. Washbrook; David Watkins; George Watkins MSc; J.W.N. Watkins; Anthony J. Watts; Dr Geoff Watts; Melvyn Westlake; Anthony White MA(Oxon), MAPhil(Columbia); P.J.S. Whitmore MBE, PhD; Professor G.R. Wilkinson; Rev H.A. Williams CR; Christopher Wilson BA; Professor David M. Wilson; John B. Wilson BSc, PhD, FGS, FLS; Philip Windsor BA, DPhil(Oxon); Professor M.J. Wise; Roy Wolfe BSc(Econ), MSc; Dr David Woodings MA, MRCP, MRCPath; Bernard Yallop PhD, BSc, ARCS, FRAS; Professor John Yudkin MA, MD, PhD(Cantab), FRIC, FIBiol, FRCP.

The General Editor wishes particularly to thank the following for all their support:
Nicolas Bentley
Bill Borchard
Adrianne Bowles
Yves Boisseau
Irv Braun
Theo Bremer
the late Dr Jacob Bronowski
Sir Humphrey Browne
Barry and Helen Cayne
Peter Chubb
William Clark
Sanford and Dorothy Cobb
Alex and Jane Comfort
Jack and Sharlie Davison
Manfred Denneler
Stephen Elliott
Stephen Feldman
Orsola Fenghi
Dr Leo van Grunsven
Jan van Gulden
Graham Hearn
the late Raimund von Hofmansthal
Dr Antonio Houaiss
the late Sir Julian Huxley
Alan Isaacs
Julie Lansdowne
Andrew Leithead
Richard Levin
Oscar Lewenstein
The Rt Hon Selwyn Lloyd
Warren Lynch
Simon macLachlan
George Manina
Stuart Marks
Bruce Marshall
Francis Mildner
Bill and Christine Mitchell
Janice Mitchell
Patrick Moore
Mari Pijnenborg
the late Donna Dorita de Sa Putch
Tony Ruth
Dr Jonas Salk
Stanley Schindler
Guy Schoeller
Tony Schulte
Dr E. F. Schumacher
Christopher Scott
Anthony Storr
Hannu Tarmio
Ludovico Terzi
Ion Trewin
Egil Tveteras
Russ Voisin
Nat Wartels
Hiroshi Watanabe
Adrian Webster
Jeremy Westwood
Harry Williams
the dedicated staff of MB Encyclopaedias who created this Library and of MB Multimedia who made the IVR Artwork Bank.

The Natural World/Contents

Lord Butler, Master of Trinity College,
Cambridge, knocks on the great door of
the college during his installation
ceremony on October 7, 1965

Preface

I do not think any other group of publishers could be credited with producing so comprehensive and modern an encyclopaedia as this. It is quite original in form and content. A fine team of writers has been enlisted to provide the contents. No library or place of reference would be complete without this modern encyclopaedia, which should also be a treasure in private hands.

The production of an encyclopaedia is often an example that a particular literary, scientific and philosophic civilization is thriving and groping towards further knowledge. This was certainly so when Diderot published his famous encyclopaedia in the eighteenth century. Since science and technology were then not so far developed, his is a very different production from this. It depended to a certain extent on contributions from Rousseau and Voltaire and its publication created a school of adherents known as the encyclopaedists.

In modern times excellent encyclopaedias have been produced, but I think there is none which has the wealth of illustrations which is such a feature of these volumes. I was particularly struck by the section on astronomy, where the illustrations are vivid and unusual. This is only one example of illustrations in the work being, I would almost say, staggering in their originality.

I think it is probable that many responsible schools will have sets, since the publishers have carefully related much of the contents of the encyclopaedia to school and college courses. Parents on occasion feel that it is necessary to supplement school teaching at home, and this encyclopaedia would be invaluable in replying to the queries of adolescents which parents often find awkward to answer. The "two-page-spread" system, where text and explanatory diagrams are integrated into attractive units which relate to one another, makes this encyclopaedia different from others and all the more easy to study.

The whole encyclopaedia will literally be a revelation in the sphere of human and humane knowledge.

Butler

Master of Trinity College,
Cambridge

The Structure of the Library

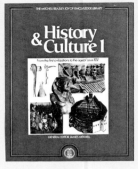

Science and The Universe	The Physical Earth	The Natural World	Man and Society	History and Culture
The growth of science	Structure of the Earth	How life began	Evolution of man	Volume 1 From the first
Mathematics	The Earth in perspective	Plants	How your body works	civilizations to the age of
Atomic theory	Weather	Animals	Illness and health	Louis XIV
Statics and dynamics	Seas and oceans	Insects	Mental health	
Heat, light and sound	Geology	Fish	Human development	The art of prehistory
Electricity	Earth's resources	Amphibians and reptiles	Man and his gods	Classical Greece
Chemistry	Agriculture	Birds	Communications	India, China and Japan
Techniques of astronomy	Cultivated plants	Mammals	Politics	Barbarian invasions
The Solar System	Flesh, fish and fowl	Prehistoric animals and	Law	The crusades
Stars and star maps		plants	Work and play	Age of exploration
Galaxies		Animals and their habitats	Economics	The Renaissance
Man in space		Conservation		The English revolution

The Natural World is a book of popular general knowledge about life on earth. It is a self-contained book with its own index and its own internal system of cross-references to help you to build up a rounded picture of nature.

The Natural World is one volume in Mitchell Beazley's intended ten-volume library of individual books we have entitled *The Joy of Knowledge Library*—a library which, when complete, will form a comprehensive encyclopaedia.

For a new generation brought up with television, words alone are no longer enough—and so we intend to make the *Library* a new sort of pictorial encyclopaedia for a visually oriented age, a new "family bible" of knowledge which will find acceptance in every home.

Seven other colour volumes in the *Library* are planned to be *Science and The Universe, The Physical Earth, Man and Society, History and Culture* (two volumes), *Man and Machines*, and *The Modern World*. *The Modern World* will be arranged alphabetically: the other volumes will be organized by topic and will provide a comprehensive store of general knowledge rather than isolated facts.

The last two volumes in the *Library* will provide a different service. Split up for convenience into A-K and L-Z references, these volumes will be a fact index to the whole work. They will provide factual information of all kinds on peoples, places and things through approximately 25,000 mostly short entries listed in alphabetical order. The entries in the A-Z volumes also act as a comprehensive index to the other eight volumes, thus turning the whole *Library* into a rounded *Encyclopaedia*, which is not only a comprehensive guide to general knowledge in volumes 1–7 but which now also provides access to specific information as well in *The Modern World* and the fact index volumes.

Access to knowledge

Whether you are a systematic reader or an unrepentant browser, my aim as General Editor has been to assemble all the facts you really ought to know into a coherent and logical plan that makes it possible to build up a comprehensive general knowledge of the subject.

Depending on your needs or motives as a reader in search of knowledge, you can find things out from *The Natural World* in four or more ways: for example, you can simply browse pleasurably about in its pages haphazardly (and that's my way!) or you can browse in a more organized fashion if you use our "See Also" treasure hunt system of connections referring you from spread to spread. Or you can gather specific facts by using the index. Yet again, you can set yourself the solid task of finding out literally everything in the book in logical order by reading it from cover to cover from page 1 to page 254: in this the Contents List (page 7) is there to guide you.

Our basic purpose in organizing the volumes in *The Joy of Knowledge Library* into two elements—the three volumes of A-Z factual information and the seven volumes of general knowledge—was functional. We devised it this way to make it easier to gather the two different sorts of information—simple facts and wider general knowledge, respectively—in appropriate ways.

The functions of an encyclopaedia

An encyclopaedia (the Greek word means "teaching in a circle" or, as we might say, the provision of a *rounded* picture of knowledge) has to perform these two distinct functions for two sorts of users, each seeking information of different sorts.

First, many readers want simple factual answers to simple factual questions, like "What is an elephant?" They may be intrigued to learn that an elephant is a mammal which in its adult prime can weigh up to 12 tons, that one of its tusks can weigh 236 lb, and that it may live as long as 77 years. Such direct and simple facts are best supplied by a short entry and in the *Library* they will be found in the two A-Z *Fact Index* volumes.

But secondly, for the user looking for in-depth knowledge on a subject or on a series of subjects—such as "What sort of animals live in Africa?"—short alphabetical entries alone are inevitably bitty and disjointed. What do you look up first—"animals"? "Africa"? "aardvark"? "anteater"? "elephant"? "hippopotamus"? "polar bear"?—and do you have to read all the entries or only some? You normally have to look up *lots* of entries in a purely alphabetical encyclopaedia to get a comprehensive answer to such wide-ranging questions. Yet comprehensive answers are what general knowledge is all about. A long article or linked series of longer articles, organized

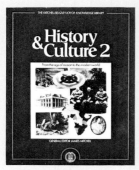

History and Culture

Volume 2 From the Age
of Reason to the
modern world

Neoclassicism
Colonizing Australasia
World War I
Ireland and independence
Twenties and the
    depression
World War II
Hollywood

Man and Machines

The growth of
    technology
Materials and techniques
Power
Machines
Transport
Weapons
Engineering
Communications
Industrial chemistry
Domestic engineering

The Modern World

Almanack
Countries of the world
Atlas
Gazetteer

Fact Index A-K

The first of two volumes
containing 25,000 mostly
short factual entries
on people, places and
things in A-Z order. The
Fact Index also acts as
an index to the eight
colour volumes. In
this volume, everything
from Aachen to Kyzyl.

Fact Index L-Z

The second of the A-Z
volumes that turn the
Library into a complete
encyclopaedia. Like the
first, it acts as an
index to the eight
colour volumes. In this
volume, everything from
Ernest Laas to Zyrardow.

by related subjects, is clearly much more helpful to the
person wanting such comprehensive answers. That is why
we have adopted a logical, so-called *thematic* organization
of knowledge, with a clear system of connections relating
topics to one another, for teaching general knowledge in
The Natural World and the six other general knowledge
volumes in the *Library*.

The spread system
The basic unit of all the general knowledge books is the
"spread"—a nickname for the two-page units that
comprise the working contents of all these books. The
spread is the heart of our approach to explaining things.

Every spread in *The Natural World* tells a story—almost
always a self-contained story—a story on the evolution of
life, for example (pages 22 to 23) or the way a cell works
(pages 26 to 27) or fossils (pages 174 to 175) or the way in
which the plant kingdom is classified (pages 38 to 39). The
spreads on these subjects all work to the same discipline,
which is to tell you all you need to know in two facing
pages of text and pictures. The discipline of having to get in
all the essential and relevant facts in this comparatively
short space actually makes for better results—text that has
to get to the point without any waffle, pictures and
diagrams that illustrate the essential points in a clear and
coherent fashion, captions that really work and explain the
point of the pictures.

The spread system is a strict discipline but once you get
used to it, I hope you'll ask yourself why you ever thought
general knowledge could be communicated in any other way.

The structure of the spread system will also, I hope
prove reassuring when you venture out from the things you
do know about into the unknown areas you don't know,
but want to find out about. There are many virtues in
being systematic. You will start to feel at home in all sorts
of unlikely areas of knowledge with the spread system to
guide you. The spreads are, in a sense, the building blocks
of knowledge. Like living cells which are the building
blocks of plants and animals, they are systematically
"programmed" to help you to learn more easily and to
remember better. Each spread has a main article of 850
words summarising the subject. The article is illustrated by
an average of ten pictures and diagrams, the captions of

which both complement *and* supplement the information in
the article (so please read the captions, incidentally, or you
may miss something!). Each spread, too, has a "key"
picture or diagram in the top right-hand corner. The
purpose of the key picture is twofold: it summarises the
story of the spread visually and it is intended to act as a
memory stimulator to help you to recall all the integrated
facts and pictures on a given subject.

Finally, each spread has a box of connections headed
"See Also" and, sometimes, "Read First". These are
cross-reference suggestions to other connecting spreads.
The "Read Firsts" normally appear only on spreads with
particularly complicated subjects and indicate that you
might like to learn to swim a little in the elementary
principles of a subject before being dropped in the deep
end of its complexities.

The "See Alsos" are the treasure hunt features of *The
Joy of Knowledge* system and I hope you'll find them
helpful and, indeed, fun to use. They are also essential if
you want to build up a comprehensive general knowledge.
If the spreads are individual living cells, the "See Alsos"
are the secret code that tells you how to fit the cells
together into an organic whole which is the body of
general knowledge.

Level of readership
The level for which we have created *The Joy of Knowledge
Library* is intended to be a universal one. Some aspects of
knowledge are more complicated than others and so readers
will find that the level varies in different parts of the
Library and indeed in different parts of this volume,
The Natural World. This is quite deliberate: *The Joy of
Knowledge Library* is a library for all the family.

Some younger people should be able to enjoy and to
absorb most of the pages in this volume on the animal
world, for example, from as young as ten or eleven
onwards—but the level has been set primarily for adults
and older children who will need some basic knowledge to
makes sense of the pages on basic cell structure or DNA,
for example.

Whatever their level, the greatest and the bestselling
popular encyclopaedias of the past have always had one
thing in common—simplicity. The ability to make even

Main text Here you will find an 850-word summary of the subject.

Connections "Read Firsts" and "See Alsos" direct you to spreads that supply essential background information about the subject.

Illustrations Cutaway artwork, diagrams, brilliant paintings or photographs that convey essential detail, re-create the reality of art or highlight contemporary living.

Annotation Hard-working labels that identify elements in an illustration or act as keys to descriptions contained in the captions.

A typical spread Text and pictures are integrated in the presentation of comprehensive general knowledge on the subject.

Captions Detailed information that supplements and complements the main text and describes the scene or object in the illustration.

Key The illustration and caption that sum up the theme of the spread and act as a recall system.

complicated subjects clear, to distil, to extract the simple principles from behind the complicated formulae, the gift of getting to the heart of things: these are the elements that make popular encyclopaedias really useful to the people who read them. I hope we have followed these precepts throughout the *Library*: if so our level will be found to be truly universal.

Philosophy of the Library
The aim of *all* the books—general knowledge and *Fact Index* volumes—in the *Library* is to make knowledge more readily available to everyone, and to make it fun. This is not new in encyclopaedias. The great classics enlightened whole generations of readers with essential information, popularly presented and positively inspired. Equally, some works in the past seem to have been extensions of an educational system that believed that unless knowledge was painfully acquired it couldn't be good for you, would be inevitably superficial, and wouldn't stick. Many of us know in our own lives the boredom and disinterest generated by such an approach at school, and most of us have seen it too in certain types of adult books. Such an approach locks up knowledge, not liberates it.

The great educators have been the men and women who have enthralled their listeners or readers by the self-evident passion they themselves have felt for their subjects. Their joy is natural and infectious. We remember what they say and cherish it for ever. The philosophy of *The Joy of Knowledge Library* is one that precisely mirrors that enthusiasm. We aim to seduce you with our pictures, absorb you with our text, entertain you with the multitude of facts we have marshalled for your pleasure—yes, *pleasure*. Why not pleasure?

There are three uses of knowledge: education (things you ought to know because they are important); pleasure (things which are intriguing or entertaining in themselves); application (things we can do with our knowledge for the world at large).

As far as education is concerned there are certain elementary facts we need to learn in our schooldays. The *Library*, with its vast store of information, is primarily designed to have an educational function—to inform, to be a constant companion and guide everyone through school,

college and other forms of higher education.

But most facts, except to the student or specialist (and these books are not only for students and specialists, they are for everyone) aren't vital to know at all. You don't *need* to know them. But discovering them can be a source of endless pleasure and delight, nonetheless, like learning the pleasures of food or wine or love or travel. Who wouldn't give a king's ransom to know when man really became man and stopped being an ape? Who wouldn't have loved to have spent a day at the feet of Leonardo or to have met the historical Jesus or to have been there when Stephenson's *Rocket* first moved? The excitement of discovering new things is like meeting new people—it is one of the great pleasures of life.

There is always the chance, too, that some of the things you find out in these pages may inspire you with a lifelong passion to apply your knowledge in an area which really interests you. My friend Patrick Moore, the astronomer, who first suggested we publish this *Library* and wrote much of the astronomy section in our volume on *Science and The Universe*, once told me that he became an astronomer through the thrill he experienced on first reading an encyclopaedia of astronomy called *The Splendour of the Heavens*, published when he was a boy. Revelation is the reward of encyclopaedists. Our job, my job, is to remind you always that the joy of knowledge knows no boundaries and can work untold miracles.

In an age when we are increasingly creators (and less creatures) of our world, the people who *know*, who have a sense of proportion, a sense of balance, above all perhaps a sense of insight (the inner as well as the outer eye) in the application of their knowledge, are the most valuable people on earth. They, and they alone, will have the capacity to save this earth as a happy and a habitable planet for all its creatures. For the true joy of knowledge lies not only in its acquisition and its enjoyment, but in its wise and loving application in the service of the world.

Thus the Latin tag "Scientiam non dedit natura, semina scientiae nobis dedit" on the first page of this book. It translates as "Nature has given us not knowledge itself, but the seeds thereof."

It is, in the end, up to each of us to make the most of what we find in these pages.

The Structure of this Book

The Natural World is a book of general knowledge containing all the information that we think is most interesting and relevant about life on Earth in all its many aspects. I haven't counted up exactly, but it must incorporate nearly 100,000 facts about nature in its 272 pages. Our intention has been to present these facts in words and pictures in such a way that they make sense and tell a logical, comprehensible and coherent story rather than appear in a meaningless jumble.

For many years we regarded the world about us as something to be exploited. Man the predator—perhaps the most predatory animal the world has ever known—burned forests to create farms, slaughtered elephants to provide billiard balls and chessmen, netted butterflies to decorate his parlours. The world, in those not-too-distant days, seemed a cornucopia. No matter how many trees were cut or burned, no matter how many animals died, there always seemed to be enough left.

But "enough" is a relative word, and it has become alarmingly apparent in the last two or three decades that man's uncontrolled cropping of the world's flora and fauna may have a snowball effect: unbalance Nature's balance and whole species begin to die out. Through a long chain, man may depend on those species for his own survival. The process was summed up a long time ago in the tale of the kingdom that was lost for want of a nail.

In The Natural World we have been conscious of that process. That is why you will find articles on endangered mammals, endangered birds, and destructive man. And we have tried to balance that picture with articles on the ways in which man is trying to curb his own destruction. Who triumphs—destructive man or constructive man—is a matter that concerns us all.

Where to start
Before outlining the plan of the contents of The Natural World I'm going to assume for a moment that you are coming to this subject, just as I came to it when planning the book, as a "know-nothing" rather than as a "know-all". Knowing nothing, incidentally, can be a great advantage as a reader—or as an editor, as I discovered in making this book. If you know nothing, but want to find things out, you ask difficult questions all the time.

I spent much of my time as General Editor of this Library asking experts difficult questions and refusing to be fobbed off with complicated answers I couldn't understand. The Natural World, like every other book in this Library, has thus been through the sieve of my personal ignorance in its attempt to re-state things simply and understandably.

If you know nothing, my suggestion is that you start with Gwynne Vevers' introduction to the subject (pages 16 to 19). He places the study of life in an historical setting and explains how it is that we have learned as much as we have about the subject and who the great discoverers and thinkers of the past and the present have been and are. If you find an historical perspective useful, I should start there. You'll find out about Gregor Mendel and Charles Darwin and Francis Crick. If you prefer to plunge straight into the facts, but don't have much basic knowledge, I suggest you study eight spreads in this book before anything else (see panel on page 14). These spreads are the "Read First" spreads. They will give you the basic facts about life and natural history and will provide you with a framework of essential information in its proper perspective. Once you have digested these spreads you can build up a more comprehensive general knowledge by proceeding to explore the rest of the book.

Plan of the Book
There are broadly three sections, or blocks of spreads, in The Natural World. The divisions between them are not marked in the text because we thought that would spoil the continuity of the book. They are:
How Life began and How We Study It.
Everyone wants to know how it all began. Primitive man explained the Creation through myth—the one most of us know is the story of the Garden of Eden—and was content to leave it at that; modern man regards the origin of life as one of the great mysteries and hence a fundamental challenge to his scientific skills.

Scientists now believe that the Earth was formed as a planet some 4,500 million years ago. About 1,000 million years later the first signs of life appeared, closely followed by the first plants, but it was not for 3,000 million years after that that the first animals to leave recognizable fossils came into being. We now have some idea of how

The Natural World, like most volumes in The Joy of Knowledge Library, tackles its subject topically on a two-page spread basis. Though the spreads are self-contained, you may find some of them easier to understand if you read certain basic spreads first. Those spreads are illustrated here. They are "scene-setters" that will give you an understanding of the fundamentals of life—its origins and continuity—and of how botanists and zoologists classify plants and animals. With them as background, the rest of the spreads in The Natural World can be more readily understood. The eight spreads are:

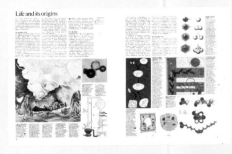

life began and what processes have moulded it and continue to mould it.

The principal process—evolution—will end only with the extinction of life itself. It acts across the depth and breadth of the living world, from the tiniest virus (it would take 250,000 of them to cover a full stop) to the blue whale which, at more than 100 tons, is the largest living mammal. A second process—heredity—ensures that the characteristics of the parents are passed on to the offspring.

Descriptions and Groupings of Living Things.
Life scientists divide living things into plants and animals. Because the plants existed before animals, The Natural World studies them first. Two spreads—"The plant kingdom" (pages 38 to 39) and "The way plants work" (pages 44 to 45)—set the scene for the seventeen spreads that deal with plants. We start with simple plants—yeasts and moulds—just as in the spreads on animals we start with the simplest animals—single-celled creatures such as the amoeba. The story then unfolds step by step through the ranks of plants without flowers, from seaweeds to conifers such as the giant California redwood. The following section deals with flowering plants (those with two seed leaves and those with one) and the section is drawn together by an explanation of how a flower works as a plant's reproductive system. The animal kingdom is divided into animals without backbones (the invertebrates) and those with backbones (the vertebrates). General anatomy, reproduction and behaviour are described before the section sets out to describe the invertebrates. We came across some fascinating facts in the course of compiling our articles on the invertebrates: for example, tapeworms can grow to more than 85 feet long; the insects that British spiders eat each year weigh more than all the people living in the British Isles; and some insects are equipped to "jam" the radar of bats.

Fish, amphibians, reptiles, birds and mammals make up the great groups of animals with backbones. The Natural World studies the enormous diversity and separate life-styles of these groups in detail. We have also tried to shed light, through recent advances in knowledge, on those aspects of vertebrate life that have long fascinated man—for example, how birds fly and migrate (pages 146 to 147),

how bats find their way in the dark (pages 160 to 161) and how whales communicate (pages 166 to 167).

Fish are the most successful vertebrates in terms of the sheer number of species. Once again, we came across some compelling facts: Pacific cod can produce eight million eggs at a time; predatory bony fish can cruise at speeds of up to six times their own body length a second; and the electric eel can generate up to 550 volts.

There are four spreads on amphibians and reptiles and then the section considers birds—how they are classified, their anatomy, how they reproduce and how they live and behave. The diversity of behaviour is enormous and ingenious: the emperor penguin can stand on ice at temperatures of below −60°C for 64 days, holding its egg on its feet to stop it from freezing; ravens can be taught to count up to seven; the honey buzzard often travels more than 6,000 miles in a single migration.

Of the 4,000 species of mammals, 42 per cent are rodents. Having discussed the behaviour of these and other mammals, including primates, the section takes a step back to consider the various ages in which life forms other than mammalian were supreme before it returns to the present—the age of mammals.

Where Living Things are Found.
The array of creatures great and small described in the first two sections would have the appearance of some museum collection if we did not show the plants and animals in their natural context. This is the function of the third section of The Natural World.

The spread called "Regions of the earth: zoogeography" (pages 192 to 193) shows how the Earth can be divided in terms of animals; the following spread, "The basis of ecology", shows how it can be divided on the basis of the plants it supports and how animal and plant communities function as food chains. With the help of panoramic drawings we then look at life in the world's grasslands, forests, woodlands, jungles and deserts; at life on mountains, tundra, polar ice caps and islands; in lakes, rivers, swamps and oceans. How to begin to describe the extraordinary beauty and wonder of this intricate web of life? Well, perhaps it is best not to spend any more time finding out how to get the most value from this book. Plunge in now—and remember when you read "Endangered

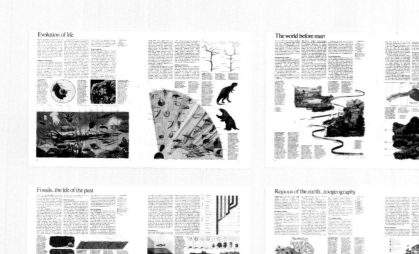

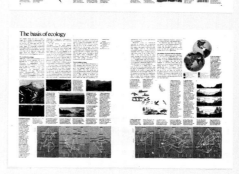

mammals", on pages 242 to 243, that we ourselves will be the most endangered mammal of all if we go on ruining our natural world with our vile pollution, our reckless elimination of the wild places of the Earth, our gross over-population, our aggressiveness and our greed.

The latest estimate is that there will be 6,000 million men, women and children alive—or half alive—on this planet by the year 2000. If we continue to mismanage nature, more than half of them might only have just enough food to survive. How much of the natural world will survive then?

The Natural World

Dr Gwynne Vevers, FLS,

Assistant Director of Science, the Zoological Society of London

Man's knowledge of life on earth has been built up over thousands of years, but it is only in the last 300 years or so that he has started to understand the basic principles that relate animals to one another and to the outside world and has collected the body of knowledge called biology. Unfortunately, in a short introduction, it is only possible to point out the important landmarks along the path of biological research.

Early man was a purely practical animal, intent on surviving in a variable, often hostile environment and on finding enough food for himself and his family. He must have known a fair amount about the habits of the animals he hunted and something about the plants around him. This knowledge cannot have stopped him being frightened when confronting the larger animals during hunting expeditions, or at night when the disturbing activities of nocturnal animals such as hyenas and owls could be sensed but not seen and fully understood. Trial and error must have soon taught him what to eat and what to avoid. He must also have found out something of the use of plants in the treatment of certain ailments, but unfortunately we know nothing of how this came about. So man's early knowledge of life on earth involved first hunting, then agriculture and the rudiments of medicine.

Great practical advances in these fields were made by some of the ancient civilizations, particularly those around the Mediterranean Sea. The rise of civilizations produced a relatively leisured class with time to observe animals and plants, to describe them objectively and to speculate on their functions, without the hocus-pocus of myths and taboos. The ancient Egyptians were skilled in certain aspects of human anatomy and also studied some of their sacred animals. For instance, they worked out the life history of the scarab beetle and knew about the metamorphosis of a tadpole into a frog. It was, however, the Greeks who really founded the natural sciences, with great numbers of men occupied in practising and teaching them. The genius of Pythagoras and Euclid endures to this day and, in medicine, Hippocrates (*c.* 460–*c.* 377 BC) laid the foundations for the study of anatomy and physiology.

Great as the contributions of these men were, it was probably Aristotle (384–322 BC) who was the greatest originator of what we now call biology or the life sciences. He was born in Stagira, a Greek settlement on the coast of Macedonia, and in middle life was tutor to the young Alexander. After moving to Athens he founded a school in 335 BC in the Lyceum, a building attached to a temple dedicated to Apollo Lycaeus. Aristotle and his pupils made many original observations on animals, particularly on the fish and other aquatic animals living in Greece and the surrounding waters. He produced a form of classification based on external and internal anatomical characteristics, and on the habits of the animals. He recognized that animals reproduce in different ways, some by sexual means, some asexually, but he thought that certain animals arose from putrefying matter by what has come to be known as spontaneous generation, an erroneous belief that lasted more than 1,900 years. Aristotle also wrote on rather more philosophical subjects such as sleep and waking, the soul and dreams, and colours and sounds. There were many other men of the period who were interested in life on earth, but perhaps none have had quite the impact of Aristotle. The Roman writer Pliny the Elder (AD 23–79) was one of these, but his works, although full of interest, were more descriptive and in the nature of compilations.

The Renaissance saw great advances in biology, as indeed in so many other fields. Leonardo da Vinci (1452–1519) had wide interests in the natural sciences as well as in the arts. He even studied fossils and decided that they were animal remains. The sixteenth century was marked by a series of great anatomists such as Andreas Vesalius (1514–64), who worked most of his life in Italy although he was actually born near Brussels. One of the most important scientific events of the seventeenth century was the discovery by William Harvey (1578–1657) of the circulation of the blood. His book on this subject was published at Frankfurt am Main in 1628, the result of many years of careful experiment and observation. It is difficult for us to realize the impact of this work on scientific thought at the time. Men like Harvey and his contemporaries in Europe were few and far between and they were wallowing in a vast sea of ignorance. It was only about this time that the Florentine physician Francesco Redi (1626–97) showed that flies did not arise from putrefying matter by a process of spontaneous generation. He carried out a controlled experiment, something that would have been unthinkable in the Middle Ages. He took two sets of dishes containing meat, leaving one set open to the air and carefully covering the other. The first developed maggots from the eggs laid by visiting bluebottles, the second produced no maggots as the flies could not reach the meat and lay their eggs.

Travel at this time was difficult but those who did manage to move far afield brought back tales that were embroidered to begin with and became even more elaborate in the retelling. As methods of transport improved, the stories of animals and plants in distant parts of the world gradually became more accurate and exploration in the sixteenth and seventeenth centuries yielded a vast body of knowledge on natural history. The minds of intelligent men and women turned from sterile myths to recording and describing these natural objects and even to thinking about how they grew and reproduced. In England the late seventeenth century saw the appearance of a small number of people who could truly be called naturalists. One of these was John Ray (1627–1705), the son of a blacksmith at Black Notley, near Braintree in Essex. About 1662, Ray started a series of tours throughout parts of England and Wales, in which he compiled detailed lists of the plants he found. A few years later, in collaboration with Francis Willughby (1635–72), a Warwickshire squire, Ray was planning a detailed description of the organic world. Willughby's early death brought this scheme to an end, but Ray continued to

Portrait of a predator—the large eyes of the snowy owl (*Nyctea scandica*) identify prey; the sharp beak kills it. Plume-like feathers protect the owl's nostrils from the freezing Arctic air.

publish works on plants, insects, fish and birds, and also made a first attempt at producing a systematic account of some animal and plant groups. He was an extremely accurate observer who may rightly be regarded as the founder of natural history in Britain.

If Ray dominated the field in the late seventeenth century, there is no doubt that the Swede Carl Linné (1707–78), usually known as Linnaeus, dominated the latter half of the eighteenth century. Linnaeus travelled widely in Europe and amassed vast collections of animals and plants. Most of these priceless finds are still preserved at the Linnean Society of London. However, he was not just a collector but the author of the system of naming animals and plants that is still in use throughout the world. In this, the binomial system, every plant and animal is given just two names, instead of a sentence or two of wordy description, which had been the method previously used. The African elephant is thus known universally as *Loxodonta africana*, the Indian elephant as *Elephas maximus*. Not only does the binomial system of nomenclature help the academic biologist to trace the relationships of species and possibly the paths in their evolution, it also allows workers in different parts of the world to share an internationally accepted code. It is interesting to reflect that the backroom workers in botany and zoology have accepted and developed this rational code of nomenclature, whereas the manufacturers of so many articles used by man, such as motor cars and electrical fittings, continue to use a chaotic multiplicity of sizes and standards.

The followers of Linnaeus in various countries extended this system and their work has greatly widened man's knowledge of the living world. There is, indeed, no doubt that the systematic approach conceived by Linnaeus was one of the first major breakthroughs in modern biology. This was greatly helped by the collections brought back to Europe by sailors and explorers, which provided material on which naturalists could practise the Linnaean system. The pirates and buccaneers of the seventeenth century had been mere collectors of wealth, whereas the explorers of the eighteenth century were intelligent, cultivated men with a full measure of natural curiosity. Captain James Cook (1728–79), the son of an agricultural labourer near Whitby in Yorkshire, was one of these. His first expedition, which left England in 1768, was concerned primarily with recording the transit of Venus, which he observed in Tahiti on 3 June 1769. His ship, the *Endeavour*, also carried a team of naturalists and artists, led by Sir Joseph Banks (1743–1820), a wealthy patron whose resources enabled him to make large collections of animals and plants at the many places visited by the *Endeavour*. The expedition was particularly valuable because it visited parts of the Pacific Ocean that had never before been investigated scientifically. Cook made two further expeditions, between 1772–5 and 1776–9. On the last of these, in February 1779, he was killed in the Hawaiian Islands.

This was still essentially a period of collecting, describing and classifying, a task that had to be done before man could go on to find out more about life on earth. In Britain the Linnean Society of London, founded in 1788, and the first national society in Britain devoted purely to biological work, played a great part in the promotion of the study of natural history. The naturalist must have felt very much in his element during this time. There was time to go out and observe wild animals and plants. At Selbourne in Hampshire, the clergyman Gilbert White (1720–93) was recording with great accuracy the natural events in his immediate surroundings. The Flintshire squire Thomas Pennant (1726–98), to whom Gilbert White wrote many of his letters, was travelling farther afield, in Britain and in many parts of Europe, and was visiting other men of letters and science. This was a time when men of science were able to travel throughout Europe without passports, carrying perhaps a few letters of introduction. Pennant did this and left a diary that is full of information on what he saw and whom he visited. There were no large museums in the modern sense of the word, but many educated people had cabinets of curiosities, containing shells, insects, stuffed birds, parts of plants and many other natural objects. Pennant's diary tells about the ones he saw, often in considerable detail.

The process of spreading a knowledge of biology was much helped by the activities of societies such as the Linnean which held meetings for the discussion of new findings and published these in their own journals. The influence of such societies increased even more in the nineteenth century. In Britain one of the major societies founded at this time was the Zoological Society of London (1826), which not only established its well-known Zoological Gardens but also published an enormous amount of zoological research, mainly in the field of classification and morphology.

About this time the young Charles Darwin (1809–82) was naturalist on the voyage of HMS *Beagle* and during its famous voyage (1831–6) he collected notes and specimens from many parts of the world. After his return, he spent several years working on the results of this expedition; in fact it was not until a good 20 years later that he published his *The Origin of Species by Means of Natural Selection* (1859), a work that revolutionized biological thought. Darwin shared the concept of evolution by natural selection with another Victorian naturalist, Alfred Russel Wallace (1823–1913), and indeed they had collaborated in presenting papers on the subject to the Linnean Society at a meeting held on the evening of 1 July 1858. In view of the later impact of their work on biology throughout the world it is a wry thought that the President of the Linnean Society in his review of the year 1858 wrote: "The year which has passed . . . has not, indeed, been marked by any of those striking discoveries which at once revolutionize, so to speak, the department of science on which they bear; it is only at remote intervals that we can reasonably expect any sudden and brilliant innovation which shall produce a marked and permanent impress on the character of any branch of knowledge, or confer a lasting and important service on mankind."

While the theory of evolution by natural selection became the subject of controversy during the remainder of the nineteenth century, the investigation of animals and plants in the field still went on at an increasing pace. In the last half of the century, for example, the study of life in the sea started to flourish as never before. Hitherto, man's interest in the sea had been largely concerned with problems of navigation and with taking fish and other marine products from coastal waters to supply human needs. Now there was a desire to know more about life in the depths of the sea. The pioneer event in this area of study was the great expedition of HMS *Challenger*. This ship circumnavigated the world in the years 1872–6 and collected large numbers of specimens by trawl and dredge. There had been a few expeditions of this type in previous years, but on a much smaller scale, and there were others in the years to follow, as for example the German Valdivia Expedition in 1898–9. Since then, work in the field of marine biology has gone on in all parts of the world and has been largely based on specialized marine biological and fisheries laboratories equipped with ships that can sample the animals and plants of an area and also the water in which they live. On a smaller scale valuable work is also being done on the flora and fauna of inland fresh waters.

The advances in biology made during the twentieth century have been quite outstanding and there is no doubt that the volume of research carried out and published in the last 50 years far exceeds anything achieved before. Many of these advances have had an important influence on the health and well-being of man and of the animals and plants that he has to varying degrees domesticated for his own use. Early work on vitamins started just before the turn of the century and our first knowledge of hormones also dates from about this time. The term "hormone" is perhaps rather restrictive, for it denotes substances such as the thyroid, pituitary and ovarian hormones that act as internal, chemical messengers within the body. There are, however, many other substances that act as chemical messengers between one animal and another – for example, the minute traces of a chemical (a pheromone) that act as a sexual attractant to moths at distances of at least 3km.

The relatively new science of genetics stems from the experiments of the monk Gregor Mendel (1822–84). This work, published in 1865 in a rather obscure Moravian journal, was completely overlooked until 1900. After this there was a veritable explosion of research into the genetics of various plants and animals, particularly of the fruit-fly. The chromosomes in the cell nucleus, which can be seen under an ordinary compound microscope, were identified as the carriers of genetic material which is arranged in linear fashion, the locus of a given genetic factor being known as a gene. Later came the exciting work on nucleic acids that has explained how hereditary material is duplicated when a cell and its chromosomes divide. This, one of the major biological discoveries of the century, was the culmination of work that had begun in the 1920s. Although the breakthrough was achieved in the Cavendish Laboratory, University of Cambridge, it was the result of American and British collaboration. The American biophysicist, James Watson, working with Francis Crick of Britain, drew on X-ray data obtained by Maurice Wilkins, of King's College, London, to construct a model of the structure of deoxyribonucleic acid (DNA) – the now famous "double helix" – the self-replicating molecule that is the basis of all living things. Watson, Crick and Wilkins shared a Nobel prize for physiology and medicine in 1962.

By the application of the results of genetic research on domesticated plants and animals the biologist has been able to provide the farmer with new strains of cereals, cattle and other animals that have increased the yields from a given area and thus helped to feed the ever-increasing human population of the world.

Much of the progress in the more physiological fields has depended upon new techniques, such as tissue culture, X-ray crystallography and the electron microscope. The latter, in particular, has provided the biologist with a new dimension, for it enables him to interpret the ultrastructure of cells, giving detail that optical microscopes cannot provide.

Observations on the behaviour of animals have also taken on a more scientific dimension. We have, indeed, come a long way from the misleading myths of the Middle Ages to objective methods of recording, and sometimes interpreting, the activities of animals in the wild. In Europe and North America the rise in the standards of natural history research from about 1800 onwards was the result of work by many different kinds of people. Up until about 100 years ago there was really no such thing as a professional zoologist or botanist. Most of the naturalists were clergymen, medical men, farmers and country squires – men who had time to spare for this pursuit. The work of such people laid the foundations of modern field studies in ecology and animal behaviour. In these fields much has been learned about the relationships of animals and plants to their environment and to one another. Some behavioural studies have provided information of use in the interpretation of human behaviour.

There is, however, a serious danger that man's attempts to exploit his environment may be proceeding at such a pace that the animals and plants sharing the earth with him may suffer. In recent years there has been increasing concern about the effects of man's activities not only in actually destroying animals and plants but also – and this is perhaps more serious – in destroying natural habitats. In the present century many animals and plants have either become extinct or are seriously threatened. The precarious state of the larger whales is a good example. Our concern for endangered wildlife must not be based on sentiment or even on the aesthetic appeal of certain species. The conservation of the fauna and flora of the earth is a matter of common sense for, as ecological research in the field has shown, the continued existence of life on earth is dependent upon the maintenance of the complex interrelationships of plants and animals, including man. Breaks in the natural chain of events have repercussions that become increasingly difficult to foresee and to control. This is why conservation is so absolutely essential.

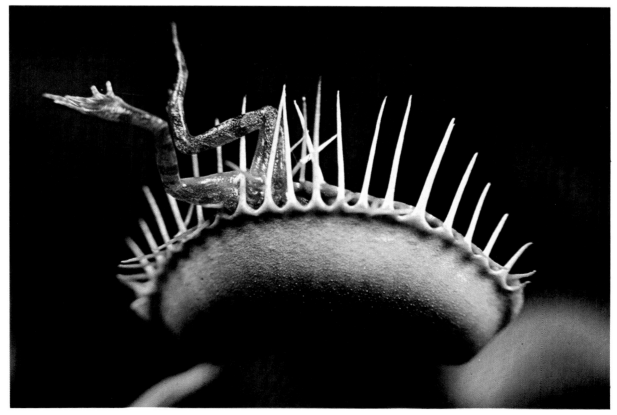

Snatching a meal– the Venus fly-trap (*Dionaea muscipula*) breaks a general rule of the natural world by being an animal-eating plant. It is unusual for it to seize a frog; insects are its normal diet.

Life and its origins

Life, it seems, should be easy to define – a horse, for instance is quite clearly alive, whereas a lump of rock is not. The nature of life is something that has puzzled biologists for centuries but the basic feature of all living organisms is their ability to produce identical copies of themselves [Key], given the correct supply of raw materials.

The essentials of life

The simplest of all living creatures [4] consist of a single living unit, the cell. More complex creatures, whether plants or animals, are made up of many hundreds, even millions, of cells but creatures that are alive all share a number of essential characteristics, as well as the all-important one of reproduction. These are movement [5], responsiveness to the environment, growth, and the ability to harness the energy sources of the environment to their own use; this they do through the action of molecules called enzymes, found within the structure of the cell.

Plants and animals, although they appear different, differ essentially only in the way in which the basic activities of life are carried out. Thus, animals move very obviously, whereas plants show complex and organized movement within their cells. Animals have sophisticated nervous systems with which to monitor their surroundings; plants are sensitive to stimuli such as light and gravity. And while plants use the sun's energy to synthesize many chemical components, ultimately all animals depend on plants as an energy source, whether they eat them directly or prey on others that do so.

For life to be maintained, a balance must exist between the energy-producing capacity of an organism and all its energy-utilizing functions – such as growth and movement and cell maintenance and repair. Within a living plant or animal, every enzyme system designed to construct new molecules in the body is balanced by other systems that break down molecules to release energy. The sum of these two systems represents the organism's metabolism.

In spite of the immense variety of living organisms, both in form and complexity, they are all fashioned out of the same kinds of molecular building blocks [6]: proteins, carbohydrates, nucleic acids and fats. Nucleic acids carry the genetic instructions that are passed from parent to offspring; proteins perform structural jobs and also take part as catalysts (enzymes) in the myriad chemical reactions that make an organism alive; carbohydrates and fats are sources of energy and building blocks for all types of organisms.

How life began

Establishing the origin of life depends on discovering how these chemicals were created. When the earth was created by cosmic events it was lifeless. Its noxious atmosphere and blazing hot temperature [1] were incapable of supporting organisms.

Before even the simplest life forms could begin to establish themselves on the maturing earth one essential step had to be taken: the evolution of the chemicals of life. This step – or, rather, countless series of random events – began the process by which the hostile atmosphere of hydrogen, methane, ammonia and water vapour of primitive earth evolved into a life-supporting cushion of oxygen, carbon dioxide and nitrogen.

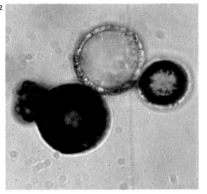

2 The most primitive cells to be formed on the early earth were probably simple structures known as protein spheroids. These have none of the detailed architecture of the single-celled organism. Laboratory experiments show that protein molecules suspended in water, heated and agitated will form tiny spheres roughly the same size as cells, with similar surrounding envelopes.

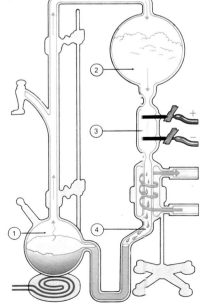

1 Some 4,000 million years ago conditions on earth were unsuitable for life. The atmosphere was composed of hydrogen, methane, ammonia and water vapour. There was very little – if any – oxygen in the atmosphere. The sun's ultra-violet rays blazed through this non-protecting layer on to the rocks and water below; volcanoes and thunderstorms were frequent; the heat was intense. The first step for the emergence of life was the creation of life-related molecules such as amino acids from the atmospheric gases. Energy was supplied by the sun, lightning, volcanoes and meteorites. The newly formed basic life molecules were polymerized by the sun's heat into primitive proteins, nucleic acid chains and carbohydrates.

3 Conditions on the primitive earth were re-created by the American scientists Miller and Urey. They mixed hydrogen, ammonia and methane gases [1]. Then these gases were further mixed with water vapour [2] and were subjected to electrical discharge [3], any liquid forming [4] being condensed back into the lower flask. This liquid was found to contain four amino acids commonly used in natural protein synthesis, some fatty acids and other life-related molecules.

The formation of carbohydrates, proteins, nucleic acids and fats probably took place as a result of favourable chemical conditions prevailing on the primitive earth. They almost certainly did not arrive on the earth conveniently pre-formed, as some Victorian scientists speculated, and there is now enough evidence to suggest that the atmosphere of the primitive earth certainly provided all the necessary ingredients to generate the more complex components of life-supporting molecules.

Scientists have managed to re-create in the laboratory the conditions they think existed on the primitive earth [3]. The first major experiment along these lines was performed in 1953 by Stanley Miller (1930–) and Harold Urey (1893–) at the University of Chicago. They passed an electric spark through a "primitive atmosphere" for one week. When they analysed the "soup" they had created they found a number of molecules characteristic of life, including four amino acids commonly present in proteins, a number of fatty acids and urea – another biologically important molecule.

Since then, representatives of all the classes of life-related molecules have been forged under harsh conditions similar to those that must once have been prevalent on our planet.

The synthetic chemistry of the maturing earth had to depend on natural energy resources such as ultra-violet light and heat from the sun, flashes of lightning, heat from volcanoes, radioactivity and the intense pressure and temperature generated when huge meteorites crashed into and became embedded in the earth's surface.

The primordial soup
Over millions of years there was a gradual build-up of vital fats, sugars, amino acids and nucleic acid components which formed "primordial soup". In order for life proper to begin, however, these substances had to be fused together.

The central part of the chemical evolution of life was the production of the nucleic acids because it is these molecules that have the ability to replicate themselves. This self-reproducing ability is crucial. Without it life would not exist or continue.

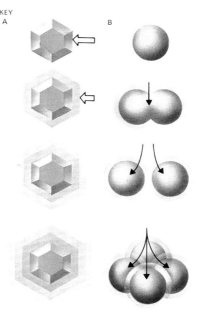

Growth and reproduction are two of the keys to life. A non-living crystal [A] can enlarge its size by aggregating more and more molecules to its surface. But this is not growth in the living sense. Organisms that are alive [B] grow via the essential process of biosynthesis: raw materials are obtained by the growing organism, broken down into simpler units, then reconstructed to fit in with the organism's demands. Non-living crystals, though they can grow, cannot perform the all-important "life" activity – unlike living cells they cannot split themselves into two identical offspring which can themselves grow and divide.

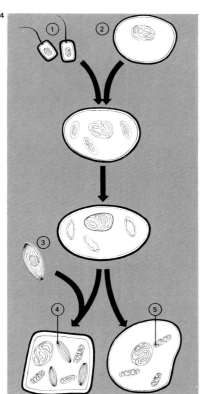

4 The first sophisticated single-celled organisms may have been produced by the aggregation of less complicated structures. The accidental engulfing of nucleic acids and enzymes (proteins) by a protein spheroid may have been the basis of a normal cell. A simple cell like this could probably carry out some basic chemical reactions to pursue simple life processes. A big advance could have been achieved if these basic cells [2] were to engulf other smaller cells such as primitive bacteria [1] or algae [3]. Some scientists believe that the sausage-shaped structures, mitochondria [5], which perform many of the energy-releasing reactions in the cells, are descendants of ancient bacteria engulfed in this way. Similarly, the green structures in plant cells, chloroplasts [4], may once have existed separately as algae.

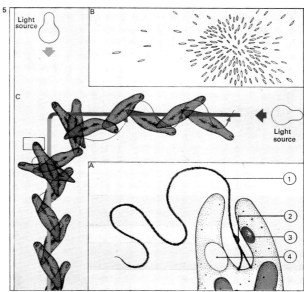
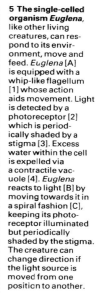

6 The molecules of life include proteins [A], carbohydrates [B] and fats [C]. Cell proteins act as building materials and as enzymes; carbohydrates and fats supply energy. Both are also structural.

5 The single-celled organism *Euglena*, like other living creatures, can respond to its environment, move and feed. *Euglena* [A] is equipped with a whip-like flagellum [1] whose action aids movement. Light is detected by a photoreceptor [2] which is periodically shaded by a stigma [3]. Excess water within the cell is expelled via a contractile vacuole [4]. *Euglena* reacts to light [B] by moving towards it in a spiral fashion [C], keeping its photoreceptor illuminated but periodically shaded by the stigma. The creature can change direction if the light source is moved from one position to another.

7 Movement of the ground substance or cytoplasm within cells is characteristic of life. In plants [A], movement of the cytoplasm sweeps the organelles within the cell into circular motion [1], a process known as cyclosis. Within an animal cell [B] the organelles, such as the mitochondria [2], may change shape or move quite independently. Or the cell nucleus [3] may spin continuously. In some human cells the nuclei rotate completely every 3.5 minutes.

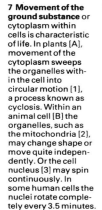
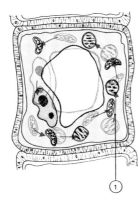
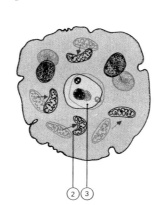

Grey — carbon
Red — oxygen
White — hydrogen
Blue — nitrogen

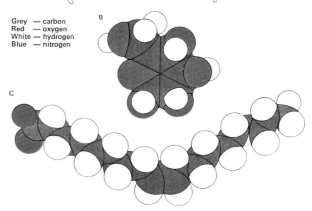

Evolution of life

The history of the earth is one of ceaseless change. This is particularly true of the plants and animals that inhabit its every corner, from mountain heights to ocean depths. The way living things have changed since life began is the story of evolution [Key, 4].

Life's beginnings

We do not know how life began, but it is almost certain that the first living organisms must have appeared in the ocean and fed on the organic molecules surrounding them, breaking them down to obtain their chemical energy without the help of oxygen. Perhaps more than a thousand million years later the important green chlorophyll pigments developed and enabled some organisms to create food substances from water and carbon dioxide using the energy of sunlight. So the first plants appeared, the "primary producers" or fixers of solar energy on which all other life forms depended.

The evidence that evolution has taken place is copious. The most irrefutable part of this evidence is the fossil record [3]. Fossils do not in themselves prove that evolution has occurred, for each one of them could have come into being quite independently. But fossils that have been recovered from successive geological eras show a distinct progression and also show that animals were adapted to cope with the conditions prevailing on earth during their lifetime [2].

Other evidence for evolution has come from the study of living animals and plants. Comparative anatomy of the limbs of animals with bony skeletons, for example, leaves little doubt that the hand of man has exact bony equivalents in the pectoral fin of a fish. Each is adapted for a particular way of life but is built to a fundamental plan that indicates fairly conclusively that fish and men share common ancestors. Embryology – the study of development – provides similar examples and so does animal behaviour. The hornbills of India and Africa, for instance, although they are different species, plaster up their nests in almost identical ways. At a biochemical level affinities of descent are shown by similarities in such features as the chemical composition of the blood proteins.

The overwhelming evidence of the fossil record that evolution has occurred as a process of gradual development by which animals and plants could compensate for adverse changes in their circumstances was not accepted until the late nineteenth century. Indeed, it was not until the eighteenth century that scientists and philosophers even thought of studying objectively the way in which species had arisen. The French philosopher Montesquieu (1689–1755) was one of the first people to put forward the idea that "in the beginning there were very few species and they have multiplied since". The adaptation of the newly discovered flying lemurs on Java suggested this to him.

Evolution by natural selection

Another Frenchman, the naturalist Georges Buffon (1707–88), was the first to suggest that ape and man have a common ancestry. Buffon, like Charles Darwin (1809–82) who followed him, incurred ridicule from his peers. It was Darwin however – and Alfred Russel Wallace (1823–1913) – who, in 1858, first drew attention to the process of evolution by natural selection, and in doing so took

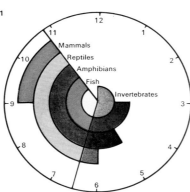

1 The time-scale of evolution can be compared to a 12-hour clock to clarify the stages involved. If midnight is regarded as the beginning of the clear fossil record about 600 million years ago, by the end of the Palaeozoic era, at 6.30, invertebrates, fish and amphibians were well established and reptiles had evolved. At the end of the Mesozoic, at 10.45, mammals had come to the fore. In the last 1.25 hours mammals have been dominant.

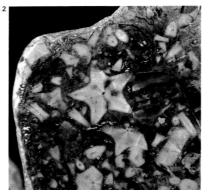

2 The star-like stem of an extinct crinoid or sea lily (Pentacrinus sp) is a common fossil in sedimentary Jurassic rocks. More than 5,000 fossil crinoids are known, the oldest of which dates from the Ordovician period over 430 million years ago. Their hard skeletons fossilized easily and most marine limestones contain fragments or even whole fossils, often in such quantities as to make up the bulk of the rock.

3 Although the primeval sea teemed with life notable absentees were the vast array of fish. Instead, during the Cambrian and Ordovician periods, there was a host of creatures without backbones. Most of these invertebrates are now extinct. Many others, including those illustrated here, have descendants still alive today. The crinoids are virtually unchanged but most of them are now extinct.

3   Graptolite   Jellyfish   Giant nautiloid   Clam   Coral   Brachiopod   Crinoid   Sea snail   Trilobite   Sea scorpion   Brittle star

the scientific world completely by storm.

Natural selection is an astonishingly simple process by which successful organisms survive while unsuccessful ones do not. Mixing of parental stock and changes (mutations) in genetic material ensure that in every species individuals begin life with small but important differences. In the battle for survival, competition for scarce resources tends to eliminate the weakest in favour of those that are strongest or most adaptable.

The species and classification

The process of evolution by natural selection is essentially a conservative one. Animals and plants have not changed over the millennia for the sake of changing but only to compensate for changes in their environmental circumstances. In species well adapted to a stable environment, most mutations are thus unfavourable and individuals carrying them soon die out because they cannot survive long enough to reproduce. In fluctuating environments adaptability is essential for survival.

The unit of evolution is the species because it represents the unit of mating.

Plants and animals of the same species can mate and produce viable offspring while organisms more distantly related cannot. The naming and grouping of species is designed not only for identification purposes but to reflect evolutionary relationships. Species that are presumed to have a single common ancestor are grouped together in a genus, and the Latin name or binomial for each consists of the generic name first, followed by the specific name. Cats are thus classified in the genus *Felis* of which there are several species such as *Felis pardalis*, the ocelot, and *Felis sylvestris*, the wild cat.

Related genera are grouped together into larger groups or families and several families together make up an order. Related orders are members of the same class and several classes form the largest practical unit of classification, the phylum. The ocelot is thus a member of the family Felidae, the order Carnivora, the class Mammalia and the phylum Chordata. This tells us a lot about its evolutionary history and relationships and the fossil record confirms it to be a descendant of a backboned carnivore.

Steps in the evolution of plants [A] have been the development of a vascular system [1], roots [2], leaves [3], cone seeds [4], frond seeds [5] and flowers [6]. Invertebrate animals [B] have evolved a gut [7], a body cavity [8], a complex internal structure [9] and segmented bodies [10]. Some are lacking external shells [11], others have legs [12].

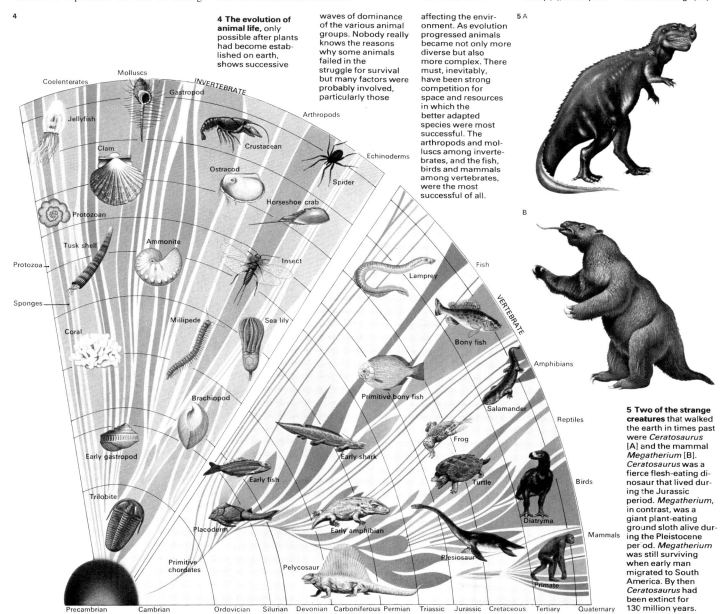

4

4 The evolution of animal life, only possible after plants had become established on earth, shows successive waves of dominance of the various animal groups. Nobody really knows the reasons why some animals failed in the struggle for survival but many factors were probably involved, particularly those affecting the environment. As evolution progressed animals became not only more diverse but also more complex. There must, inevitably, have been strong competition for space and resources in which the better adapted species were most successful. The arthropods and molluscs among invertebrates, and the fish, birds and mammals among vertebrates, were the most successful of all.

5 A

B

5 Two of the strange creatures that walked the earth in times past were *Ceratosaurus* [A] and the mammal *Megatherium* [B]. *Ceratosaurus* was a fierce flesh-eating dinosaur that lived during the Jurassic period. *Megatherium*, in contrast, was a giant plant-eating ground sloth alive during the Pleistocene period. *Megatherium* was still surviving when early man migrated to South America. By then *Ceratosaurus* had been extinct for 130 million years.

23

The world before man

The story of the earth's evolution, unfolding over a vast span of millennia, begins with an empty stage. Although the fabric of the earth's crust is some 4,600 million years old, the first stirrings of life did not disturb the barren expanses of its surface until about 1,000 million years after its formation. A further 3,000 million years were to elapse before the appearance of creatures that left recognizable fossil evidence.

The time scale of evolution

From their studies of the earth's crust, scientists have distinguished three broad geological eras following the long awakening of the Precambrian. They are the Palaeozoic (Greek for "ancient life"), Mesozoic ("middle life") and Cenozoic ("recent life"). Each is divided into richly diverse periods and the Cenozoic, spanning about 65 million years, is further subdivided into epochs.

Although the origin of life remains a subject for continuing speculation, it was not until the publication of Darwin's theory of evolution in 1859 that the argument became the province of scientists as well as philosophers. Modern palaeontologists, aided by technology, have confirmed much of his intelligent guesswork by accurate measurement. Great advances in particular have been made in the dating of fossil remains. The recent discovery of the basic genetic material known as deoxyribonucleic acid (DNA) has also increased our understanding of two contrasting mechanisms in evolution. One is the way in which species reproduce themselves faithfully and the other is the process by which new species of animals and plants come into being. The second involves mutation – minute changes to DNA instructions.

During the earliest period, the Precambrian, the earth was devoid of life for possibly 4,000 million years. However, although there was no oxygen in the atmosphere, the primeval oceans of that desert world already contained the basic constituents of life. Primitive organic structures such as bacteria and algae were the first to evolve, and their appearance, more than 3,500 million years ago, was the turning point in the history of earth, which had become inhabited.

Following on from early soft-bodied forms, the shelled creatures of the Cambrian period provided the earliest yield of fossil remains, the most numerous being the many-legged trilobites. It was not until the Ordovician period that the first fish-like vertebrates began to appear [1]. By the time that jawed fish had evolved, towards the end of the Silurian, marine plants were reaching the shore.

The first land dwellers

At last, as the Devonian period opened, there were living things on land as well as in the sea. This was a time of great topographical change. The crust of the earth rose and fell, throwing up huge mountain ranges, and the oceans advanced and receded several times, exposing mud that was rich in organic materials. As lush vegetation grew up to carpet bare rock, the first insects appeared. Then came the first vertebrate to emerge from the sea – the lungfish – and, by the end of the Devonian, amphibians had evolved.

In the Carboniferous period the reptiles developed. These new animals had better brains and physical systems than the amphi-

1 The age of fish began in the mid-Palaeozoic era about 400 million years ago. Before that time, there was neither ozone nor oxygen in the earth's atmosphere. Then strong-stemmed plants spread across the landscape, changing the environment for the lungfish and primitive amphibians that evolved from Devonian fish stock. These creatures were the first vertebrates to begin colonizing the land.

1 Dipterus
2 Pterichthys
3 Drepanaspis
4 Pteraspis
5 Ichthyostega

2 The reptiles were the dominant life form during the Mesozoic era between 225 and 65 millions years ago. At that time the land supported a lush vegetation of ferns and conifers, which were gradually replaced by a more modern flora of broad-leaved trees and flowering plants. Primitive birds, as well as reptiles, had evolved by the Jurassic. Early mammals were present but were insignificant.

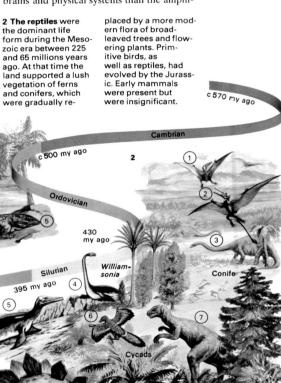

3 The major stages of life's evolution can be depicted as a ribbon of life spanning an interval of 600 million years. During that time a remarkable and ever-changing succession of life forms has populated the earth. Geological periods are the major divisions of this vast time span. They are shown to scale along the ribbon that winds through reconstructions [illustrations 1, 2 and 5] of three of the most significant eras, the Palaeozoic, Mesozoic and Cenozoic. The latter is the latest geological era, dating back 65 million years. The four most recent epochs are the Miocene, Pliocene, Pleistocene and Recent. During the Miocene the mountains of the Alps, Himalayas and Rockies were uplifted, temperate and polar regions cooled and grasslands replaced the forests. Grazing animals spread over the plains. In the Pliocene the world continued to cool, so tropical plants and animals retreated to lower latitudes. Camels, horses, antelopes and masto-dons lived on the plains of North America and Asia. During the Pleistocene an ice shield covered northern latitudes of America and Eurasia, advancing and receding over the plains four times. Men began farming, after the final retreat of the ice, about 10,000 years ago. Many gave up their former ways of life as wandering hunter-gatherers and set about establishing permanent agricultural settlements on the fertile land of the plain.

Protolepidodendron
Cyclostigma
Barrandeina
Psilophyton
Zosterophyllum
Sphenophyllum

c 570 my ago
Cambrian
c 500 my ago
Ordovician
430 my ago
Williamsonia
Silurian
395 my ago
Devonian
Carboniferous 345 my ago
Permian 280 my ago
Triassic 225 my ago
195 my ago Jurassic
135 my ago
Cretaceous
Conifers
Cycads

1 Pterodactylus
2 Rhamphorhynchus
3 Diplodocus
4 Plesiosaurus
5 Pelonerstes
6 Archaeopteryx
7 Antrodemus
8 Oligokyphus
9 Stegosaurus

Palae-ocene	Eocene	Oligocene	Miocene	Pliocene	Pleis-tocene
65 my ago		38 my ago	26 my ago	c 7 my ago	c 2 my ago
		Tertiary			Quaternary

bians from which they had evolved and, furthermore, did not have to return to water to lay their eggs. The cotylosaurs, or stem reptiles, were a simple group that gave rise to many new forms, the most significant development from them being the mammal-like reptiles that sprang up in the Permian. These finally evolved into the first mammals in the desert conditions of the Triassic.

Curiously, it was the thecodonts – small in size, but one of the most successful reptile groups – that evolved into some of the biggest creatures that ever lived, the dinosaurs [2]. Not all members of the vast dinosaur family were giants, however: the meat-eating *Podokesaurus*, for example, was only the size of a chicken. But among the long-necked vegetarians of the late Jurassic and early Cretaceous periods were the 25m (82ft) long *Diplodocus*, and *Brachiosaurus* which, weighing more than 50 tonnes, was the heaviest land animal of any era.

Recent theories tend to suggest that the dinosaurs and other large reptiles, including the flying pterosaurs, were warm-blooded and more like mammals than reptiles in behaviour. Certainly they gave rise to warm-blooded descendants, such as birds, which may have evolved directly from one of the two orders of dinosaurs. The first mammals in this age of reptiles were possibly monotremes – the egg-laying mammals.

The end of reptile dominance
Towards the end of the Mesozoic era, sweeping geological changes altered the face of the earth. Gradually the single large continent had been breaking up into separate land masses [Key]. But now the slow-moving drama of evolution suffered a bewildering change of cast: for no apparent reason, the dinosaurs and their bizarre relatives, the great swimming and flying reptiles, died out. With their demise came the birth of a new era, the Cenozoic, and the way became clear for the proliferation of mammals [5].

The catalogue of mammalian life, studded with such oddities as the first horse – the size of a fox – was enlivened above all by the arrival, a million years ago, of *Homo erectus*, the first man. It had taken more than 4,000 million years for him to appear.

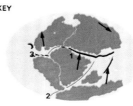

correspond to a network of cracks formed 200 million years ago. Pangaea began to split in two major blocks 180 million years ago. The southern block later subdivided.

The shifting continents began as a single "supercontinent", Pangaea. The outlines of the present continents

1 Trenches.
2 Mid-ocean ridges
3 Transverse faults
→ Direction of drift

The rift between the eastern and western continents reached northwards splitting Greenland from North America about 65 million years ago.

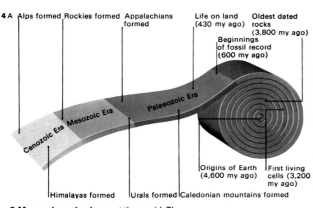

4A Alps formed Rockies formed Appalachians formed Life on land (430 my ago) Oldest dated rocks (3,800 my ago) Beginnings of fossil record (600 my ago)

Cenozoic Era Mesozoic Era Palaeozoic Era

Origins of Earth (4,600 my ago) First living cells (3,200 my ago)

Himalayas formed Urals formed Caledonian mountains formed

4 Fossils preserved in rocks provided a fragmentary record of life on earth in past ages. There are few traces of life before 600 million years ago, but the rocks reveal how huge supercontinents shifted over the face of the earth [A]. The time chart [B] traces the steps in evolution from the earliest evidence through the appearance of the first fish and land dwellers to the age of mammals.

5 Mammals evolved as warm-blooded offshoots of reptilian stock and became predominant in the Cenozoic. Before this flowering plants also overtook other forms of vegetation and spread throughout the world. The relatives of many modern mammals can be seen among these early Tertiary forms.

1 *Indricotherium*
2 *Uintatherium*
3 *Moeritherium*
4 *Hyracotherium*
5 *Diatryma*
6 *Brontotherium*
7 *Arsinoitherium*
8 *Andrewsarchus*

Era		Period	Began my ago	Length my	Development of Life
Cenozoic		Quaternary	2	2	Dominance of mammals Spread of man
		Tertiary	65	63	Flowering plants dominant Hoofed mammals and primates appear
Mesozoic		Cretaceous	135	70	Flowering plants appear. Mammals and birds become numerous
		Jurassic	195	60	The age of reptiles. Primitive birds appear. Coniferous forests widespread
		Triassic	225	30	Worldwide deserts. First mammals. Reptiles numerous
Palaeozoic		Permian	280	55	Modern insects appear Much life in the sea and freshwater
		Carboniferous	345	65	First reptiles. Winged insects appear. Ferns and horsetails common
		Devonian	395	50	Fish abundant. First amphibians
		Silurian	430	35	Seaweeds abundant. First land plants. Jawed fish and sea-scorpions common
		Ordovician	500	70	Corals and trilobites abundant
		Cambrian	570	70	First abundant fossils. Sea urchins and graptolites common
		Precambrian	4600	4030	Earliest traces of life, algae and bacteria.

The cell in action

Cells are the basic units of life, the blocks of which all living things are built. Most are minute structures – measuring a few thousandths of a millimetre in diameter – and in a human being, for instance, there are roughly 100 million million of them existing together in organized harmony. But whether a cell lives or dies independently, as do bacteria and protozoans (which are single-celled organisms), or as part of a more complex multicellular organism, such as a horse or a human, it has the basic potential to utilize raw materials and to reproduce itself.

A closer look at the cell

Cells can be thought of as simple sacs packed with the molecules (particularly proteins, nucleic acids, fats and carbohydrates) needed for life. This was the picture that biologists had of cells until they developed sophisticated techniques for examining their structure and activities more closely. Only then did they realize that the internal architecture of a cell is organized in a complex way to perform its many functions.

There are many varieties of cell shape [2,

5]. Animal cells, for example, may be roughly spherical, as in the liver; spiky as in bone; flat, as in the skin surface or elongated like nerve cells which may send long fibres from one part of the animal to another. Despite these differences, which reflect differences in function, there is an underlying pattern in cell construction just because cells share all the properties and requirements of living things.

At its simplest [Key] the cell can be seen as a sphere with a thin outer membrane (the plasma membrane) containing a smaller, denser sphere (the nucleus) suspended in a jelly-like substance (the cytoplasm). Use of the electron microscope, however, has advanced our knowledge considerably, and revealed a high level of organization.

When analysed in more detail the cell's plasma membrane [4] is found to be a sandwich with layers of protein forming the "bread" and fat molecules providing the "filling". The membrane is not merely a boundary wall but is actively involved in cell operations. For instance, materials are constantly passing in and out of the cell [7, 8, 9] and there seem to be special areas in the

membrane that work as selective transport channels. These are constructed in such a way that passage is permitted only to those substances that are needed.

The plasma membrane is the meeting point between the inside and the outside of the cell and is thus involved in communication. Some neighbouring cells establish closer contact with each other by means of minute filaments (desmosomes) which overlap between the cells. The plasma membrane also possesses special receptors on its outer surface with which other chemical messengers, such as hormones, interact.

Information and energy

Inside nearly all cells the most prominent structure is the nucleus, without which the cell dies. It is there that all the genetic information (coded in the genes) is stored in structures called chromosomes as deoxyribonucleic acid (DNA), which has the remarkable ability to replicate itself. Also inside the spherical nucleus is the nucleolus which is involved with the synthesis of proteins within the cell.

CONNECTIONS

See also
20 Life and its origins
28 The genetic code
30 Principles of heredity
44 The way plants work

In other volumes
152 Science and The Universe
154 Science and The Universe
156 Science and The Universe
144 Science and The Universe
60 Science and The Universe
114 Man and Society

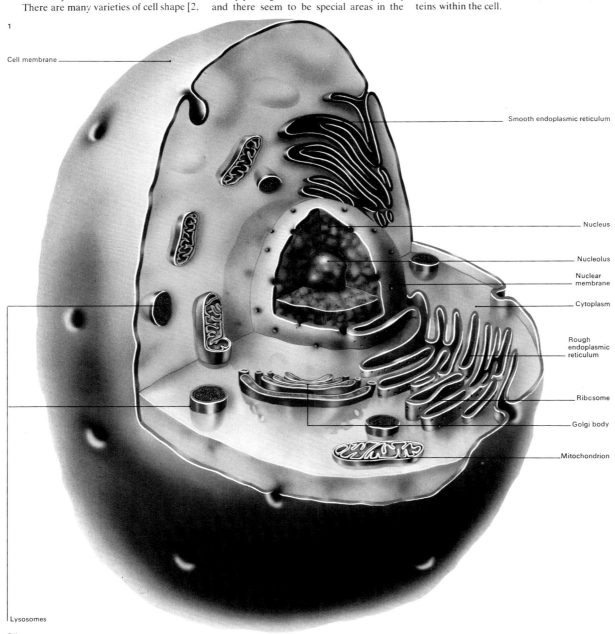

1

Cell membrane

Smooth endoplasmic reticulum

Nucleus

Nucleolus

Nuclear membrane

Cytoplasm

Rough endoplasmic reticulum

Ribosome

Golgi body

Mitochondrion

Lysosomes

1 Animal cells are all built to the same basic pattern. The nucleus is a membrane-bound sac containing the genetic material of the cell. The genetic information is coded for in deoxyribonucleic acid (DNA). This is combined with proteins to form chromosomes. The nuclear membrane is perforated by pores that may be important in controlling exchange of substances between the nucleus and cytoplasm. The cytoplasm contains numerous small structures called organelles. Prominent are the mitochondria, small sausage-shaped bodies that are responsible for the energy production. Scattered in the cytoplasm are several multi-layered membrane systems: the smooth endoplasmic reticulum, the rough endoplasmic reticulum and the Golgi body. The smooth endoplasmic reticulum is concerned with the manufacture of lipid (fat) molecules while the rough endoplasmic reticulum manufactures proteins destined for export from the cell. The granular nature of the rough endoplasmic reticulum is created by the presence of globular ribosomes on the surface of its membranes and it is on the ribosomes that proteins are assembled. The Golgi body, amongst other things, is thought to modify some of these proteins. Sacs of enzymes, called the lysosomes, are concerned with breaking down some large molecules that enter the cell.

There is a continual flow of messages from the nucleus to the cytoplasm carrying instructions for the manufacture of specific proteins. These messages are translated into proteins on globular structures known as ribosomes. Proteins destined for export from the cell are made on the ribosomes situated on an extensive network of membranes, the rough endoplasmic reticulum, designed to effect rapid removal of newly made proteins to the exterior through the plasma membrane. Proteins for internal use are made on ribosomes floating free in the cytoplasm.

Proteins play an enormously important part in cell activities. As enzymes they work as catalysts to promote the many reactions that go on in every living cell. They can also work as hormones, in which case they may be modified before their export. Modification and transport is thought to be effected by another membrane structure, the Golgi apparatus [6].

Scattered throughout the cell cytoplasm are sausage-like structures, the mitochondria [3], which are packed with enzyme systems designed to metabolize fatty acids and other energy-releasing molecules. Because the mitochondria produce all the cell's energy they are known as its powerhouses. Other enzyme-packed structures dispersed in the cell are lysosomes. The enzymes they contain are responsible for digesting many materials – including noxious ones – entering the cell.

The cells of plants

Plant cells [Key] differ from animal cells in a number of ways and have, for example, a tough cellulose coat surrounding the filmy plasma membrane. This coat is punctured in a number of places to allow for inter-cellular communication. An important feature of most plant cells is their green colour and this is produced by the substance chlorophyll contained in packets known as chloroplasts. It is the chloroplasts that are responsible for exploiting the sun's energy to manufacture carbohydrates in photosynthesis.

Many plant cells possess one or more spaces or vacuoles. They contain cell sap and sometimes, as in *Spirogyra*, the nucleus is suspended in the centre of one of the vacuoles on strings of cytoplasm.

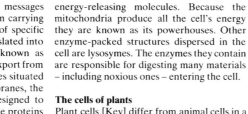
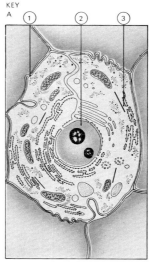
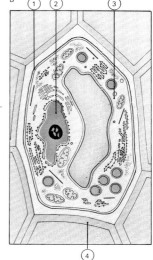

Both animal [A] and plant [B] cells have a cell membrane [1], nucleus [2] and cytoplasm [3]. But in plants the cell wall is covered with a rigid coating of cellulose [4].

2 The function of muscle cells is to contract and so they are packed with special protein fibrils, made of actin and myosin. These interact to make the cell shorten. The extra energy needed to effect contraction is provided by the activity of an abundance of mitochondria.

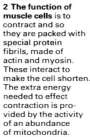

5 Nerve cells send messages from one area of the body to another and are therefore elongated. Messages are received by small fibrils or dendrites at the end nearest the cell body (the central dark area) and pass down the long fibre, the axon, to other nerves or perhaps to a muscle or gland.

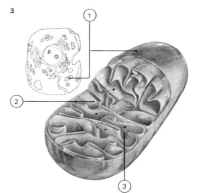

3 Mitochondria [1] are the cell's powerhouses: enzymes within them metabolize nutrients to release energy in a form, ATP (adenosine triphosphate), that can be used for synthesis of cell materials. The mitochondrion has two membranes: a smooth outer one and a much-folded inner one [2]. Scattered spherical granules in the mitochondrion [3] collect up essential calcium ions.

4 The cell membrane [A], magnified two million times, appears layered. Models of the membrane show fat globules [B], fat layers [C, E] and a protein channel [D].

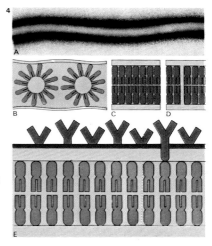

6 Large particles can enter the cell via an infolding of the membrane [A, B]. These vesicles [4] fuse with lysosomes [3] containing digestive enzymes. Digestion products [5] are absorbed and waste products are ejected [C]. Materials made by the endoplasmic reticulum [1] may leave via the Golgi body [2].

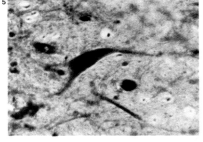

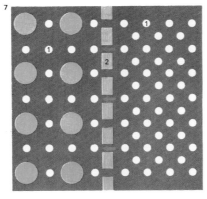

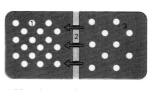

7 Molecules may enter cells in different ways depending on the environment that exists within the cell and outside it. Transfer of molecules [1] may occur by diffusion when their concentration is higher outside the cell membrane [2] than it is inside it. This is a passive process and continues until the concentrations are equal on each side of the membrane. Molecules move into the cells of the alimentary tract by diffusion during the digestive process.

8 Many large molecules such as proteins [1] are permanently enclosed by the cell membrane [2]. The more concentrated they become the more they attract water [3]. This water movement is called osmosis.

9 Cells can take in molecules from low to high concentrations by active transport requiring energy. Carrier molecules [1] in the membrane bind to the incoming molecules [2] on one side of the membrane and release them on the other.

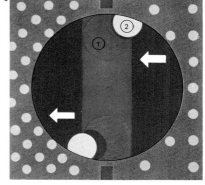

The genetic code

Inside every living cell, in its nucleus, is a master plan that governs both the minute-to-minute activities of the cell itself and those of the whole organism. The nucleus also contains the genetic information passed from parent to offspring in the reproductive process. The life-plan is encoded in a molecular substance called deoxyribonucleic acid (DNA). The DNA molecule has two characteristics essential to its role at the centre of life: it has the ability to store information and it can make exact copies of itself. The information in the DNA is translated into proteins vital to the life of the cell.

The code secret

The DNA molecule is a long strand too fine to be seen even with the most powerful optical microscope. It is arranged in the form of a twisted rope ladder with millions of rungs – the double helix [Key]. The struts are made up of alternating units of phosphate and deoxyribose sugar. Each rung contains a linked pair of chemical compounds called nucleic acid bases. There are only four of these bases – adenine (A), thymine (T), cys-

tosine (C) and guanine (G). Because of their differing but complementary structures, A can link only with T while C can link only with G. The order in which the bases present in the nucleotide units of DNA (a nucleotide unit consists of a nucleic acid base with a sugar and a phosphate attached) are arranged on one side of the ladder therefore precisely determines the order of bases on the other side. When a DNA strand is divided it will always re-form in the same pattern, although the pattern is different for every individual.

On the DNA blueprint there is a specific code for each type of protein manufactured in the cell. The codes are contained in genes – long segments of the DNA ladder, each containing from several hundred to a few thousand rungs. It is the precise sequence of base units on these rungs that makes up a protein code and one gene is thought to code for one polypeptide chain (part of a protein). Other genes, not directly involved in protein manufacture, are concerned with the machinery for translating genetic information [2] into protein manufacture.

This translation mechanism from code to

protein is extraordinarily elegant. A protein is built out of a long chain of amino acids. These acids, together with the energy needed to synthesize them into protein, are supplied to the cell by food. Once inside the membrane of the cell (in the cytoplasm) they are collected up according to the instructions of the DNA and brought to the cell's manufacturing plant which is called the ribosome.

To achieve this, certain genes in the DNA order a substance called ribonucleic acid (RNA) to deliver to the ribosome a blueprint for the protein required [4]. Part of the DNA ladder temporarily untwists and separates down the middle. The messenger RNA is a substance chemically similar DNA. The main differences between them are that the RNA is single stranded, contains the sugar ribose and has the base uracil (U) instead of thymine. In the synthesis of messenger RNA [5] one strand of the DNA ladder acts as a template and the RNA is made as a complement to it. The messenger RNA then travels to the ribosome where it serves as a mould for the assembly of a protein. Other RNA molecules, called transfer RNA, are each

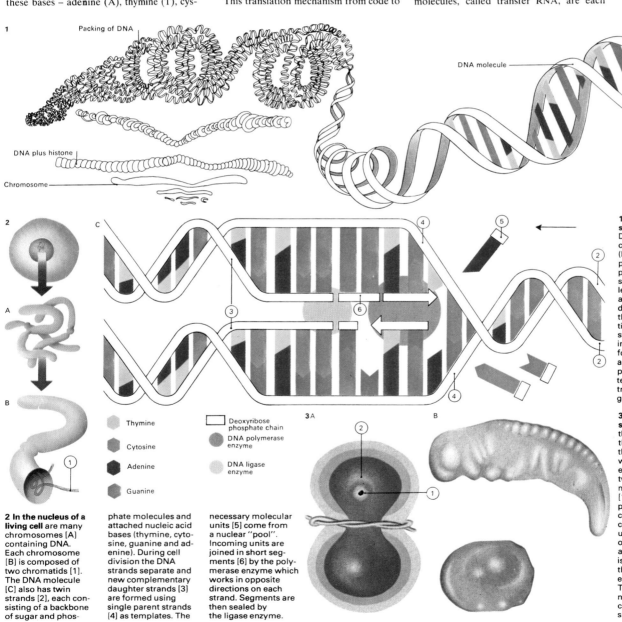

1 Packing of DNA

DNA molecule

DNA plus histone

Chromosome

1 In the chromosomes, the long DNA strand is associated with proteins (histones) and tightly packed so that it occupies only one ten-thousandth of its unwound length. Scientists are still trying to discover exactly how the DNA helix is so tightly packed. The structure probably involves regular folding of the helix and its associated proteins. These proteins also help to control the activity of the genes in the DNA.

3 Experiments with salamander eggs show the importance of the nucleus and something of the way it works. The fertilized egg is split into two [A] leaving the nucleus in one half [1] plus a greyish patch, the grey crescent [2]. This crescent, formed under the influence of the nucleus just after fertilization, is as essential as the nucleus to development [B]. The half with neither grey crescent nor nucleus shrivels and dies.

Thymine

Cytosine

Adenine

Guanine

Deoxyribose phosphate chain

DNA polymerase enzyme

DNA ligase enzyme

2 In the nucleus of a living cell are many chromosomes [A] containing DNA. Each chromosome [B] is composed of two chromatids [1]. The DNA molecule [C] also has twin strands [2], each consisting of a backbone of sugar and phosphate molecules and attached nucleic acid bases (thymine, cytosine, guanine and adenine). During cell division the DNA strands separate and new complementary daughter strands [3] are formed using single parent strands [4] as templates. The necessary molecular units [5] come from a nuclear "pool". Incoming units are joined in short segments [6] by the polymerase enzyme which works in opposite directions on each strand. Segments are then sealed by the ligase enzyme.

coded to collect a specific amino acid and bring it to the right place on the assembly mould. The code word for each amino acid is composed of three base units of the DNA. Different triplet sequences (called codons) of any of the four base units – such as GUG or GAA – can produce 64 possible code words. Some amino acids may have more than one code word while other code words seem to represent control instructions.

Protein manufacture
Protein synthesis takes place as the ribosome moves along the messenger RNA mould and the correct transfer RNAs, loaded with their specific amino acids, move in to recognize the triplet sequence; the amino acids are then released from their RNA carriers and are linked together in the specified order. The proteins so formed may be structural, as in the case of the collagen found in the skin. or more actively functional, as with enzymes.

Other hormones are themselves among the chemical signals that trigger genes to dispatch particular messenger RNAs. Similarly, other chemical signals switch off genes and

stop further protein production. The structural genes that carry the code information for specific proteins do not exist in isolation. Each is associated with controlling elements in a genetic unit. Countless genetic units are linked together on the long DNA molecular strand. And the DNA strand itself is wrapped up in association with certain proteins [1]. At cell division this nuclear material forms chromosomes. In humans, there are 23 pairs of chromosomes and, when dividing, these are thick enough to be seen with the aid of an ordinary microscope.

DNA duplication
The precise base-pairing in the double helix explains the ability of the DNA to duplicate itself accurately during cell division and pass on a complete copy of its genes from one cell to the next. When a cell divides in two, the DNA ladder splits neatly down the middle [3] and two new halves are synthesized by an enzyme called DNA polymerase. Each cell contains one old and one new DNA strand, ensuring that the new ladder has the same sequence as the parent one.

KEY

The **double helix** is the thread of life and is the name given to a ladder of de-oxyribonucleic acid (DNA) that forms a long molecular coil in our chromosomes. The chemical structure in this ladder encodes genetic instructions in cells of every single living organism. A model of the DNA strand shows the characteristic winding of the double helix. Its structure was discovered in England in 1953 by James Watson and Francis Crick at Cambridge University and by Maurice Wilkins at the University of London. The significance of their work in advancing knowledge of biology was recognized by the award of a joint Nobel prize in 1962.

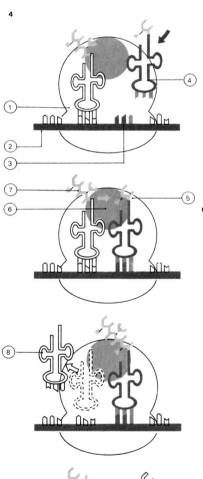

Uracil
Cytosine
Adenine
Guanine
Enzyme

4 Proteins are assembled on particles known as ribosomes [1]. The code for the protein structure is contained in messenger RNA [2]. Each amino acid in the protein is coded for by a triplet of bases

called a codon [3]. Molecules of transfer RNA [4] bring the required amino acids [5] to the ribosomes, where the enzyme peptidyl transferase [6] links the amino acid to the growing peptide chain [7]. Once its amino acid has been transferred the transfer RNA moves away [8] and a new one arrives [9]. The strand of messenger RNA moves relative to the ribosome so that new codons are continuously being presented for translation.

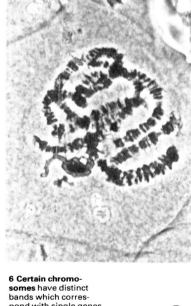

6 Certain chromosomes have distinct bands which correspond with single genes. When these become active, synthesizing messenger RNA, the band expands to form chromosome puffs. With a high-powered microscope, a gene can be seen in action.

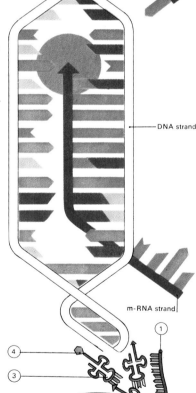

DNA strand

m-RNA strand

5 Messenger RNA molecules are made by using one DNA strand as a template. RNA is very similar to DNA but is single-stranded, the sugar backbone is modified and the base uracil replaces thymine. RNA is transcribed from the DNA base sequence. The information encoded in the chromosome is required for the synthesis of protein structures in the cytoplasm of the cell. The messenger RNA [1] therefore moves to the cytoplasm where it joins up with ribosomes [2] so that proteins can be made from amino acids [4] transported by transfer RNA [3]. As the new polypeptide chain emerges [5] it begins to fold into its characteristic shape [6].

Uracil
Thymine
Cytosine
Adenine
Guanine
Ribose phosphate chain
Deoxyribose phosphate chain
RNA polymerase enzyme

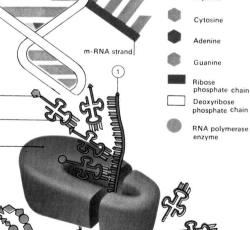

Principles of heredity

Heredity is the mechanism by which characteristics are passed from parent to offspring. These characteristics include everything from obvious features such as eye colour or stature to the hidden parameters of body metabolism such as structure and quantity of enzymes. Such characteristics are said to be inherited by the offspring from the parents. The detailed study of the mechanism of heredity is called genetics.

Mendel and genetics

Genetics is a relatively young science but the forces of heredity have been appreciated for a very long time. Ten thousand years ago, when man was on the threshold of organized agriculture, he quickly realized that better and better crops could be raised by actively crossing the good strong plants that had arisen by accidents of nature. But until the middle of the nineteenth century nobody was concerned with how selective breeding worked nor with the factors involved in the passing on of characters through the generations. The great breakthrough was achieved by an Austrian monk, Gregor Mendel

(1822–84) [2], but for a number of reasons his work was ignored for many years.

In his monastery Mendel investigated the inheritance of characters, or traits as they are often called, using pea plants as his experimental material [3]. The traits he examined included flower colour, texture of the pea seed and the size of the plant. The first stage of his experiments required two different strains of pea to be crossed, forming what is called a hybrid. By looking at the resulting hybrids he obtained some idea of the interactions of characters. For instance, whenever he crossed tall plants with short ones, no matter which plant donated the pollen, all the hybrids were tall. He said that the factor for "tallness" was dominant to that for "smallness" in these plants; the factor for "smallness" was said to be recessive.

In a series of such experiments Mendel demonstrated dominance and recessiveness in other characters. For instance, red flowers are dominant to white; smooth peas are dominant to wrinkled; and "stem" or axillary flowers are dominant to terminal flowers. The next step was to allow the hybrid plants

to self-fertilize, a technique that allowed the recessive traits to reappear: for instance, self-fertilized hybrid tall plants yielded some small and some tall offspring. But the important thing was that there were always three times as many with the dominant trait as there were with the recessive trait.

The importance of genes

To explain his results Mendel postulated the existence of "factors" or genetic elements (now known as genes), a pair of which was needed for each character. In the case of stem size, for example, there may be two genetic elements for tallness (TT) or one for tallness (T) and the other for shortness (t). Because tallness is dominant, a Tt plant has a tall stem. The only way a plant can have a short stem is to double the dose of the shortness gene (tt).

When the two members of a gene pair are identical, as in TT, the organism is said to be homozygous for that character; a mixture, such as Tt, is a heterozygote. Once a gene pair is recognized, it identifies the genotype of an organism, whereas the actual appearance of the organism is its phenotype: if the genotype

1 The basis of sexual reproduction is the fusion of gametes from the male and female (sperm and egg). In this way genetic material from two individuals intermingles to create a new individual. The result is a tremendous variation between members of the same species. In humans, although there has been an estimated one hundred thousand million people born, no two have been alike (apart from identical twins). Gametes are produced by reduction division or meiosis, resulting in cells with only half the normal number of chromosomes. When two gametes fuse at fertilization the correct number is restored.

[A] In a cell destined to form sperm the chromosomes appear, when cell division starts, as thin strands.

[B] The chromosomes become thicker and arrange in pairs: one from the mother, one from the father.

[C] Later each member splits itself lengthwise with each part connected at one point only.

[D] "Crossing over" occurs with an exchange of genetic material, important in generating diversity.

[E] The chromosome pairs align themselves around a spindle structure; they then separate.

[F] The paired chromosomes are parted by contractile fibres and migrate to the two poles of the spindle.

[G] The centrioles (the spindle forming bodies) now separate to form four poles.

[H] Unlike the earlier division there is now no exchange of genetic material.

[I] After the new spindles are formed the chromosomes migrate in 4 groups.

[J] The 4 sperm are produced, each with half the normal number of chromosomes.

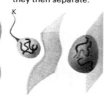
[K] A similar reduction division occurs in the female, only one cell becoming an egg.

[L] New cells have the correct number of chromosomes and mixed parental genes.

2 Gregor Mendel is the father of genetics. He was profoundly interested in science and performed many experiments on inheritance in pea plants. By analysing mathematically the outcome of crosses between different strains of pea he was able to establish the basic laws of inheritance. Mendel sent his results to a Swiss botanist who thought them unimportant. The value of the work, published in an obscure journal, was not realized until after Mendel's death.

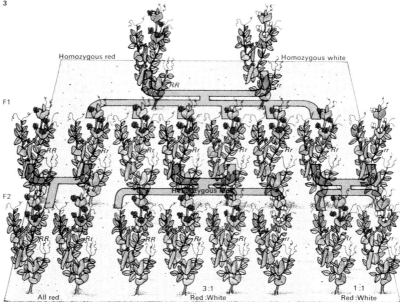

Homozygous red

Homozygous white

F1

F2

All red

3:1
Red:White

1:1
Red:White

3 Mendel used garden peas to demonstrate dominant and recessive inheritance. Pure-breeding plants were assumed to carry two genes (to be homozygous) for red (*RR*) or white (*rr*) flowers. The first (F₁) generation of progeny were all red, suggesting that the red gene was dominant. All the members of this F₁ generation were thus *Rr* in constitution. This was proved in the second (F₂) generation backcrosses between the two parental types and the F₁ offspring. All the progeny of red-red crosses were red. Of the red-white crosses half were red and half white. Crosses between F₁ progeny gave a 3:1 ratio of red to white flowers.

of a pea plant is heterozygous *Tt* its phenotype will be tall.

Mendel's experiments laid the foundations for understanding inheritance. We now talk of genes, not genetic elements; we know that characters are usually governed by a collection of genes, not just one; and we know that genes are part of the chromosomes located in the nucleus, a discovery made by an American, Walter Sutton (1877–1916), 30 years after Mendel's death. Humans have 46 chromosomes in each cell, made up of 23 pairs; one set of the pair comes from the mother and the other set from the father [Key]. This means that for every character a gene (or genes) from the mother interacts with the equivalent gene of the father to produce the phenotype of the offspring.

Sex chromosomes and sex linkage

The double set of chromosomes is achieved when a sperm fertilizes an egg (ovum) [1]. The sperm contains just one half of the male set while the ovum contains half of the female set. When they join to create an embryo the full complement is restored. Of the 23 chromosome pairs, one is different in that the two chromosomes in the pair are not always identical in size. This is the sex chromosome pair: the sex chromosomes are denoted X, which is big, and Y, which is small. A person whose genotype has two X chromosomes is female, an XY is male.

The difference in size between X and Y chromosomes gives rise to a number of so-called sex-linked diseases of which haemophilia [5] and colour blindness [6] are best known. The gene for haemophilia, a disease in which the blood clotting mechanism is defective, is recessive and is located on the X chromosome. For a female to have this disease, therefore, she must have a double dose of the haemophilia gene, one on each of the X chromosomes. Now, because the Y chromosome is short, it has no room for the haemophilia (or the anti-haemophilia) genes. A male will therefore suffer from haemophilia if the culprit gene appears on the X chromosome because there is nothing on his Y chromosome to combat it. Harmful recessive traits are more likely to be expressed phenotypically with inbreeding.

The chromosomes are the carriers of genetic information. The human chromosomes are 46 in number and consist of 23 pairs. In each pair one chromosome comes from each parent. One of the pairs is responsible for sex determination. In the male, they are dissimilar and are known as X and Y.

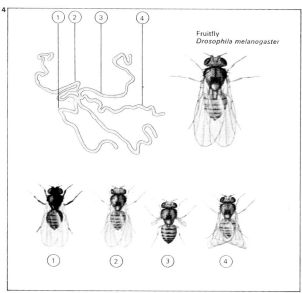

4 The fruit fly (*Drosophila melanogaster*) is a favourite of geneticists because it takes up little space, reproduces every 10 to 15 days and its cells contain only four pairs of chromosomes. Geneticists have mapped many genes on these chromosomes: for example, genetic damage at 1 produces bar-shaped eyes; at 2, sepia bodies; at 3, lack of wings; and at 4, curly wings. *Drosophila* also has giant chromosomes whose basic structure can be seen by looking at cells in the salivary glands. They have bands across them and each band is now thought to correspond exactly to a gene.

5 Queen Victoria's family numbered among its male members several haemophiliacs. It seems that Queen Victoria herself received from one of her parents a gene for haemophilia that had changed spontaneously from the norm – that is, it had mutated. Her father was normal and there is no evidence to suggest that her mother was a carrier of the gene. By various marriages this haemophilia gene was spread through many European royal families. It affects mainly males because the gene is carried on the X chromosome. Males have an XY complement of sex-determining chromosomes so that if the haemophilia gene is present on the X, the individual will have haemophilia. Victoria's family also share other obviously inherited characteristics such as those of facial features.

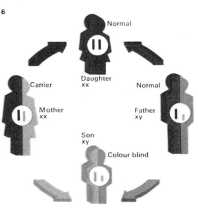

6 Colour blindness is a sex-linked trait. The recessive gene for colour blindness (*c*) is carried on the X chromosome. This means that a woman may carry one such gene but not show it. Because she has two X chromosomes the second will probably bear the dominant gene for normal colour vision (*C*), making her *Cc*. If her son inherits the X chromosome with the *c* gene, he has no dominant gene for normal colour vision on his Y chromosome, and is colour blind.

7 In Mendel's early experiments he dealt with traits that were either one thing or the other – red flowers or white; wrinkled peas or smooth. But because most characters are influenced by more than one gene they tend to appear in a kind of spectrum. Height in humans spreads from short to tall, but because there is a pool of genes contributing to this factor there is a so-called normal distribution with an average in the middle between extremes.

Evolution: classical theories

The word evolution means an unfolding, a gradual development, and it is a word that has become closely linked with the origin of animal and plant species as they exist today. Scientists now believe that through millions of years of the earth's life, simple organisms have developed into more complex ones, each better adapted to prevailing conditions than its predecessor. It is generally accepted that the ultimate survival of a new form is determined by an effect known as natural selection, or the survival of the fittest – a theory intimately associated with the name of Charles Darwin(1809–82) [Key].

Ideas about the origin of species
Darwin published his theory of evolution in 1859 in a book whose full title is *On the Origin of Species by Means of Natural Selection, or the Preservation of Favoured Races in the Struggle for Life*, but which was popularly abbreviated to *The Origin of Species.* He suggested that the development of species is a continual process, and implied that man himself must have evolved from an ape-like stock. This was considered by many to be an heretical proposal at the time. The first edition of the book, some 1,200 copies, was sold out on the day of publication, an indication of the prevailing interest in the subject.

After much critical examination by scientists, and a great deal of public scorn and derision concerning man's supposed origins in apes, the theory of natural selection was assimilated and has remained ever since the central pillar of our ideas on evolution.

The modern conception of species as such began with John Ray (1627–1705) in the seventeenth century, and was established fully by Carl Linnaeus (1707–78) in the following century. It was the Linnaean classification system that really emphasized the relationship between similar species, the variations on a theme that so impressed Darwin in his observations.

One theory about the origin of species that held sway for a long time, and encompassed the Christian view on the creation of man, suggested that species were created spontaneously, perhaps sequentially, after a series of disasters. But the obvious relationships between species that emerged from studying Linnaeus's classification of organisms forced people to look for the origin of species in a process of gradual change.

It was Georges Buffon (1707–88) who first seriously suggested that the environment had an important effect on the evolution of species. This idea was developed by Jean Baptiste Lamarck (1744–1829), and it is his name that is attached to the suggestion that species inherited characteristics that were adaptations to the environment.

The basis of Lamarck's theory
Basically, Lamarckism is simple and attractive. It proposes that changes in external conditions create new needs in the species living there [6]. These new demands lead to new patterns of behaviour that may involve modifying the use of existing organs, thereby altering their structure. It is these altered structures that the offspring are thought to inherit. For instance, Lamarckian theory proposes that giraffes have developed long necks as a result of their repeated attempts to reach food in high trees.

Lamarckism attracted a lot of support,

1 On 27 December 1831 Darwin set sail on board HMS *Beagle* at the beginning of a five-year expedition. The *Beagle* explored extensively around South America, and as the ship's naturalist, Darwin made an enormous collection of samples of plants, animals and rocks. It was during the voyage of the *Beagle* that Darwin amassed the data that led to his theory of natural selection.

The Galápagos Islands

Pinta (*T.e. abingdoni*)

Marchena (*T.e. Darwini*)

San Salvador

Fernandina

Santa Cruz

Darwin's route

San Cristobal (*T.e. chathamensis*)

Isabela (*T.e. microphyes*)

Santa Maria (*T.e. elephantopus*)

Española (*T.e. hoodensis*)

N

2 While on the Galapagos Islands, Darwin discovered specimens that convinced him that a new explanation was needed of the origin of species. He was especially intrigued and puzzled by the similarities and differences between the giant tortoises, *Testudo elephantopus,* a different sub-species of which inhabited each island. Shells of the tortoises named next to their islands are shown in illustration 3.

3 The giant tortoises can all be regarded as varieties (sub-species) of *Testudo elephantopus,* a species believed to have originated from a South American tortoise (*Testudo tabulata*). The shells vary from island to island. Unfortunately many of the tortoises are now becoming rare because introduced animals are destroying their food, eggs and young. A conservation programme is now under way.

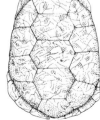
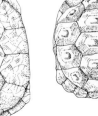

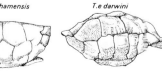

T.e. abingdoni   *T.e. microphyes*   *T.e. elephantopus*   *T.e. hoodensis*   *T.e. chathamensis*   *T.e darwini*

some of it after Darwin's natural selection theory was published, and it is easy to see why. But in fact the laws of inheritance simply do not allow Lamarckian inheritance to operate. As we now know from modern genetics, characters are passed from parent to offspring by means of the genes in the germ cells (the ones that form gametes), and structural changes in distant parts of the body do not modify the genetic constitution of the germ cells or indeed of any cells.

It was against the background of Lamarckism that Darwin produced his theory of natural selection. He set sail in HMS *Beagle* [1] as the expedition's naturalist and returned five years later a confirmed evolutionist, the transformation having taken place on the Galapagos Islands.

The fruits of Darwin's voyage

In all his explorations during the five-year voyage, Darwin was impressed by the subtle variation between species [3], particularly among the finches on the Galapagos Islands [5]. Darwin noticed that in almost all organisms there is a massive production of poten-

tial offspring (whether eggs or spores), and that only a few survive. Life, then, was a struggle for existence. The next important step in the development of Darwin's ideas was his recognition of the great individual variation within populations.

The combination of these two points produced a third: those variants that survived to adulthood in the struggle for life were, presumably, the ones most fit to do so. Darwin supposed that individual variation could be inherited by offspring from their parents. He therefore saw evolution operating through the natural selection of inheritable variations.

Darwin first developed this theory as early as 1838 but felt unable to publish it, perhaps because it went so much against his father's beliefs. Eventually he was virtually forced into publishing when Alfred Russel Wallace (1823–1913) sent Darwin a short paper on his theory on evolution, a theory that matched Darwin's own exactly. The two men presented a joint paper to the Linnean Society in 1858, and Darwin published *The Origin of Species* a year later.

Englishman Charles Robert Darwin developed the theory of natural selection in evolution.

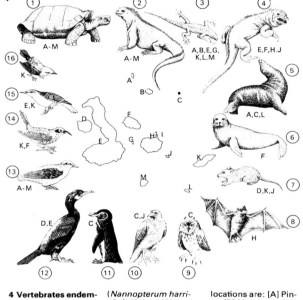

4 Vertebrates endemic to the Galapagos include: [1] giant tortoise (*Testudo elephantopus*); [2] marine iguana (*Amblyrhynchus cristatus*); [3] lava lizard, (*Tropidurus*); [4] land iguana (*Conolophus subcristatus*); [5] sea lion (*Zalophus californianus wolleback*); [6] fur seal (*Arctocephalus australis galapagoensis*); [7] mouse (*Oryzomys*); [8] bat (*Lasiurus*); [9] short-eared owl (*Asio flammeus*); [10] hawk (*Buteo galapagoensis*); [11] penguin (*Spheniscus mendiculus*); [12] flightless cormorant (*Nannopterum harrisi*); [13] dove (*Nesopelia galapagoensis*); [14] mocking thrush (*Nesomimus melanotis*); [15] golden warbler (*Dendroica petechia*); [16] scarlet flycatcher (*Pyrocephalus nanus*). Their island locations are: [A] Pinta; [B] Marchena; [C] Genovesa; [D] Fernandina; [E] Isabela; [F] San Salvador; [G] Pinzón; [H] Santa Cruz; [I] Baltra; [J] Santa Fé; [K] San Cristóbal; [L] Española; [M] Santa Maria.

5 The finches of the Galapagos Islands provided Darwin with an important clue to his theory of evolution. They are a perfect example of how small, localized populations can evolve. All the modern finches are thought to have descended from a single line of birds that flew from South America. When they arrived there were no indigenous finches on the islands. The invading birds were all seed eaters. Some adapted to tree living, others to a cactus habitat and still others to the ground. The different populations adapted to different foods and this is reflected in the shape of the beak. The ecological and geographical separation allowed divergent evolution until separate species developed which are unable to interbreed. Today there are 14 different species on the islands classified in six separate genera.

A *Geospiza* sp
B *Platyspiza* sp
C *Camarhynchus* sp
D *Cactospiza* sp
E *Certhidea* sp
F *Pinaroloxias* sp

Ancestor: seed-eating ground finch from South America

Food source
Seeds
Buds and fruit
Insects

Habitat
Trees
Ground
Cacti

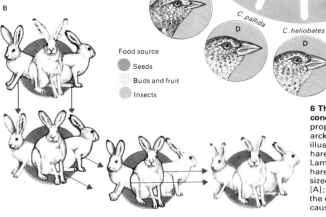

6 The contradictory concepts of evolution proposed by Lamarck and Darwin are illustrated by Arctic hares. According to Lamarck, all the hares have the same sized ears initially [A]; the influence of the environment causes them to become shorter and this is passed on through the generations. Darwin said that all the hares originally had ears of different lengths [B] but that those with the shortest ones survived better and were thus the ones to breed.

33

Evolution in action

Since its publication in 1859, Charles Darwin's theory on the origin of species has enjoyed phases of popularity, and sometimes it has been almost totally eclipsed by contemporary ideas. But, modified and improved by new data and new ideas, the theory has survived in essence and remains central to current views on evolution.

The key word in Darwin's theory is variation: he continually emphasized the subtlety of the variations within a species. This phenomenon is known as continuous variation. In contrast, there are often distinct differences between species and these are termed discontinuous variations. Emphasis on these forms of variation led to the first challenge to Darwinism.

The challenge to Darwin's theories
In 1894 William Bateson (1861–1926) published a book, *Materials for the Study of Variation*, in which he pointed out with some force the difference between continuous and discontinuous variation. It was then an easy step to suggest that evolution may have progressed through jumps provided by discontinuous variation, rather than the smooth slides of continuous variation.

Bateson's implication was taken up enthusiastically by Hugo de Vries (1848–1935) in observations of cultivated plants. He noticed that in a population of plants there occasionally appeared a variant strikingly different from the rest. Such variations, he insisted, were quite outside the range of continuous variation. De Vries named this phenomenon "mutation", a totally new concept, to explain the appearance of discontinuous variations.

Out of all this emerged the mutation theory of evolution, much of which de Vries published in the first years of the twentieth century. Evolution, it suggested, was mediated through sudden spontaneous changes rather than depending on natural selection acting on continuous variations. At about that time the work of Gregor Mendel (1822–84) on inheritance was rediscovered, after almost 40 years of obscurity. Genetics as a real science was then born and, superficially, it seemed to support the mutation theory as against orthodox Darwinism.

With the reality of chromosomes and the notion of genes now before them, it was easy for geneticists to envisage evolution in terms of major jumps. For instance, X-rays and certain drugs could inflict change within a chromosome, and rearrangement of the chromosomes themselves also occurred, both of which could induce physical changes.

The results of chromosomal re-ordering
One dramatic example of a new species emerging as a result of chromosomal change was a plant, *Spartina townsendii*, more commonly known as rice grass. A count of chromosomes on this grass, which was first noticed in 1870 on mudflats in the south of England, totalled 126: this grass was clearly a hybrid between two other species of *Spartina*, one with 56 chromosomes and the other with 70. The new species thrived and multiplied phenomenally, a good example of a new species created by discontinuous variation filling an ecological niche and surviving. This type of process, although probably not uncommon in plants, is rare in animals of any complexity because the mechanics of sexual

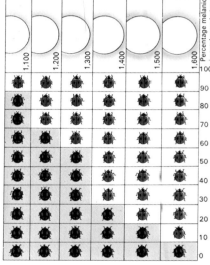

1 The Peppered moth, (*Biston betularia*) [A], of Europe originally lived on lichen-covered trees where its light colour gave some protective camouflage. During the 1800s industrial pollution killed much of the lichen and covered trees with soot. The light coloured moth was now more easily seen by its predators and two darker forms, *insularia* [B] and *carbonaria* [C], evolved. The darker moths were now better camouflaged and soon became abundant in industrialized areas. [A] now thrives in non-industrial areas and the darkened, but not black form [B] is found in semi-industrial zones. This is an example of evolution in action.

2 The typical two-spot ladybird has two black spots on a red background, but melanic forms have the colours reversed. The melanic forms or morphs appear more frequently if there is little sunshine and are more active at lower temperatures, as shown here. A ladybird of each form is chilled for 30 minutes at 5°C (41°F) [1]. The activity of the two forms is recorded as the number of squares each enters in a set time [2]. The melanic form is found to be more active [A]. The frequency of each form in the population is plotted against the hours of sunshine. In areas of low sunshine and consequently temperature, melanic forms are more common [B].

Yellow 1–5 bands

Pink 1–5 bands

Brown 1–5 bands

Yellow

Pink

Brown

3 The shells of the snail *Cepaea nemoralis* can be of three different colours: pink, yellow or brown. A series of bands may be superimposed on the shells. The proportion of colours and banding varies, depending on the ecology of the habitat and the ability of the predator thrush to recognize them. In short turf [A] the greenness and uniform nature of the background favours the yellow unbanded snail. In beech wood [B] the degree of banding depends on the leaf carpet. In rough herbage [C] yellow banded snails are best camouflaged. Leaf litter [D], in a deciduous wood, favours brown shells, without bands.

reproduction tend to prevent such chromosome assemblage [5].

Mutational events are undoubtedly important in evolution, and as geneticists discovered more and more about the structure of genes, they came to realize how subtle such changes could be. The hereditary information in genes is written in a code of nucleotide bases, triplets of which represent single amino acids. If one of these bases is altered, a new amino acid may be coded, slightly altering the structure of the protein for which the gene carries the total information. A changed amino acid in an unimportant part of the protein molecule has no effect on the organism's phenotype, and it is this on which the forces of natural selection act.

A mutation in a crucial area of the molecule, however, may make itself felt. And the degree of modification in the phenotype depends on the number of genes that are responsible for that particular character. If the character is determined by a single gene, mutation in that gene may have a dramatic effect, in other words producing a discontinuous variation; but if a battery of genes

combine to shape that character, mutation in one of them may have only minimal effect.

It is therefore clear that a sharp distinction cannot be made between orthodox Darwinism and the mutation theory – they overlap. The principle of selection remains sound, and the new synthesis of ideas has become known as neo-Darwinism.

The problem of altruism and the individual

By its nature, selection acts on the individual: it is the survival of individuals that overall makes a group of them (a population) successful; it is on the individual that the forces of the environment act. And yet this emphasis on individual rather than group selection has thrown up the problem of explaining altruism in animals, a trait that undoubtedly exists. For example, a worker bee will sting an invader to defend the hive, and die in the act. How can suicide be favoured by natural selection and survival of the fittest? The answer to this puzzle can be seen in the drive to ensure the future survival of genes identical with those possessed by the now dead individual.

KEY

10 million years old

3 million years old

A series of fossil shells from successive layers of rock illustrate how small, almost imperceptible changes in structure can give rise to large changes that are obvious at first glance. Despite this, all the shells are members of the same species, a fact that would not be known without such a full fossil record. They range from three to 10 million years old.

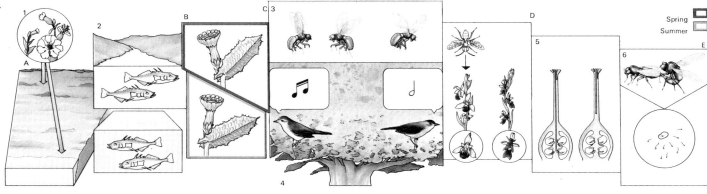

Spring
Summer

4 Species are kept distinct in a number of ways. Isolation can be due to habitat [A]. Hybrids of sea campion (*Silene maritima*) and bladder campion (*S. vulgaris*) are viable [1], but one grows on the sea-shore and the other in meadows. Three-spined sticklebacks (*Gasterosteus aculeatus*) have two populations in Belgium, one marine the other freshwater [2]. Of two species of *Lactuca* [B] one flowers in spring, the other in summer. The chances of cross-fertilization are thus reduced. Isolation is also due to behavioural differences [C]. Fruit flies [3] have a complex courtship so that related species are unable to mate; if they do [E] the gametes are usually incompatible or the egg is unattractive to the sperm [6]. In blackcaps (*Sylvia atricapilla*) and garden warblers (*S. borin*) [4] the song is highly specific. Mechanical isolation [D] exploits the senses of insects, flowers having different colours and structures. If transferred, pollen often does not grow a tube long enough to reach the ovule [5].

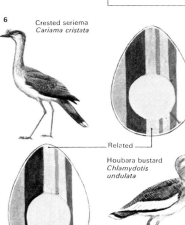

Crested seriema
Cariama cristata

Related

Houbara bustard
Chlamydotis undulata

Not related

Secretary bird
Sagittarius serpentarius

5 Sometimes species (and even subspecies) are kept distinct by incompatability of genetic material after mating. The mule [A] produced by a mare and a male donkey is sterile, as is a hinny, the cross between a stallion and a female donkey. If a fertile (viable) hybrid is formed from different species and is then mated with one of its parents, weak or sterile offspring arise, as in certain cotton plants [B] and fruit flies [C]. Related species of grain (*Triticum aestivum*, wheat and rye *Secale cereale*) [D] and different races of the North American leopard frog (*Rana pipiens*) [E] can produce hybrid seeds or eggs respectively but the embryos fail to develop.

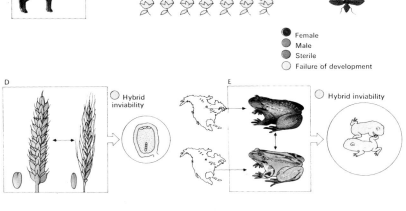

Hybrid infertility

Hybrid breakdown

Female
Male
Sterile
Failure of development

Hybrid inviability

Hybrid inviability

6 Many species of birds are similar in appearance due to convergent evolution, but by referring to the proteins of the white of the eggs their true relationships can be established.

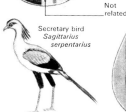

Bacteria and viruses

Bacteria and viruses are the smallest forms of life on earth; about 250,000 of them would cover the full stop at the end of this sentence. They are simple creatures, but they are not primitive [1]. They have a capacity for survival in inhospitable places and an ability to adapt to new conditions that puts them among the most successful and advanced of living things.

Although associated with disease, death and decay, the role of bacteria in returning essential nutrients to the soil, and in making complex foods that can be used by other living creatures, means that they are essential to the balance of nature.

Viruses and their effects

Viruses have been described as "living chemicals". They neither feed, breathe, grow nor move and are never found living free in nature. Yet once within a living cell [4] they can take control of it and subvert it from its ordinary activities to the production of new viruses. All living cells are susceptible to attack by viruses – even bacteria fall victim to a special virus, the bacteriophage ("bacteria-

eater") [5]. Virus "particles", seen under the microscope, are the infectious agents, inert bundles of chemicals travelling between the last "victim" cell and the next.

Viruses comprise a core of hereditary material surrounded by an envelope of protein. Plant virus particles [2] enter living cells through damaged cell walls, while animal virus particles [3] are taken up by the cell as if they were morsels of food. Bacteriophages inject themselves into their victims. Once inside, the viral hereditary material takes control, diverting its victim's cellular machinery to new virus production.

Viral infection is usually specific. A particular virus will attack only one organism, or only one part of that organism. Virulent viruses destroy the cells they attack with the release of the new viruses – these cause diseases such as yellow fever, poliomyelitis, influenza, the common cold and smallpox. Latent viruses do not destroy their victims immediately, but may coexist with them for long periods.

Antibiotic drugs are comparatively ineffective against viruses, but living cells

produce antibodies that give immunity to viral attack, and vaccination with killed virus can be used against diseases such as poliomyelitis. Interferon, a substance produced naturally by living cells in response to viral attack, prevents viral multiplication, but has proved difficult to use medically.

The ubiquitous bacteria

Unlike viruses, bacteria are cellular, but their cells are very much more simple than those of higher organisms and vary greatly in size and shape [6, 7, 8]. Bacteria are found everywhere and can live in conditions that defeat more complex organisms. They have been found 9km (6 miles) below the surface of the ocean, floating in the upper atmosphere, attached to rocks washed by the spray from boiling springs and surviving in small pockets of liquid water in otherwise frozen soil. When conditions worsen some bacteria produce a resistant resting stage, the endospore, the most resistant living thing known. Even boiling will not kill some endospores.

Of all places to live, the body of another organism is the most dangerous. Bacteria

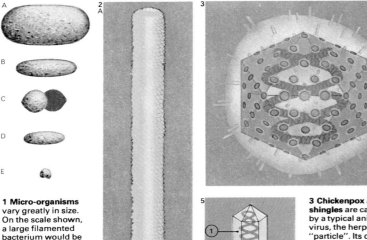

1 Micro-organisms vary greatly in size. On the scale shown, a large filamented bacterium would be too large for the page, while the smallest virus would scarcely be visible. Micro-organisms are measured in micrometres or microns (μ). One μ equals a millionth of a metre. [A], [B] and [C] are cellular bacteria. *Haemophilus influenzae* [D] is a large virus and [E] is a mycoplasma, the smallest known free-living cell.

2 Tobacco mosaic virus "particles" each take the form of a long, hollow rod [A], measuring 300 millimicrons in length and 18 millimicrons in diameter. Ribonucleic acid (RNA) forms a helical spine [1] around which the protein units [2] are arranged. This virus was the first to be fully separated from its host plant and purified. Infected leaves were ground up and the virus separated. This showed that a living organism could be handled like a chemical.

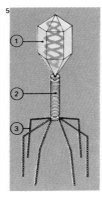

5 The bacteriophage, the most complex of all viruses, infects bacteria. It consists of a head [1], a protein tail surrounded by a retractable sheath of protein units [2] and tail fibres [3].

3 Chickenpox and shingles are caused by a typical animal virus, the herpes "particle". Its deoxyribonucleic acid (DNA) core is surrounded by 162 protein units in the shape of a solid with 20 plane faces.

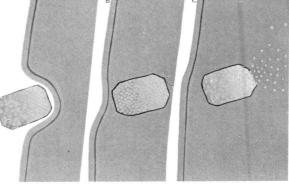

4 Viruses can live and multiply only within living cells. The viral hereditary material is simply a blueprint for the creation of more viruses. In multiplying they may cause disease. Cowpox, for example, is caused by *Vaccinia* virus, one of the largest viruses. It can enter the host cell only by courtesy of its victim. [A] A pocket forms in the host's cell membrane which the virus particle enters and is then ferried across [B]. At the inner surface the viral core is released into the cell contents hereditary material escapes from the core [C] and the production of new virus begins.

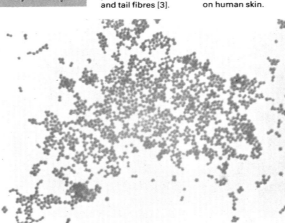

6 Typhoid is caused by the rod-shaped bacterium named *Salmonella typhi.*

7 Round staphylococci form cell clusters. These are from a boil on human skin.

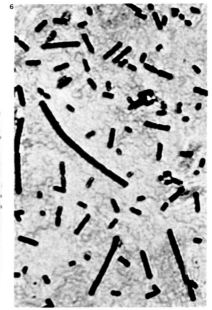

gain entrance through wounds. Once inside they must resist their victim's defence mechanism, chiefly phagocytes (cells that would engulf them) and antibodies that would destroy them. Some bacteria have slimy capsules on their outer surface that phagocytes cannot attack. Others are able to live within the phagocyte after being engulfed; some bacteria produce camouflage substances that give the invaded cell the illusion that no invasion has taken place and therefore no antibodies are produced.

Harmful and helpful

Bacteria can harm in three ways: by clogging vital passages through sheer numbers; by producing poisonous substances – tetanus poison (toxin) produced by the soil organism *Clostridium tetani* is one of the most poisonous substances known to man – and by producing an allergic reaction in their victims. Antibiotics have been effective against bacterial infections for some time, but bacteria are appearing that are resistant to these chemicals. Bacteria multiply rapidly, dividing once every ten minutes in good conditions, so

the chances of mutants arising, resistant to particular antibiotics, are high.

Bacteria living within other creatures are not always harmful. The intestines of cows, sheep and goats all contain a special sac, the rumen, which contains bacteria and helps the animal to digest plant cellulose [12].

The mycoplasmas, smallest of all cellular creatures and perhaps an interim stage between the viruses and the bacteria, can exist in nature by living on sewage, but may also live in animals where they cause diseases such as a form of arthritis in swine.

Bacteria break down dead bodies and return the vital substances contained within them to the soil. Without the continual recycling of these vital building blocks, life could not continue. Man makes use of bacteria in sewage works to render organic waste harmless, in composting [13] and in making cheese, butter and vinegar.

Between bacteria, viruses and other creatures a fine balance is constantly maintained in which coexistence is the dominant theme. Invasion leading to disease and death is the exception rather than the rule.

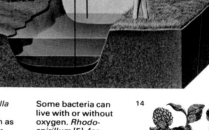

A bacterium is bounded by a rigid wall [2] outside which may be a slimy capsule [1]. A fine membrane [3] surrounds the cell contents. The

hereditary material [4] lies free within the cell, together with the machinery that manufactures the cell substance [5]. Some bacteria

appear hairy [6] and some bear flagella [9]. Granules of storage material [7, 8] are also found in some bacteria such as *Chlorobium* spp.

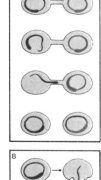

8 Cystitis and other urinary tract infections may be caused by *Proteus vulgaris*, a rod-shaped bacterium bearing waving flagella. The shape of the bacterial cell is important to its survival. Spherical cocci are more resistant to dehydration than rods or spirals, while rods are better able to abstract food from their surroundings. Spirals and rods are the most common bacterial shapes.

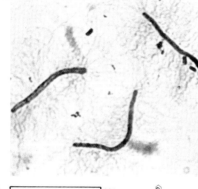

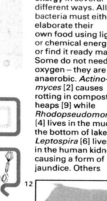

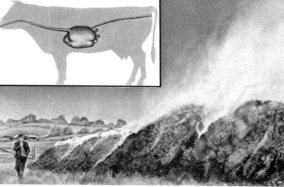

9 Bacteria acquire energy in several different ways. All bacteria must either elaborate their own food using light or chemical energy, or find it ready made. Some do not need oxygen – they are anaerobic. *Actinomyces* [2] causes rotting in compost heaps [9] while *Rhodopseudomonas* [4] lives in the mud at the bottom of lakes. *Leptospira* [6] lives in the human kidney, causing a form of jaundice. Others

such as *Gallionella* [3] need oxygen, while some, such as *Rhizobium* [1], fix nitrogen in the root nodules [8] of leguminous plants. *Bacillus anthracis* [7] causes anthrax in hoofed animals.

Some bacteria can live with or without oxygen. *Rhodospirillum* [5], for example, obtains its energy from chemical reactions when oxygen is available and from light when oxygen is sparse.

10 Mating behaviour in bacteria can be similar to sexual reproduction [A]. The "male" forms a tube through which part of the hereditary material passes to the "female". The resulting cell shows some of the characteristics of the "male". The process is called conjugation. More unusually, a bacterial cell may

break down, releasing its hereditary material, parts of which can be picked up by a nearby cell which gains those characteristics of the destroyed cell and is said to be transformed [B]. However, bacteria commonly reproduce by splitting in half. Their hereditary material divides first and then the whole cell splits to produce two identical daughter cells [C]. This is known as fission.

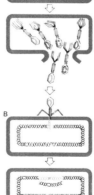

11 Viruses multiplying in a cell can pick up a part of their victim's hereditary material by mistake [A] and carry it to a second victim where it becomes incorporated in the new host's hereditary material. Specific genes can be so transferred [B]. This is transduction.

12 Cattle, sheep and goats live on grass yet without bacteria they cannot digest the cellulose and so release the sugars essential to energy provision. A special section of their gut, the rumen, contains vast numbers of bacteria and other

organisms that break down the complex plant cellulose structure into simpler sugars.

13 Compost heaps usually comprise soil, dead plant matter and excreta. Bacteria break these materials down into

simpler substances, many of them nitrogenous, which can be used as fertilizer. As a result of bacterial activity energy is released as heat and a compost heap may become very hot and burst into flames when air enters.

14 All living things need nitrogen but cannot assimilate nitrogen gas. Small nodules on the roots of leguminous plants contain bacteria that help to "fix" nitrogen gas and convert it into a usable form.

The plant kingdom

Traditionally, the plant kingdom is studied in two distinct halves, the lower or non-flowering plants and the higher or seed-bearing plants. The lower plants include algae, mosses, ferns, horsetails and club mosses, while the higher plants consist not only of the well-known profusion of flowering plants but also most trees and shrubs.

Many plants have two "forms", one sexual, the other asexual or vegetative. This is known as alternation of generations. In the simplest plants the sexual form is the dominant – in land plants there has been a shift of emphasis to the asexual phase, culminating in the flowering plants where the sexual phase is reduced to a few cells.

Early plant types

Today, more than 400,000 species of plants have been described and it is thought that, through evolution by natural selection, all are descendants of a small fraction of the progeny of earlier species [Key]. Those early species, less fitted for competition under natural conditions, mostly became extinct. Ancestry of the major groups of plants can be traced back to simple single-celled plants similar to *Euglena* [1], which reproduce asexually by mitosis, the process whereby a nucleus reproduces itself without any change in the number of chromosomes.

Relatively early in the evolutionary tree, fungi [3], mosses and liverworts [4] appeared. The latter group, called Bryophyta, has developed multi-cellular sex organs and shows affinities with the ferns and horsetails The fungi developed independently of green plants as a highly specialized group containing no chlorophyll (the green matter that uses the sun's energy to manufacture food) and were therefore unable to obtain food by the photosynthesis characteristic of all green plants. Fungi are either parasites on other living creatures, or saprophytes that obtain energy from dead organic matter. However, the absence of chlorophyll has not impeded the fungi's success as a group – they have evolved to produce a great number of species diverse in form and function and are of importance in the economy of the living world: with the bacteria, they form the most important decomposing agents and they are the scavengers of the plant kingdom. The fungi and algae combined to form the lichens.

Ferns, clubmosses and horsetails

Among the next group of plant types are the pteridophytes – ferns [5], clubmosses and horsetails [6]. These include about 10,000 species and as a group show much variety of form – in particular the specialization in which cells for conducting water up the stem lose their chlorophyll to form a central tube of dead tissue. These tubes may then split into parallel tubes and branches. With the development of the vascular system and a growth tissue (cambium) plants of larger and more complex shape became possible. The primitive life cycle with two generations, one sexual (gametophyte) and one asexual (sporophyte) became more complex. In ferns the gametophyte generation is suppressed and gives rise to the dominant sporophyte.

The range of seed plants

A major development was the evolution of seed plants or Spermatophyta – distinguished by the formation of seed during reproduction

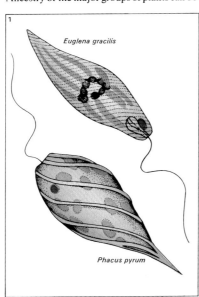

Euglena gracilis

Phacus pyrum

Red limestone seaweed
Corallina officinalis

1 Euglenoids are some of the simplest of all plants. *Euglena* spp are microscopic, single-celled freshwater algae containing chlorophyll, which live by photosynthesis – the process whereby sunlight is converted to chemical energy by green plants. Euglenoids have a worldwide distribution in fresh and salt water and these are just two of about 10,000 living species of single-celled plants.

2 Red algae are mostly deep-water seaweeds. They show great diversity of form, from unicellular to ribbon species.

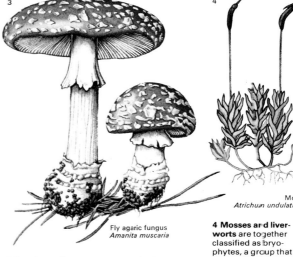

Fly agaric fungus
Amanita muscaria

3 Fungi contain no chlorophyll and must live as parasites on green plants or as saprophytes on dead or decaying matter.

Moss
Atrichum undulatum

4 Mosses and liverworts are together classified as bryophytes, a group that has multicellular sex organs and no true conducting or vascular system. There are about 25,000 species occurring in all parts of the world, but most individual plants are small and inconspicuous. They are important in the formation of soils in many barren regions.

Coast redwood
Sequoia sempervirens

Hart's tongue fern
Phyllitis scolopendrium

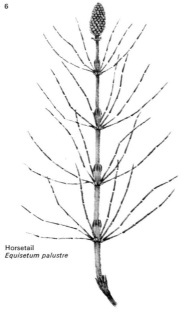

Horsetail
Equisetum palustre

5 Ferns, unlike bryophytes, are mostly large plants. Anatomically they differ from bryophytes in having special water-conducting cells. They also have a region of growing cells – cambium – which allows the development of larger plants such as the tree-ferns of tropical regions. Ferns have an interesting life cycle. The typical "fern plant" such as this one is the asexual stage. It produces spores which grow into the small and inconspicuous sexual stage.

6 Horsetails are allied to ferns and have cylindrical "leaves", borne in whorls, and jointed stems. The spores are produced in cone-like structures at the tips of fertile stems. They are "living fossil" relatives of giant Devonian trees.

7 Gymnosperms are one of the two main groups of seed-bearing plants. They are the group bearing naked seeds and have naked ovules with two or more cotyledons. No gymnosperm has sepals or petals and each flower is unisexual. The largest class of gymnosperms is the cone-bearing conifers. The conifers are wind-pollinated and seed fertilization can be a complex process, taking many years for the development of the pollen, fertilization and finally for the maturation of the seed.

[7] and [8]. These now form the main vegetation over most of the freshwater and land surfaces of the earth and some specialized plants have even adapted to colonize such inhospitable regions as salt marshes, tundra and deserts. The number of seed plant species is estimated at more than 250,000 and included in them is a fascinating variety of form and size. The smallest plant in the group is rootless duckweed *Wolffia*, a tiny green aquatic globular plant less than one millimetre in diameter; the largest are the giant forest redwood trees such as *Sequoia sempervirens* which can grow to 100m (330 ft) or more in height.

All the plants in this group produce seed, almost all by sexual reproduction. Pollen grains are produced by male organs and egg cells form in the female ovule. A pollen grain then fuses with an egg cell to produce an embryo in a seed. The cycle from seed to seed-producing plant can vary greatly in length and many annual weeds have very rapid life cycles – shepherd's purse (*Capsella bursa-pastoris*) can complete a life cycle in 21 days. After producing seeds shepherd's purse

dies but perennial plants can live to a great age. Some specimens of the bristle-cone pine (*Pinus aristata*) may survive to be nearly 5,000 years old.

Seed plants may be distinguished from other types by their flowers [10], which normally have conspicuous petals or sepals (the leaves enveloping the flower) surrounding the reproductive organs. There are some seed plants apparently without flowers, but all have obvious pollen-producing organs and ovules. Bark surrounding the stems of woody seed plants is another characteristic of the group, as is a large, extensive root system.

Vascular plants have special pores or openings in leaf surfaces called stomata [12]. These normally have two kidney-shaped guard cells capable of shrinking and swelling to close or open the pores and by this means the rate at which water droplets transpire from the surfaces of the plant can be controlled. This is most clearly seen in plants with special adaptations to drought conditions; plant species growing in deserts have stomata that are also structured with hairs to aid water conservation.

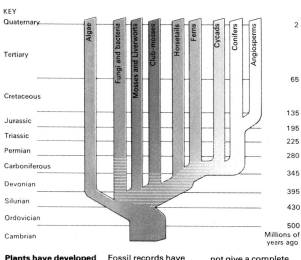

KEY

Plants have developed through the ages from simple unicellular algae to the complex flowering plants that exist in the world today.

Fossil records have been found of most major plant groups. Although these records provide a broad outline of plant evolution, they do

not give a complete picture. Because of the relative scarcity of plant fossils the relationships of the various plant groups is still uncertain.

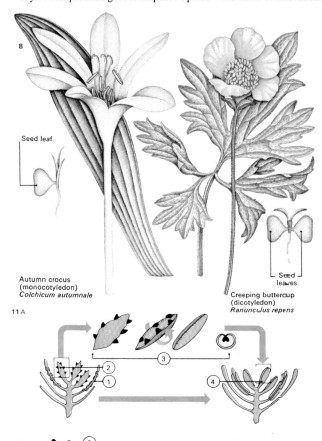

Seed leaf

Autumn crocus (monocotyledon)
Colchicum autumnale

Seed leaves

Creeping buttercup (dicotyledon)
Ranunculus repens

8 Angiosperms are seed-bearing plants with ovules enclosed in ovaries. They are divided into monocotyledons (one seed leaf; parallel-veined leaves) and dicotyledons (two seed leaves; net-veined leaves).

9 Angiosperm fruits show an evolutionary sequence from one ovary with many ovules, as in the sugar apple (*Anona* sp) [A], to an ovary with one ovule, as in wheat (*Triticum* sp) [E], or a number of ovaries with many ovules, as in the tomato (*Lycopersicum* sp) [F]. The cherry (*Prunus* sp) [B], walnut (*Juglans* sp) [C] and orange (*Citrus* sp) [D] represent intermediate stages with one ovary and ovule or many ovaries and one ovule.

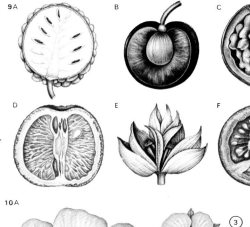

9 A

B

C

D

E

F

10 A

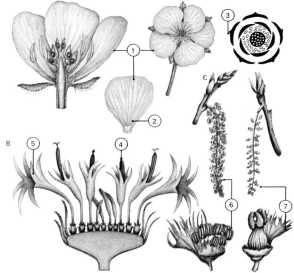

B

10 Flowers contain the sex organs of higher plants. Some such as the buttercup (*Ranunculus* sp) [A] have large petals [1] and honey-guide markings [2]. A flower-plan is shown [3]. Daisy family flowers comprise a number of florets (small flowers). In cornflowers (*Centauria* sp) [B] these are central and fertile [4] or sterile and marginal [5]. Poplar (*Populus* sp) [C] has small, open male [6] or female [7] flowers.

11 A

B

11 Two major theories exist on the way flowering plants evolved. The euanthemum theory [A] suggested that the ancestral flower was fern-like with separate male [1] and female [2] parts or sporophylls. These then folded over [3]

to produce parts of the flower [4]. The pseudanthemum theory [B] supposes that flower evolution began with a complete flower head with both male [5] and female [6] flowers, which gradually [7, 8] became more and more condensed.

12

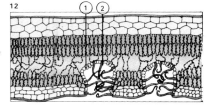

12 A stoma is a pore in the surface layer of cells of leaf or stem, characteristic of seed plants and of the ferns. Each stoma is surrounded by guard cells [1] or hairs [2] which control retention or loss of water as vapour from the plant.

13

13 Outer cells of leaves may form hairlike projections, some of which are specialized secretory cells. The section through the epidermis of a *Thymus* leaf shows a club-shaped gland containing the scented oil of thyme secreted by the plant.

Yeasts and moulds

Moulds and mildews, rusts and smuts are all fungi, members of a vast group of organisms which, like animals, cannot manufacture their own food and, like plants, have no organs of locomotion to go in search of it. All fungi are therefore either parasites, forced to live in or on other living things, or saprophytes [6] living on dead and decaying organic matter. The fungi, a group comprising 100,000 or more distinct species, include not only the microscopic single-celled yeasts [11] but also the familiar mushrooms, toadstools and puff-balls.

Harmful and helpful fungi
Although fungi are chiefly the enemies of the green plant, certain species will attack animals: some trap and kill nematode worms [8], while others cause disease in insects [9]. Athlete's foot and ringworm in man are also fungal diseases. These are minor irritations, however, compared with the enormous problem that fungal diseases cause agriculturists. Roots and shoots, foliage, leaves and fruits of plants – all are subject to attack by various fungi [Key].

Fungi do have their uses, however. Yeasts yield alcohol by causing the fermentation of sugars, the central process in the making of wine and beer. Bakers use yeast in their dough – the fungus produces bubbles of carbon dioxide, so making the bread "rise". Many fungi produce antibiotics, substances that can be used to control bacterial diseases in man and other animals. The members of the genus *Penicillium* [4] are among the most important to man. Species of this genus are not only the source of the antibiotic penicillin (produced by *Penicillium notatum*), but are also used in cheese-making. Many forest trees enjoy mutually beneficial liaisons (mycorrhiza) with fungi – the fungus receives nourishment from the tree which, in turn, receives extra mineral salts from the fungus.

With the exception of the single-celled yeasts, all fungi are constructed of fine filaments called hyphae. These are surrounded by a stout wall made not of cellulose, as in green plants, but of another carbohydrate, fungal chitin, which is similar to the material found in the shells of insects and crustaceans. These filaments ramify through the sub-

stance on which the fungus lives, forming a loose network called a mycelium. In some species elaborate and substantial fruiting bodies are produced by the combination of many thousands of fungal filaments.

Fungal classification
Because fungi cannot search out food, their only hope of finding new sites for colonization is to produce reproductive spores and distribute them widely. The methods of sexual reproduction and spore dispersal so developed are very distinctive and form the basis of fungal classification. Those fungi in which sexual reproduction has never been observed are lumped together in a rag-bag category called "fungi imperfecti". Among them are the grey moulds, the ringworm fungus and the wilt fungi. Most other fungi are grouped together as either phycomycetes, ascomycetes or basidiomycetes depending on the way in which they produce their spores.

There are about 1,800 species of phycomycetes; members of the group are simple fungi that do not produce complex

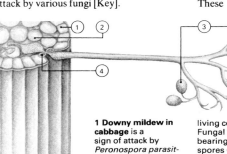

1 Downy mildew in cabbage is a sign of attack by *Peronospora parasitica*, a fungus that enters the plant through the epidermis [1] then ramifies, sending forked feeding tubes into living cells [2]. Fungal filaments bearing dispersal spores (conidia) [3] cause the white mildewy appearance. The spore-bearing filaments emerge through the stomata [4] of the plant.

2 The cup fungus *Peziza* lives saprophytically on dead wood and organic matter in the soil. As its name suggests it forms a cup-like, brightly coloured, fruiting body [A], above the ground. This bears [B] the fertile hymenium layer [A], consisting of sterile hairs [2] and spore-containing asci [3], which point towards the light. The eight ascospores are shot 2—3cm (1in) out into the air.

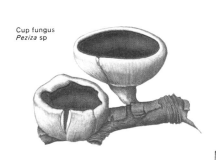

Cup fungus
Peziza sp

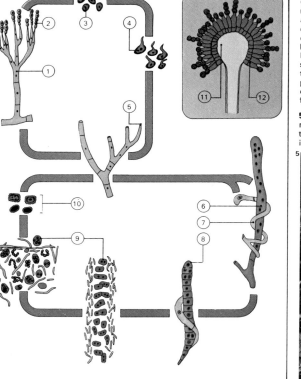

3 Black rust of wheat is caused by *Puccinia graminis*, a fungus that divides its complex life cycle between two hosts. In early summer, spores on an infected wheat plant [A] are dispersed to infect other wheat. Later, new resting spores [C] are produced. These survive the winter on the wheat stalk and germinate [D] in the spring, producing spores to infect the barberry plant [E], on which a form of sexual reproduction [F] produces another wheat infector.

4 The blue mould (*Penicillium* sp) [A] reproduces asexually and sexually. Chains of asexually produced conidia [2] are borne on a hypha [1]. Released [3], the spores germinate [4] to produce new hyphae [5]. Sexual reproduction follows the formation of a female hypha [7] which spirals around a male one [6] and fuses with it [8]. The contents of the hyphae mix and asci [9] containing ascospores [10] are produced. *Aspergillus* [B] produces a vesicle [11], from which stalks bearing conidia (spores) eventually emerge [12].

5 Slime moulds, more like animals than fungi, resemble individual amoebae at one stage. Later the cells join to form a large sporing structure.

fruiting bodies. They include many water moulds, downy mildews [1] and bread moulds [6] and the chytrids, many of which parasitize algae and other fungi. These species produce dispersal spores asexually, either within a special structure, the sporangium (spore-producing body), or at the end of an erect hypha (the externally produced spores are called conidia). They also have a form of sexual reproduction: two identical-looking specimens of bread mould may, in fact, come from different mating strains. When these strains meet, chemical substances pass between them, there is a recombination of nuclear material and, in favourable conditions, a resistant spore suitable for dispersal to new environments is produced.

Ascomycetes are the largest class of fungi and some form stable relationships with algae to produce lichens. Most ascomycetes can reproduce asexually by conidia, but their chief characteristic is the sexually produced ascus, a sac-like cell in which ascospores (generally eight) are produced. Some ascomycete fruiting bodies are edible – good examples are the common morel and the truffle. The group includes the cup fungi [2], flask fungi, powdery mildews and yeasts.

The toadstools, bracket fungi, puff-balls and stink horns are basidiomycetes, a group that also includes the rusts [3], smuts and the commercially significant fungus *Serpula lacrimans,* the cause of dry rot in wood. This fungus produces rhizomorphs – long, creeping strands of fungus that can even grow over dry brick to spread the infection farther. The basidiomycete species all produce four basidiospores on short stalks at the end of a short, club-like basidium. The wind must catch the discharged spores if they are to be dispersed effectively in large numbers.

Settling in with the host
When fungal spores land on a suitable host plant they germinate in the surface moisture, producing a germ tube that gains entry either by piercing the host's outer covering or epidermis, or by growing in through open stomata [1]. Soil fungi can usually gain entrance through the delicate root cells, but some fungi cannot attack living tissue and must enter through wounds or dead matter.

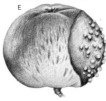

KEY
A   B   C   D   E

Victims of fungal attack among green plants include barley [A], susceptible to loose smut, and rye [B], here bearing ergots of the dangerous *Claviceps purpurea.* Potato crops may be ruined by the potato blight (*Phytophthora infestans*) [C] which attacks the growing leaves of the plant. Wart disease of the potato tuber is caused by *Synchytrium endobioticum* [D]. Apples that become infected with *Sclerotina fructigena* turn brown and rot [E].

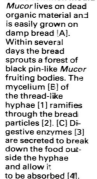

6 The common mould *Mucor* lives on dead organic material and is easily grown on damp bread [A]. Within several days the bread sprouts a forest of black pin-like *Mucor* fruiting bodies. The mycelium [B] of the thread-like hyphae [1] ramifies through the bread particles [2]. [C] Digestive enzymes [3] are secreted to break down the food outside the hyphae and allow it to be absorbed [4].

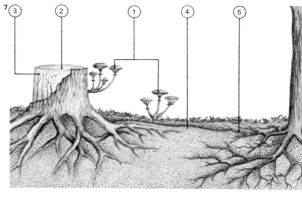

7 The honey fungus (*Armillaria mellea*) is a basidiomycete with conspicuous fruiting bodies [1]. These occur on dead tree stumps [2] where they grow as saprophytes. From the hyphae [3] in the dead stump rhizomorphs – black, root-like strands [4] – creep out through the soil to attack the roots of live trees [5]. The fungus kills living tissue, forming a suffocating network below the bark of the tree.

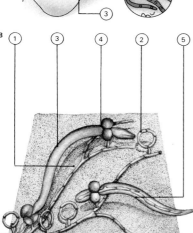

8 Some fungi can trap, kill and digest eelworms. The fungus *Dactylella bembicodes* grows as a mass of threads [1] in rotting wood, which is a favourite haunt of nematode worms. The worm trap [2] consists of a small ring of cells on a short stalk like a lasso. Once in the "noose", the victim [3] is trapped by the suddenly inflating cells [4]. Ramifying fungal hyphae then penetrate [5] the victim which is eventually absorbed by the fungus.

9 A fly dies from fly cholera caused by the phycomycete *Entomophthora muscae* – grim proof that some animals, mostly insects, fall victim to fungal diseases. Spores of the fungus adhere to the fly's body and a germ-tube penetrates its skin. The fungus [B] grows inside the fly and eventually plugs its entire body with hyphae. Spore-bearing filaments break out all over the dead fly, squirting spores in a great halo around the corpse [A].

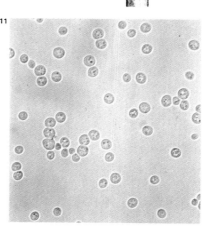

10 Tumours and galls are abnormal growths on plants caused by parasites. Fungi, bacteria and viruses, as well as some small animals, all cause plant galls. These are seen as rapid and uncontrolled cell growth, but rarely kill the plant. It seems certain that substances produced by the invader set off the gall-making process, but their exact composition remains a mystery.

11 Yeasts are among the most commercially important of the fungi. These ascomycetes are simple, unicellular structures. Ubiquitous in nature, they live in weak sugar solutions (such as are found on fruit surfaces). They multiply by budding.

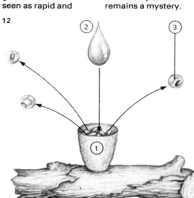

12 Carriage by rain is a novel method of spore dispersal. The basidiomycete bird's-nest fungus (*Crucibulum vulgare*) has a fruiting body like an open vase [1]. Rain [2] bouncing off the inside of the cup removes spore-bearing tissue [3].

Mushrooms and toadstools

Mushrooms and toadstools, sprouting from the earth with their diminutive, fleshy "umbrellas", are among the most familiar fungi known to man. Botanists classify them in the Basidiomycetes – a name they have given to them because their spores are formed on special structures called basidia.

Most people use the term "mushroom" when referring to edible members of the class and "toadstool" when describing poisonous fungi. Botanists or more specifically mycologists (botanists who study fungi) make no such distinction. The term "mushroom" is used as a collective name to include all the larger Basidiomycetes: the gill-bearing fungi (agarics), fleshy fungi with pores (boletes), bracket fungi (polypores), fairy clubs, puff-balls and their allies, stinkhorns, jelly fungi and others. The term "toadstool", on the other hand, is lapsing into disuse amongst botanists.

The life cycle of an agaric
The life cycle of an agaric [Key] begins with the germination of two basidiospores. Each of these carries half the hereditary (genetic) material of the species. When they germinate each basidiospore produces an extensive system of branching threads (hyphae), collectively known as a mycelium. This penetrates the material on which the fungus grows. Usually, before any agaric fruitbodies such as the typical mushroom "umbrella" can be produced, there must be a fusion of the complementary mycelia. The fruitbodies are thus produced from a mycelium with complete hereditary material. A microscopic examination of the resulting fruitbodies shows that the gills (on the underside of the "umbrella") are densely covered with bodies known as basidia – each with four spines (sterigmata), at their apex, producing a single basidiospore to repeat the cycle of growth.

Methods of spore dispersal
Mushrooms produce an enormous number of spores during their very short life-span; an ordinary field mushroom about 10cm (4in) in diameter will produce some 16,000 million spores in a period of about six days.

Except in puff-balls and their relatives basidiospores are discharged explosively from the basidium and although they move only a fraction of a millimetre, this is sufficient to carry them clear of the fertile surface. They then fall free of the fruitbody and are dispersed by the breeze.

Raindrops help to disperse the spores of puff-balls [1], bird's nest fungi and earth stars (*Geastrum* spp) [3].

A number of puff-balls that grow in deserts are released from the sand when ripe and are blown far away by the wind, scattering their spores as they go.

The spores of some mushrooms, especially those that grow in dung, such as *Coprinus* spp, are dispersed by animals. Small rodents such as squirrels may carry agaric fruitbodies for some distance before partially eating them. Slugs and snails also disperse spores over shorter distances. The stinkhorns [2] and their allies are a group of fungi specifically adapted to spore-dispersal by insects. *Aseroe rubra*, for example, attracts insects not only with its powerful smell of putrefying flesh, but also by means of the bright colours of its flower-like, red body.

Mushrooms, even when greenish, lack

1 Puff-balls provide a familiar example of the kind of fungi whose spores are dispersed by the action of raindrops. When ripe, the fruitbody is made up of a thin, perforated, papery wall enclosing a mass of powdery spores that are held between fine cottony threads. When a raindrop falls on the fungus, the wall is momentarily depressed and functions like a pair of bellows as it puffs out a small cloud of spores.

2 Stinkhorns start as a whitish "egg" [1] that is about 5cm (2in) across. Inside its thin skin is a thick layer of jelly surrounding the immature fruitbody (receptacle). When ripe, the receptacle elongates rapidly. It ruptures the egg and carries the glutinous olive spore mass [2] on its conical reticulated cap [4]. Flies, attracted by the putrid smell, eat the gluten [3] and fly away with spores in their gut or on their bodies.

Stinkhorn fungus *Phallus impudicus*

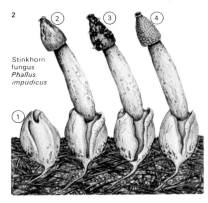

3 The spores of earth stars are dispersed by raindrops in the same way as those of true puff-balls. The fungus, which is at first globular or bulb-shaped, eventually splits into a number of rays. These are grouped round a central thin, papery spore sac that has a distinct opening through which the spore clouds are released. In *Geastrum triplex* the rays curl under the fungus and push it clear of leaf litter, in this way exposing it more effectively to the rain. The rays of other species are hygroscopic: they curl over to enclose and protect the spore sac in dry weather and open again to expose it during rainy periods.

4 Fairy rings [A, B] are produced by a mycelium that grows out equally from a central point. Fungal threads advance and break down the proteins in dead vegetation. As a result, food substances are returned to the soil and an outer ring of lush, green grass springs up. Within this is a ring of poor grass that is thought to be caused either by the mycelium's clogging of the air spaces in the soil – thus preventing water from reaching the grass roots – or else by the fungus's parasitic action on the roots. The third and innermost ring is another lush region that is probably produced by the release of new food materials on the death and decomposition in the soil of the old fungal mycelium.

4 Luxuriant zone — A zone _____ Depressed zone · Normal zone

A

B

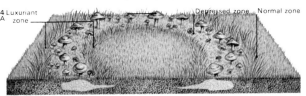

5 The bracket fungus (*Laetiporus sulphureus*) grows chiefly on old oaks. It is easily recognizable when young by its juicy tiers of sulphur yellow or orange fruiting bodies, but the colours fade with age.

6 The cauliflower fungus (*Sparassis crispa*), which measures up to 25cm (10 in) across, grows at the base of conifer trunks and forms intricate masses of small, crisped or wavy lobes. The popular name comes from the fruitbody, which resembles a large waxy cauliflower head, has a spicy smell and is edible. Several species growing in woods in Eurasia and North America are all very similar. These species are also edible. Some polypores of deciduous woodlands are somewhat similar but have far fewer and thicker lobes.

chlorophyll and as a result are unable to manufacture their own carbohydrates by means of photosynthesis. Instead they obtain these substances from the breakdown of complex organic substances in humus and other vegetable debris, or by living parasitically on trees as many bracket fungi do. But many woodland agarics form a special kind of association with the roots of various trees. These fungi form a sheath of mycelium round a tree's rootlets and from this numerous branching threads penetrate between or into the cells of the roots. Deep penetration of the root is usually prevented by a reaction from the tree that inhibits the fungal growths. Such a relationship is known as a mycorrhiza and infected roots are believed to be better able to absorb nutrients from poor soils thus benefiting the tree. The fungus, in its turn, obtains various organic substances.

Some mushrooms, the bracket fungi [5] in particular, cause losses in forestry by spoiling or killing trees. Among these are the birch polypore (*Piptoporus betulinus*), a familiar sight wherever birches grow, and the large perennial, brown, woody brackets of *Gan-*

doderma applanatum, whose stratified fruit-bodies with tube-layers are commonly found on beech and other trees.

Poisonous and non-poisonous mushrooms

Many mushrooms are edible and tasty [6, 7], but some kinds are deadly and there is no infallible way of distinguishing between edible and poisonous species in the kitchen. The only safe way is to learn from an expert. The most dangerous fungus is the death cap [8]. Any mushroom with a white cap, white gills and a stem bearing a ring and a sheathed base should be avoided. The yellow staining mushroom [8], easily confused with the field mushroom, is mildly poisonous. In general all boletes are safe to eat except for those with red spores and red marks on the stem. Most puff-balls are also edible if eaten while still white inside, but the earth balls (*Scleroderma* spp) should be avoided.

Some agarics, notably species of *Psilocybe* found in Mexico, cause hallucinations when eaten; eating such fungi can be dangerous because they are not readily distinguishable from deadly species.

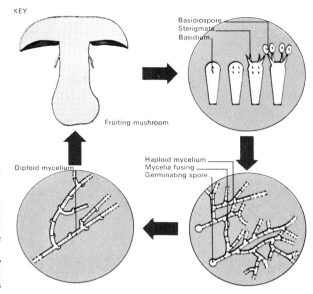

KEY

Basidiospore
Sterigmata
Basidium

Fruiting mushroom

Diploid mycelium

Haploid mycelium
Mycelia fusing
Germinating spore

The life cycle of a field mushroom is typical of higher fungi, but the mycelia may also produce an asexual (mould) stage.

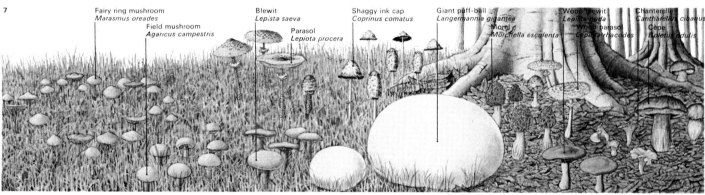

7 Fairy ring mushroom
Marasmius oreades

Field mushroom
Agaricus campestris

Blewit
Lepista saeva

Parasol
Lepiota procera

Shaggy ink cap
Coprinus comatus

Giant puff-ball
Langermannia gigantea

Morel
Morchella esculenta

Wood blewit
Lepista nuda

Wood parasol
Lepiota rhacodes

Chanterelle
Cantharellus cibarius

Cèpe
Boletus edulis

7 Edible mushrooms are numerous. The fairy ring mushroom (*Marasmius oreades*) is rather leathery, with a bell-shaped cap that varies from reddish tan when wet to buff when dry. It can be dried and used for flavouring soups. The common field mushroom (*Agaricus campestris*) can be recognized most easily by the colour of the gills. These are pink at first but eventually turn purplish brown. The blewit (*Lepista saeva*) and wood blewit (*L. nuda*) are closely related but the former grows in pastures while the latter is a woodland species. The parasol (*Lepiota procera*) is different from the wood parasol (*L. rhacodes*) in that it grows in pastures, has brown, snake-like markings on the stalk and flesh that does not turn reddish when it is bruised. Shaggy ink caps (*Coprinus comatus*) have distinctive cylindrical white caps resembling closed umbrellas. When ripe, the gills turn black and the entire cap drips away as an inky liquid from below. Giant puff-balls (*Langermannia gigantea*) may reach 30cm (12in) across and weigh more than 1kg (2.2lb). Morels (*Morchella esculenta*) have a characteristic conical honeycomb-like cap. Chanterelles (*Cantharellus cibarius*) are top-shaped, fleshy, have a smell of apricots and thick, irregularly branched gills. *Hygroporopsis aurantiaca* is similar but less fleshy, has regularly forked gills and does not smell. The cèpe (*Boletus edulis*) has a cap resembling a "penny bun", its spores are whitish and its stem is pale.

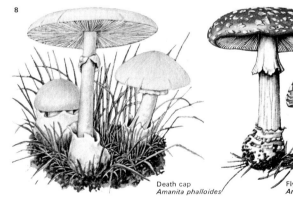

Death cap
Amanita phalloides

Fly agaric
Amanita muscaria

Yellow staining mushroom
Agaricus xanthodermus

Devil's boletus
Boletus satanas

Common earth ball
Scleroderma sp

8 Poisonous mushrooms are not always easily recognizable, even by experts. Fungi of the genus *Amanita* develop a membrane that forms a kind of envelope or veil covering the entire plant. One species, the death cap, is so highly poisonous that it is usually lethal to human beings. It is most commonly found under oak and beech and has a yellowish green, indistinctly streaky cap about 9cm (3.5in) across. The gills are white, and the stem, which is also white, bears a pendulous ring and has a sheathed base. Fly agarics (*A. muscaria*) grow under birch trees with which they form a mycorrhizal association. Their red caps with pyramidal white scales are unmistakable. The yellow staining mushroom can easily be confused with the field mushroom because it grows in similar places. It turns bright yellow when cut or bruised. The plant is only mildly poisonous and some people are unaffected by it. The devil's boletus (*Boletus satanas*), which has a pale cap, red pores and a red network on the stem, grows in chalk beechwoods in southern England. The common earth balls (*Scleroderma* spp), unlike true puff-balls, have a thick rind and a purplish-black spore mass.

The way plants work

Green plants and a few specialized bacteria manufacture all the food that sustains life on earth. Their raw materials are water, carbon dioxide, mineral nutrients and sunlight; plants must have adequate supplies of those essentials and they have evolved special structures to ensure that they do.

Photosynthesis in green plants
All green plants, which range in size from microscopic single-celled algae to trees 100m (328ft) in height, have at least two features in common. First, the cells of all plants are surrounded by a firm wall of cellulose [1]; it may be less than 0.001mm thick but it gives the plant structural strength. Second, all green plants contain the pigment chlorophyll, which gives them their colour and which is vital for photosynthesis, the process by which they manufacture their own food. Chlorophyll is responsible for capturing energy from the sun, which is used for the chemical processes of food manufacture.

Even plants that look brown or red – some seaweeds for example – contain chlorophyll, although its greenness is masked by the presence of other pigments. A few plants have no chlorophyll and must either live parasitically on other plants (for example dodder which parasitises nettles) or strike up a mutual dependence with them (a symbiosis) as the bird's nest orchid does with a fungus. Leaves on parasitic plants are often reduced to mere scales.

Although virtually any part of a plant can contain chlorophyll, the leaves [3] are specialized for photosynthesis and are especially rich in the pigment. In the process of photosynthesis the sun's energy is used to split water in the plant into oxygen and hydrogen. The oxygen is released, but the hydrogen is used to convert carbon dioxide from the atmosphere into sugars (carbohydrates) through a series of complex chemical reactions involving enzymes. Sugars so produced can then be converted into all the other chemicals found in the plant by a further series of reactions.

Food manufactured in the leaves of plants is transported for use in other parts of the plant through tissue called the phloem. This is a series of elongated cells arranged in a system of "veins" running through the plant. The cells are separated from each other at their ends by perforated plates called sieve plates. Fine strands of living material pass through these plates and some researchers believe that these strands transport the food.

The water system
Water moves through a green plant's conducting or vascular system in xylem cells which partner the phloem cells in the veins. Together, the xylem and phloem are called vascular bundles. Xylem cells form a series of long tubes strengthened by stout cell walls. Sometimes these are wide, mutli-cellular structures. To offset evaporation from the plant surface, water is absorbed by the roots and passed up to the leaves [5] by forces that include osmosis (a special type of diffusion) and capillary action (the raising of water in very fine tubes) and which can raise water to the top of the highest tree.

Most plants, whether aquatic or terrestrial, remain anchored in one place, although some microscopic algae move through the water by thrashing whip-like flagella or hair-

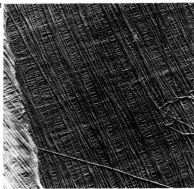

1 The plant cell wall is formed by two layers of cellulose fibres. These fibres lie parallel to each other in the inner layer but criss-cross in the outer layer. Cells communicate with one another through strands of living tissue that pass through fine holes in the cell wall. The holes may aggregate to form pits, in the centre of which is sometimes suspended a disc or torus.

2 All the hereditary information needed to create a whole plant is contained in the genetic material of a single cell. Cells of a carrot plant [1], taken from the branch root [4], storage root [3] or leaves [2] and grown in coconut milk, will all give rise to plantlets [5]. From these plantlets entire carrot plants will grow eventually. Known to botanists as totipotency, this phenomenon occurs only rarely.

4 A single chloroplast [A], typical of those found in the palisade layer of cells [B] of a land plant, is contained in a membrane [1]. Its internal structure comprises a series of minute plates called lamellae, each a few molecules thick. Chlorophyll molecules are found chiefly on these membranes in areas called grana [3]. Between the grana is a granular mass called the stroma [2].

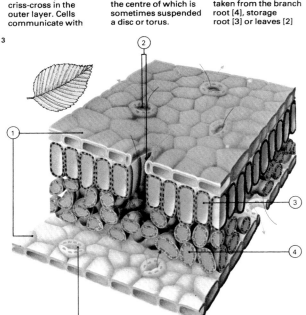

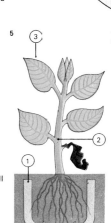

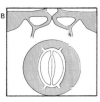

3 Both surfaces of a leaf are protected by epidermal cells [1], covered with a waterproof cuticle that prevents excessive water loss. These impermeable surfaces are pierced by small holes called the stomata [2], found mostly on the leaf's lower surface. Through them water vapour, oxygen and carbon dioxide enter and leave the leaf. Below the epidermis on the upper side is the palisade [3], which consists of cylindrical cells filled with chloroplasts where photosynthesis chiefly takes place. Below the palisade layer is the spongy mesophyll [4] where the products of photosynthesis are stored before being transported to other parts of the plant.

5 Water is absorbed by tiny hairs on the roots of a plant [1]. It is passed upwards in the xylem tissue [2] and evaporates through the stomata [3]. This transpiration process helps to cool the leaves and aids absorption of mineral salts.

6 Stomatal apertures are bordered by special door-like guard cells. The stoma usually open in light [A] to allow photosynthesis and close in darkness [B] to prevent water loss. Modified stomata of desert plants aid water retention.

like cilia. Plants are designed to trap as much sunlight as possible while resisting damage from inclement weather. Water plants are flexible and frond-like to ride with the currents. Land plants orientate their green leaves so that the flat surfaces face the sun. The roots of land plants are firmly embedded in the soil. The shape and form of a plant can vary enormously with its environment – a plant living at the top of a mountain can look superficially quite different from the same species living in a valley.

Unlike animals, plants have no sophisticated nervous system. Their growth is controlled by hormonal or growth substances. These are produced in one part of the plant and transported to another where they exercise their influence. Some hormones, including the auxins, the gibberellins and the cytokinins, promote growth. Of those that inhibit growth, the most common is abscisic acid. Others influence processes such as the ripening of fruit.

All the plant hormones affect plant growth in various ways. Auxin, for example, is chiefly responsible for the elongation of plant cells, but it also plays a part in the initiation of rooting and by counteracting the effects of abscisic acid, in controlling abscission – the process by which a plant "cuts off" dead leaves and ripe fruits before they are shed. Gibberellins play various roles in plant growth but are especially known for their ability to stimulate growth in dwarf plants [7]. Cytokinins are important in the regulation of plant cell division [13].

The importance of sunlight

Sunlight, vital to plants for photosynthesis, also helps to regulate plant growth. Plants grown in the dark are spindly and yellowish [12] because there is no destruction, by light, of auxin and no formation of chlorophyll. Phytochrome, a chemical sensitive to light, is part of the internal clock that tells a plant how long the day is and thus what time of year it is. Some plants flower regardless of the length of day, but others will flower only when days are long or when days are short [11]. This is important ecologically because the best time for flowering and setting seed differs in various parts of the world.

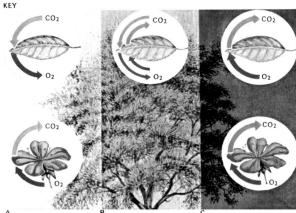

Trees, like all plants, breathe continuously, using oxygen (O_2) and releasing carbon dioxide (CO_2). During the day, photosynthesis takes place in the leaves faster than respiration and only the uptake of carbon dioxide and the release of oxygen can be detected [A]. In twilight, photosynthesis slows down and all oxygen given off is used in respiration while all the carbon dioxide given off in respiration is used in photosynthesis. This is called the compensation point [B]. At night, photosynthesis is halted and it is easy to detect carbon dioxide release and oxygen intake [C].

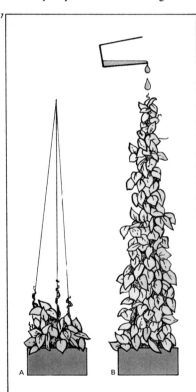

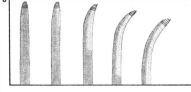

7 A dwarf bean [A], after treatment with a solution containing gibberellic acid, grows as tall as a normal bean [B]. This hormone is one of several able to modify inherited growth patterns.

8 Plant stems bend towards the light because the growth hormone auxin, produced in the tip, is diverted to the shaded side of a plant where it stimulates markedly faster growth.

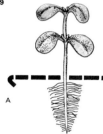

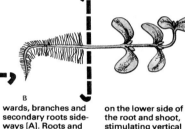

9 Plant response to gravity is called geotropism. Roots normally grow downwards, stems upwards, branches and secondary roots sideways [A]. Roots and shoots of a plant on its side [B] will try to regain their former positions. Deposited starch grains promote auxin accumulation on the lower side of the root and shoot, stimulating vertical growth in the shoot but inhibiting it in the root, which then grows downwards.

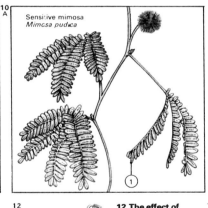

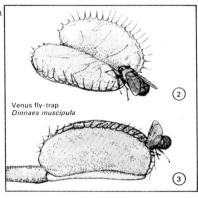

10 Movements of plant parts are sometimes caused not by growth but changes in the structure of the plant's cells. In the sensitive mimosa [A], leaflets [1] are attached by special cells, the pulvini. If a leaf is shaken these cells lose water and the branch may suddenly sag. The Venus fly-trap [B] is a carnivorous plant. When an unwary insect touches long hairs on the leaf [2] an effective closing trap is rapidly sprung [3].

Sensitive mimosa *Mimosa pudica*

Venus fly-trap *Dionaea muscipula*

11 Some plants will flower only on long days, some only on short days. The flowering tobacco plant [A] was grown where daylight hours were long, while the non-flowering plant [B] was not.

Flowering tobacco *Nicotiana tabacum*

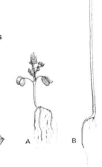

12 The effect of light on the growth of plants is called photomorphogenesis. One mustard seedling has been grown in the light [A]. Another of the same age and genetically identical has been grown in darkness [B]; it is tall and spindly and its leaves and roots have not developed. Light is essential to all green plants and a seedling will use up all the stored food in its seed-leaves in an attempt to grow towards the light.

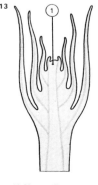

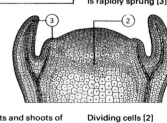

13 New cells are manufactured chiefly at the tips of the roots and shoots of plants by a process called apical growth. The shoot apex of a vascular plant [1] consists of a small group of cells, the meristem, which divides to produce more cells, so the tip constantly grows away from the roots. Dividing cells [2] produce leaf initials [3] and in the axil of each leaf is a bud that could develop into a new branch. The apical bud may prevent the growth of lateral buds by producing an inhibiting hormone; this is apical dominance.

Algae and seaweeds

The simplest plants are those in which a single cell is capable of independent existence. These, together with their multi-cellular relatives, are called algae and there are about 20,000 species of them in the world today. Algae are technically distinguished from higher plants by the absence of a sterile cell layer round the sexual cells and, although some algae can be large and complex, they never attain the degree of anatomical differentiation of flowering plants and ferns.

The classification of algae

Algae are a rather loose assemblage of not very closely related organisms that can be classified into 11 major groups based on differences in pigmentation, cell-wall composition, nature of reserve food materials and type of organelles – parts of cells doing special work – present in cells. These groups are the blue-green algae (Cyanophyta), red algae (Rhodophyta), diatoms (Bacillariophyta), yellow-green algae (Xanthophyta), golden-brown algae (Chrysophyta), dinoflagellates (Pyrrophyta), cryptomonads (Cryptophyta), brown algae (Phaeophyta), euglenoids (Euglenophyta), green algae (Chlorophyta) and stoneworts (Charophyta).

All the groups of algae possess chlorophyll "A", the green pigment primarily responsible for photosynthesis, the process by which the plant converts light energy into chemical energy. Indeed the green algae have the same pigments in the same proportions as the higher plants and are thought to be the ancestral stock from which these have evolved over millions of years. In other groups the accessory pigments differ in nature or in proportions, giving them their characteristic colours. These may, however, be much modified both in different species and according to environmental conditions.

Blue-green algae [7] are in many respects more like bacteria than like other algae. They lack true organelles, have little or no power of movement and no truly sexual reproductive processes. In all the other groups each cell has a true nucleus (an organelle with its own surrounding membrane), which controls important cell processes such as division, and one or more plastids – organelles that contain the photosynthetic pigments. In most groups (except the red algae) there are at least some species or stages bearing flagella – thin, flexible, hair-like structures that beat rhythmically and serve as the principal means of locomotion. In some groups – the brown algae [4, 5], for example – they are restricted to reproductive cells. Differences in number, position and kind of flagella occur among different major groups and help to distinguish them. Some nonflagellate algae, particularly certain diatoms, are capable of quite rapid gliding movements whose mechanism is not fully understood.

Natural habitats of algae

Algae occur in almost all possible habitats to which adequate light penetrates. They may colour permanent snow and ice fields red or green, may multiply rapidly in small temporary puddles (thus often giving rise to stories of coloured rain) and have been found in hot springs at temperatures of up to 85°C (185°F). They abound in brackish and marine waters and are found even in saline pools where the salts are crystallizing out.

1 Chlamydomonas [A] has a cup-shaped plastid [1], two equal flagella [2] and a red eyespot [3]. It usually reproduces by division of the cell contents [B] into 4 to 16 motile cells (zoospores) that then disperse. In sexual reproduction the divided products (gametes) fuse together in pairs [C, D] after release, forming a non-motile resting stage [E] that produces four zoospores [F] when conditions are favourable. The organism lives mostly in freshwater, especially stagnant pools, where it swims about in response to the light. The red eyespot is sensitive to light and determines the direction of its movement.

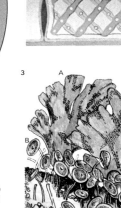

2 Kinds of algae include *Volvox* [A] with motile *Chlamydomonas*-like cells; single-celled desmids [B], varied and beautiful; and *Spirogyra* [C], whose filaments consist of cells joined end to end.

3 The sea lettuce or green laver (*Ulva lactuca*) is a familiar seaweed [A] that is occasionally eaten. *Acetabularia* [B] is one of a group of lime-encrusted green algae from tropical and sub-tropical seas. It resembles a delicate toadstool and has been widely used and studied by scientists because the single-celled stalk, up to 5cm (2in) high, can be cut into two living parts with and without a nucleus.

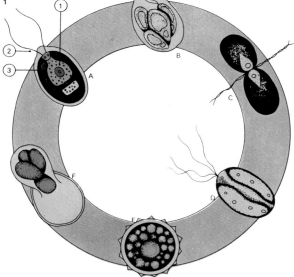

4 The life history of *Fucus* is often used as an example for the brown algae because species of this genus are common. The most usual form of reproduction is a specialized sexual process. On the tips of certain fronds are swollen areas known as receptacles [1]. Male and female reproductive organs are produced on separate plants in cavities in the swollen areas. Male reproductive organs (antheridia) are released in spring as orange slime [2]. Female reproductive organs (oogonia) [3] are similarly released as a green slime. When the seaweed is covered by the tide the antheridia [4] and oogonia [5] burst, releasing the reproductive cells into the water. The male gametes [6], 64 per antheridium, have two unequal flagella which they use to propel themselves. They fuse [8] with non-motile egg cells [7], eight per oogonium in the water. In a few other groups the gametes are as in *Fucus*, but in most groups all the gametes produced are motile. In all these groups non-sexual reproduction also occurs: the contents of some cells divide and are released as motile flagellate zoospores, which later develop into new plants.

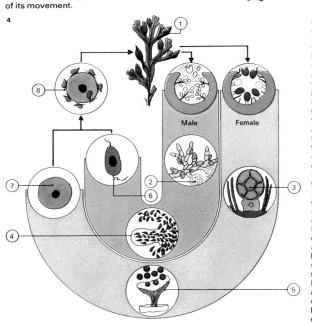

5 Brown algae range from small filamentous types to large plants with elaborate structures. *Dictyota* [A] is anchored by rhizoids and has thin flat fronds. New plants are all of this kind. But in many brown algae two different kinds of plants are produced. In furbelows, (*Saccorhiza polyschides*) [B], the zoospores produce microscopic filaments, which in turn produce gametes. Fusion of gametes results in the growth of a plant with a frilled stem and fan-like frond.

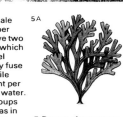
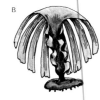

Although the smaller forms are individually inconspicuous they may multiply rapidly enough to colour the water. The Red Sea was so named because of the frequency in its waters of a reddish blue-green alga, *Trichodesmium*. Such water blooms may cause fish and other aquatic organisms to suffocate, taint drinking water and, as Red Tide organisms (certain dinoflagellate species) do, poison the waters with powerful toxins. Algae also play an important role in purifying sewage effluents by removing dissolved substances and keeping the waters oxygenated.

Some algae remain suspended in the upper layers of the water and together with small drifting animals form the plankton that is a rich food source for higher animals. Others, which are known as benthic forms, occur on or among bottom deposits or become attached to plants and animals on the bottom. These are limited in development by their need for adequate light for photosynthesis, but in very clear waters may be found at depths of as much as 200m (656ft). The richest vegetation on the sea coast occurs just

below the level exposed to air by receding tides, but a wide variety of species live between tides, often forming distinctive zones at different levels characterized by the species.

Algae also inhabit non-aquatic regions. Large numbers grow in the upper layers of soils (even in deserts), on buildings, trees and other terrestrial plants, and even in animals such as some sponges, flatworms and shellfish. Algae even grow on corals, where they play a fundamental role in the immense productivity of coral reefs.

Fossilized deposits

In most of the major groups of algae some or nearly all members have lime or silica deposited either in the cell walls or as a form of external shell. These hard parts are readily preserved as fossils and, indeed, fossil diatom deposits 914m (3,000ft) thick have been found. Such fossilized forms can give an indication of the ages of the groups to which they belong. Among red algae the calcareous corallines were well developed 500 million years ago and arose much earlier. Ancient calcareous green algae are also known.

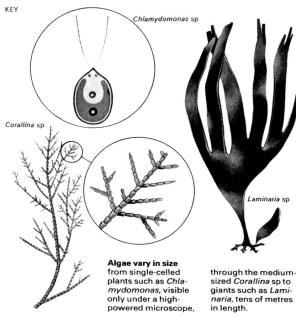

Algae vary in size from single-celled plants such as *Chlamydomonas*, visible only under a high-powered microscope, through the medium-sized *Corallina* sp to giants such as *Laminaria*, tens of metres in length.

6 The red algae are mainly marine. Dulse (*Palmaria palmata*) [A], is a common edible species, as is Irish moss (*Gigartina stellata*) [B] and laver (*Porphyra umbilicus*) [H]. *Dumontia incrassata* [C] is common in shallow inter-tidal pools. It dies in late summer but new plants appear in spring. *Phymatolithon polymorphum* [E] is a common encrusting coralline species, both inter-tidally and sub-tidally. The beautiful crimson fronds of *Heterosiphonia plumosa* [F] grow best sub-tidally while the stiff, dark red fronds of *Ptilota plumosa* [G], are most frequently found growing on the stripes of the oarweed (*Laminaria hyperborea*) [D].

8 The uses of seaweed are many and varied. Traditionally they have been used as manure and for food, for which they are specially cultivated in Japan. They once provided a source of iodine and soda (for glass-making). Nowadays, jellylike substances are extracted – agar from red algae and alginates from brown algae. These are colourless, tasteless and odourless and are used for growing microbes, food canning, making non-fat creams (edible and cosmetic), emulsifiers for paint, beer, ice-cream, tablets, capsules, for film emulsion and artificial thread. Fossil diatoms provide abrasives and are used in making toothpastes, filters, absorbents and insulating material.

7 The blue-green algae are a primitive but successful group well represented in terrestrial, marine and freshwater habitats, being particularly abundant in muddy places such as salt marshes and paddy fields. They consist of simple cells aggregated into loose or regular colonies such as *Dermocarpa*, or filaments – branched or unbranched. The colonies may be embedded in a gelatinous matrix. Reproduction is vegetative or by non-motile spores. Some filamentous types such as *Anabaena* have thick-walled cells and are capable of converting atmospheric nitrogen into ammonia.

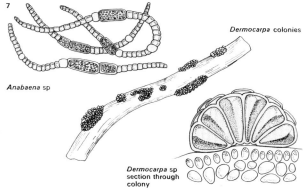

The lichens

Lichens occupy some of the most forbidding regions of earth, establishing themselves in environments where few other species of living things are able to survive. They are found farther south and farther north than any other type of plant. In the Himalayas they have been found at altitudes of more than 5,600m (18,460ft).

Distribution and structure

Lichens can adapt themselves to almost any kind of surface. They will grow on sun-baked rock in arid deserts, on the backs of weevils or in the bleached skulls of dead animals. One species (*Verrucaria serpuloides*) is continuously submerged in freezing Antarctic waters; another (*Lecanora esculenta*) travels on the wind. And although lichens are on the whole highly sensitive to industrial pollution, *Lecanora conizaeoides* has actually increased in areas of high industrial pollution.

There are some 15,000 known species of lichens, classified into three main groups on the basis of their characteristic growth forms. The foliose (leafy) lichens [3] flourish in regions where the rainfall is heavy. Crustose

(crust-forming) lichens [2], which cling close to a surface, are resistant to drought and prevail in deserts. The fruticose (stalked) lichens absorb water in the form of vapour and are particularly suited to humid climates. As a result, lichens vary greatly in both size and appearance. Some have filaments that reach lengths of 2.75m (9ft) or more. Other lichens are smaller than a pinhead.

Lichens arise from the association between two distinct plant types, an alga and a fungus, and provide one of the most successful examples of the symbiotic relationship called "mutualism" – a term used to describe the beneficial partnership that can develop between two dissimilar organisms.

The algal partner of a lichen belongs to either the blue-green algae (Cyanophyta) or to the green algae (Chlorophyta). Most fungal members belong to the fungal group known as Ascomycetes, which derive their name from the ascus or sac in which they form their reproductive cells. With very few exceptions, a lichen is the product of a single species of fungus and a single species of alga. The most common alga (found in more than

50 per cent of lichens) is the single-celled green alga *Trebouxia*.

Lichens are extremely slow-growing plants. Most crustose types rarely grow more than 1mm (0.04in) a year. Other types grow a little faster but rarely exceed more than 1cm (0.4in) a year. It follows, therefore, that the larger lichens are extremely old – some that occur in the Arctic are thought to be more than 4,000 years old. A technique known as lichenometry uses lichens for dating rock surfaces. The ages of glaciers and of the giant megaliths (large stone structures) that are found on Easter Island in the South Pacific have been determined by this technique.

The roles of fungus and alga

The great ages attained by lichens indicate that they have evolved a highly organized and well-balanced relationship between the algal and fungal partners. The exact nature of that relationship, however, is still not entirely clear. The alga is a simple green plant which, like other green plants, manufactures its food by the process of photosynthesis – the conversion of the carbon dioxide and water

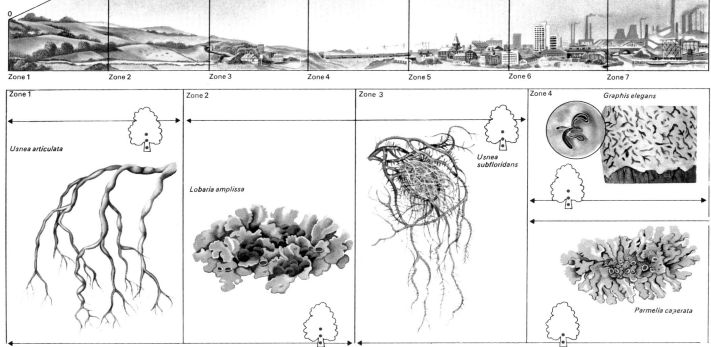

1 By mapping lichens it is possible to pinpoint areas of high and low pollution in a given region and to assess what importance factors such as wind direction, elevation and ground shape have on the sources of pollution. The diagram shows the monitoring of such a region, based on specific lichens and of the green alga (*Pleurococcus viridis*) found on oak trees compared with a graph indicating the varying levels of the poisonous gas sulphur dioxide (SO_2). Few lichens are found where the sulphur dioxide concentration is very high: the result is a "lichen desert". Outside this is a "struggle zone", which merges into the normal lichen flora. Species like *Lecanora conizaeoides* have increased in distribution as the concentration of sulphur dioxide has increased over the British countryside. Some species of lichen occur in more than one zone. Species present in zone 1 are *Usnea articulata, Lobaria amplissima*; zone 2 – *L. amplissima, Usnea subfloridans*; zone 3 – *Usnea subfloridans, Parmelia caperata*; zone 4 – *Graphis elegans, Parmelia caperata, Parmelia saxatilis*; zone 5 – *Lecidea scalaris, Parmelia saxatilis*; zone 6 – *Lepraria incana, Pleurococcus viridis*; zone 7 – *Pleurococcus viridis, Lecanora conizaeoides*.

Location on oak tree
Branches
Trunk
Base

SO_2 mg/m³

SO_2 mg/m³ 180
140
100
65
45
40
30
0

Zone 1   Zone 2   Zone 3   Zone 4   Zone 5   Zone 6   Zone 7

Zone 1
Usnea articulata

Zone 2
Lobaria amplissa

Zone 3
Usnea subfloridans

Zone 4
Graphis elegans
Parmelia caperata

vapour contained in the air into carbohydrates by exposure to light. On the other hand, the fungal partner does not have any of the green matter (chlorophyll), which is essential for photosynthesis.

The simple carbohydrate formed in the algal layer is excreted, absorbed by the fungus and transformed by it into a different carbohydrate. This carbohydrate flow is the basis of the symbiotic relationship that has developed in the lichen. The transfer of nutrients from alga to fungus is swift: fungi have been found to convert sugars derived from algae into fungal products within three minutes of the start of photosynthesis. The alga may also provide the fungus with vitamins and stimulate the growth of the lichen. For its part, the fungus takes in water vapour, which accelerates the rate of photosynthesis in the alga, while the fungus provides shade for the alga underneath – some algae such as *Trebouxia* dislike strong light.

The uses of lichens

Lichens have many uses – as food for animals (reindeer obtain two-thirds of their food supply from lichen [4]), as nesting material for birds and as shelter and homes for hundreds of species of small invertebrates such as mites, beetles, snails and moths. Man also has derived benefit from lichens. For more than two centuries Icelandic moss (*Cetraria islandica*) has been prescribed for chest and cough ailments. Usnic acid, present in some lichens, is used to combat infection from superficial wounds and also in the treatment of tuberculosis. Modern medical research has revealed the existence of antibiotic substances effective against diseases such as scarlet fever and pneumonia and lichens are used in other industrial processes. Litmus, a chemical indicator that turns red in acid solutions and blue in alkaline ones, is derived from *Roccella* sp of lichen, which grow in North Africa and the Canary Islands.

Lichens are particularly sensitive indicators of atmospheric pollution [1]. Radioactive derivatives of strontium and caesium are absorbed and retained in high concentrations. This poses a great threat to the reindeer and – ultimately – to those (Eskimos and other northern peoples) who eat them.

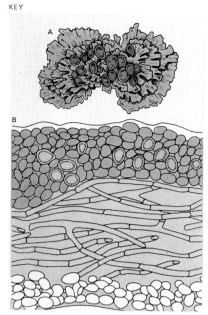

Lichens arise from the mutually beneficial partnership that can develop between an alga and a fungus. *Xanthoria parietina* [A] is commonly found along coasts on rocks and on the walls and roofs of many buildings. The bright orange, saucer-shaped fruiting bodies (apothecia) differ little in structure from those of an isolated fungus. A section through a lichen [B] shows a thin upper layer of tightly packed fungal strands. Embedded in this upper layer are isolated green algal cells. The main body of the lichen is made up of enmeshed fungal strands, below which is another thin layer similar to the upper one.

4 Reindeer moss (*Cladonia rangiferina*) [B] is the staple diet of reindeer [A] in the long months of the Arctic winter when there is little other green vegetation available for consumption.

Other deer and musk oxen also graze large quantities of it. Lichens contain a high proportion of carbohydrates which provide the large amounts of energy required in such extreme conditions.

However, lichens are low in other nutrients. A diet based exclusively on lichens produces symptoms of protein deficiency. In Scandinavia some species of *Cladonia* are used to make alcohol and sugar.

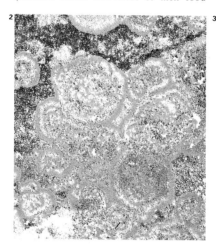

2 Encrusting lichens cling close to the surface to which they attach themselves. The species shown here is *Caloplaca heppiana*. It is frequently to be found growing on walls and tombstones. (This and similar species of lichen can be used for dating purposes.) Such lichens are often brightly coloured – the pigment being derived from the fungal partner.

3 A typical leafy lichen is the tree lungwort (*Lobaria pulmonaria*), one of 70 species which are mostly subtropical. The tree lungwort occurs in the northern temperate zone. Its "leaves" can grow to lengths of 10cm (4in). When wet they are bright green; when dry, yellowish brown. Reproduction is by means of spores produced in cup-shaped structures on the "leaf" surface.

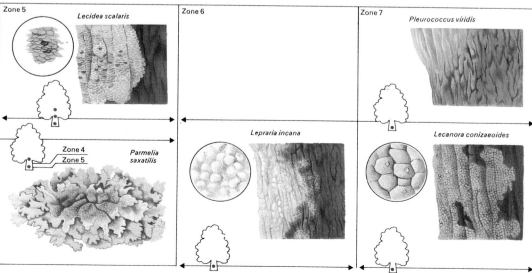

Zone 5 — *Lecidea scalaris*

Zone 4 / Zone 5 — *Parmelia saxatilis*

Zone 6 — *Lepraria incana*

Zone 7 — *Pleurococcus viridis* / *Lecanora conizaeoides*

5 Lichen extracts were once used to colour the cloth for Scottish kilts. The lichens coloured the cloth yellow, brown, red and purple. Intermediate colours were obtained by overdyeing. Lichen extracts are now used in the manufacture of an antibiotic, a perfume and litmus.

Mosses and liverworts

Mosses, liverworts and hornworts – collectively known as the bryophytes – are among the simplest and most primitive of land plants. The mosses total some 14,000 species, the liverworts about 8,000, the majority of both groups being tropical. Most species favour moist, shady habitats in soil or dead leaves or on bark or rocks. Some mosses, however, are able to survive many months of drought and desiccation.

The structure of bryophytes

It is generally assumed that the bryophytes evolved from green algae and they still retain many primitive features that betray this ancestry [Key]. Mosses and their allies are structurally very simple. Some bryophytes bear a superficial resemblance to flowering plants but although they appear to have stems, leaves and roots these structures are much simpler than in higher plants. Liverworts are so called because they are often roughly liver-shaped and in the Middle Ages were reputed to be good for liver disease.

Leafy liverworts, such as *Lophocolea*, a common liverwort found on damp logs and tree stumps, display "leaves" carried on a central "stem". In the mosses, this adaptation is taken a step further, and the thallus shape is lost completely.

The differences between mosses and leafy liverworts can be seen externally. The leaves of liverworts never have a central midrib and are nearly always in two or three rows. Moss leaves are never divided in shape and are rarely in more than one row. The rhizoids – the fine hair-like structures that root the plant in the soil and absorb nutrients – are made up of a single cell in the liverworts, but have many cells in the mosses. They also differ in the shape of the capsule – the spore-containing structure – and in the way that the spores are dispersed. In liverworts, the capsule is oval or spherical and the spores when mature are dispersed through four longitudinal splits which develop in the capsule wall. In mosses, the capsule is usually cylindrical and often has a lid that is shed when the spores are mature, exposing a circle of "teeth" round the open "mouth" [1].

Unlike higher plants, most bryophytes have no specialized cells to conduct water and food round the plant (some mosses are an exception), nor do they possess woody tissues which could support a large body. Thus all bryophytes are comparatively small.

The need for water

Despite all these primitive features, bryophytes show significant adaptations to life on land compared with their algal ancestors, which are nearly all confined to water. The most significant adaptation to the land habit is a pattern of alternation of generations in which the plants producing sexual organs are dominant and the plants producing asexual spores are virtually parasitic on the dominant form [1].

Alternation of generations is the rule in virtually all plant species. It implies that there are two distinct stages in the plant life cycle, one which has only one complete set of chromosomes and another, formed as a result of sexual reproduction, which has two sets of chromosomes, one from each parent. In the algae, both generations may look identical, but in the bryophytes each is distinct.

The reason for separate generations is

CONNECTIONS

See also
38 The plant kingdom
52 Ferns and horsetails
44 Algae and seaweeds
176 Plants of the past

In other volumes
136 The Physical Earth

1 The life cycle of a moss, *Polytrichum,* begins with the dispersal of the tiny spherical spores by a censer mechanism through gaps in the teeth guarding the apex of the capsule [1]. The spores [2] germinate to form a branched filament, the protonema [3]. The upper parts turn green and buds appear on those filaments that grow into a plant [4]. Mature male [5] and female [7] *Polytrichum* plants are similar. Male sex organs [6] are borne in a flower-like open cup at the shoot apex. Female sex organs [8] are borne in a cluster, also at the shoot apex. Male sperms [9] swim from male to female plant and fertilize the egg [10]. This develops into an embryo [11] that gives rise to the moss sporophyte generation. The sporophyte borne on the female [12] has a capsule holding spores [13] produced by meiosis.

2 The thallus of the liverwort *Marchantia* in cross-section [A] shows unspecialized cells [1] bounded by surface layers [2]. Scales and rhizoids [3] are found on the lower surface. The moss "stem" [B] shows a central cylinder of cells [4] that conduct water up the stem. The surrounding cells [5] conduct food. Unspecialized cells [6] lie between these and the outer epidermal layer.

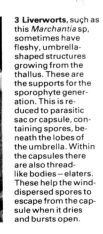

3 Liverworts, such as this *Marchantia* sp, sometimes have fleshy, umbrella-shaped structures growing from the thallus. These are the supports for the sporophyte generation. This is reduced to parasitic sac or capsule, containing spores, beneath the lobes of the umbrella. Within the capsules there are also thread-like bodies – elaters. These help the wind-dispersed spores to escape from the capsule when it dries and bursts open.

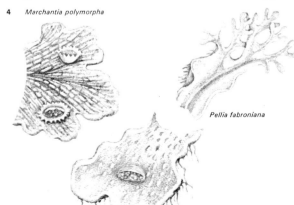

4 *Marchantia polymorpha*

Pellia fabroniana

Lunularia cruciata

4 Dispersal units, gemmae, enable mosses and liverworts to regenerate entire adult plants vegetatively from small pieces of plant tissue. Gemmae may be spherical or plate-like and are borne in small cups or envelopes on the thallus surface. *Marchantia* produces large disc-shaped gemmae, while those of *Lunularia* are biconvex and just visible to the naked eye. In *Pellia*, the shoots and the margin of the thallus separate to form new plants.

that bryophytes need water for sexual reproduction. The male sperm must swim – using two whip-like flagella – to the female egg. Because the male reproductive organs and the female reproductive organs may be borne on separate plants, there has to be an unbroken film of water extending from the male organs to those of the female. On land, this occurs only during rainfall or after a heavy dew, so the evolution of sexual reproduction was more risky and less dependable as a means of colonizing new environments. The bryophyte answer was to find a way of making each mating much more effective; the fertilized egg became not a mature bryophyte plant but a spore-producing organism, or sporophyte, dependent on the mature plant for its food. Each of the thousands of spores produced as a result of one mating could give rise to a new plant.

In both mosses and liverworts, the sporophyte consists of a foot embedded in the tissues of the bryophyte plant. This foot usually develops a long stalk and a capsule which contains and disperses the spores. *Anthoceros*, the hornwort, is distinct from all other bryophytes because its spore-producing structure consists of only a foot and a capsule.

Bryophytes can also reproduce vegetatively, often from small detached fragments. Many have developed specialized deciduous shoots or groups of cells called gemmae from which the entire adult can regenerate [4].

Important bryophytes

Sphagnum [6], the moss of the peat bogs, is another untypical bryophyte as well as being the only one which is of any commercial importance, being widely used in horticulture. Because it is able to soak up and retain large amounts of water it can be used as a surface mulch for water conservation. Dried and partially decomposed it may be added to heavy clay soils to improve their texture. Peat is dug and burned as a fossil fuel in some parts of northern Europe. Bryophytes also play a significant role in soil ecology. They prevent erosion and retain surface water, and they help to break down rocks to soil.

Bryophytes, however, are believed not to have given rise to any other plant groups.

KEY

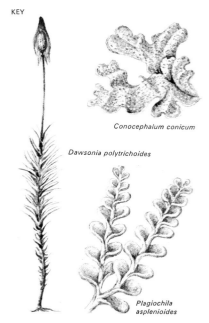

Conocephalum conicum

Dawsonia polytrichoides

Plagiochila asplenioides

Mosses and liverworts are almost certainly descended from algae and show stages in the development of a land habit. The thallose liverwort *Conocephalum conicum*, for example, with dark green, ribbon-like branches, looks rather like a thalloid green alga. *Plagiochila asplenioides*, a leafy liverwort, shows a new development: the thallus has become divided to give the appearance of a central "stem" bearing large rounded "leaves". This development is most marked in mosses (*Dawsonia polytrichoides* is an Australian example) where the plant grows in an upright manner and the "leaves" and "stem" are seen to be well developed.

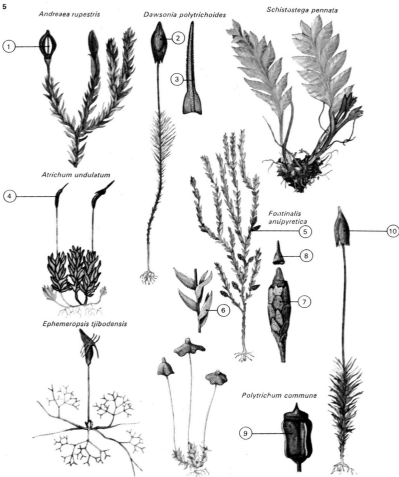

5

Andreaea rupestris

Dawsonia polytrichoides

Schistostega pennata

Atrichum undulatum

Fontinalis antipyretica

Ephemeropsis tjibodensis

Polytrichum commune

Splachnum luteum

5 Mosses vary in growth and colour according to species. *Andreaea rupestris* is dark reddish-brown with tiny capsules [1] that open by four longitudinal slits. *Dawsonia polytrichoides* is an Australian species with a distinctive capsule [2] and thin, pointed leaves [3]. *Schistostega pennata*, a European moss, has flattened, translucent leaves. *Atrichum undulatum* is common on heaths and in woods, and has a capsule [4] with a long, pointed cap. *Fontinalis antipyretica* is an aquatic moss [5] whose boat-shaped leaves have a sharp keel [6]; the capsules are oblong or cylindrical [5, 7] and there is a pointed cap [8]. *Ephemeropsis tjibodensis* has an extensive protonema, while the mature plant is small. *Splachnum luteum* has a spore-producing body that resembles an umbrella. *Polytrichum commune* is extremely common and has a capsule [9] that looks like a four-sided box. It bears a long, golden brown cap [10] which is released before the spores are dispersed.

6

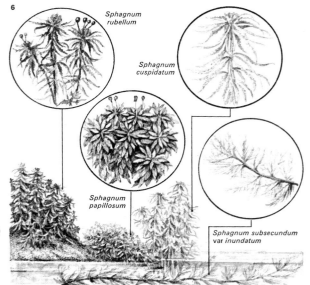

Sphagnum rubellum

Sphagnum cuspidatum

Sphagnum papillosum

Sphagnum subsecundum var *inundatum*

6 The peat mosses (*Sphagnum* spp) can absorb and retain enormous amounts of water – over 25 times their own dry weight. They can reproduce asexually extremely efficiently by separating off branches. As a result, open stretches of water that are colonized by *Sphagnum* species are drained by the ever encroaching moss, while the level is raised by dead and decaying moss tissue accumulating on the surface. At an early stage in the formation of a peat bog, the water surface is covered with a dense mat of *Sphagnum* and other species, forming a treacherous "quaking bog". These bogs are capable eventually of supporting large trees and in time the swamp is entirely replaced by woodland.

7

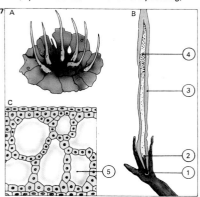

7 The hornwort (*Anthoceros* sp) [A], is a rare plant found in moist, shady places. Long sporophytes [B], growing from the upper surface of the plant, are fed from the gametophyte [1] by a foot [2]. Fertile spores [3] develop from the tissues round the sterile central column [4] and are released when the capsule splits. The plant [C] often has cavities [5] containing colonies of blue-green algae.

Ferns and horsetails

Ferns, horsetails and club-mosses, the pteridophytes, are the most primitive members of the plant kingdom and have a system of tubes (a vascular system) to conduct water and food around the plant [2]. This characteristic they share with flowering plants and, like the mosses and liverworts, the pteridophytes have a life cycle [1] that shows well-marked alternation of generations.

The pteridophyte life cycle

The dominant part of the pteridophyte plant's development is known as the sporophyte stage and is involved with the production of asexual spores. When they germinate, fern spores give rise to small, insignificant plants (prothalli) that look rather like liverworts. These are called gametophytes because they produce the sex cells. For male and female sex cells to come together, fuse and give rise to a new spore-producing plant, water is essential because without it the male cells could not "swim" to the female cells. This need for water explains why most ferns and horsetails can exist only in damp environments and also why they are

called cryptogams, a name that means "hidden marriage". To early botanists the sexual lives of ferns were a mystery because they did not produce easily visible structures such as pollen, fruits and seeds in the way that flowering plants do.

Where ferns are found

Ferns and horsetails have become greatly reduced in numbers from their abundance in the Carboniferous period [Key], but this does not mean that they are no longer of importance in the plant population. One has merely to see a hillside of bracken turn golden-brown in autumn to realize that some species of fern are still very common indeed. The key role of water in the reproduction of ferns means that most species occur in warm, damp regions. In forest and jungle, the enormous complex fronds of tropical ferns, quite unsuited to dry or exposed habitats, are ideally suited to capturing the maximum amount of light available in the forest shade. Ferns also occur in many other habitats – even as far north as the Arctic tundra – and can grow in amazing abundance.

There are about 10,000 species of ferns. The greatest number and diversity (about 2,500 species) are found in South-East Asia where ferns and horsetails [7] cover much of the jungle floor and also grow in the boles of the trees. In contrast there are only about 150 species of ferns and fern allies to be found growing in the countries of Europe.

Useful and ornamental species

Few pteridophytes are of economic importance but some are eaten, especially in Japan. Spores of the club-moss *Lycopodium*, which form a very light, mobile powder, were once used for demonstration purposes in physics experiments to reveal the vibration patterns on an object, such as a drum skin, that was generating sound. An extract of the male fern (*Dryopteris* sp) was once commonly used as a highly effective and relatively safe vermifuge or "worm powder" for human beings – other substances used for this purpose such as yew leaves were rather hazardous. Bracken (*Pteridium aquilinum*) is one of the very few ferns regarded by man as a "weed". With its rapid growth and persistence it is responsible

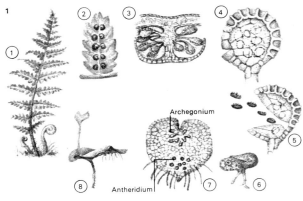

1 The dominant plant in the life cycle of *Dryopteris*, the male fern, is the spore producer [1]. From brown patches, the sori [3], on the leaf undersides [2] asexual spores [5] are produced from masses of tissues known as sporangia [4]. The spores float dust-like on the breeze and germinate in damp places [6]. Each forms a heart-shaped prothallus [7] that measures 1–2cm (0.5–0.75in) long. On the underside of the prothallus male (antheridia) and female (archegonia) organs containing sex cells (gametes) are formed. The motile male gamete swims to and fuses with the female. Thence a new sporophyte forms [8].

2 Ferns have more primitive systems for conducting water and mineral salts (vascular systems) than do flowering plants. The xylem or water-conducting tissue [1] is made up of tracheids, single elongated cells with stiffening. Instead of the xylem and phloem (food-conducting tissue) [2] being arranged in bundles as they are in higher plants there is a more primitive arrangement. The simplest fern stem [A] consists of a solid mass of xylem completely surrounded by phloem, pericycle [3] and endodermis [4]. This is a protostele. Another arrangement is the solenostele [B], which may be hollow or contain pith [5].

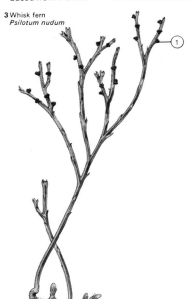

3 Whisk fern *Psilotum nudum*

3 The whisk fern is a primitive fern relative belonging to a group known as the Psilotales. It grows over much of the tropical and subtropical world but only "squeezes" into Europe. It has a green stem bearing bilobed spore leaves (sporophylls). Each of these has a three-chambered sporangium (spore bearer) [1] on it. The plant has no roots but instead bears a swollen underground stem.

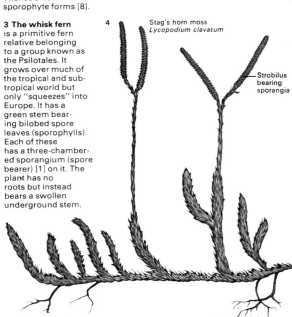

4 Stag's horn moss *Lycopodium clavatum*

Strobilus bearing sporangia

Root

4 Club-mosses are widespread descendants of a plant group that reached its peak in the Carboniferous, over 280 million years ago, and whose fossil remains are found in coal. The types that remain today like stag's horn moss grow from tropics to tundra.

5 *Selaginella* is a relation of the club-mosses but differs in that its leaves have a small outgrowth at the base, known as the ligule. Many species are restricted to tropical rain forests where they flourish in the perpetual humidity.

for ruining large areas of pasture land in Britain.

Many ferns are grown in gardens and greenhouses because, once established, they are highly tolerant of conditions unsuitable for most flowering plants. During the reign of Queen Victoria (1837–1901) there was a craze for fern collecting and certain types of fern, especially the so-called filmy ferns [11] with very delicate leaves only one cell thick, were collected and grown in Wardian cases (sealed glass boxes). This led to a serious depletion of some ferns in the wild in both Britain and Europe. Certain other genera such as *Woodsia*, a fern of high mountain regions, were also over-collected.

Many forms of various fern species found in the wild are grown in nursery gardens. Some of these forms have such names as *Phyllites scolopendrium* "palmatum" (palmate, or hand-shaped, harts-tongue) and *Asplenium trichomanes* "cristatum" (crested maidenhair spleenwort). Many of the more common fern species can still be grown in shady spots in gardens where the soil is poor.

Certain ferns, such as *Salvinia* and *Azolla*

spp [8], live totally in water. These two genera are native to tropical and subtropical regions but have been widely introduced into Europe. Water ferns can form large mats over ponds and *Azolla* often turns crimson.

The horsetails (*Equisetum* spp) are fern allies that vary from small creeping plants to scramblers reaching up to about 9m (30ft). Many of the species have much silica (sand-like material that is very rough and scratchy) in their stems. One of them, the "Dutch rush" (*Equisetem hyemale*), was once used for scouring pans and can still be bought in music shops for shaving down and smoothing the reeds of clarinets and saxophones. Horsetails are almost impossible to eradicate.

True ferns can often be recognized even when they do not resemble ferns (pillwort looks like a fine grass) by the characteristic way in which the new shoot tips unroll. In ferns they appear to unwind like a spring, a phenomenon known as "circinate vernation". Thick-stemmed ferns in the early stages of shooting resemble crosiers, or fiddle heads, and hence have been given these common names.

KEY

Tertiary
Cretaceous
Jurassic
Triassic
Permian
Carboniferous
Devonian
Silurian

The true relationships between pteridophytes is not known due to a lack of fossil evidence. The number of living species is small compared to the abundance of forms in the Carboniferous period which was truly the age of ferns. It was in this period that enormous beds of plants were fossilized to form coal seams. The modern groups that remain are the Psilotales [1], club-mosses [2], the selaginellas [3], quillworts [4], the horsetails [5] and lastly the ferns [6].

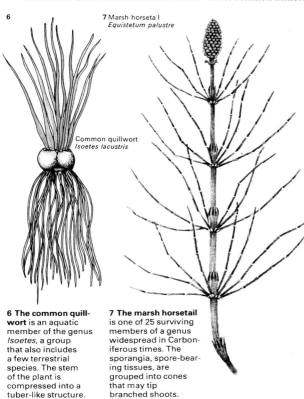

6 Common quillwort
Isoetes lacustris

7 Marsh horsetail
Equistetum palustre

8

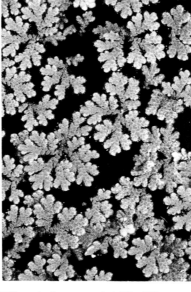

9 Royal fern
Osmunda regalis

6 The common quill-wort is an aquatic member of the genus *Isoetes*, a group that also includes a few terrestrial species. The stem of the plant is compressed into a tuber-like structure.

7 The marsh horsetail is one of 25 surviving members of a genus widespread in Carboniferous times. The sporangia, spore-bearing tissues, are grouped into cones that may tip branched shoots.

8 The water fern (*Salvinia* sp) occurs in Africa and grows so rapidly that it may completely choke lakes and ponds. During the construction of the Kariba dam, two species were allowed to meet and hybridize. This hybrid showed so much vigour that in a short while this "man-made" species, named *Salvinia molesta*, lived up to its title and caused serious economic problems in the area.

9 The royal or flowering fern flourishes in wet, peaty areas all over the world. It is common in western Europe but because it is often collected and moved to gardens its European represent-atives are becoming rare. The fronds of the fern show a marked difference between those that are sterile [1] and those that are spore-bearing [2]. The fronds may grow to 3.5m (10ft) long.

10 The tree ferns, such as *Dicksonia*, build up a trunk of leaf bases like palms. Most tree ferns are tropical but this species (*Dicksonia antarctica*) is found in eastern Australia.

11 The delicate, humidity-loving filmy ferns have fronds that are only a single cell thick. In Victorian times they were much collected and grown with great care in so-called Wardian cases or in greenhouses.

11

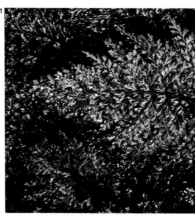

12

Stag's horn fern
Platycerium bifurcatum

12 The stag's horn fern has flat, spreading basal leaves which are often brown. These may create a "plate" from which forked, fertile fronds project. Its distribution is from tropical Africa to India and Australia.

The cone bearers

A significant proportion of the world's vegetation is classified as gymnosperms – plants with naked seeds. The group – which includes the conifers, whose seeds are normally borne in structures called cones – consists of about 66 genera with 600 species that are woody plants, mostly with bark; they have primitive male and female flowers, and all have two or more seed leaves.

Development of gymnosperms
Gymnosperms were of great importance in prehistoric times. They were a highly varied group that dominated the world's flora in the Mesozoic era more than 220 million years ago. Today they are represented only by a relatively small remnant of those earlier species, but they still form a very important section of the world's flora, including well-known trees such as firs, pines and yews.

The seed plants differ from the ferns, or pteridophytes, in the alternating generations between sexual and asexual, or spore-producing, stages. As the male and the female cells of the sexual generation became specialized in ferns and primitive gymnosperms, the male cells were motile, with whiplash flagella that enabled the sperms to move through water or a film of moisture. This happens in cycads and ginkgo, but the "swimming" stage has disappeared in the pine and all flowering plants, thus releasing them from the restriction of needing water for the last stage of the fertilization process.

In conifers fertilization is a complex process that takes a long time. In the pine (*Pinus* sp) it takes three years for the pollen grains to mature. Each grain has two wings to facilitate wind pollination and a mature tree can produce about 1kg (2.2lb) of pollen each year, which may be blown for some hundreds of kilometres downwind. The pollen grain landing on a female cone can take a year or more to grow in and fuse with the cell inside, and two more years may pass before the cone is mature and the seeds enclosed within it are ripe and ready for dispersal.

Groups of gymnosperms
There are five main sub-groups of gymnosperms. The Cycadales are tropical, palm-like trees known as tree-ferns or cycads [12]. The cones of cycads can be very large, and in one species, *Encephalartos caffer*, cones weighing 42kg (92lb) have been reported – the largest cones known on any plant. Another species, the Australian *Macrozamia peroffskyana*, has cones 60cm (2ft) long.

The Ginkgoales group contains only one living genus and one species, namely the maidenhair tree (*Ginkgo biloba*) [10]. It was thought that this species, too, had become extinct as a wild plant, but in recent years Chinese botanists have found some wild maidenhair trees, thus confirming that it does still exist in the wild state. It is a common avenue tree in American cities. The group that includes yews and podocarps is known as the Taxales and all its members have fleshy fruits. They are the only gymnosperms that do not produce cones. The fourth group of gymnosperms, the Gnetales, include the *Welwitschia* [11], a small, specialized class.

Most important of the gymnosperms are the Coniferales, the group that contains 48 out of the total of 66 gymnosperm genera. These are mainly evergreen, temperate trees with needle-like or scale-like leaves [Key]

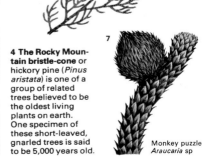

2 The wood of coniferous trees is called softwood because it is easily worked. Characteristic of gymnosperms is the absence of vessels in the vascular or water- and mineral-conducting system. In cross-section the softwood shows a single type of cell tracheid or fibre [1]; when magnified this looks square in outline. Large tracheids laid down in spring alternate with smaller summer tracheids forming annual rings [2]. The tubes serve mainly as space for the transport of sap through minute openings called pits [3]. At right-angles to the rings, rays [4] radiate from the centre for movement of sap and water.

1 The reproductive cycle of the Norway spruce starts when pollen grains [2] from under the scales [3] of the male cone [1] are blown on to two unfertilized eggs [5] on scales [6] of the female cones [4]. The pollen grain nucleus passes down the tube [7] and fuses with the female nucleus [8] to form a winged seed on each scale [9]. The seeds disperse and the embryo [10] germinates [11, 12].

3 The California redwoods (*Sequoia sempervirens*) are magnificent examples of cone-bearing trees. The tallest tree in the world, growing in Redwood Creek Valley, California, is 112m (367ft) tall and has a girth of about 14m (46ft) measured at 1.5m (5ft) above the ground.

4 The Rocky Mountain bristle-cone or hickory pine (*Pinus aristata*) is one of a group of related trees believed to be the oldest living plants on earth. One specimen of these short-leaved, gnarled trees is said to be 5,000 years old.

Willow-leaf podocarp
Podocarpus saligna

Mediterranean cypress
Cupressus sempervirens

Monkey puzzle
Araucaria sp

5 Podocarps or yellow-woods are native to tropical mountains. They bear berry-like cones, some of which are edible. This willow-leaf podocarp grows in the Andes of southern Chile.

6 True cypress trees are evergreen conifers of North America, Europe and Asia. The Italian form of the Mediterranean cypress (*Cupressus sempervirens stricta*) is commonly planted in formal gardens.

7 The Chilean pine, better known as the monkey puzzle, has characteristically stiff, dark green leaves on tiered branches. Resin from the bark is used in medicinal preparations and seeds of the large female cones are edible.

8 Common juniper *Juniperus communis*

8 Junipers are chiefly Northern Hemisphere evergreens. The fruits, although commonly called berries, are true cones with overlapping scales that look like berries. The fruits of one species are used to flavour gin.

and they often form the dominant vegetation in colder parts of the world, being especially abundant in temperate forests both north and south of the Equator. Economically they are of great importance as timber trees. Conifers are often grouped together as "firs" and most have softwood timber containing resin. This is formed in minute canals that run through the wood giving the characteristic turpentine smell to freshly sawn softwood and is the commercial source of many useful resins.

The recognition of conifers

The "firs" comprise a diverse group of trees, most of which can be identified by the shape of their cones and the number and grouping of their leaves or "needles". In the pines, for example, there are only two, three or five needles together. In cedars (*Cedrus* spp) the needles are in bundles or clusters of many individuals, firm and evergreen. The larches (*Larix* spp) [9], also have many needles in clusters but these are deciduous.

When the young twigs remain green through the summer the conifer may be a yew (*Taxus* sp) if the needles are long but short-stalked; Japanese cypress (*Cryptomeria* sp) if the leaves are small scales grouped in spirals, or a Californian big tree (*Sequoiadendron*). Where the needles are arranged in two rows on short shoots that are soft to the touch and fall in autumn, the swamp cypress (*Taxodium* sp) has these short shoots arranged singly, while the dawn redwood (*Metasequoia* sp) has the short shoots opposite in pairs on the stem. Another distinction is a clear stalking of the needles; where this occurs and the tree top overhangs, it is a hemlock (*Tsuga* sp). The Douglas fir (*Pseudotsuga taxifolia*) has sharp-pointed buds and lemon-scented needles. The silver firs (*Abies* spp) have characteristic round scars where needles are pulled off, while a spruce (*Picea* sp) needle takes a small piece of stem with it.

The conifer trees are mostly evergreens. One of the exceptions is the larch, which is deciduous. Most historic among the gymnosperms is the dawn redwood. Like the ginkgo it is a "living fossil" and it was only in 1941 that Chinese botanists found specimens of a tree new to them. Later expeditions found 25 trees of this species.

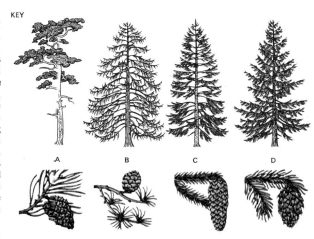

Conifers can be identified by their shapes, needles and cones. The Scots pine (*Pinus sylvestris*) [A] has a flattened crown and triangular cones. The larch (*Larix decidua*) [B], as its name suggests, is one of the few deciduous conifers. It has open cones. The Norway spruce (*Picea abies*) [C] has long, pendulous cones and is the European Christmas tree. The Douglas fir (*Pseudotsuga taxifolia*) [D] has cones with pointed scales and long, soft needles.

9 The larch genus includes about a dozen species that are native to the cool, temperate regions of the Northern Hemisphere. Some of them are among the most commonly planted timber trees. Larches grow tall and straight, and their coarse-grained, hard, heavy, rot-resistant wood is used in the making of ships' masts, telegraph poles and pit props. But many people also grow larches purely for ornament because of their pleasing shape. In forests [A] the trees grow rapidly, and disease-resistant types can be selected. In the open [B] larch trees grow a fine regular crown. Larches are deciduous and sessile cones on bare twigs appear in early spring. [C] Male flowers are small, silvery and inconspicuous; female flowers, the familiar red-purple "larch roses", are followed by green cones.

C Larch
Larix sp

Female cone

'Larch rose'

Male cone

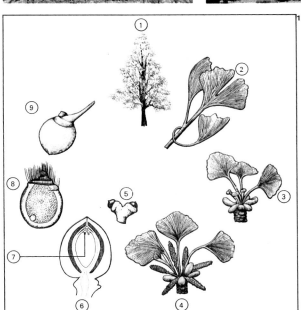

11 *Welwitschia* sp

10 The maidenhair tree (*Ginkgo biloba*) [1] has typically lobed leaves [2] with parallel veins. The sexes are separate: female trees bear ovules [3], male ones [4] produce pollen. The young ovule [5] matures into a structure [6] equipped to receive pollen. On landing on the ovule, pollen is drawn into the pollen chamber [7]. Two large motile spermatozoids [8] are produced, fertilization takes place and a naked seed is formed [9].

11 The *Welwitschia* is a remarkable woody plant, the sole member of its family. It is adapted to desert life by restricted water loss, and it grows only in parts of South-West Africa.

12 Cycads have a long history dating back to Triassic times. They look like palms or ferns and today are found in the tropics. *Cycas revoluta*, from Japan, is widely grown in warm areas, and in conservatories in cooler regions.

Seed fern (Cycad)
Cycas revoluta

Pines and other conifers

The conifers, as their name suggests, are plants with cones. But what makes them most remarkable is their size and longevity. Conifers are by far the largest plants on earth and at their tallest – the maximum is 112 m (367ft) – literally "reach for the sky". For man a century is a remarkable lifespan, but some conifers live for a thousand years.

The conifers are classified into eight families, and of the 50 genera 33 are confined to the Northern Hemisphere. Nearly all are evergreens with tough, dark-green needle-like or scale-like leaves and they form dense and deeply shaded forests.

Man has used conifers for hundreds of years, particularly for timber and for paper-making, and these trees have the added advantage of being able to thrive on poor, thin soil where arable farming is impossible. Conifers add wood to their stems faster than broadleaves do and can thrive even where the sun's rays are weak.

Conifers are also ornamental and have graced gardens and palaces since ancient times. Their upstanding, formal appearance adds elegance to man-made architecture.

CONNECTIONS

See also
54 The cone bearers
206 Northern pine forests
222 Mountain life

In other volumes
218 The Physical Earth

1 Tamarack *Larix laricina*

2 **The white spruce** grows in the far north of Canada and like other spruces is a prized timber tree. It reaches a height of 23m (75ft).

2 White spruce *Picea glauca*

1 The Tamarack or American larch of eastern North America can grow to a height of 18m (60ft). Like other larches this conifer is unusual in being deciduous and not evergreen. All are classified with the pines (family Pinaceae). They have resinous, weather-resistant timber.

3 Lawson cypress *Chamaecyparis lawsoniana*

5 **The Scots pine** (*Pinus sylvestris*) is seen here in the setting from which its name is derived on the shores of a Scottish loch. This tree, which can grow 0.6 to 0.9m (2 to 3ft) a year for 20 to 30 years is the best known of the family Pinaceae. All pines (*Pinus* spp) have two kinds of leaves; there are brown, scale leaves on the long shoots and needle-shaped ones on the short shoots.

6 **The Norway spruce** (*Picea abies*) is the commonest species of *Picea* and Europe's tallest native tree, reaching a height of 54m (180ft). It is seen here growing in the Italian Tyrol, in the foothills of the Alps. The tree may live more than 1,000 years and may not reach maturity (as shown by the production of cones) until it is 30 years old. It can tolerate shade and thus forms thick forests.

4 **The cypresses** are well known as ornamental garden trees and include golden varieties. This species, *Cupressus macrocarpa*, which grows to a height of 45m (150ft) is seen here in a Mediterranean setting. All 15-20 species of cypress are distinguished by their scale-like leaves with which the twigs are densely covered.

3 **One of the tallest of conifers**, the Lawson cypress, can grow to a height of 60m (200ft) and live for 600 years. It is also known as Oregon cedar and is grown both for its attractive looks and its timber. This member of the family Cupressaceae is native to California and Oregon. There are six related species in Japan and North America.

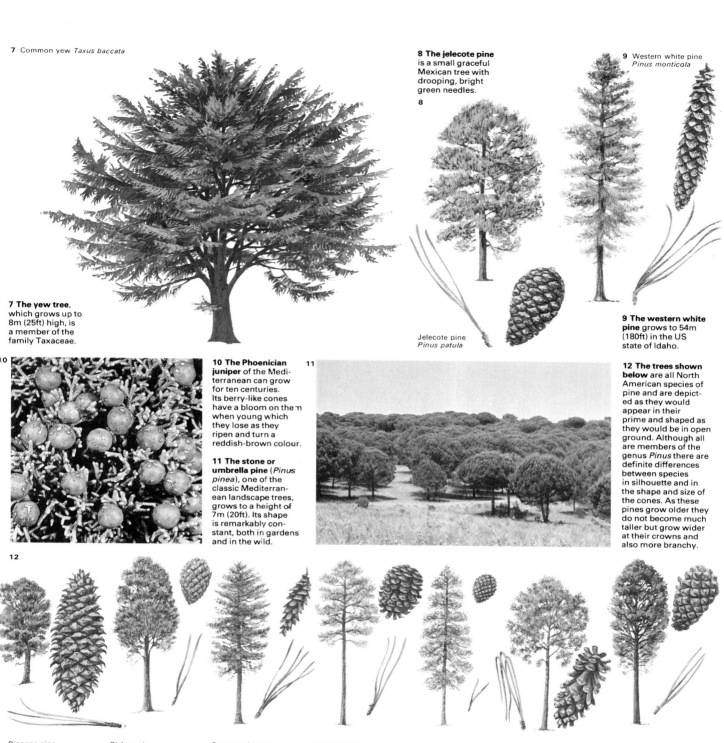

7 Common yew *Taxus baccata*

7 The yew tree, which grows up to 8m (25ft) high, is a member of the family Taxaceae.

8 The jelecote pine is a small graceful Mexican tree with drooping, bright green needles.
8

Jelecote pine
Pinus patula

9 Western white pine *Pinus monticola*

9 The western white pine grows to 54m (180ft) in the US state of Idaho.

10 The Phoenician juniper of the Mediterranean can grow for ten centuries. Its berry-like cones have a bloom on them when young which they lose as they ripen and turn a reddish-brown colour.

11 The stone or umbrella pine (*Pinus pinea*), one of the classic Mediterranean landscape trees, grows to a height of 7m (20ft). Its shape is remarkably constant, both in gardens and in the wild.

12 The trees shown below are all North American species of pine and are depicted as they would appear in their prime and shaped as they would be in open ground. Although all are members of the genus *Pinus* there are definite differences between species in silhouette and in the shape and size of the cones. As these pines grow older they do not become much taller but grow wider at their crowns and also more branchy.

12

Bigcone pine
Pinus coulteri

Bishop pine
Pinus muricata

Eastern white pine
Pinus strobus

Loblolly pine
Pinus taeda

Lodgepole pine
Pinus contorta

Longleaf pine
Pinus palustris

Monterey pine
Pinus radiata

Pitch pine
Pinus rigida

Ponderosa pine
Pinus ponderosa

Red pine
Pinus resinosa

Shore pine
Pinus contorta

Slash pine
Pinus elliottii

Sugar pine
Pinus lambertiana

Whitebark pine
Pinus albicaulis

Flowering plants: two seed leaves

Most of the flowering plants belong to a group that botanists call the dicotyledons because within the seeds of these plants the embryo has two separate seed leaves or cotyledons attached to it which often act as food supplies. Such plants have a characteristic internal anatomy [Key]. The many woody shrubs and trees among dicotyledons provide useful timber, but the non-woody dicotyledons are just as valuable to man because many of them are food plants.

Worldwide dicotyledon distribution

Non-woody dicotyledons occur almost everywhere; each continent has a large population, including the Arctic and the Antarctic. These plants fill many different ecological niches and colonize all types of habitat except permanent snow and ice, from natural regions such as marshes, forests, rocky cliffs and pastures to totally man-made areas such as walls, roofs, roadsides and demolition or clearance sites. Some thrive in the driest deserts, others grow totally immersed in fresh or salt water. This last type of habitat, salt water, often causes a curious state known as physiological drought, in which the plant behaves rather like a desert cactus in retaining its internal moisture.

Ornamentals and food providers

Almost all families of dicotyledons provide plants that are used for food or grown as ornamentals. Many buttercups [1] and their relatives (family Ranunculaceae) are grown as ornamental garden flowers: these include anemones, pasque flowers (*Pulsatilla* spp), delphiniums and clematis. In the crucifer family (Cruciferae) a few species such as stocks, wallflowers, honesty and alyssum are grown as garden flowers, but much more important are the cabbage and its many relations of the genus *Brassica*. Cabbage, brussels sprouts, cauliflower, broccoli, curly kale and many other cabbage-like vegetables are surprisingly all different forms of a single plant species. Other crucifers are rape, an important source of vegetable oil, black and white mustard, swedes and turnips.

Another important crop family is the Solanaceae or nightshade family, which includes the potato, tomato and many drug sources such as deadly nightshade and henbane. The daisy family (Compositae) [7] provides not only many decorative flowers but also artichokes (Jerusalem and globe), lettuce, chicory, endive and oil crops such as sunflower. The rose family (Rosaceae) is represented by orchard fruits (apples, pears, plums, peaches), blackberries, raspberries, strawberries and garden flowers throughout the temperate world. Among the latter are geums, lady's-mantles, *Kerria*, *Dryas* and, of course, roses. Besides food and ornamental plants, certain families, such as the dead nettle family (Labiatae) with its many aromatic species, have been used from ancient times as flavourings. These include mint [11], thyme, sage, basil and marjoram. Fruits of certain families are also aromatic and are known as spices. Examples from the carrot family (Umbelliferae) are dill, fennel, coriander and caraway seeds. The roots of a few species are also used as spices. Among these are horseradish and liquorice.

Other plant families yield fibres for clothing or cordage. The mallow family (Malvaceae) is represented by the cotton plant,

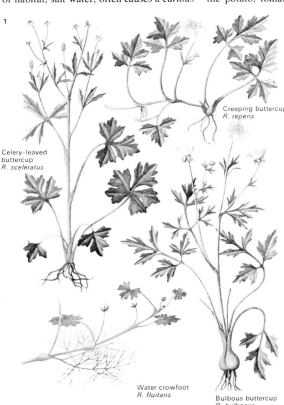

1 Buttercups are common plants of the family Ranunculaceae that vary greatly in shape and size and occur in many ecological niches from water to meadows. The crowfoots are generally aquatic and have white flowers. Their leaves are of two kinds – floating and submerged. Other buttercups of different species grow on dry land, but each occupies a separate habitat and has distinctive anatomical characteristics. *Ranunculus bulbosus*, for example, as its name suggests, has an underground bulb. Both the creeping *Ranunculus repens* and bulbous buttercups are prolific weeds of meadows and gardens.

Creeping buttercup
R. repens

Celery-leaved buttercup
R. sceleratus

Water crowfoot
R. fluitans

Bulbous buttercup
R. bulbosus

4 Grass of Parnassus is not a grass but is related to the saxifrages. It occurs in damp grassland in Europe and throughout temperate Asia. The pretty five-petalled white flower has gleaming golden pinheads called staminodes that entice insects into visiting the plant in search of nectar. This is the only means the plant has of attracting insect visitors, which then pollinate it. The flower stalk has a heart-shaped leaf, whose base surrounds the stem.

Grass of Parnassus
Parnassia palustris

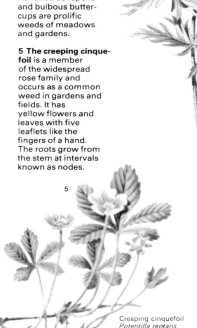

2 The meadow cranesbill is a beautiful plant common on roadside verges. There is a rare variety that has white flowers. Closely related is the frequently cultivated South African genus, *Pelargonium*, which is also known as the geranium. The name "cranesbill" derives from the shape of the fruit.

Meadow cranesbill
Geranium pratense

5 The creeping cinquefoil is a member of the widespread rose family and occurs as a common weed in gardens and fields. It has yellow flowers and leaves with five leaflets like the fingers of a hand. The roots grow from the stem at intervals known as nodes.

Creeping cinquefoil
Potentilla reptans

6 The rue-leaved saxifrage is a rare plant of rocks, walls and dry soils. It is a member of the widespread family Saxifragaceae, whose name, which in Latin means "breaker of rocks", illustrates the kind of stony habitat that this plant and its relatives occupy. Saxifrages are frequently grown as rockery plants – examples are the mossy saxifrage (*Saxifraga hypnoides*) and the hardy hybrid London pride (*Saxifraga* x *geum*).

Deptford pink
Dianthus armeria

3 The Deptford pink is a member of the family Caryophyllaceae and is closely related to garden carnations. It has a wide range and is found on sandy soils throughout Europe; it has also been introduced into North America. Although it is a rare plant, local populations may be quite extensive.

Rue-leaved saxifrage
Saxifraga tridactylites

and flax belongs to its own family (Linaceae). Hemp, or *Cannabis sativa* (family Cannabiaceae) provides the drug cannabis and also yields the familiar cordage fibre. Some herbaceous dicotyledons grow in places where the supply of nutrients from the environment is very poor, such as bogs.

Insectivorous plants

Plants have evolved to fill low-nutrient niches and in order to obtain the missing nutrients (especially nitrogen) some resort to trapping, digesting and absorbing insects. These insectivorous species belong to several families. They include the sundews, butterworts and bladderworts of Europe and the pitcher plants of the genera *Sarracenia* and *Nepenthes,* which are found in the tropics and subtropics. These plants have many different ways of catching their food. Sundews (*Drosera* spp) trap insects by means of "flypaper". They have short tentacles coated with sticky "glue" and insects passing close to the plants or landing on them get stuck. The plants then produce digestive enzymes that dissolve the insect and the plants absorb the nutrient fluid. Butterworts (*Pinguicula* spp) behave similarly, but produce the glue over the whole leaf surface.

Bladderworts (*Utricularia* spp) grow in water and have traps, each one of which is a small vesicle with a door and signal hairs growing near it. When a small water creature touches these hairs, the door opens and both water and creature rush into the trap where digestive enzymes break down the food. The pitcher plants are basically pitfall traps that attract insects and other small creatures to the top of a container, often by the promise of nectar; the victim eventually slips on the smooth surface and falls into a soup of enzymes where hairs directed downwards trap it until it drowns. The best known insectivorous plant is probably the Venus fly-trap (*Dionaea muscipula*), the leaf of which has a hinged trap at its tip with touch-sensitive hairs. These will not respond to a single touch, but require a short burst of touches, such as that given by a moving insect, to trigger off a response, bringing the two halves of the trap together with a sudden snap. The trapped prey is then dissolved and digested.

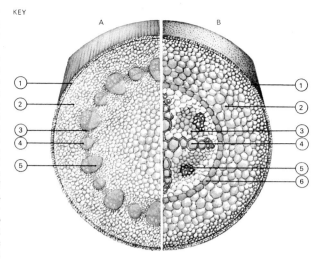

The hallmarks of a dicotyledon show in its stem [A] and root [B]. Both are covered by epidermis [1] surrounding a cortex [2], inside which the growing layer or cambium [3] gives rise to vascular bundles containing water-conducting xylem [4] and food-conducting phloem [5]. In the root a central cross-shaped pattern can be seen and there is an endodermis layer [6].

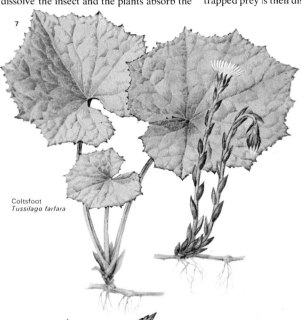

Coltsfoot
Tussilago farfara

7 Coltsfoot is a well-known composite whose yellow dandelion-like flower appears before the leaves. Coltsfoot is widespread on basic soils and its leaves are used as a herbal cough cure. Its name comes from the shape of its leaves, which is similar to a horse's hoof.

8 The knotweeds (family Polygonaceae), of which the black bindweed is one, are a group of 900 species which include in their numbers dock, sorrel and many other common weeds. The name of the family comes from the knot-like formations that grow on the plant stems. The plants are found worldwide.

Black bindweed
Polygonum convolvulus

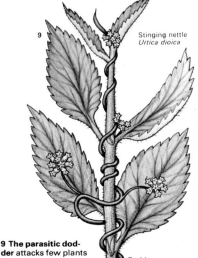

Stinging nettle
Urtica dioica

Dodder
Cuscuta europaea

9 The parasitic dodder attacks few plants except the stinging nettle. It feeds by haustoria [1], modified "roots" that enter the host plant.

Opium poppy
Papaver somniferum

10 The opium poppy has been cultivated since the Middle Ages. The drug opium and its derivatives, morphine and heroin, are extracted from the latex of the seed pods. The seeds themselves are used as cattle food and as a source of oil. The poppy is probably European in origin.

Mint *Mentha* sp

11 Mint has long been used as a food flavouring. Many species of mint are known and they show subtle differences in the aromas they discharge. Crosses between water mint (*Mentha aquatica*) and spearmint (*Mentha spicata*) are the basis of cultivated peppermint (*Mentha x piperita*), which is commercially grown on a wide scale in the USA and used to flavour gums, toothpaste and a variety of drugs.

Flowers of wayside and hedgerow

Of all the world's flowering plants the non-woody members of the dicotyledon group (plants with two seed leaves) are the most numerous. Of the total of 200,000 dicotyledon species, which are grouped into over 250 families, more than half are non-woody herbs. Many of these herbs are cultivated in gardens and greenhouses in temperate regions, where they are treated as ornamental plants or as vegetables.

The largest families of dicotyledons include both familiar and useful herbs, such as the buttercups (Ranunculaceae); the brassicas (Cruciferae); the spurges (Euphorbiaceae); the roses (Rosaceae); the legumes (Leguminosae); the carrots (Umbelliferae); the nettles (Labiatae); the potatoes (Solanaceae); and the dandelions (Compositae).

Although many of these herbs are employed by man as foods and flavourings, many others are weeds that hamper his agriculture, absorbing nourishment that he would prefer his crops to take. Most weed plants have extremely efficient reproductive mechanisms and many of them spread vegetatively as well as by means of seed.

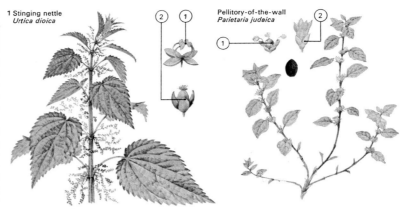

1 Stinging nettle
Urtica dioica

Pellitory-of-the-wall
Parietaria judaica

CONNECTIONS

See also
54 Flowering plants: two seed leaves
38 The plant kingdom

In other volumes
172 The Physical Earth
212 The Physical Earth
190 The Physical Earth

1 The stinging nettle and the pellitory-of-the-wall are two of the few species of the family Urticaceae that grow in temperate regions. Each plant bears both male [1] and female [2] flowers. There are 35 species of *Urtica* and all of them have bristle-like stinging hairs, which are long, hollow cells. The tips of these are toughened with silica and they are easily broken off. When the plant is touched the hairs penetrate the skin like surgical needles, the tips are lost and the poison contained in the cells is released. The pellitory-of-the-wall is essentially a nettle that does not sting. Both it and the stinging nettle are used in herbal medicine as cures for rheumatism. Nettles are sometimes cooked and eaten as a vegetable, or used to make nettle beer.

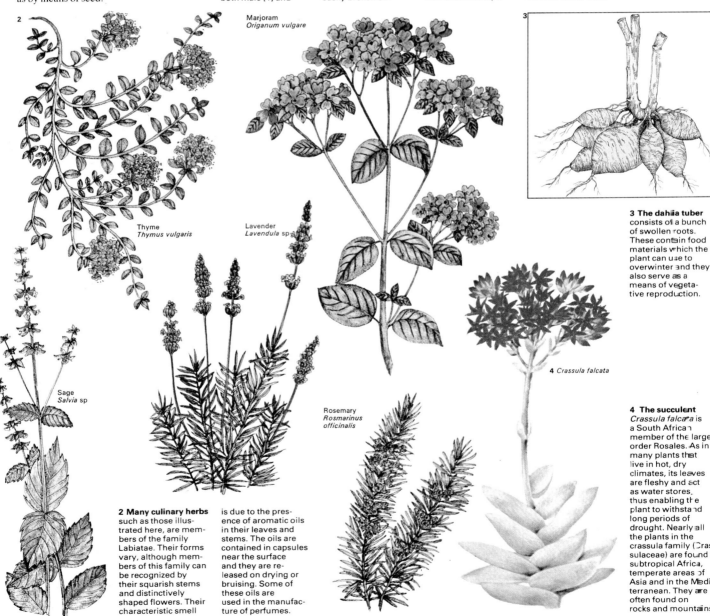

Marjoram
Origanum vulgare

Thyme
Thymus vulgaris

Lavender
Lavendula sp

Sage
Salvia sp

Rosemary
Rosmarinus officinalis

4 *Crassula falcata*

2 Many culinary herbs such as those illustrated here, are members of the family Labiatae. Their forms vary, although members of this family can be recognized by their squarish stems and distinctively shaped flowers. Their characteristic smell is due to the presence of aromatic oils in their leaves and stems. The oils are contained in capsules near the surface and they are released on drying or bruising. Some of these oils are used in the manufacture of perfumes.

3 The dahlia tuber consists of a bunch of swollen roots. These contain food materials which the plant can use to overwinter and they also serve as a means of vegetative reproduction.

4 The succulent *Crassula falcata* is a South African member of the large order Rosales. As in many plants that live in hot, dry climates, its leaves are fleshy and act as water stores, thus enabling the plant to withstand long periods of drought. Nearly all the plants in the crassula family (Crassulaceae) are found in subtropical Africa, temperate areas of Asia and in the Mediterranean. They are often found on rocks and mountains.

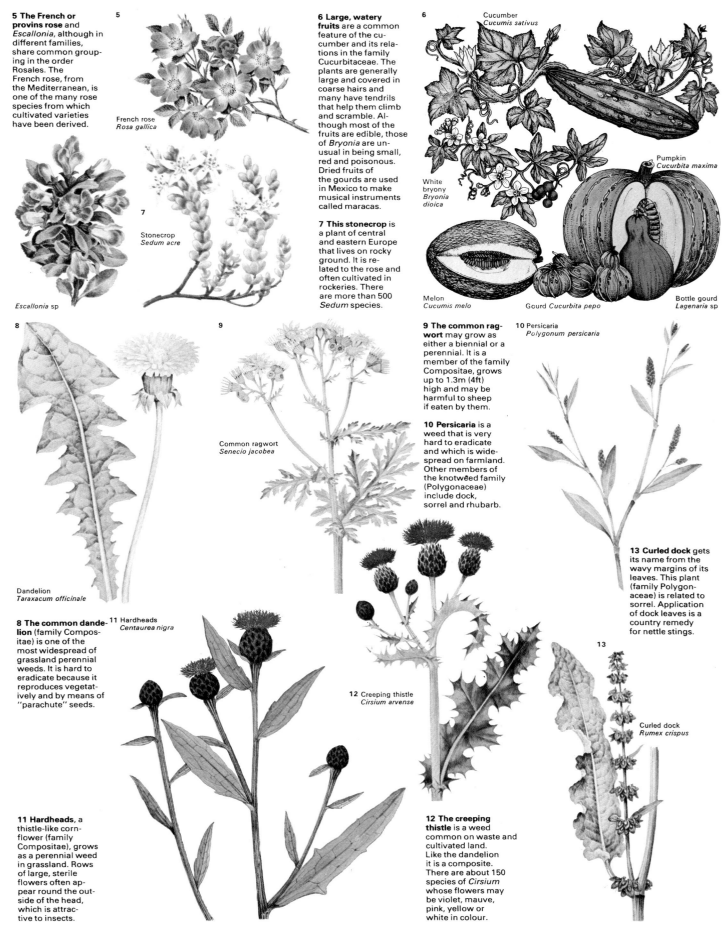

5 The French or provins rose and *Escallonia*, although in different families, share common grouping in the order Rosales. The French rose, from the Mediterranean, is one of the many rose species from which cultivated varieties have been derived.

5

French rose
Rosa gallica

Escallonia sp

7

Stonecrop
Sedum acre

6 Large, watery fruits are a common feature of the cucumber and its relations in the family Cucurbitaceae. The plants are generally large and covered in coarse hairs and many have tendrils that help them climb and scramble. Although most of the fruits are edible, those of *Bryonia* are unusual in being small, red and poisonous. Dried fruits of the gourds are used in Mexico to make musical instruments called maracas.

7 This stonecrop is a plant of central and eastern Europe that lives on rocky ground. It is related to the rose and often cultivated in rockeries. There are more than 500 *Sedum* species.

6

Cucumber
Cucumis sativus

Pumpkin
Cucurbita maxima

White bryony
Bryonia dioica

Melon
Cucumis melo

Gourd *Cucurbita pepo*

Bottle gourd
Lagenaria sp

8

Dandelion
Taraxacum officinale

8 The common dandelion (family Compositae) is one of the most widespread of grassland perennial weeds. It is hard to eradicate because it reproduces vegetatively and by means of "parachute" seeds.

9

Common ragwort
Senecio jacobea

9 The common ragwort may grow as either a biennial or a perennial. It is a member of the family Compositae, grows up to 1.3m (4ft) high and may be harmful to sheep if eaten by them.

10 Persicaria is a weed that is very hard to eradicate and which is widespread on farmland. Other members of the knotweed family (Polygonaceae) include dock, sorrel and rhubarb.

10 Persicaria
Polygonum persicaria

13 Curled dock gets its name from the wavy margins of its leaves. This plant (family Polygonaceae) is related to sorrel. Application of dock leaves is a country remedy for nettle stings.

13

Curled dock
Rumex crispus

11 Hardheads
Centaurea nigra

11 Hardheads, a thistle-like cornflower (family Compositae), grows as a perennial weed in grassland. Rows of large, sterile flowers often appear round the outside of the head, which is attractive to insects.

12 Creeping thistle
Cirsium arvense

12 The creeping thistle is a weed common on waste and cultivated land. Like the dandelion it is a composite. There are about 150 species of *Cirsium* whose flowers may be violet, mauve, pink, yellow or white in colour.

61

Woody flowering plants

The world's flowering trees, although they appear to be quite different from the rest of the earth's flora, do not in fact form a distinct botanical category. Most of them are simply flowering plants (angiosperms) that possess woody stems. All living trees, with the exception of the gymnosperms such as conifers and ginkgo, are flowering trees. These are divided botanically, on the basis of the number of seed leaves or cotyledons they possess, into the monocotyledons, which have one seed leaf, and the dicotyledons, which have two seed leaves.

Relatively few monocotyledons are flowering trees and those that are, unlike dicotyledonous trees, produce no wood. The most familiar are the several hundred palm species, such as the date (*Phoenix dactylifera*) and the coconut (*Cocos nucifera*). Most tree species are dicotyledonous and the majority grow in the world's tropical forests. Many valuable wood trees make up this group including teak, ebony and mahogany. In contrast the temperate regions support few tree species. Oak, ash, beech and birch are the four most common temperate trees. They grow in mixed woodlands, often accompanied by other species such as the Spanish chestnut, or (less usually) form single-species forests, as in the case of the black beech [6] of New Zealand.

Size and growth of trees

The key to the size of trees lies in a continuous ring of cells within the trunk known as the cambium [1]. The multiplication of these cells, which lie just below the bark, results in growth. The cells produced by the cambium rapidly become specialized as elements of the transport system within the trunk, by which water, food substances and essential mineral salts are carried through the tree's trunk, branches and leaves.

The tree's structural and mechanical strength – endowed by its woodiness – is a consequence of the fate of the cells produced by the cambium. As the cells of the water-transporting system, known collectively as the xylem, enlarge and age they gradually become invested with layers of lignin, a tough substance related chemically to cellulose. Lignin endows a rigidity that helps trees to stand erect – the root system also aids in this.

Division of the cambial cells does not proceed at a steady pace. In temperate trees growth is much faster in summer than in winter. As a result the woody cells that arise in the summer months appear as lighter-coloured rings when the tree trunk is cut across. Because this is a yearly cycle of growth the tree's age is easily determined by counting the number of these annual rings. In tropical trees growth patterns follow the cycle of wet and dry seasons.

The bark that covers the tree is a product of a special tissue, the cork cambium. This protective covering for the all-important cambium varies greatly in thickness and consistency from species to species. The bark of the silver birch, for example, is thin and papery, while that of the cork oak is extremely thick and porous. In some savanna trees the bark is even fire-resistant.

Leaves and photosynthesis

Trees, like all other green plants, produce the food they require by photosynthesis. This is a process in which energy from the sun is

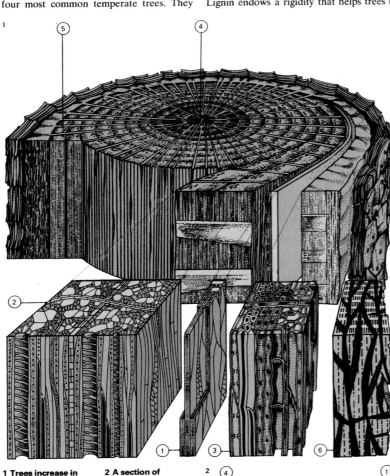

1 Trees increase in girth by rings of new wood produced annually in temperate zones but less often in the tropics. The cambium [1] produces xylem [2] and phloem [3]. They are alive but the heartwood [4] is dead. The medullary rays [5] allow the transport of food across the trunk. Bark [6] is a protective outer coating.

2 A section of trunk is cut here to show how buds are held in reserve in case the main shoot is lost. After a mature branch has been broken off [1] a "suppressed" bud [2] begins to grow. A totally suppressed bud [3] grows just enough to keep pace with the thickening of the trunk. A short shoot [4] may grow into a branch.

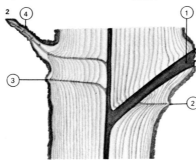

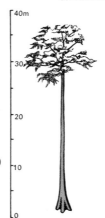

root system such as that of an oak (*Quercus* sp) has roots that spread outwards, branching repeatedly until the soil is threaded with a mat of root fibres that often extends far beyond the reach of the crown of the tree. The red mangrove (*Rhizophora mangle*) rises above its swampy habitat on stilt roots which enable it to survive changes in the mud level. The white

Oak
Quercus robur

Red mangrove
Rhizophora mangle

White mangrove
Avicennia officinalis

3 Roots anchor a tree securely to the ground and take up from the soil the essential minerals and the huge quantity of water it needs – several hundreds of litres a day is not unusual. A fibrous

Tropical forest tree

mangrove (*Avicennia officinalis*) sends roots upwards in the airless mud to enable them to obtain oxygen. Root systems develop in different ways. The shallow roots of tall rain-forest trees concentrate in the humus-rich topsoil and grow flange-like buttresses that support the trunk. These can grow to 4m (13ft) up and out. The common alder (*Alnus glutinosa*) is another resident of waterlogged places that has supporting roots.

trapped by the green pigment chlorophyll in the leaves and used to build food materials from the simple chemicals water and carbon dioxide. The leaves of trees show a profusion of shape and size and may be simple or complex. Near the extremes of the scale are the small, simple leaves of the birch measuring up to 5cm (2in) in length and the giant leaves of the rubber plant, which can be more than 30cm (12in) long.

Each leaf on a tree has a limited life, which depends on climatic factors. In many temperate trees all the leaves are shed every autumn and reappear in spring [4]. These trees are described as deciduous. In contrast, evergreens are those in which the leaves die and are replaced at random every two or three years. Large numbers of tropical trees are evergreen but those that endure a dry season are often deciduous.

In temperate regions the flowers of trees are generally unspectacular, often to such a degree that they appear to the casual observer to be non-existent. The reason for this lack of floral flamboyance is that these trees are wind pollinated and do not need to attract living pollinators such as insects. Those woody plants that are pollinated by insects – most shrubs, climbers and tropical trees – produce large, colourful, often scented flowers [7, 8].

The results of successful pollination – the fruits of trees – are as varied as the flowers. The winged fruits of the sycamore and the ash, the huge nuts of tropical trees such as the coconut, the fleshy fruits of apples and plums and the pea-like pods of the acacia and laburnum are some familiar examples.

The future of flowering trees

The distribution of flowering trees over the earth has varied greatly since these plants made their first appearance. Analysis of pollen remains shows that preglacial European woodlands contained not only the familiar modern trees but also those regarded as alien species, for example magnolias and rhododendrons. Were it not for their removal by man, forests would still cover huge areas of the globe. Only in recent years has man made any attempt to conserve the trees of the world's temperate regions.

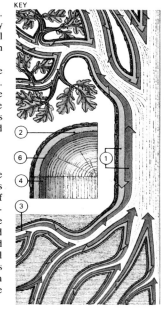

Sap circulates in a tree [1]. Food made in the leaves is carried in the phloem [2]. Water and dissolved minerals [3] for food production are taken from the roots in the sapwood or xylem [4]. Shoots, root tips [5] and cambium [6] are the growing points.

4 Plant hormone (auxin) [1] is produced by the leaf. As the leaf ages the auxin production slows and an abscission layer [2] forms. The leaf falls off leaving stalk [3] and stalk vein scars [4].

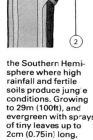

Black beech
Nothofagus solandri

the Southern Hemisphere where high rainfall and fertile soils produce jungle conditions. Growing to 29m (100ft), and evergreen with sprays of tiny leaves up to 2cm (0.75in) long, black beech forms vast one-species forests. Waratah is the floral emblem of New South Wales. A small shrubby plant with blooms up to 10cm (4in) long, it grows in acid soils in near-desert conditions in the Western part of the State.

Waratah
Telopea sp

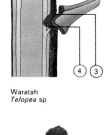

6 Variety of form characterizes those flowering trees that can grow in extreme climates. Black beech is a close relative of the familiar northern beech but grows in

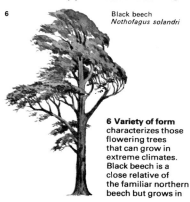

7 Passion flower *Passiflora quadrangularis*

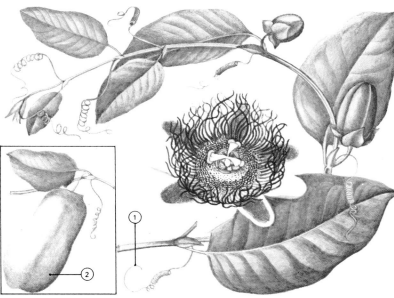

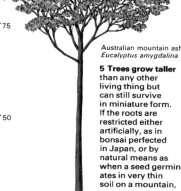

Australian mountain ash
Eucalyptus amygdalina

5 Trees grow taller than any other living thing but can still survive in miniature form. If the roots are restricted either artificially, as in bonsai perfected in Japan, or by natural means as when a seed germinates in very thin soil on a mountain,

a fully formed tree only a few centimetres high will result. The California redwood, the tallest tree, is closely rivalled by a eucalyptus such as the mountain ash of Australia. The eucalyptus can grow as much as 14m (45ft) in two years. All parts of this tree are rich in oil. It is also grown for its hard timber. The English oak is one of 450 species of oak that grow as trees, bushes and shrubs. It enlarges slowly – about 4.5m (15ft) in 10 years – but produces wood of prodigious strength. *Espeletia* grows on snowy ledges over 400m (1,300ft) up in the Sierra Nevada.

English oak
Quercus robur

Frailejon
Espeletia sp

Bonsai tree

7 The passion flower is a climbing woody plant. It makes its way by means of twining tendrils [1]. Some species produce large, edible fruits [2].

8 The honeysuckle twines its stem around its tree hosts. Saplings can be marked for life by its spiraling stem; such wood is much prized for walking-sticks. In extreme cases the host is completely strangled as the honeysuckle stops sap flow. The honeysuckle is

Honeysuckle
Lonicera periclymenum

insect-pollinated and the flowers have

both male and female parts.

Trees, shrubs and climbers

Most of the world's flowering trees are noted more for their shape and form than for their flowers. Nearly all of the broad-leaved species (as distinct from narrow-leaved conifers) are deciduous and are members of the group of plants known as dicotyledons.

During the evolution of the flowering trees, shrubs and climbers the great step forward came with insect pollination rather than wind pollination. One advantage of this is that there is no need for trees to grow in large stands in order for pollination, and thus fruit formation, to be successful.

Most trees with significant, attractive flowers belong to one of three plant families, the Magnoliaceae (magnolias), the Rosaceae (roses) and the Leguminosae (peas). Of these the rose relatives are the most important economically because they often bear edible fruits – apples, pears, plums and cherries – for which they are cultivated. The fruits of many other tree families are eaten by man and include the olives (Oleaceae), figs (Moraceae), walnuts and pecans (Juglandaceae), chestnuts (Fagaceae) and oranges, limes and lemons (Rutaceae).

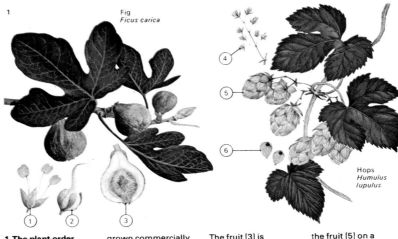

Fig
Ficus carica

Hops
Humulus lupulus

1 The plant order Urticales contains a number of diverse families as widely different as the elms, figs, hops and stinging nettles. Most of these are grown commercially. The fig is grown for its fruit in the warmer areas of Europe. Male flowers [1] and female flowers [2] are found on the same tree. The fruit [3] is shown in section. Hops are widely cultivated in southern England and are used to flavour beer. Shown here are the male flower [4], the fruit [5] on a branch and the seeds [6]. The other member of the Urticales that is commercially important is the elm, which is grown for its versatile timber.

4 The limes or lindens (family Tiliaceae) have some of the largest leaves found on deciduous trees. Of these the largest – 30cm (12in) long – are on the American basswood of North America.

4 American linden
Tilia americana

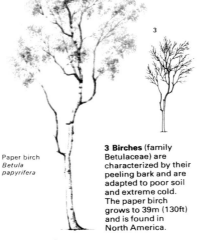

Elder
Sambucus nigra

2 The elder (family Caprifoliaceae), native to Europe, grows to a height of 12m (40ft). In early summer the tree bears clusters of white flowers.

Paper birch
Betula papyrifera

3 Birches (family Betulaceae) are characterized by their peeling bark and are adapted to poor soil and extreme cold. The paper birch grows to 39m (130ft) and is found in North America.

5 The leaves of the ash (family Oleaceae) are distinctive in being split into many small leaflets giving the impression of fine foliage. The white ash of eastern North America is a timber tree up to 41m (135ft) in height.

5 White ash
Fraxinus americana

6 European willow
Salix caprea

6 The willows (family Salicaceae) are fairly small trees, the European willow growing to 10m (35ft). The seeds are light, wind dispersed, and contain little food. They often grow by water where rapid germination is effected.

7 Holly
Ilex aquifolium

7 The holly (family Aquifoliaceae), with its thick foliage of evergreen thorned leaves, is planted as a windbreak and for ornamental purposes. Its red, winter berries create a striking contrast to the dark foliage. It grows to 21m (70ft).

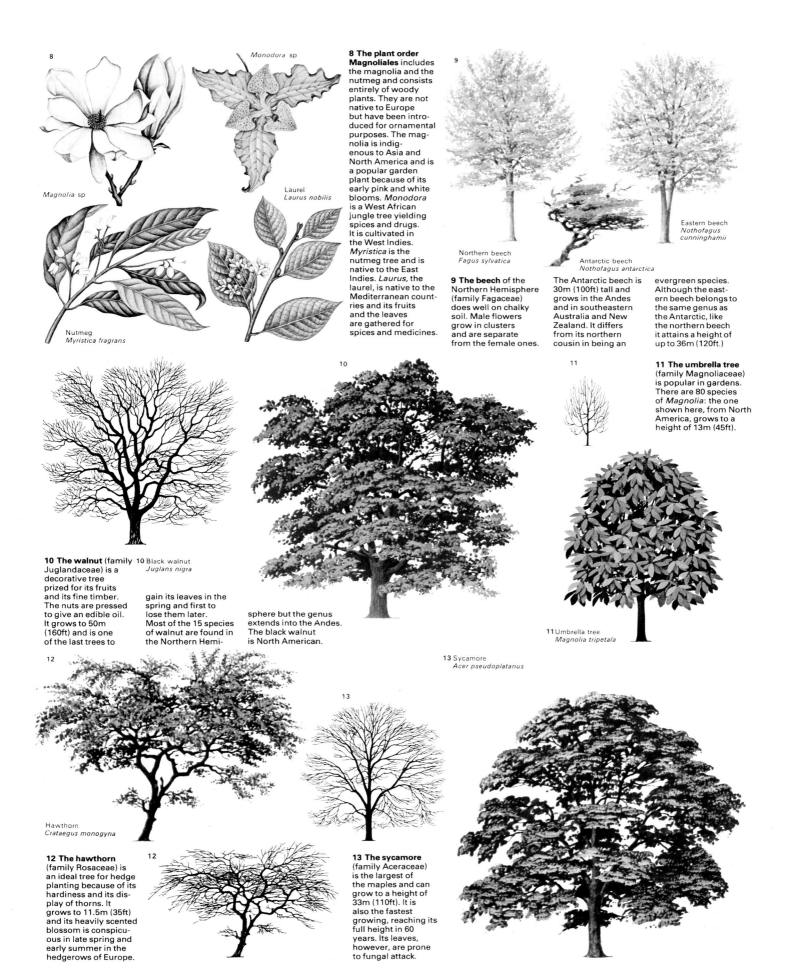

8 *Monodora* sp

Magnolia sp

Laurel
Laurus nobilis

Nutmeg
Myristica fragrans

8 The plant order Magnoliales includes the magnolia and the nutmeg and consists entirely of woody plants. They are not native to Europe but have been introduced for ornamental purposes. The magnolia is indigenous to Asia and North America and is a popular garden plant because of its early pink and white blooms. *Monodora* is a West African jungle tree yielding spices and drugs. It is cultivated in the West Indies. *Myristica* is the nutmeg tree and is native to the East Indies. *Laurus,* the laurel, is native to the Mediterranean countries and its fruits and the leaves are gathered for spices and medicines.

9 The beech of the Northern Hemisphere (family Fagaceae) does well on chalky soil. Male flowers grow in clusters and are separate from the female ones.

9

Northern beech
Fagus sylvatica

Antarctic beech
Nothofagus antarctica

Eastern beech
Nothofagus cunninghamii

The Antarctic beech is 30m (100ft) tall and grows in the Andes and in southeastern Australia and New Zealand. It differs from its northern cousin in being an evergreen species. Although the eastern beech belongs to the same genus as the Antarctic, like the northern beech it attains a height of up to 36m (120ft.)

11 **11 The umbrella tree** (family Magnoliaceae) is popular in gardens. There are 80 species of *Magnolia*: the one shown here, from North America, grows to a height of 13m (45ft).

10 The walnut (family Juglandaceae) is a decorative tree prized for its fruits and its fine timber. The nuts are pressed to give an edible oil. It grows to 50m (160ft) and is one of the last trees to

10 Black walnut
Juglans nigra

gain its leaves in the spring and first to lose them later. Most of the 15 species of walnut are found in the Northern Hemisphere but the genus extends into the Andes. The black walnut is North American.

10

13 Sycamore
Acer pseudoplatanus

11 Umbrella tree
Magnolia tripetala

12

Hawthorn
Crataegus monogyna

12 The hawthorn (family Rosaceae) is an ideal tree for hedge planting because of its hardiness and its display of thorns. It grows to 11.5m (35ft) and its heavily scented blossom is conspicuous in late spring and early summer in the hedgerows of Europe.

12

13

13 The sycamore (family Aceraceae) is the largest of the maples and can grow to a height of 33m (110ft). It is also the fastest growing, reaching its full height in 60 years. Its leaves, however, are prone to fungal attack.

Flowering plants: one seed leaf

Flowering plants – called angiosperms – are the most dominant form of vegetation on earth. The angiosperms are divided botanically into two groups the monocotyledons and the dicotyledons. The names of these two groups indicate the number of seed leaves or cotyledons contained within the seed – all the monocotyledons have one seed leaf and all the dicotyledons have two seed leaves.

Distinguishing features

The monocotyledons and dicotyledons are distinguished by several other features, apart from the number of seed leaves they possess. In the monocotyledons the food and fluid transport system is grouped into closed, scattered, vascular bundles [1]. The leaves have parallel veins and usually no distinct central vein or midrib [4]. The flower parts of monocotyledons are arranged in threes, usually with three or six petals. Those few species that are woody have no bark. Pollen grains of monocotyledons bear only a single furrow.

The food and fluid transport system of the dicotyledons is made up of vascular bundles arranged as a single ring within the stem. The leaves have a network of veins and a distinct midrib. Flower parts have a five-, a four-, or more rarely, a two-fold arrangement. There are many woody species all of which have bark. The pollen grains of dicotyledons have three furrows.

Throughout the world there are some 55,000 species of monocotyledons, forming about a quarter of all flowering plants and including the grasses, palms, and orchids. They show a considerable range of form, and leaf size in this group can vary from a few millimetres to the 20m (65ft) long leaf of the raffia palm (*Raphia ruffia*). The flowers of many families also show considerable variety and include the lilies (Liliaceae) [8], orchids (Orchidaceae), pineapples and other bromeliads (Bromeliaceae) as well as the palms (Palmaceae), sedges (Cyperaceae) and grasses (Gramineae). More examples of monocotyledons are on the following pages.

Flowers of vivid beauty

The flowering monocotyledons include many of the most beautiful and ornamental garden plants, as well as decorative aquatic plants [5]. The lilies are widespread garden plants, together with species of narcissus (Amaryllidaceae) and irises (Iridaceae). Other striking examples of flowers in this group are the bird of paradise flowers and canna flowers [11].

The orchid family [10] displays a particular wealth of brilliant blooms, but even the most irregular and intricately shaped of orchid flowers still retain the characteristic three-fold monocotyledon arrangement of the flower parts, although this is sometimes hard to detect. Another large group of versatile and curious monocotyledon plants is found in the pineapple family. Most of these grow on other plants especially trees (as such they are called epiphytes), a few are terrestrial plants. Members of the pineapple family grow throughout the American tropics, in deserts, jungles or on salt-saturated beaches, and from sea-level to mountain slopes. A characteristic of many plants in this family is that the colour of the leaves changes when they flower.

Monocotyledons are widely distributed round the world. They range from palms,

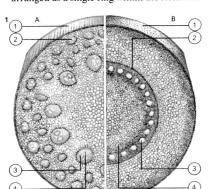

A Stem cross-section
1 Epidermis
2 Fibre bundle
3 Phloem
4 Xylem

B Root cross-section
1 Epidermis
2 Endodermis
3 Xylem
4 Pith

1 Monocotyledonous stems have an irregular arrangement of the food- and fluid-carrying tubes, the phloem and xylem. In the stem [A] they are scattered, but in the root [B] they are central.

Date palm
Phoenix dactylifera

Screw pine
Pandanus sp.

2 The date palm, with its graceful column-like trunk, can grow up to 30m (100ft) in height and live for 200 years often in arid conditions. The tree bears either male or female flowers and its fruit has been a staple food in North Africa and India for centuries.

3 Screw pines, with their characteristic stilt roots [1] above ground, are a feature of tropical Asia, the islands of Polynesia and the Indian Ocean. A few kinds grow in Africa. In some areas they form the dominant vegetation. Their tough, pliable leaves are made into mats and baskets.

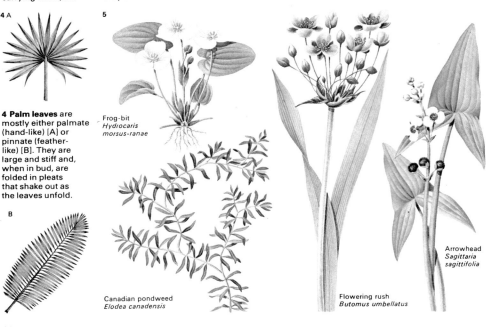

4 Palm leaves are mostly either palmate (hand-like) [A] or pinnate (feather-like) [B]. They are large and stiff and, when in bud, are folded in pleats that shake out as the leaves unfold.

Frog-bit
Hydrocaris morsus-ranae

Canadian pondweed
Elodea canadensis

Flowering rush
Butomus umbellatus

Arrowhead
Sagittaria sagittifolia

5 Aquatic monocotyledons clearly show a parallel leaf venation and those with conspicuous flowers, such as frog-bit, flowering rush and arrowhead, have the typical three petals. The plants can be bottom-rooted or free-floating and may grow their leaves and flowers above the surface. Canadian pondweed, which is free-floating and submerged, has tiny flowers that grow on long stalks to the surface. Other species have small flowers that open under water, some consisting only of an ovary and stamens.

Bog Cotton
Eriophorum sp

6 Bog cotton, a sedge found in areas of northern peat bog, is recognized by its white cotton-plumed, nodding seed heads.

which flourish in the tropics, to the bog-cottons [6], which cover vast expanses of the Arctic tundra. Many of these single-seed leaf plants provide food, such as grain crops useful to man, and are therefore of considerable economic value.

Valuable food plants

The palms are of immense economic importance to man because they produce food, timber, wax, oil, sugar, wine and fibre. Species of palms vary greatly in size and form. A few are slender climbers, but most have a characteristic appearance of woody trunks without branches, topped by a crown of large, stiff leaves. They have small individually inconspicuous flowers, usually in large clusters, and the variable fruit may be berries (the date) or drupes (the coconut).

The coconut palm (*Cocos nucifera*) can be used in many ways and is considered to be the most valuable of all palms. Another valuable species, the date palm (*Phoenix dactylifera*) [2], has flower clusters that may contain as many as 10,000 blossoms. To ensure good yields of fruit, growers sometimes hang bran-

ches of male flowers on the flowering female tree. Dates are especially nutritious and in northern parts of Africa they are a staple food of man and other animals.

Members of the grass family are probably among the most important of all the plants. The family is a large one, containing more than 600 genera and 10,000 species distributed throughout the world. Grasslands flourish in regions where rainfall totals 25–75cm (10–30in) annually. Many grass species are of vital importance to man and animals in the form of crops such as grain or green pasturage as well as being used for a variety of products such as fibres and building materials.

Bamboos [9] are grasses and with a few exceptions are tropical. The mountain species are able to withstand the cold of high localities. The woody, hollow stems, which are tough, stringy and braced by solid cross joints, have great strength. The bamboo grows profusely in South-East Asia and its stems can even be used for reinforcing concrete. Bamboo shoots are also the staple diet of the giant panda.

Meadow grass
Poa pratensis

Meadow grass is an important hay and green pasture grass in Europe and North America and as such it is an economically valuable member of the large and widespread grass family (Gramineae). The flower of the grass is a minute spikelet, usually arranged in open branching clusters known as panicles. The flowers are cross-fertilized by the wind and the single ovule then develops into a seed or grain. Grassland will evolve readily wherever forest or scrub cover is sparse and where there is sufficient moisture and nutrients in the soil. Vast areas of the world are natural grasslands, for example the steppes of Asia and the North American prairies.

7

Bluebell
Endymion non-scriptus

10

Californian lady's slipper orchid
Cypripedium californicum

8

Climbing lily *Gloriosa simplex*

7 The bluebell is a spring flowering bulb that is found in woodlands throughout Europe. Most bulbous plants for example tulips and daffodils, are monocotyledons. The bulbs are food storage organs for the plants and provide a means of vegetative reproduction.

10 The Californian lady's slipper orchid is so named from the distinctive slipper shape of the large concave lower petal or lip. There are 50 species of lady's slipper orchid, in a variety of showy colours and shapes, growing in tropical and temperate areas of the world.

11 Flowers of the banana group Zingiberales, such as the bird of paradise, Indian shot, banana and ginger flowers, are often spectacular and colourful. Although it is hard to detect at a glance, these irregular and intricately shaped flowers have the three-fold monocotyledon arrangement.

8 The climbing lily is a native of Africa. The lily family includes cultivated species and is widespread and varied. It includes many of the most fragrant and beautiful of all the flowering plants.

9 The bamboos are tropical members of the grass family Gramineae. They vary greatly in height, with giant species growing up to 33m (120ft), and are often found in dense, impenetrable thickets.

9

11

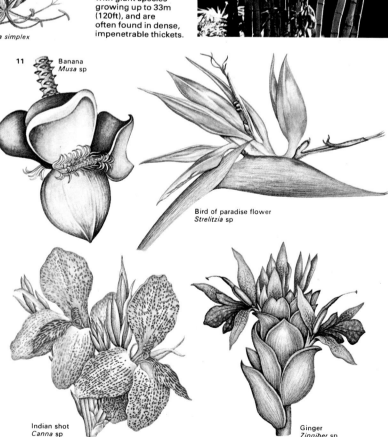

Banana
Musa sp

Bird of paradise flower
Strelitzia sp

Indian shot
Canna sp

Ginger
Zingiber sp

Grasses, reeds and rushes

The group of plants that includes the grasses, palms, lilies, orchids, reeds and rushes is known as the monocotyledons and comprises some 55,000 species grouped into 31 families. The monocotyledons, mostly herbs and found worldwide, include some of the earth's most useful and attractive plants. From the grass family [2] (Gramineae) come many of man's staple food crops, including wheat (*Triticum* sp) and sorghum (*Sorghum* sp), as well as luxury foods such as sugar from the sugar cane (*Saccharum* sp).

Bamboo [7], another grass, is a building material widely used in Asia and the papyrus [6] (family Cyperaceae) was used as a writing material by the ancient Egyptians. Other monocotyledons are used to brighten man's homes and gardens. Some of the most fascinating of these are the bromeliads (Bromeliaceae) [1] and the tradescantias [4] (Commelinaceae) – trailing plants that are widely grown. Other monocotyledons are found in such diverse habitats as desert, where the century plant [3] (*Agave* sp) thrives, and marshland where bulrushes [5] (family Typhaceae) are common.

1 The bromeliads (family Bromeliaceae) come from the New World tropics. Some, such as *Aechmea fasciata,* are epiphytes – that is, they grow on trees, large cacti, rocks and other supports, but do not obtain nourishment from them. The leaves of many bromeliads form a "vase" that catches and holds water and dissolved minerals from which nourishment is obtained. In the vase entire life cycles of small animals and plants may occur.

1 Bromeliad
Aechmea fasciata

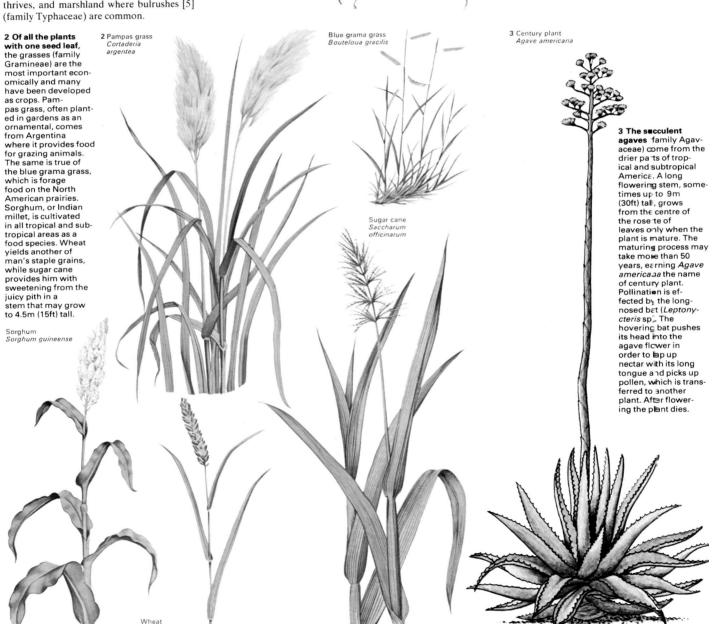

2 Of all the plants with one seed leaf, the grasses (family Gramineae) are the most important economically and many have been developed as crops. Pampas grass, often planted in gardens as an ornamental, comes from Argentina where it provides food for grazing animals. The same is true of the blue grama grass, which is forage food on the North American prairies. Sorghum, or Indian millet, is cultivated in all tropical and subtropical areas as a food species. Wheat yields another of man's staple grains, while sugar cane provides him with sweetening from the juicy pith in a stem that may grow to 4.5m (15ft) tall.

Sorghum
Sorghum guineense

2 Pampas grass
Cortaderia argentea

Blue grama grass
Bouteloua gracilis

Sugar cane
Saccharum officinarum

3 Century plant
Agave americana

3 The succulent agaves (family Agavaceae) come from the drier parts of tropical and subtropical America. A long flowering stem, sometimes up to 9m (30ft) tall, grows from the centre of the rosette of leaves only when the plant is mature. The maturing process may take more than 50 years, earning *Agave americana* the name of century plant. Pollination is effected by the long-nosed bat (*Leptonycteris* sp). The hovering bat pushes its head into the agave flower in order to lap up nectar with its long tongue and picks up pollen, which is transferred to another plant. After flowering the plant dies.

Wheat
Triticum aestivum

4 Boat lily
Rhoeo discolor

Wandering Jew
Zebrina pendula

4 The tradescantia family (Commelinaceae) comprises over 300 species, which are mostly found in the tropics and subtropics throughout the world. Most are creeping plants but some, including the boat lily, have upright habits. Wandering Jew is one of the many tradescantia species cultivated for the beauty of its leaves, which are variegated. The flowers are often small, few and inconspicuous, but in some species are large with violet-coloured petals and six bright yellow, pollen-bearing anthers. Vegetative reproduction is assisted by the formation of roots at the leaf bases.

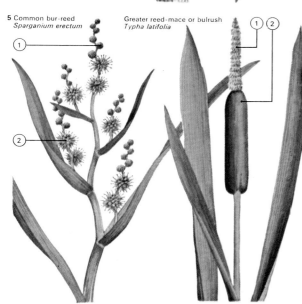

5 Common bur-reed
Sparganium erectum

Greater reed-mace or bulrush
Typha latifolia

5 The bur-reeds and reed-maces (or bulrushes) of the families Sparganiaceae and Typhaceae are usually found in swamps and at water margins. Both male [1] and female [2] flowers are borne separately on the same plant. In the reed-mace they occur on the same head, known as a spadix.

6 The sedges and club-rushes (family Cyperaceae) are a widespread group of plants usually found in waterlogged acid soils. Pendulous sedge grows in large clumps in damp, shady woods while sea club-rush is a plant of salt marshes. Papyrus, from which an ancient writing material was made, is a member of the same family.

6 Pendulous sedge
Carex pendula

Female flowers   Male flowers

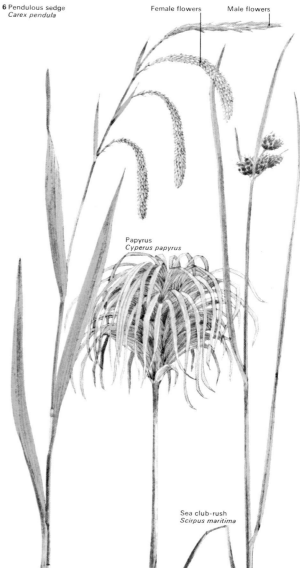

Papyrus
Cyperus papyrus

Sea club-rush
Scirpus maritima

7 The bamboos are woody members of the grass family that vary in height from a few centimetres to several metres. Bamboos often grow in dense, impenetrable clumps in tropical regions, but some, such as *Arundinaria alpina*, grow on mountain sides and can withstand cold conditions. Because of their abundance, strength and unusual stem structure [1] (woody and hollow with solid, stiffening cross joints), bamboos are used by man for many purposes, including the building of houses and furniture. Large canes are used as pipes in irrigation schemes. Young shoots can be eaten and are regarded by many people as a delicacy.

7 *Arundinaria japonica*

Arundinaria alpina

69

How flowering plants reproduce

The profusion of rainbow coloured, richly scented and curiously shaped flowers that exists throughout the world is an essential part of the plant kingdom's equipment for sexual reproduction, although nearly all plants can reproduce asexually and sexually. Asexual reproduction [9] can take place by simple budding, by means of bulbils (bundles of swollen leaves), rhizomes (underground stems) or tubers (swollen parts of roots or stems). Or, in a few so-called apomictic species such as dandelions, seed can be produced without sexual fusion. Sexual reproduction is effected through specialized organs in which a male cell fuses with a female cell. This gives rise to a new plant that combines the characters of both parents.

Sexual reproduction

Plants that reproduce sexually may either have the male and female organs in the same flower (bisexual), or in different flowers on the same plant (monoecious) or on different plants (dioecious). As the flower bud is formed and the flower parts develop, so the male and female organs differentiate. The female part consists of the ovary, in which there is the ovule (embryo seed) and from which projects the stigma (the pollen receiver) on a stem or style of varying length. The male organ is the stamen, made up of an anther, bearing pollen grains, and the stalk-like filament. Most pollen grains are carried by wind or animals – frequently insects – to the ovule which contains the egg nucleus and in which, after fertilization, the young embryo subsequently develops to give rise to a seed and eventually to a new plant [4].

Pollen grains, too small to be seen individually by the human eye, can be detected as "dust" when shaken from the anthers of flowers. But when magnified, pollen is seen to vary greatly in size and shape and electron microscopy has revealed remarkable structuring of ridges and hollows in the outer coat of each pollen grain. In many flowering plants there is a physical affinity between the shape of the pollen grain and the surface cells of the female stigma, contributing to successful fertilization. In the primrose, for example, there are two distinct flower types [Key] but the structure of the pollen grains is such that those from one type fit the stigmas of the other and vice versa.

The evolution of flowers

In the earliest and most simple types of flowers, pollen was transferred from stamens to stigmas by wind. Many trees have wind-pollinated flowers, often forming catkins. The male flowers are simple and open in structure to facilitate pollen dispersal. Pollen is produced in very large quantities so that from the cloud of pollen released as catkins toss in the breeze there is a high chance of some grains at least falling on ripe stigmas of a nearby female flower of the species. The full range of diversity of flower adaptation is found in flowers pollinated by animals [1]. The direct transfer of the pollen by a creature provides greater efficiency in fertilization, but to ensure that it will occur the flower must attract the required insect (or other animal) pollinator. To this end many flowers have bright colours and studies of the sight capacity in bees and other insects have shown them to be particularly receptive to certain colours. The petals of flowers are modified

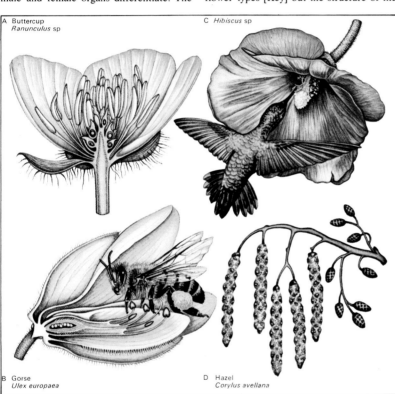

A Buttercup Ranunculus sp
C Hibiscus sp
B Gorse Ulex europaea
D Hazel Corylus avellana

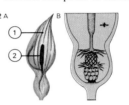

1 Flowers are adapted to different pollination methods. These are: non-specialized simple flowers [A]; bee-specialized flowers [B]; an hibiscus pollinated by humming-birds [C]; wind-pollinated catkins [D].

2 The flower of the cuckoo pint (*Arum maculatum*) [A] is remarkably adapted for pollination. As the spathes [1] unfurl [B] the exposed spadix [2] heats up, releasing a carrion scent attractive to

owl-midges. These insects fly to the scent, crawl down and pollinate the female flowers and become trapped in a ring of hairs [C]. The midges are dusted with pollen from male flowers and escape as the hairs wither [D].

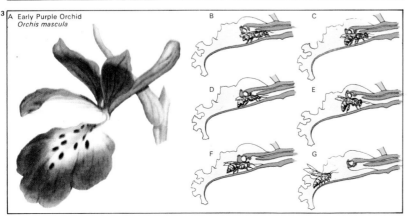

A Early Purple Orchid Orchis mascula

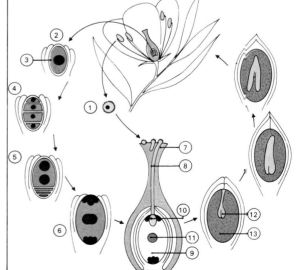

3 In the pollination of the early purple orchid [A] the bee enters with a bundle of pollen (a pollinium) from another flower. As it backs out, the pollinium is left on the stigma. The bee's back is smeared with liquid to which the pollinium sticks [B-G].

4 Flower maturation involves pollen production [1] and ovule development [2]. The ovule mother cell [3] divides repeatedly [4, 5] to form an embryo sac [6] containing eight nuclei. When pollen lands on the stigma [7] a pollen tube [8] forms

and the male nucleus from the pollen grain passes down it, dividing in two as it does so. In the ovule [9] one fertilizes the egg nucleus [10], while the other fuses with one polar nucleus [11]. The embryo [12] develops within nutritious endosperm [13].

leaves – in wind-pollinated flowers they are usually small and mostly green.

In insect-pollinated flowers the petals are often large and showy and the green pigment found in leaf cells is replaced by pigments that provide a new colour range. Yellow and white pigments are primitive colours found in the simpler flowers while red, purple and blue pigments are found in more advanced flowers. Orange is an uncommon natural flower colour and completely black petals are unknown in the wild, although "black" petals have been produced in some cultivated species such as tulips.

Flower shape and scent
Many intricate flower shapes have been evolved as a result of the plant/insect relationship [8]. In each insect-pollinated flower there are glands producing nectar and scents enticing the insect to visit [2]. In the petal markings of most flowers there is a pattern of honey guidelines, that are sometimes black, to direct the feeding insect to the parts of the flower necessary for pollination. Simple flowers that are open and saucer-

shaped have nectar at the petal bases in the centre and can be pollinated by any visiting insects. More specialized flowers are tube-shaped and the tube often fits one particular type or species of insect exactly, so that pollen is "wiped" off the insect's body.

Orchids have a most complex pollination mechanism with flowers often shaped, marked and scented to resemble an insect and some even have furry-textured petals [3]. Equally specialized are long-tubed flowers whose nectar can be reached only by long-tongued butterflies and moths, birds with long, curved bills [1] or even bats. Where the evolutionary lines between plants and pollinators run parallel, animals are able to feed from and pollinate flowers through mutual adaptation.

After fertilization, flower petals and stamens wither and the ovary develops into the ripening fruit containing the seed. There are many adaptations for seed dispersal including wings, parachutes [5] and explosive mechanisms [7] to ensure that the seeds are carried far enough from the parent to colonize new soil.

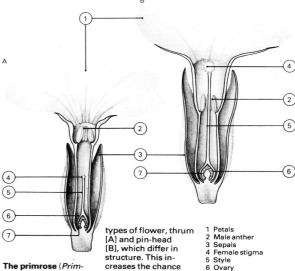

The primrose (*Primula vulgaris*) has two types of flower, thrum [A] and pin-head [B], which differ in structure. This increases the chance of cross-pollination.

1 Petals
2 Male anther
3 Sepals
4 Female stigma
5 Style
6 Ovary
7 Ovules

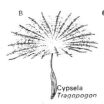

5 Some plant seeds are shaped for wind dispersal. Air currents lift the winged sycamore samara [A], spinning the fruit some distance away. The parachuted cypsela of goat's beard [B] floats on the breeze.

6 Man and animals assist seed dispersal when fruits with hooked spines become caught on fur or clothing. These trouser legs have captured fruits of burr-marigold, cleavers and agrimony.

7 After fertilization a flower's petals fade and drop, leaving the ripening fruits. In the geranium family each fruit is divided into five single-seeded parts attached by a stalk or pedicel [1]. These are at first joined to a central column [A]. After ripening, and in dry air, the outer cells of each pedicel contract causing the fruits to spring up and away from the plant [B]. In this way seeds can be thrown several metres.

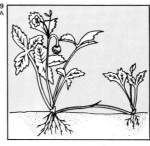

8 Berries, pollen and nectar can form the staple diet of many birds and small mammals, and often there is intricate interdependence between the plants and animals. In Australia, parrots [1] and parakeets [2] swinging in the gum tree branches [3] to seek out fruit will shake the tasselled flowers, and pollen grains that dust their feathers (or are caught on their tongues) will be carried to new plants. Honeyeaters [4] with long, curved bills and honey possums [5], with narrow noses and long tongues, use these adaptations for feeding on the rich nectar of the *Banksia* flowers [6–9]. Others such as the dormouse possum [10] carry pollen grains caught in their fur or whiskers.

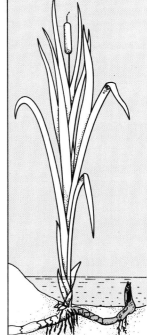

9 New plants may arise by a variety of asexual methods. In the strawberry [A] they spring from creeping stems, in the bulrush [B] from swollen underground stems or rhizomes and in the wild onion [C] from bulbils in the flower head.

The animal kingdom

All animals are made up of a cell or cells with a nucleus that contains the cell's genetic specifications. The main groups into which the animal kingdom is divided are known as phyla and the most primitive phylum is that of the Protozoa, animals that are made up of just one single nucleated cell. This means that each individual protozoan must be completely self-sufficient, with special structures (organelles) to perform not only those functions that are essential to all cells, but also those that in many-celled animals are shared between different tissues. Protozoans may be plant-like, or animal-like.

From sponges to shells

Although there are protozoans that live in colonies, the great step that separates them from all other animal forms is that which led to the development of interdependent groups of cells with a distinct form. The phylum Porifera, which contains the sponges, is the most primitive multi-cellular phylum and sponge cells retain a high degree of independence. The Coelenterata are more complex and the members of this phylum have differentiated tissues. These animals are composed essentially of a gut and a ring of tentacles. Sessile forms, such as *Hydra* and sea anemones, may also have a foot. Others, such as the jellyfish [3], are free-swimming.

Apart from the small phylum of Ctenophora – the sea walnuts or comb jellies – which are very like the coelenterates, the next three major phyla in the evolutionary tree consist of worms – flatworms (Platyhelminthes) [4], roundworms (Aschelminthes) [5] and the Annelida [6]. They are all hermaphrodites, the most widely known of the segmented annelids being the common earthworm.

The phylum Mollusca contains more than 80,000 species, which makes it second in number among the invertebrates only to the arthropods. Molluscs are most conspicuous for their external shells; the marine gastropods and bivalves provide the varied and fascinating sea shells that litter the world's beaches. The most important class of molluscs other than the Gastropoda [7] and the Bivalvia are the Cephalopoda – the squids and octopuses. In some of these the shell has become internal and in *Octopus* [8] it has disappeared altogether. Because of the elaboration of their mode of locomotion and their carnivorous habits, cephalopods have relatively highly developed nervous systems.

External and internal skeletons

There are about one million known species of arthropods, of which the most numerous fall into three classes: Crustacea – the lobsters [9], shrimps and crabs; Arachnida – spiders [10] and scorpions; and Insecta [11]. The arthropods have colonized every conceivable ecological niche on land, in the air and in water. They, like annelids, are characterized by a segmented body, but this has been adapted to such an extent that it is often only clear during the larval phase; but more importantly they feature an external protective skeleton. Most also have compound eyes and complex nervous systems.

The Echinodermata are marine animals which, although invertebrates, have an internal skeleton. The most familiar of these are the starfishes [12] and sea-urchins. There are many minor invertebrate phyla but the

1 Amoeba is a single-celled animal that belongs to the vast phylum Protozoa. The protozoans flourish wherever there is moisture and they range in size from the microscopic *Amoeba* and the blood parasite *Pseudoplasmodium* to the freshwater ciliate *Spirostomum*, which is 3mm (0.1in) long. Because it is difficult classifying single-celled organisms as animals (Protozoa) or plants (Protophyta), they are often grouped as the Protista.

Sarcodine
Amoeba proteus

African buffalo
Syncerus caffer

2 The African buffalo is one of the large mammals, made up of millions of cells organized into specialized tissues. With a height of 1.5m (4.9ft), the buffalo is many millions of times larger than the largest protozoan. Mammals have colonized most environments in which protozoans can flourish; and some protozoans, such as *Trypanosoma*, colonize mammals. *Trypanosoma* causes sleeping sickness.

3 Jellyfish belong to the Coelenterata, the first unambiguously metazoan or multi-celled animal phylum in the evolutionary tree, sharing with the Ctenophora the feature of radial symmetry.

4 Parasitic tapeworms are members of the phylum Platyhelminthes. All are flattened and bilaterally symmetrical. Parasitic flukes and free-living flatworms are also in the phylum.

Jellyfish
Chrysaora hyoscella

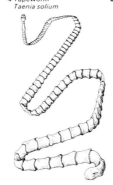

4 Tapeworm
Taenia solium

5 Threadworm
Oxyuris vermicularis

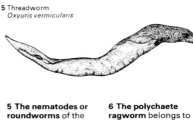

5 The nematodes or roundworms of the phylum Aschelminthes are probably the most numerous metazoans in the animal kingdom. The 10,000 species live anywhere from the desert to the deep blue sea.

6 The polychaete ragworm belongs to the phylum of segmented worms which, at their largest, are larger than any other worm-like invertebrate. Their segments are essentially identical.

6 Ragworm
Nereis diversicolor

7 The freshwater winkle, *Viviparus viviparus*, belongs to the largest class of molluscs, the Gastropoda, with more than 35,000 species. Bivalves and cephalopods are also in the order Mollusca.

8 The octopus, a mollusc that has lost its shell in the course of evolution, belongs to the cephalopods. This group includes the 20m (65ft) giant squids, largest of all the invertebrates.

9 The lobster belongs to the arthropod phylum and is one of the jawed (mandibulate) species. It represents the Crustacea, the only primarily aquatic arthropod class, which is made up of more than 30,000 distinct species.

9 Norway lobster
Nephrops norvegicus

Octopus
Octopus vulgaris

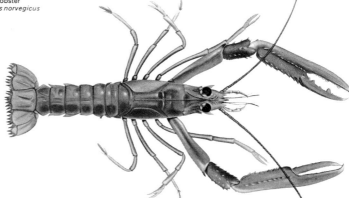

majority of the remaining invertebrates belong to the Hemichordata, which is a sub-phylum of the Chordata and contains mostly marine worms. About 95 per cent of animal species are invertebrates and numerically arthropods are the most important group, because they constitute three-quarters of all animal species. The vertebrates constitute just one sub-phylum of the chordates and all have backbones. The other small sub-phyla include the tunicates (sea squirts) and the lancelets. They have no true backbones.

Fish, amphibians, reptiles and mammals
The group Pisces, the fish, is much the largest group of vertebrates and is also the oldest in evolutionary history. The two principal kinds of fish are the cartilaginous, or shark-like fish, and the bony fish [13] which comprise the larger and more advanced group.

The Amphibia, the frogs, toads [14], newts and salamanders and a few legless worm-like species – the caecilians – are the survivors of the first group of vertebrates to develop legs, enabling them to emerge from the waters in which life began. But they have not quite abandoned the aquatic habit: their eggs are still laid in water and the larvae (tadpoles) breathe through gills.

The reptiles [15] made the final break with the water through the development of eggs with shells, for protection and preservation of moisture, and yolks for nourishing the embryo. The living reptiles are the snakes, lizards and turtles, which are mainly tropical species because their cold-bloodedness makes them too sluggish to survive in colder climates. To a zoologist birds [16] seem little more than "glorified reptiles". They have an essentially reptilian physiology but a warm-blooded metabolism, light, hollow bones and feathers. Their complex behaviour, including song, is unknown in reptiles.

The mammals [2], the hairy animals, are the only other warm-blooded class of animals. They are defined by their protracted care of the young which they nurture internally (except for the egg-laying monotremes such as the platypus) and feed with milk from their mammary glands for a varying period after birth. Behavioural complexity and intelligence reach their peak in the mammals.

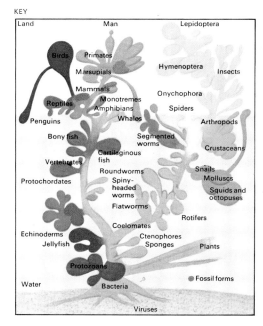

KEY

Land · Man · Lepidoptera
Birds · Primates · Hymenoptera · Insects
Marsupials
Mammals · Onychophora
Monotremes
Reptiles · Amphibians · Spiders
Whales · Arthropods
Penguins
Segmented worms
Bony fish · Crustaceans
Cartilaginous fish
Vertebrates · Roundworms · Snails
Protochordates · Spiny-headed worms · Molluscs
Squids and octopuses
Flatworms
Rotifers
Coelomates
Echinoderms · Ctenophores · Plants
Jellyfish · Sponges
Protozoans
Water · Bacteria · ● Fossil forms
Viruses

10 Tarantula
Aphonopelma sp

10 Spiders, scorpions, mites and ticks are arachnids and belong to the chelicerate or jawless division of the Arthropoda. Among arthropods, Arachnida are second to Insecta in numbers.

11 The butterfly is one of the most beautiful of the Insecta, which is a huge class of arthropods containing some 1,000,000 known species, more than all the other animal species put together.

11 Asian swallowtail
Papilio philoxenus

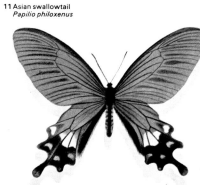

12 · Common seastar
Asterias rubens

12 The starfish has hard projecting tubercules, forming part of its internal skeleton. This is the most important feature characterizing the phylum to which it belongs, the Echinodermata. The name means "spiny-skinned". The starfish has a central mouth located on the underside of its body. It eats mussels, snails, oysters and clams, and is found in all seas except near the poles.

13 Catfish
Silurus glanis

13 Fish were the first vertebrates to have jaws. They belong to a sub-phylum (Pisces) of the Chordata and an evolutionary landmark from the point of view of the emergence of man.

14 The toad belongs to the order Anura, whose members are the commonest of the three amphibian orders. The Amphibia were the first of the tetrapod vertebrate classes – the four-footed land animals.

14 European green toad
Bufo viridis

15

Boa constrictor
Constrictor constrictor

15 The boa constrictor belongs to the legless Serpentes sub-order of Reptilia, the other sub-orders of which – as the first vertebrate class to become completely independent of water – have legs.

16 The stonechat, one of the 8,600 species of the vertebrate class Aves, belongs to the most diverse vertebrate group after Pisces. Almost all the special features of bird physiology can be traced to the requirements of flight.

16

Stonechat
Saxicola torquata

Animal anatomy

All animals, from the microscopic protozoan amoeba to the huge African elephant, share the same essential requirements. They must be able to obtain a supply of food and oxygen, get rid of unwanted waste products, reproduce themselves, move and be capable of monitoring and responding to the environment in which they live. In the single-celled amoeba these tasks are easily fulfilled; thus oxygen is obtained by simple diffusion.

Layers of cells

During evolution, animals were able to become more complex only because their body plans were organized in a way that allowed the essential life processes to continue. The first big jump from the protozoan body plan came with the appearance of the coelenterates – sea anemones and their relations. In these multicellular creatures the body is a simple tube bounded by two layers of cells – the ectoderm and the endoderm.

True organs did not arise until animal anatomy had taken another step forward – the development of a third layer of cells, the mesoderm, between the ectoderm and the

endoderm. The simplest animals to have three cell layers are the platyhelminths, of which the flatworm is a member. From this mesoderm layer there arise muscles, reproductive organs and excretory cells.

The ectoderm forms the nervous tissue, along which coded information is conveyed to and from different body parts. The simple flatworm body is the first on the evolutionary tree to have a recognizable "head" containing clusters of nerve cells or ganglia and also primitive "eyes".

The three-layer body plan provided the potential for enormous evolutionary advance and this arrangement is found in all the remaining groups of animals. Despite this, animal bodies can still be thought of as tubes, and in all animals from the sea anemone onwards digestion takes place in the innermost tube – the gut – which is lined with endoderm.

The gut also contains bands of muscular tissue made from mesoderm cells. The purpose of these muscles is to move food from mouth to anus. In the flatworm this process is accompanied by waves of contraction down

the whole of the animal's body. This is very restrictive and little evolutionary advance was possible until the development of a fluid-filled cavity, the coelom [Key], separating the gut from the rest of the body.

The simplest animals to have a coelom are the annelids – the earthworms and their allies – and in these creatures the coelom is easily visible in a cross-section. In higher animals the coelom is barely detectable because it is filled with numerous glands and organs.

The segmented body plan

The most striking feature of annelid anatomy is its division into segments [1]. A closer, internal study shows that each of these is essentially identical, apart from those at the "head" end [6] where there is some concentration of nerve tissue, and at the tail. For annelids, the greatest advantage of the segmented body plan is as an aid to movement. In other invertebrate animals, however, different body segments have become specialized to do different jobs. This is particularly so in the arthropods [2], a group that includes the insects. In these animals the

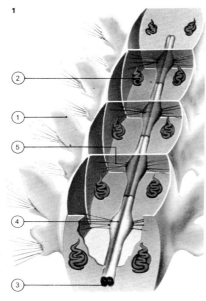

1 Except at head and tail the segments of an annelid worm such as the ragworm *Nereis* are virtually identical. From the sides of each segment project "side feet" or parapodia [1] that are used as paddles in swimming. Also in each segment are paired excretory organs [2]. Running centrally is the ventral nerve cord [3] and every segment contains a mass of nerve cells, the segmental ganglion [4], from which originate segmental nerves [5]. These provide a system of communication within the body. While the nervous system transmits coded information, the blood system conveys oxygen to all the body cells.

2 Although totally different in evolutionary origin, man and the crayfish depend on similar anatomical systems to circulate blood, digest food and perform other body functions. But the crayfish has a

hard external skeleton covering its body while man, in common with other vertebrates, has an internal skeleton to which muscles are attached. The segments of the crayfish body show some specialization.

- ■ Blood system
- Nutrition
- Reproduction
- Excretion
- Respiration
- Nervous system
- ■ Skeleton
- ■ Endocrine system

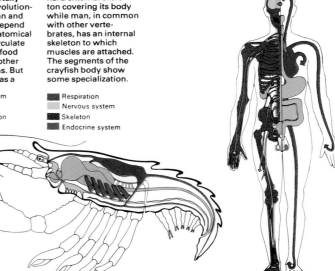

3 Fish, reptiles, birds and mammals are all vertebrates with essentially similar skeletons. The fish [A] and the lizard [B] both possess similar skeletal features such as a skull [1], a backbone [2], ribs [3] and limb support bones [4]. The lizard's foot [5] is an evolutionary development of the fish's fin [6] and has the 5-toed pattern typical of vertebrates.

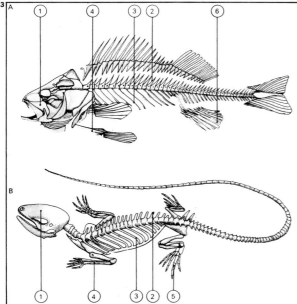

4 The differences between the skeletons of a bird [A] and a mammal [B] reflect the differences in their ways of life. The limb bones of birds [1] are reduced and fused as are those of the backbone [2]. The mammal, having no need for lightness, tends to have larger and heavier bones.

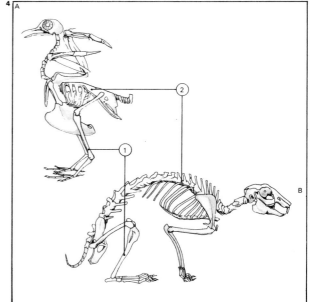

segments are equipped with jointed appendages but these are modified to form structures such as legs and mouthparts. Internally, sex organs, a heart to push blood round the circulatory system and other organs are positioned in particular segments and there is a head with a simple brain and most have large compound eyes.

External and internal skeletons

Small, soft-bodied animals such as worms have little need of any body supports, and have no skeleton, but the arthropods have developed an external skeleton that serves the dual purpose of protection and the provision of rigid material to which muscles can be attached. Such an arrangement does have its problems, however, for it means that the animal cannot grow without moulting this skeleton. Not until the skeleton was formed internally did animal anatomy take another big step forwards. This occurred with most significance in the primitive chordates. These animals possessed a notochord, the forerunner of the backbone.

The chordate body, like that of the annelids, is divided into well-defined segments. These are related to the development of the backbone because each one contains bands of muscles – myotomes – attached to the notochord and whose contraction enables the animal to swim.

It is only a small step from this body plan to that of a fish. Like all animals with backbones [3, 4] the fish has a body divided into head, trunk and tail, the latter lying behind the anus. It also has a chambered heart and a blood system divided into veins, arteries and capillaries. In the fish, oxygen is obtained from the surrounding water by way of gills well supplied with blood vessels.

The next really significant anatomical advance came with the emergence of life on land. For the first time animals breathed through lungs and had internal skeletons developed to form the struts of four well-defined limbs, each with a five-toed foot, on which the body could be raised off the ground. Other advances on the vertebrate plan [3, 4, 5] are essentially adaptations to different modes of life and include the wings of birds and the placentas of mammals.

KEY

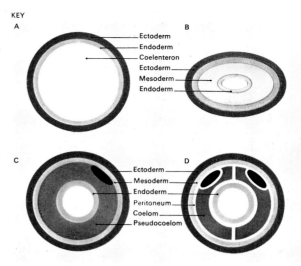

Sections cut across animal bodies show increasing complexity. The 2-layered coelenterate body [A] and the 3-layered flatworm body [B] are both simple tubes. All higher animals [D] have an additional cavity, the coelom. This differs from the pseudocoelom of roundworms [C] in its lining.

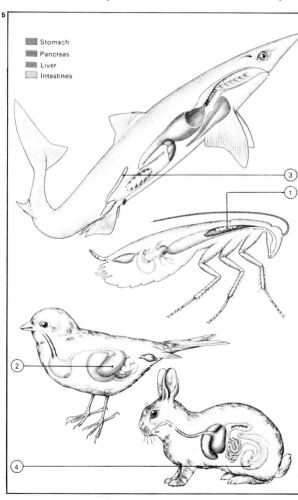

5 The tubular digestive system consists of an intake opening, the mouth, leading to a storage and processing chamber, the stomach. This is followed by the intestines where the absorption of digested food takes place. Associated with the intestines and involved in digestion are the liver and the pancreas. Differences in animal digestive systems are the result of dietary dissimilarities. The insect has a gizzard [1] with "teeth" for grinding food. A bird's gizzard [2] is a modified stomach with a horny lining. It contains small stones that are used to grind food. The fish has a rectal gland [3] to excrete excess salt. The rabbit has an enlarged caecum [4] that is the site of cellulose digestion.

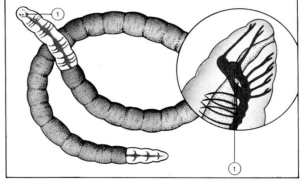

6 The nerve cord of the earthworm is specialized at the "head" end of the animal to form cerebral ganglia [1]. This development makes possible a range of co-ordinated responses to the environment such as those used, for example, in finding food, in burrowing and in mating. The outer covering of the body of the earthworm bears receptors that respond to light, temperature changes, chemicals and touch.

7 The frog, like other vertebrates, has a well-defined head containing a lobed brain. The cerebrum is concerned with learned behaviour, the optic lobes with vision, the olfactory lobes with interpreting smells and the cerebellum with balance and the co-ordination of movement. No one area of the frog's brain shows any dominance over the others, but compared with a fish the cerebrum is larger.

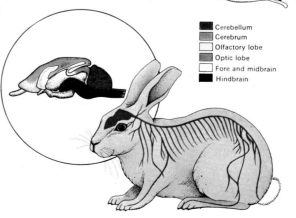

Cerebellum
Cerebrum
Olfactory lobe
Optic lobe
Fore and midbrain
Hindbrain

8 The brain of the rabbit, like that of other mammals, is remarkable for the size of its cerebrum and its cerebellum. These are associated, respectively, with the large range of voluntary movements the animal performs and with the very fine degree of muscular co-ordination achieved. The size of the optic and olfactory lobes is an indication of the importance of the senses of sight and smell. In man the cerebrum overshadows all the other parts.

Sex in the animal world

In the natural world all sexual activity is directed to one end: the fusion of male and female sex cells or gametes to form the first cell of a new organism. The means to that end are many. In flowering plants, for example, the male sex cells or pollen may reach the female sex cells by being borne on the wind or carried by animals. In the animal kingdom the male and female gametes are, respectively, spermatozoa and eggs [Key] and their fusion may take place either inside or outside the female body. In all living organisms vastly more sex cells are produced than are ever shed or conjugate and this seems to be a natural insurance policy for fertility.

The evolution of sex

Why did sex evolve? Many primitive animals reproduce simply by budding and in principle there is no reason why humans should not do the same. But sex, by bringing together the inherited characteristics of two parents in a new combination, makes for greater variety in the offspring. This variety is not just the spice of life: in evolutionary terms it is the essence of survival, for variety offers a wide choice on which the forces of natural selection can operate. It provides the material for an infinite number of genetic variants within a species and for vigorous offspring and it allows for the "storage" of genes that are not immediately useful. These genes, inherited from one parent, are dominated or "disguised" by genes from the other.

Bringing the sexes together

The first requirement for mating is to bring the two animals together at the right moment [1]; in many species this is achieved by the secretion of chemicals known as pheromones by the female. The male traces her by sense of smell, sometimes exquisitely sensitive, as in moths which are able to detect one or two molecules of female pheromone borne on the wind [5]. Other animals may rely more on the other senses. Male frogs, for example, have special mating croaks with which they attract females. The tones of the male's croak develops with his sexual maturity and only reaches the right pitch when he is old enough to mate. Until then, females ignore him.

The most spectacular ways of bringing the sexes together are the elaborate visual displays of some animals – from the "dance" of the red-bellied male stickleback in the mating season to the courtship rituals of some birds, which may flaunt their plumage like the peacock or even build a decorative love-nest like the bowerbird.

Many of the mating calls and visual displays of animals are extremely elaborate and serve a dual function. Not only do they bring the two sexes of the same species together, but they also ensure that the two sexes in different species are kept apart. The behavioural ritual has to be exactly right or the female will not respond. The same applies to pheromones: the chemicals must be exactly the right ones, sometimes even in exactly the right proportions, or the male will not court the female. This is why so much evolutionary energy has been expended on extravagant or subtle ritual and decoration in the mating season: it plays a vital part in the process of reproductive isolation by which similar species can live in the same territory and yet remain separate.

There are several mechanisms, once male

1 Sexual reproduction in one form or another has evolved in most life forms. Some mosses [1] have male and female sex cells on different plants; sperm is carried to the egg on a film of moisture. Many plants, such as foxglove [2], have female eggs and male pollen on one plant. Earthworms [3] are also hermaphrodite (with both sex organs) but nonetheless mate with a reciprocal exchange of sperm. In single cells [4], mating involves only the exchange of small pieces of genetic material rather than a complete package. Sticklebacks [5] and frogs [7] fertilize their eggs internally. In many other creatures such as dragonflies [6], chaffinches [8], field mice [9] and rabbits [10] fertilization is internal. Coupling is belly-to-back, whether on the ground, in water or on the wing. Fertilization in all mammal species takes place internally.

2 Cuttlefish *Sepia* sp

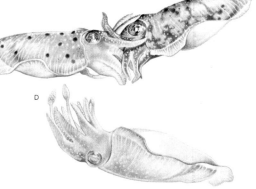

2 The male cuttlefish courts a female [A] by displaying to her, often blushing red all over as he does so. The two animals then join [B], either side by side or head-to-head, with their tentacles entwined. The male then transfers a package of sperm – a spermatophore – to the female's mantle cavity using a modified tentacle, the hectocotylus, which may then drop off. The male then dies [D]. So does the female, but not until she has laid her eggs on the sea-bed [C].

and female have been brought together, for the final step of fusing the genetic information of the two. Some single-celled organisms have a special structure on the cell surface that effects the transfer of small pieces of DNA (the material containing the genetic blueprint) from one organism to another. In hermaphrodites, such as the earthworm or the snail, the behaviour of the two partners is identical, each transferring sperm to the other. In terrestrial species with distinct sexes, fertilization is usually internal, the male inserting a penis or some analogous organ into the female to deposit the sperm. Sometimes, as in most aquatic species, the eggs are laid first and fertilized externally afterwards, while cuttlefish and squids, in which fertilization is internal, use specialized tentacles to transfer sperm to egg [2].

Socio-sexual relations
The social aspects of sex range far beyond courtship and mating. The hormones that burnish the plumage of the male mallard and bring a scarlet blush to the belly of the stickleback are also the ones that in many species

cause aggressive behaviour between males. Rutting deer engage in ritual encounters which, although they may not represent fighting in the strict sense, certainly appear hostile from a human point of view. The vigorous defence of territory by some male birds is also hormone-dependent.

In insects socio-sexual relationships can be extremely complex. The honey-bee queen monopolizes the attentions of the male drones, whose sole function is to fertilize her eggs. The workers, which build, clean, forage and act as nursemaids for the entire colony, are females prevented from reaching sexual maturity by the food they receive and by chemicals (pheromones) which they lick from the queen.

In higher animals living in colonies or herds it is usual for a single male to dominate other males and gain exclusive access to a large number of females. In deer and seals, for example, a male gathers around him a harem of many females [6]. In seals the younger bulls may often be left out in the cold and not acquire harems for many years after they have reached sexual maturity.

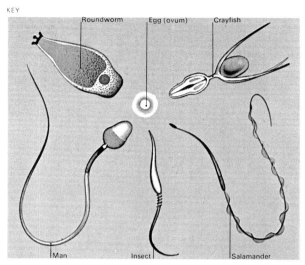

At the moment of fertilization, the *raison d'être* of sexual activity, the sperm from the male partner enters the egg from the female, each carrying genetic material. Sperm vary in shape and size from species to species but can be grouped into flagellate and non-flagellate types.

3 The courtship and mating ritual of the Uganda kob begins when the male circles the female with a prancing gait, his head held high to display the white patch on the throat [A]. He sniffs her vulva [B], causing her to squat, and can detect by her smell that she is in heat. He then tests her willingness to mate with a sharp kick [C] probably a relic of fighting behaviour. If she is ready the male quickly mounts her [D] to ensure rapid intromission and ejaculation.

3 Uganda kob
Kobus kob tomasi

4 The pairing of the wandering albatross (*Diomedea exulans*), which breeds on islands in the Subantarctic, takes place by way of an elaborate ritual involving both auditory and visual signals. This ritual establishes a pair bond. The sequence begins with neck stretching and bill clapping [A] which are followed by bowing of the heads in a preening posture [B]. The birds then stretch towards one another [C] with more bill clapping, "nibbling" at one another's bills. The male then circles the female with wings open; she turns on the spot, always facing the male [D], until finally both stand with necks stretched and wings open [E] as the preliminary to mating.

5 A

B

C

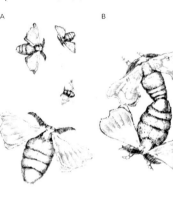

5 A male moth detects a female [A] by a "sex scent" or pheromone she secretes from a gland on her leg. He "smells" this scent by means of sensory receptors on his antennae [D] which enable him to detect very small quantities of the pheromone. After copulation [B] and fertilization the female seeks out a suitable larval food-plant, such as mulberry, on which to lay her eggs [C].

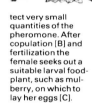

6 The South American sea lion (*Otaria byronia*), like other members of its order (Pinnepedia), must breed on land. The males come ashore to establish territorial rights over a section of beach, often fighting their rivals fiercely. As the females come ashore the bulls try to gather a harem of 9–10, which precipitates more inter-male rivalry. Soon after the females have landed they give birth to single pups (conceived the previous year) after which the bulls mate with them. For most seals and sea lions the gestation period is from 250–365 days. It is so variable because delayed implantation is known to occur in some species.

Principles of animal behaviour

Animals must be observed in their natural habitats if their behaviour is to be understood, because much of their behaviour is innate and has evolved to enable them to survive, feed and reproduce in those habitats [1]. Instinctive patterns of animal behaviour have been moulded by the demands of the environment in the course of evolution and reflect the animal's evolutionary history in just the same way that its skeleton does. Insects provide excellent study material for investigating animal behaviour because, unlike mammals, their actions are not modified by a higher intelligence.

The basis of behaviour

An animal's action – a cat pouncing on a mouse, a peacock spreading its tail to the hen, a spider spinning its web – is determined by three factors. One is the external stimulus – the mouse or the peahen. The second is its own sense organs and nervous system which determine what the animal can see, hear and feel and also what behaviour patterns it can produce in response. The third is the state of its body chemistry, such as its state of hunger

or, for example, the level of its sex hormones.

Not only do the sense organs of animals give them widely varying amounts and types of information about the world about them, but studies of animal behaviour have made it clear that, for each species, there are certain stimuli that have special significance. These, known as sign stimuli, may be visual, like the red belly of the male stickleback during the breeding season, which induces the female to spawn, or chemical, like the chemical released by the female moth to attract the male. Chemical sign stimuli are known as pheromones. Sign stimuli trigger what are known as innate releasing mechanisms (some sign stimuli are called releasers) which lead to the expression of stereotyped behaviour, sometimes known as fixed action patterns. Parental behaviour, for example, is triggered in many species by the round face and large foreheads of many infant animals [Key].

Feeding behaviour

Both the stimuli and the behaviour released by them vary enormously in complexity. One of the most primitive patterns is that involved

in the feeding behaviour of the frog [4]. Frogs perceive as prey any relatively small moving object in their field of vision and the detection of such an object releases a rapid dart of the tongue in its direction. But while the feeding behaviour of the frog is barely more than a reflex, the feeding behaviour of some insects involves vastly complicated social signs. Most famous of these is the dance of the worker honey bee [2], which won its discoverer, Karl von Frisch (1886–), a Nobel prize in 1973. A worker bee that has found a food source returns to the hive and performs a "dance" that conveys to other workers information on the whereabouts and nature of the food. The dance causes the workers to leave the hive in the direction encoded in the forager's movements.

A fact of which beekeepers are only too aware is that honey bees become enraged at the approach of a thunderstorm, the result of their sensitivity to changes in the earth's aerial electrical stress field. Recent studies suggest that they begin to show the first signs of annoyance as the electrical oscillations approach 10kHz when a thunderstrom is on

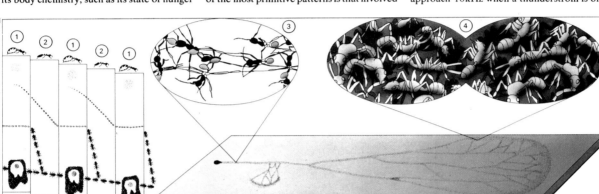

1 **Army ants** (*Eciton burchelli*) of the tropics are found in colonies made up of soldiers and two kinds of worker ant, with a queen whose function is to lay eggs and produce more ants. Army ants undergo an alternating rhythmic cycle of static [1] and nomadic [2] phases. In the static phase, eggs are laid by the queen and the colony remains in one bivouac for two or three weeks. During the nomadic phase, raiding swarms leave the bivouac but instead of returning, set up fresh bivouacs each night. The advancing front of a swarm on the move is a dense mass of ants [4], while the bivouac is made by ants which are attached to each other to form chains and bridges [3].

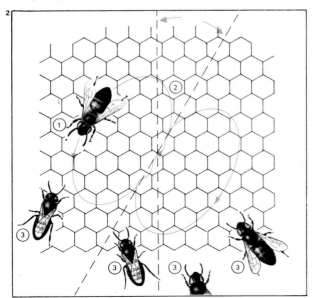

2 **The dance of the honey bee** stimulates worker bees at the hive to go out and find food sources whose whereabouts is encoded in the messenger bee's movements. The messenger [1] moves in a figure eight [2]; the angle at which it dances indicates the position of the food source in relation to the sun and the duration of abdominal flicks during the straight run is proportional to the distance of the food from the hive. When other workers [3] go out for food they follow the messenger's directions and, furthermore, gather nectar only from flowers that have the same scent as that returned to the hive by the messenger bee.

3 **Locusts** are types of grasshoppers that respond to overcrowding by migrating in huge swarms. In the solitary (*solitaria*) form [A], the locust is relatively inactive. The *gregaria* form is distinguished by the darker colour of the immature insects, or nymphs. A *gregaria* female [B] is shown laying eggs, or ovipositing. The eggs hatch into flightless nymphs which in crowded conditions gather into an army and set off in search of food. The nymphs mature en route and take to the air as a swarm, a relentless plague that devastates everything in its path and often ruins vast areas of cropland.

the way, but it was noticed also that other influencing factors are increased atmospheric moisture and a fluctuating concentration of negatively charged ions. Without these, it was observed, electricity alone is not enough to infuriate the insects.

The gregarious feeding behaviour of some insects often has terrifying consequences. At a given level of overcrowding, African locusts [3] gather into a vast, rapacious swarm.

In army ants [1] social behaviour is as rigid as that of honey bees, with different forms performing different and highly organized functions in the colony. Army ants bivouac in a solid column formed from the linked bodies of their fellows in a cylinder up to 1m (39in) across, with the queen and larvae in the centre. During the day, foraging parties of ants disperse in fan-shaped swarms. For most ants, the social signals that control co-operative behaviour are chemical.

The voracious behaviour of ants and locusts is the evolutionary outcome of the need to feed. In the caddisfly larva [5], this necessity has given rise to an amazing chain

of behavioural responses resulting in its clothing itself in a private fortress. In this case, each stage of the building triggers off the next stage, until the protective home is finished.

Instinct and learning

In higher animals, behaviour is such an intricate mixture of the learned and the instinctive that the two components become almost impossible to separate, but in insects, where instinct is predominant, the contribution of learning can clearly be defined. Digger wasps, for example [6], deposit their eggs in holes in the ground stocked with prey and then set out on a hunting expedition in which they capture further prey to take back to the next hole. The choice of prey is entirely innate and is based partly on visual clues but also on smell (for digger wasps that capture bees cannot distinguish bees from other insects of about the same size on sight alone). The return to the next hole after a hunting trip, however, entails learning. The wasp circles the hole for a few seconds before leaving and remembers its appearance so that it recognizes it after a trip lasting an hour or more.

KEY

The young of many species, from gulls [A] to cats [B], tend to have round heads, large foreheads and short noses, regardless of the grown face. The face shape is a sign stimulus for parental actions.

Other examples of sign stimuli include the recognition of particular sorts of flowers by some insects [C].

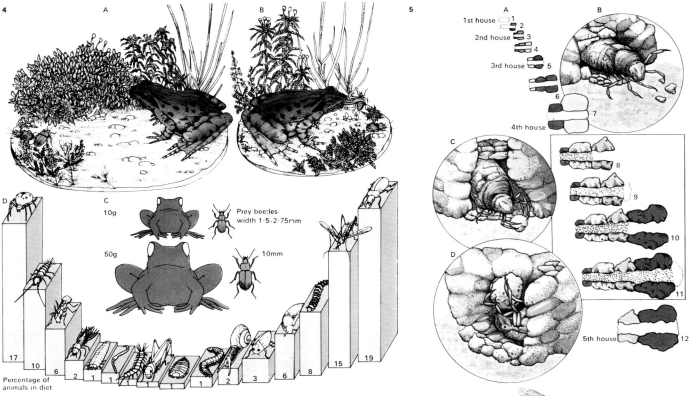

4 The motionless hunter, the common frog (*Rana temporaria*), feeds by catching prey only when it comes in reach of its tongue [A, B]. The size of prey the frog will eat depends entirely on its own size – whether it can open its mouth wide enough to accommodate the meal [C]. The usual diet of the frog includes insects of all kinds [D], according to their seasonal availability, but beetles make up the greatest proportion.

5 Caddisfly larvae build a total of five houses [A] through their larval development, one for each of five moults and each bigger than the previous one. Cross-sections show the series of houses [1–7] and how they are enlarged by stones [8–12]. In between, during moults, the larva blocks off the opening. The larva filters out small stones by raking [B]; it rejects large stones [C] by touch. The stones are fitted together [D] by silk spinning.

6 Digger wasp *Sphex* sp

6 The female digger wasp learns the landmarks around a nest hole [A, B]. If the conspicuous pine cones [C] are moved, the wasp searches in vain for the nest [D].

Single-celled animals

The single-celled organisms that are regarded as being the simplest forms of animal life are protozoans, members of the large group or phylum of living creatures known as the Protozoa. They number more than 30,000 species and nearly all of them are invisible to the naked eye, growing no larger than the size of a pinhead.

In common with all other animals the majority of Protozoa are mobile and take in complex food materials that are broken down inside the body to provide energy. In addition they all need water to survive and although the majority are either marine or freshwater species a significant number live as parasites within higher animals, including man, where they may cause a variety of diseases.

Methods of locomotion

Protozoa are divided into four main classes depending on the ways in which they move. Members of the class Mastigophora bear one or more of the thread-like structures — flagella – whose beating action propels the animal forwards. All the Ciliata are equipped with many hair-like projections known as cilia [3]. These lash in a synchronized pattern and in doing so produce movement. The members of the class Sarcodina move by means of pseudopodia, extensions of the "ground substance" or protoplasm with which their cells are filled. The Sporozoa all lack specialized locomotory structures.

Varieties of Protozoa

The Protozoa with flagella include members of the genera *Euglena* and *Chlamydomonas*, all of which contain chlorophyll and thus photosynthesize, plus dinoflagellates in a cellulose capsule, which are found in the plankton. *Trypanosoma* is a parasite that lives in man, where it causes sleeping sickness. The trumpet-shaped *Stentor* is a remarkable ciliate which, when feeding, attaches itself to the surface of a water plant.

The sarcodines are also a diverse group. *Amoeba proteus* is a common free-living representative, others such as *Entamoeba* sp live within the human gut, some of them as parasites. The plankton contains foraminiferans, marine species that secrete calcareous shells round their single-celled bodies.

Radiolarians and heliozoans form comparable shells composed of silica, which are collected and used as abrasives.

As exceptions to the rule of diversity, all sporozoans show a high degree of uniformity. They are all parasites and lack the cell parts (organelles) necessary for locomotion and feeding – they have no need to move and can take in food ready-digested. They also have unique life cycles incorporating both asexual and sexual reproductive stages in which spores, each bearing hundreds of offspring, are produced.

Protozoa reproduce asexually or "vegetatively" by splitting in two, a process called binary fission. When fully grown, an equal division of cytoplasm and nucleus takes place and "daughter" organisms are formed [5]. In adverse conditions some flagellate (Mastigophora) and some amoeboid (Sarcodina) species secrete a hard and impervious protective cyst with which they surround themselves and in which the cell may duplicate itself. When the environment is again favourable the cyst breaks open and releases the offspring, which reproduce asexually.

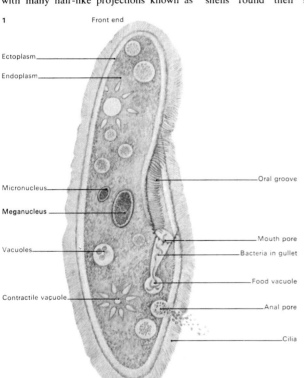

1 **Paramecium** is a specialized single-celled animal. The outer layer of the cell, the ectoplasm, is bounded by a tough skin through which many tiny hairs or cilia protrude. Regular beating actions of the cilia produce locomotion. The oral groove extends into the inner granular endoplasm, forming a blind-ended gullet. Food particles are pushed into the gullet by ciliary action and enclosed in vacuoles. As food vacuoles circulate through the endoplasm their contents are digested by enzymes. Undigested and unabsorbed materials are expelled via the anal pore. The water content of the cell is controlled by the action of the two contractile vacuoles. Of the two nuclei present, the larger or meganucleus controls all day-to-day activities of the cell, while the smaller micronucleus is concerned with cell reproduction.

Front end
Ectoplasm
Endoplasm
Micronucleus
Meganucleus
Vacuoles
Contractile vacuole
Oral groove
Mouth pore
Bacteria in gullet
Food vacuole
Anal pore
Cilia
Rear end

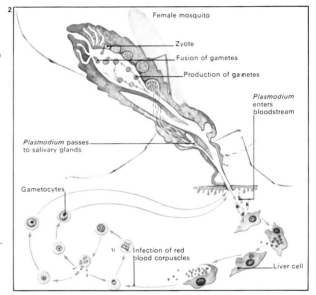

Female mosquito
Zyote
Fusion of gametes
Production of gametes
Plasmodium enters bloodstream
Plasmodium passes to salivary glands
Gametocytes
Infection of red blood corpuscles
Liver cell

2 **The malaria-producing protozoan** *Plasmodium vivax*, when injected into the human bloodstream by an infected female *Anopheles* mosquito, rapidly invades the liver cells and multiplies. As cells rupture the plasmodia escape and infect other cells. In blood cells numerous cycles of multiplication, cell rupture and reinvasion occur before male and female sex cells (gametocytes) appear. If another mosquito takes up the infected cells gametocytes divide within its stomach, forming gametes. These fuse to produce zygotes, which release plasmodia to enter salivary glands and repeat the cycle.

3 **Protozoa** move about in three main ways. Sarcodines move by means of streaming protoplasmic extensions [A]. The forward flow of inner plasmasol, coupled with a continuous reversible change of fluid plasmasol into the outer, jelly-like plasmagel, produces movement. Mastigophora are equipped with flagella whose whip-like beating actions [B, C] pull the organism along. Ciliata have numerous tiny beating cilia [D] which propel them.

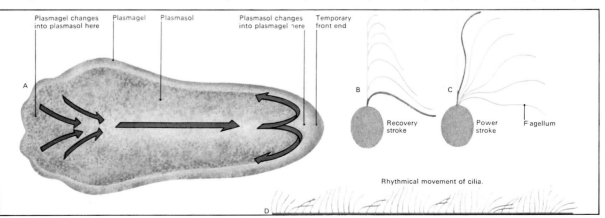

Plasmagel changes into plasmasol here
Plasmagel
Plasmasol
Plasmasol changes into plasmagel here
Temporary front end
Recovery stroke
Power stroke
Flagellum
Rhythmical movement of cilia.

Sexual reproductive methods of the Protozoa are numerous. *Paramecium* [1, 4] generally reproduces sexually by conjugation; two individuals of different strains fuse side by side and, after nuclear divisions and interchange of nuclear material, they separate. Further division of each organism takes place, producing a total of eight daughter organisms, four from each, equipped with nuclei of mixed parentage.

Sponges are simple creatures formed through the aggregation of many cells. They are classified in the phylum Porifera.

The structure of sponges

All sponges are sedentary aquatic animals and are divided into three main classes. Calcareas, such as *Sycon, Leucosolenia* and *Grantia*, bear an internal skeleton built up of "needles" of calcium carbonate. The "glass-sponges", or Hexactinella, are equipped with skeletons containing silica. Most representatives of the third class, the Desmospongia, possess a skeleton composed of the protein spongin, as in the case of the common bath sponge (*Spongia mollissima*) [6], but other species bear a skeleton of spongin and silica.

The basic structure of sponges consists of a sac with a large opening at the top and numerous perforations in the side walls. The outer layer of the sac is formed by a mosaic of flattened "covering" cells. The inner layer is composed of "collar" cells with whiplash flagella like those of the flagellate protozoa. Numerous cells, some like amoeba and others employed in the secretion of skeletal substance, lie sandwiched between inner and outer strata, forming the middle or mesenchyme layer. Spanning all three layers are many tube-shaped "pore" cells.

Sponges reproduce asexually by budding. Buds may separate from the parent and grow into new individuals or remain attached to form a branched colony. Sexual reproduction in sponges begins with the formation of sperm and eggs from undifferentiated mesenchyme cells. Although most sponges are hermaphrodite self-fertilization is rare. Sperm released from one sponge is carried by water to the eggs of another and fertilization follows. Larvae develop from fertilized eggs, subsequently escape and grow into sponges.

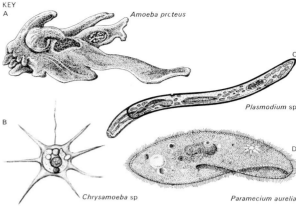

Amoeba proteus

Plasmodium sp

Chrysamoeba sp

Paramecium aurelia

Protozoans can be divided into four classes on the basis of the way in which they move about. Sarcodines, such as *Amoeba proteus* [A], move by means of pseudopodia, extensions of their protoplasm. Pseudopodia are also used to form food cups for the capture of prey. Most sarcodines are free-living protozoans. Locomotion in Mastigophora [B] is produced by whip-like flagella. Sporozoa, which include the malaria-producing *Plasmodium* sp [C], are all parasitic and lack specialized locomotory structures. The class Ciliata contains a majority of free-living (non-parasitic) species such as *Paramecium* [D] that move by ciliary action.

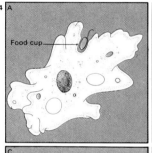

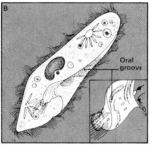

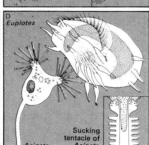

4 Like other animals protozoans feed on complex organic food materials to obtain energy. *Amoeba* spp [A] engulf materials by forming food cups around them with special pseudopodia. Food is later broken down within vacuoles by the action of digestive enzymes. *Paramecium* spp [B] feeds primarily on bacteria drawn into the oral groove by the cilia. *Trichonympha* spp [C] are symbiotic protozoans that live in the guts of termites. They engulf wood particles that the termites cannot digest themselves. *Acineta* spp [D] feed only on specific species of ciliate. These are often several times larger than themselves.

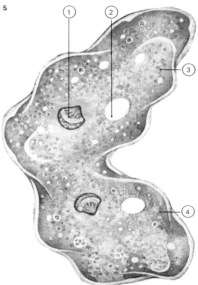

5 Asexual reproduction in protozoans takes place by binary fission. In *Amoeba* this simply involves the splitting of one cell into two equal parts. The process starts with the appearance of chromosomes in the nucleus, which become shorter and thicker. Chromosomes, each made up of two chromatids, line up across the cell while contractile vacuoles [2], having previously replicated, begin to separate. The chromatids are drawn apart [1] and the cytoplasm starts to split into equal halves. With chromosome division complete, the cytoplasm finally divides. The two resulting daughter amoebae [3, 4] are identical.

6 Sponges are members of the phylum Porifera, the "pore bearers". They vary from the vase-shaped *Sycon* spp [A] to the more complex and advanced species such as the common bath sponge (*Spongia mollissima*) which bear elaborate, highly branched internal systems of canals [B, C], rather than a simple cavity. The continuous stream of water [D] that passes through pore cells [1] and into the central cavity brings with it microscopic food particles. These are captured [E] by flagellate collar cells [2] and digested within food vacuoles [3]. Amoeboid mesenchyme cells [4] then carry the digested food from one place to another within the walls of the sac.

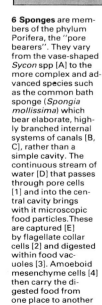

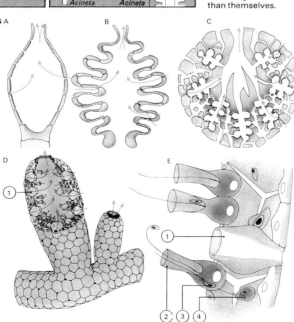

7 Man has long gathered sponges, as shown, for cleaning and polishing purposes making use of their soft, water-absorbent, yet tough and durable texture. Commercial sponges are found on the sea bed in tropical and subtropical waters. In deep waters they are collected by divers, but in shallow coastal areas they can simply be pulled up from the ocean floor by using trident-like poles. Once collected, the sponges are dried to kill off the thin layer of protoplasm that surrounds the horny skeleton. Further cellular material, and the shells and skeletons of the organisms trapped within the fibrous remains, are removed by a repeated pounding of the sponge before it can be used domestically.

Sea anemones, hydras and corals

Flower-like sea anemones, rock-like corals and translucent, tentacled jellyfish are some of the most attractive creatures in the ocean. And diverse though they may seem, these animals are all coelenterates [Key], a group numbering more than 9,000 species, all of which are aquatic and most of which inhabit marine shallows.

The feature that qualifies corals, jellyfish and freshwater hydras alike for coelenterate membership is the possession of a large, central body cavity or coelenteron (hence their name). The body comprises a concentric arrangement of body parts and the organization of body cells into rudimentary tissues in which the cells work in co-operation as elements of the whole, not as independent members of loose cell aggregates such as those found in the sponges.

The coelenterates are the first animals on the evolutionary tree to show this level of organization and they all share a similar pattern of tissue arrangement. There is an outer layer of cells, the ectoderm, and an inner layer, the endoderm, separated by mesogloea, a gelatinous material that may be little

more than a thin film – as in *Hydra* – or that may form the bulk of the animal as in the many species of jellyfish.

Coelenterates such as sea anemones are solitary animals but others, such as the plant-like *Obelia*, are made up of a colony of several sub-individuals or polyps. When these polyps are not identical the animal is said to be polymorphic – some marine colonies have separate polyps for feeding, for protection, for reproduction and even some have such polyps for swimming.

Phases of development

The life history of many coelenterates shows two distinct phases – a free-swimming or medusoid stage followed by a sessile stage of attachment and growth – which means that some species can be both bottom-living and reside in open water. In different coelenterates, however, there is a different emphasis of these two phases and this explains why these animals show such a variety of form. In *Obelia* [5], for example, the medusoid phase is relatively brief and followed by a longer, predominant period of attachment, a cycle

typical of the coelenterate group, the Hydrozoa. When mature, the *Obelia* colony gives rise to reproductive polyps which then produce medusae of their own.

Among the Scyphozoa the situation is reversed and it is the medusa that predominates. In the third coelenterate sub-group, the Anthozoa, which includes the corals [6–8] and sea anemones [4], the attached phase predominates so much that there is no medusoid phase at all. In these types, eggs and sperms are shed directly from the gonads – situated on certain areas of the endoderm lining the polyp's coelenteron – and pass out through the mouth. The fertilized eggs then divide to form balls of cells that settle and grow into new individuals.

There are, however, some exceptions to this systematic grouping, especially among the Hydrozoa. *Hydra* [1], for example, has no medusa and its life history resembles that of the sea anemone – except that the sperm and eggs develop on the outside instead of on the inside of the polyp. And there are even a few hydrozoans – to complete the range – which have a dominant medusa and a polyp

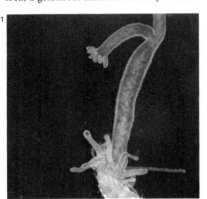

4 The dahlia anemone (*Tealia felina*) belongs to the group of coelenterates known as sea anemones. They can be found stuck firmly to rocks below the upper tide line. When left exposed by the receding tide the anemone reduces itself to a compact, blob-like mass. When covered by water the tentacles expand, ready to trap any small crabs or fish that stray within reach.

1 Species of Hydra are found throughout the world with the exception of frigid zones. When feeding, *Hydra* normally narrows its body and allows its extended tentacles to trail freely. If a water-flea or some other potential food source brushes against the tentacles, the nematocysts then discharge, partially paralysing the prey. Next, the tentacles contract and thus draw the victim towards the mouth.

2 The stinging hydroid (*Lytocarpus philippinus*) is one of the colonial coelenterates. The colony arises from a single individual that has produced buds in much the same way as its solitary cousins – with the difference that the new polyps remain attached to their parent. The coelenteron forms a continuous cavity and the end result is a branching collection of polyps.

3 Budding is one method of asexual reproduction. When conditions are good, a small bulge [1] forms on the parent *Hydra*. This grows into a new individual and eventually detaches itself.

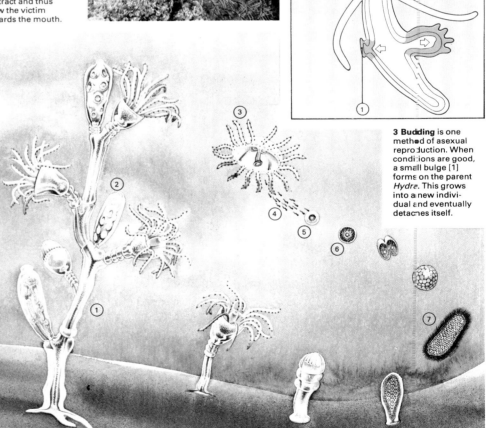

5 A mature colony [1] of *Obelia* bears reproductive polyps [2] concerned solely with production of the free-swimming or medusoid form [3] which carries *Obelia's* reproductive organs. Once separated from the parent colony the medusae shed their sperm [4] or eggs [5] into the water. The fertilized egg [6] subdivides to form a ciliated larva [7] that swims until it finds a suitable surface on which to settle. Here it grows and buds asexually.

stage that is either minute or non-existent.

Compared with the complexities of their sexual lives, the asexual reproduction of coelenterates is a relatively straightforward affair – a *Hydra* can bud [3] a new individual from its body, a sea anemone divides itself in two. Asexual reproduction may also result in the formation of colonies of individual polyps that are linked by a continuous inner cavity (coelenteron). This ability of the coelenterates to reproduce asexually means that they have considerable powers of regeneration – a mere fragment broken from an animal may be able to grow into a new individual with full powers of sexual reproduction.

How coelenterates feed

In most coelenterates feeding is assisted by the tentacles that surround the mouth. Liberally armed with stinging "cells" (nematocysts) [9], these tentacles paralyse the prey and draw it towards the captor. All the tentacles co-operate, bending round to attach themselves to the victim and thrusting it into the coelenteron. The mouth then closes and the endoderm cells secrete digestive enzymes

into the central gastric cavity. These break down the food, either into soluble products available for immediate absorption or into small particles that can be engulfed by the endodermal cells. The residue of the meal is finally expelled – by means of contractions of the body – through the reopened mouth.

Powers of movement

All coelenterates can move, although motion may be restricted to mere bending of the tentacles and shape changes. All movements are made possible by the muscle fibres in the cells of both the ectoderm and endoderm. In addition, the base of the sea anemone is richly endowed with muscles that enable it to move over rocks with a kind of gliding action. *Hydra* can perform similar movements, but it may also change position more rapidly by a kind of somersaulting action.

Even the simplest of coelenterate movements requires some degree of co-ordination. This co-ordination is the responsibility of a diffuse network of nerve cells that runs through the tissue and forms a primitive nervous system.

KEY

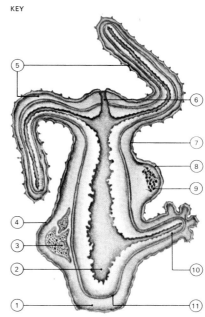

A hydroid polyp – shown here in schematized longitudinal section – illustrates some of the characteristics of coelenterates. The mouth [6], surrounded by a ring of tentacles [5] bearing stinging cells, opens directly into the digestive cavity or coelenteron [2]. The body wall is constructed of an outer layer of cells which is known as the ectoderm [7]. This is separated from the inner layer or endoderm [11] by a jelly-like layer, the mesogloea [1]. Coelenterates may reproduce by budding [10] or by a conventional sexual process in which sperm [9] generated by a testis [8] fertilize the egg [3] shed by an ovary [4].

6 Staghorn coral is a widespread species of coral reef. At first glance it may not be obvious that corals are living animals, since the individual polyps are enclosed in stony, protective cups. The individual polyps of coral resemble small sea anemones but these warm-water species tend to live colonial rather than solitary existences.

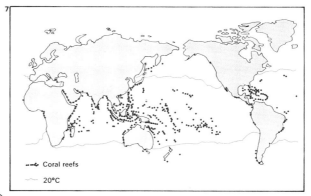

→→ Coral reefs

— 20°C

7 Reef-building corals – like those of the Great Barrier Reef fringing the Australian coast – grow only in the tropical oceans bounded by the 20°C (68°F) isotherm shown on the map. As the polyps bud and divide, so the colony expands upwards and outwards. The resulting masses of limestone long outlast the polyps that once secreted them. Corals can be exquisitely coloured but most are white.

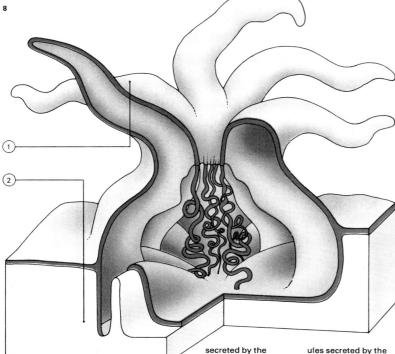

8 A coral polyp [1] embeds itself in its protective cup [2]. In stony corals – the type shown here – the cup is made of calcium carbonate (limestone) secreted by the ectodermal cells of the polyp's outer surfaces. In other groups, the skeleton is formed of horny or calcareous spicules secreted by the cells in the central mesogloeal layer and is thus to be found inside the polyps rather than surrounding them.

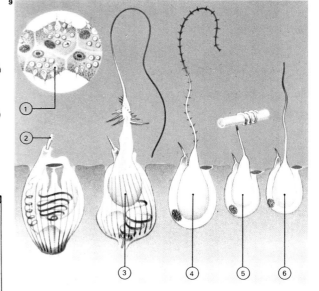

9 Nematocysts or stinging capsules [1] are the coelenterates' weapons of offence and defence. Located in greatest concentration over the tentacles, they play an essential role in feeding. Some [3] inject a paralysing poison into the prey; some [4, 6] secrete a sticky substance; while others [5] have coiling threads. Projecting from one side of each nematocyst is a small hair [2], which acts as a trigger. When the hair is touched by a passing animal the nematocyst fires. The mechanism of firing is not fully understood, but is thought to depend on a sudden pressure increase of the fluid within the capsule. Each nematocyst fires only once, after which it is discarded.

The jellyfish

The fearsome Portuguese man-of-war might seem to have little in common with *Hydra*, the sea anemones and corals, but in fact all of them are coelenterates. Jellyfish belong to the classes Scyphozoa (true jellyfish) and Hydrozoa (colonial jellyfish). The Scyphozoa are typically free-swimming and bellshaped while the conspicuous hydrozoan jellyfish are colonies of animals that behave as an individual. Each member of the colony is modified to perform a certain function such as feeding or swimming. The colonies are always free-floating or swimming instead of being permanently attached to a rock.

The scyphozoan group

A scyphozoan jellyfish [Key] and a simple polyp such as *Hydra* are both built with the same kind of body plan. In all coelenterates, the ectoderm (outer layer of cells) is separated from the endoderm (inner layer) by jelly-like mesogloea. There is so much mesogloea in the jellyfish that it makes up most of the animal's bulk.

Jellyfish come in diverse shapes and sizes, but all share the basic "umbrella" form, fre-

quently with a margin fringed with tentacles. The mouth lies at the centre of the subumbrella surface, usually separated from the gastric cavity proper by an extremely short tube, the manubrium. The corners of the mouth are often drawn out into trailing fronds, the mouth lobes. The fronds consist of membrances, folded and narrowing to a point. In *Aurelia* [1], a jellyfish that swims in large numbers round the British coast, these mouth lobes bear ciliated (hair-fringed) grooves surrounded by nematocysts (stinging cells) that paralyse the small prey on which *Aurelia* feeds. The cilia sweep the prey up through the mouth and into the gastric cavity where abundant stinging capsules paralyse any live creature that has been captured. The process of digestion then takes place.

Many jellyfish are bulky and need an efficient system of moving food, oxygen and waste products round their bodies. This is provided by a branching series of pouches and radial canals that link the central gastric cavity with a narrow circular canal running round the margin of the bell. The ciliated lining of these canals keeps a current of water

circulating freely to carry materials along.

The free-swimming medusoid (jellyfish) stage plays a highly important part in the life of the Scyphozoa. Far from acting as a mere vehicle for the gonads (sex organs), the medusa is the form in which the animal passes most of its life.

Jellyfish swim by a kind of jet propulsion, alternately opening and shutting the bell and so thrusting themselves through the water. The jelly counteracts the movement of the muscles, restoring the animal to the open shape when the muscles relax.

An active free-swimming existence clearly demands more sophisticated sense organs than those that would meet the needs of an animal permanently attached to rocks or seaweed. Many jellyfish have light-sensitive areas near the margin of the bell which detect light intensity and thus the surface of the water. A series of ingenious statocysts (balancing organs) permits the animal to stay upright in the water. These two systems are important because jellyfish are slightly denser than seawater and, as a result, sink unless they keep on pulsating. A jellyfish

1 The life cycle of the jellyfish *Aurelia* includes the adult medusoid form [A] with four violet, horseshoe-shaped gonads round its mouth. Testes and ovaries are on separate individuals. Sperm emerges from the mouth and enters the female's gastric cavity. Each ferti-lized egg grows into a ciliated larva [B] which is released to settle on a rock or similar surface [C]. It grows into a small polyp called a scy-phistoma [D] and at the right temperature divides into eight-armed buds called ephyrae [E]. These break free [F] to become adult jellyfish.

2 The colonioal hydrozoan *Velella* is made up of a colony that lives at the sea surface. It is supported by an oval float bearing a vertical sail, which allows it to be wind-driven. There are three kinds of polyps: a large feeding polyp just below the float [1]; an outer ring of protective polyps with stinging cells [2]; and between these two there is a cluster of reproductive polyps [3].

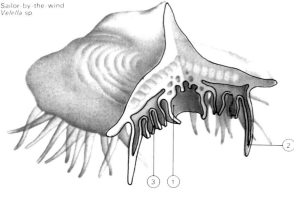

2 Sailor-by-the-wind *Velella* sp

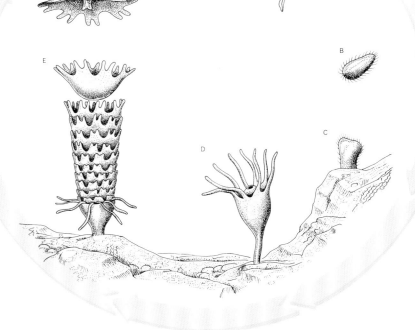

3 All jellyfish [A] orientate themselves by means of balancing organs called statocysts [B]. Each is a fluid-filled sac [1] enveloping a solid hanging granule called a statolith [2]. When the jelly-fish tilts, the granule swings against an adjacent sensory process [3] which sends nerve impulses to the contractile cells of the umbrella. Extra contraction on one side causes the jellyfish to right itself.

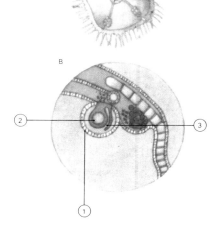

that was unable to distinguish "up" from "down" could easily find itself swimming directly towards the sea bottom, and perhaps away from the habitat of its normal prey.

The hydrozoan group

The second group of jellyfish, the unattached colonial hydrozoans, show an extraordinarily advanced degree of polymorphism. *Velella* [2] illustrates several of the polyp types, but it is by no means the most complicated. Its feeding, reproductive and protective polyps hang from the underside of a gas-filled float, itself thought to be a modified medusoid. Medusoids contribute to the swimming bells that propel surface-dwelling hydrozoans, which do not rely on the wind for movement. *Physophora* has both a float and swimming bells which drag the rest of the polyps behind them – an arrangement found in *Muggiaea* except that this has just one large swimming bell at the head of the colony and no float.

Physalia, the Portuguese man-of-war, [4] is without doubt the best known of the colonial hydrozoans and its appearance is justifiably frightening to human swimmers. Its sting is extremely painful and the appearance of large fleets of *Physalia* off bathing beaches can be regarded as a local disaster. The man-of-war paralyses fish – its normal food – with its long, stinging tentacles and then pulls its inert prey in towards the waiting mouths of the feeding polyps.

The comb jellies

The Ctenophora [5] or comb jellies are a group that is separate from the coelenterates but has much in common with them. Both share the inner and outer layers separated by a jelly-like mass. The typical ctenophore tends to be rounded with a gastrovascular cavity opening from a mouth that is located at the lower pole. At the opposite end – the apical pole – is a small sense organ that keeps the animal in balance and corresponds to the statocyst of the jellyfish. If a ctenophore has tentacles, they are armed with adhesive cells that are used in capturing prey. Ctenophores are hermaphrodites and shed their eggs and sperm into canals underlying the ciliary bands. Fertilization takes place in the sea to yield a free-swimming larva.

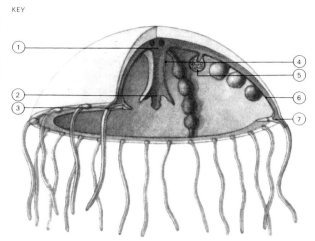

The diagram of a scyphozoan jellyfish shows the gastric cavity [1] opening to the exterior through a mouth [2] at the end of a stalk-like organ (manubrium) [4]. The cavity is continuous, via a series of radial canals [6], with a circular canal [7] round the margin of the bell. The margin bears tentacles and sense organs connected by nerve rings [3]. The gonads [5] are situated below the umbrella and open into the gastric cavity.

4 *Pelagia noctiluca*

4 Many beautiful species of jellyfish float in the world's oceans. The Portuguese man-of-war (*Physalia* sp) is not a true jellyfish but a colony of hydrozoans. The tentacles may grow up to about 18m (60ft) long. It feeds mainly on fish up to 30cm (12in) in length. Although an inhabitant of tropical waters, it is sometimes found in temperate seas. Species of *Chrysaora*, *Rhizostoma* and *Cyanea* are found in the Atlantic Ocean but they are rarely seen close inshore. The largest species of cyanea (*C. arctica*) has been recorded as having a bell measuring 3.6m (12ft) across, with tentacles more than 30m (100ft) long. A jellyfish that lives in warm seas and is responsible for luminescence in the water is *Pelagia noctiluca*. The sea wasps of the tropical Pacific are among the most dangerous animals known, despite their small size – an overall length of about 30cm (12in). They possess a poison that can kill a man in 10 minutes (and often much sooner). *Porpita* is a colonial species like *Velella*.

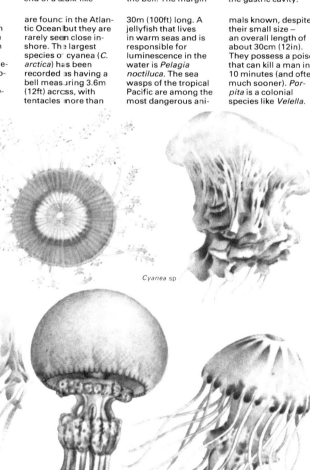

Portuguese man-of-war
Physalia physalis

Porpita mediterranea

Cyanea sp

Sea gooseberry
Pleurobrachia sp

5 The sea gooseberry, *Pleurobrachia*, is a member of the Ctenophora. It swims by co-ordinated beating of the rows of cilia that lie in eight bands stretching almost the whole distance from pole to pole. It has long tentacles that sweep the water for small animals. These are periodically "wiped" off near the mouth located at the lower pole. Sea gooseberries are common in coastal waters all over the world and may grow up to 3.8cm (1.5in) in diameter.

Sea wasp
Chironex sp

Rhizostoma sp

Compass jellyfish
Chrysaora hyoscella

Flatworms, flukes and tapeworms

Flukes and tapeworms are two groups of parasitic animals with deadly potential to both man and his domestic animals. These creatures whose effects, worldwide, have considerable economic significance, are placed zoologically together with the free-living flatworms in the phylum Platyhelminthes. All of them have simple bodies and all are bilaterally symmetrical – that is, the two halves that lie on either side of a line drawn from head to tail are mirror images.

The life of flatworms

The free-living flatworms [1] are simple, ribbon-like creatures with no circulatory system and only a single opening into the gut. The inner and outer cell layers, the endoderm and ectoderm, characteristic of the coelenterates (the sea anemones and their relatives) are separated, however, by a third mass of cells – the mesoderm. The mesoderm is responsible for the formation of muscles and reproductive organs. Indeed, the appearance of true organ systems is in itself a further advance over the coelenterate body plan.

Most of the free-living flatworms are aquatic; they move by creeping (effected by muscle contraction) or by beating of the hair-like cilia with which their surface layers are usually covered. The flatworm, which is carnivorous, uses its pharynx (the wide cavity that links the mouth with the intestine) to feed. The pharynx is pressed on to the food, then with muscular movements particles are torn off and passed into the intestine. Indigestible material is voided back through the pharynx and ejected.

Flatworms are the most primitive group of animals to possess a proper excretory system; tubes running down either side of the body and opening to the exterior by excretory pores serve to link a series of "flame cells". These cells, so called because of the bundle of hair-like cilia constantly flickering within them, are thought to regulate the worms' water content.

A concentration of nervous tissue at the front of the flatworm [2] forms a brain into which run the nerve bundles from two primitive eyes. Most species tend to shun the light and track down food largely by the use of organs sensitive to chemicals (chemorecep-tors). Flatworms respond rapidly to water-borne chemicals released by a potential food source. As soon as they become aware of even a weak solution of these chemicals, they move towards the higher concentration.

The life of flukes

The simplest form of parasitism is attachment to the outside of the host's body and a number of flukes have adopted this style of life. *Gyrodactylus*, for example, can be found attached by suckers and hooks to the gills of the fish on which it feeds. But more important to man are the internally parasitic flukes such as *Schistosoma* [5], the genus responsible for bilharzia or schistosomiasis, and the liver fluke *Fasciola* [4] which commonly infects the bile of sheep.

Apart from the obvious need to attach themselves to their host and feed, the requirements of internal parasites are fairly simple. Thus the organs of movement, digestion and sense are reduced and the surface cilia are replaced by a tough cuticle. Reproductive organs, on the other hand, must be enlarged because the offspring of the

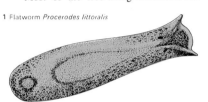

1 Flatworm *Procerodes littoralis*

1 The free-living flatworm named *Procerodes littoralis* grows to a length of 2cm (0.75in) and lives on the rocks of the upper and middle seashore. It belongs to a group called the Turbellaria, most of which are aquatic, and many marine. The ribbon-like body – which ensures that oxygen and waste products have only a short distance to diffuse – is clearly advantageous for an animal with no circulatory system.

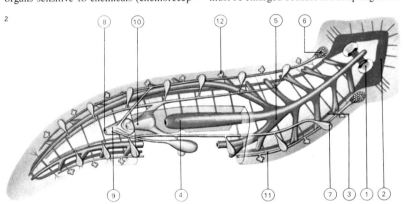

2 The turbellarian flatworm has sensory, reproductive and digestive systems. The head has eyes [1] over a brain [2] from which run a pair of nerve cords [3]. A pharynx that can turn inside out [4] opens into a branching gut [5]. All species are hermaphrodite with ovaries [6] and testes [7]. A genital pore [8] leads to the genital chamber [9] in which lies the penis [10]. On passing down the oviducts [11] the eggs are fed by yolk glands [12].

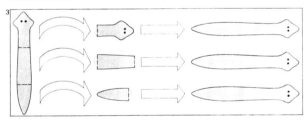

3 The flatworms have remarkable powers of regeneration. Cut into three, each segment will form a whole new worm. Indeed, free-living flatworms often divide asexually by pulling themselves in two. The adaptability of the group is further demonstrated by starving them; when food is in short supply they not only shrink in size, but actually digest their own organs. Later, when times are better, the missing organs are regenerated.

Liver fluke *Fasciola hepatica*

4 The liver fluke (*Fasciola hepatica*) is parasitic in the bile ducts of cattle and sheep. Its powers of reproduction are enormous; it may produce over 40,000 eggs that leave the host with its faeces. The eggs hatch to form free-swimming larvae that enter the intermediate host, a freshwater snail. Further changes take place until a tiny, tailed version of the adult fluke bores its way out of the snail and crawls up the stem of a plant before being eaten by another sheep.

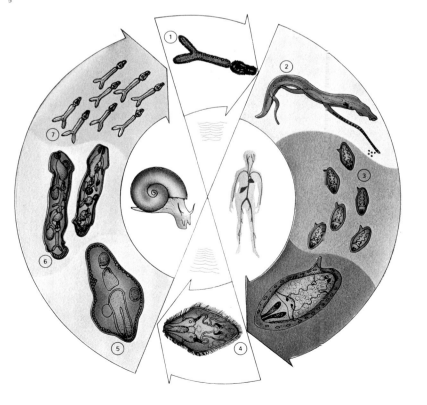

5 Adult human blood flukes of each sex [2] live in the blood vessels of the gut, causing abscesses and internal bleeding (schistosomiasis, common in Africa and the East). The eggs [3] leave via bladder or intestine and on reaching water hatch to release free-swimming larvae [4] that penetrate the tissues of a snail. Further development [5, 6] gives rise to other larvae [7, 1] that escape into water and reinfect another man by boring through his skin.

parasite face severe hazards as they search for a new host. All flukes, therefore, have massive and prolific organs of reproduction which ensure that even if 99 per cent of the eggs are lost a few will survive to enter the body of another host.

The animal harbouring the adult fluke is said to be the "final" host. To facilitate transmission between final hosts there may be one or more intermediate hosts. *Schistosoma* [5] uses just one intermediate; the Chinese liver fluke (*Clonorchis*) uses two – a snail followed by a fish. These intermediates are themselves used as staging posts in which the larval parasites pause in their reproductive progress while continuing to multiply their numbers still further.

The life of tapeworms

In the tapeworm [6] simplification of the anatomy is carried still further. A digestive system is absent – for an animal that lies bathed in the predigested contents of its host's gut has no need to do more than absorb the necessary food. Only the head of the tapeworm [8], armed with hooks and suckers,

is attached to the host. For the rest, the animal is simply a succession of separate segments or proglottides [9], formed by budding of the upper end of the tape, trailing free in the host's intestine.

Fertilized tapeworm eggs, surrounded by yolk and a protective shell, are stored in the tapeworm's uterus. At the end of the tape each segment is effectively a sac of eggs; a few segments at a time become detached and pass out with the host's faeces. Like the flukes, tapeworms [7] may have one or more intermediate hosts. *Diphyllobothrium*, the fish tapeworm found in the intestines of men, dogs and cats and which may be up to 27m (89ft) long, has two intermediates; *Taenia solium*, the pork tapeworm, has only one. The eggs of *Taenia* shed from the human gut hatch only if eaten by a pig. The pig's digestive juices dissolve the shell and a small six-hooked embryo emerges to bore its way through the intestine, enter the blood-stream and so reach a region of muscle. There it encysts as a bladder worm [10], so remaining until eaten by another human when the life cycle starts again.

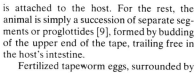

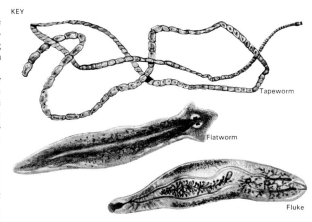

KEY
Tapeworm

Flatworm

Fluke

Three groups of animals – free-living flatworms, parasitic flukes and tapeworms – comprise the Platyhelminthes. Some of the best known of the Platyhelminthes are parasitic and these offer an excellent illustration of the modifications most characteristic of a parasitic way of life. Thus the flukes have hooks and suckers, complex reproductive systems and reduced sensory equipment, while the segments of the tapeworm are devoid even of a gut, because food exists in their environment ready made. These creatures are little more than bags containing reproductive organs.

6 The pork tapeworm (*Taenia solium*) is parasitic in man and may reach a length of over 4m (14ft). The intermediate host is the pig, which becomes infected through eating human faeces. Reinfection of man can be avoided by cooking pork well.

6

Pork tapeworm
Taenia solium

8 The head or scolex of a tapeworm is the point of attachment to the host. Hooks and suckers hold the worm firmly to the wall of the gut, making it extremely difficult to remove, even by drugs designed for this purpose. And as long as the head remains, a new tape can always grow.

7

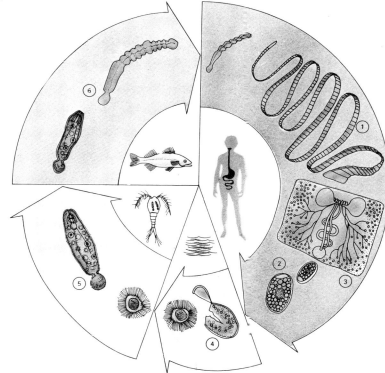

7 The broad tapeworm (*Diphyllobothrium latum*) is a human parasite with two intermediate hosts; the water flea and many freshwater fish of Europe, the Americas and the Far East. The adult worm [1] lives in man's intestine where it may grow to a length of many metres. The terminal segments – shown enlarged [3] – detach and as many as 13 million eggs [2] may be discharged every day in the faeces. In water the eggs hatch into embryos [4] and are eaten by water fleas; here they develop into the first larval form [5]. If the flea is eaten by a fish the larvae penetrate its tissues to form a secondary larva [6]. If an infected fish is then eaten uncooked by a man these secondary larvae are released. Using their small hooks they attach themselves to the wall of the gut and develop into adult worms in about three weeks, whereupon the cycle restarts.

8

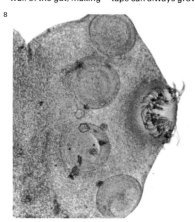

9

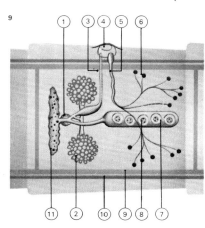

9 Apart from the excretory canals [9] and nerve cords [10], a tapeworm segment is largely reproductive organs. A branching testis [6] leads into a sperm duct [5]; together with a vagina [3], this opens into a genital pore [4]. Eggs from the ovaries [2] receive the products of shell [1] and yolk [11] glands. The uterus [7] stores young embryos [8].

10 "Measly meat" is infested with large numbers of cysts of tapeworm larvae. Cooking kills them.

10

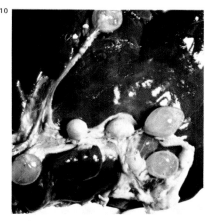

87

Earthworms, ragworms and roundworms

Ragworms and earthworms, the segmented worms, are grouped together zoologically with the leeches in the large group or phylum Annelida. These worms number more than 9,000 species and although they are found all over the world, in fresh water, in seawater and on land, they are far less widespread than the 13,000 species of roundworms [Key]. Almost every habitat has been occupied by round or nematode worms, which are formally in the class Nematoda, the largest subdivision of the phylum Aschelminthes.

The life of roundworms
Nearly all vertebrate animals, and many invertebrates, have parasitic roundworms living within them. The roundworms most important to man are parasitic species that cause disease. The intestinal inhabitant *Ascaris lumbricoides* sometimes causes pain and diarrhoea, usually in children [4], while *Wuchereria bancrofti* is the causative agent of elephantiasis. Hookworms, guinea worms [6], whipworms and giant kidney worms are all nematodes that cause disease in man. Other nematodes are free-living, but

wherever they occur they do so in large numbers. A single handful of soil may contain several million of them.

Adult roundworms are generally spindle-shaped [1]. They have thin, inflexible body walls which restrict their movements to whip-like body contortions. Most possess a simple, tubular alimentary tract and a primitive excretory system consisting of an H-shaped arrangement of canals that opens via a pore at the front of the body. The high pressure of body fluids probably forces excreta into the canals and out of the body. The nervous system is also built on an H-shaped plan and although the system is a simple one, round-worms have sensory structures that respond to touch, chemical and light stimuli.

In many roundworms, including the human (and animal) parasite *Ascaris lum-bricoides*, the two sexes are separate and males are smaller than the females [4]. Males are equipped with only a single testis, females with a pair of ovaries. During copulation the male injects sperm into the female. The sperm crawl like amoebae towards the eggs and fertilization takes place within the

female. The female *Ascaris* can lay up to 200,000 eggs a day and these are eliminated in the faeces of the host. Characteristically, parasitic nematode worms have very complex life cycles, often involving more than one host and several developmental stages.

In hermaphrodite roundworms, where both testis and ovaries are present, sperm are produced first and then stored. As eggs are released stored sperm fertilizes them. In all species fertilization produces a zygote, which becomes enclosed in a protective cyst. Juvenile worms grow inside these cysts but on escaping they pass through as many as four larval stages before becoming adults.

Bristleworms, earthworms and leeches
Segmented worms are divided taxonomically into three classes. The polychaetes – the rag-worms or bristleworms – are predominantly marine and have bristle-covered "paddles" on the body, the parapodia. Bristleworms include the burrowing lugworms (*Arenicola* spp), tube-forming fan worms [8], such as *Sabella* spp and *Serpula* spp, and the more mobile ragworms (*Nereis* spp) [10] and

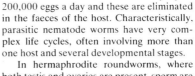

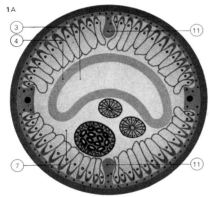

1 Roundworms and earthworms show distinct differences. An earthworm [B] has a true coelom [1], a fluid-filled body cavity formed by the splitting of the meso-derm into inner and outer layers. The in-ner "splanchnic" mesoderm gives rise to a muscular layer [2] around the intestine [3] while the outer "somatic" layer has longitudinal [4] and circular [5] muscle layers. There is also a coelomic lining [6]. A female *Ascaris* [A], a parasitic roundworm of man, has a pseudo-coelom [7], a cavity formed by the coalescing of fluid-filled spaces within the mesoderm. The intestine [3] lacks a muscular coat and only longitudinal muscle bands [4] are present. Blood vessels [8, 9] and developed excretory organs [10] are present only in the annelids. Nerve cords [11] are present in both groups.

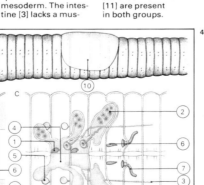

2 The earthworm is an hermaphrodite organism. However, self-fertilization is rare. The earth-worm's internal reproductive organs consist of many paired structures (side B, top C). In the testes [1] male germ cells are produced. These sub-sequently develop into sperm within the seminal vesicles [2]. The vasa deferentia [3] conduct sperm from testes sacs [4] to the male openings on the lower surface (segment 15). In receptacles (segments 9, 10) partners' sperm are stored [5] be-fore fertilization occurs. In the ova-ries [6] eggs are produced and these pass to the female pores (in segment 14), via the oviducts [7]. Externally [A] the openings of oviducts [8] and the vasa defer-entia [9] are visible as well as the cli-tellum [10], the region of glandular epidermis that secretes the cocoon. During mating the clitellum rests against the area of genital openings of the partner. After exchange of spermatozoa fertilization occurs in the cocoon.

3 Mating in earth-worms begins with the formation of a mucous slime tube around the front portions of two worms close together. Sperm, released from the vasa deferentia, pass along grooves in the body to the sperm receptacles of the annelid partner. The worms then separate. The clitellum then secretes a mucous sheath, which glides down the body picking up ripe eggs from the oviducts and sperm from sperm receptacles. Once the sheath leaves the worm it becomes sealed, forming a cocoon. Fertilization then takes place and zygotes are formed. From these, small worms rapidly grow.

4 The parasite *Ascaris lumbricoides* inhabits human and pig intestines where it is harmless except in large numbers, when it blocks the gut. The female roundworm grows to 30cm (12in); the male is shorter.

5 The sea mouse or aphrodite (*Aphrodita aculeata*) is an unusual polychaete annelid worm found on the sea-bed just offshore in temperate waters where it feeds on dead animals. These creatures, which measure 7.5–15cm (3–6in) in length, are occasionally washed up on beaches after storms. Partly concealed among its irides-cent hairs are dark, hollow spines thought to contain poison. If spines penetrate and break off in the skin they cause injury.

palolo worms. The class Oligochaeta contains earthworms (*Lumbricus* spp) and freshwater worms such as *Tubifex* spp. These lack parapodia and possess only a few bristles.

Earthworms are nocturnal, burrowing animals and come to the surface only after heavy rain, when the water has cut off their underground air supply. Earthworms feed on decaying plant and animal matter, and soil. Their burrowing aids soil cultivation by allowing more air and water to reach the soil.

The third class of segmented worms, the Hirudinea or leeches, are flattened animals. Some are parasitic but many others are free-living. Parasitic leeches live outside their hosts and suck their blood. Leeches [7] always have bodies divided into 33 segments. In parasitic species the gut is merely a reservoir for readily digested food.

The bodies of all annelid worms consist of rows of identical segments. The body wall is equipped with circular and longitudinal muscle layers and the intestine is surrounded by a thick muscular coat [1]. There is a simple blood circulatory system and a digestive system consisting of a tubular tract modified

according to the feeding habits of the species. The excretory and nervous systems are also in segments. The nervous system is composed of a nerve cord along which are spaced concentrations of nervous tissue – the ganglia – within each segment. The ganglia of the front segments, however, have become combined together to form a primitive "brain". This is concerned mainly with relaying messages from the touch, pressure, temperature and light receptors on the body wall.

Reproductive behaviour

The reproductive behaviour of annelids is varied [3]. In bristleworms sexes are separate and testes or ovaries are present in most body segments. Gametes are released as the body splits open and fertilization occurs externally.

Earthworms and leeches are hermaphrodite and gonads are present only in specific segments. Self-fertilization is rare and copulation involves a pairing up of worms during which exchange of sperm takes place. Subsequently sperm and eggs are released into a mucous bag and, once this has left the parent worm, fertilization occurs.

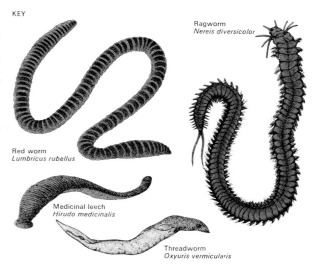

KEY

Ragworm
Nereis diversicolor

Red worm
Lumbricus rubellus

Medicinal leech
Hirudo medicinalis

Threadworm
Oxyuris vermicularis

The annelids and roundworms comprise over 22,000 animal species. The marine ragworm, the red earthworm and the leeches are all representatives of the phylum Annelida. This round-or threadworm is a parasitic species of the phylum Aschelminthes and often inhabits the human intestine.

6 Guinea worms are parasitic nematodes (*Dracunculus medinensis*) that live under the surface of human skin [A]. The male [1] fertilizes the female, then dies. The female [2] lives on, producing millions of active larvae. As those form the female stops eating and migrates to nearer the surface, usually in the lower leg, where a blister develops. The blister bursts when the leg is immersed in water, during swimming or bathing, and the larvae are discharged [3]. They readily curl up [4] and are swallowed by the crustacean *Cyclops* [B] in which larvae develop [5, 6]. These will infest a man who drinks the water.

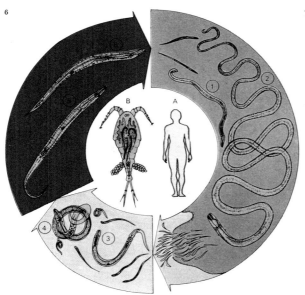

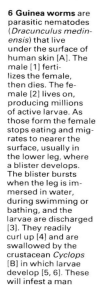

7 Leeches are common in tropical and temperate regions in freshwater habitats and feed on a variety of small animals. Many leeches feed as external parasites and suck the blood of animals. They were once used for medicinal bleedings. Leeches attach themselves with suckers at each end of the body. The anterior sucker surrounds the mouth (right) which, in many leeches, is armed with three saw-like teeth. As the leech inflicts a wound, as on this human arm, it injects a blood anticoagulant – hirudin. At a single meal a leech can take ten times its weight in blood, securing food for up to nine months.

8 Feather worms are polychaetes of the order Sedentaria – so named as they live permanently within tubes which they build in the sand. The front portions of their bodies are above the surface and bear many feather-like structures that function as respiratory gills and food collectors. Beating cilia covering the gills direct small organisms towards the mouth. The worms shown are *Spirographis* sp.

9 Giant earthworms such as this *Megascolides australis* live only in moist tropical regions. They can grow to as much as 3.6m (12ft) in length. Also shown are two common earthworms which rarely grow to more than 15cm (6in) in length.

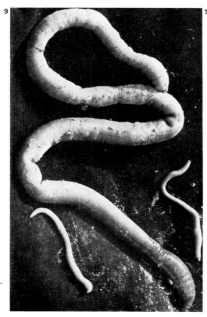

10 Ragworms (*Nereis* spp) are marine polychaetes. They live in burrows in the mud or in fine sand on the seashore, occasionally pushing their heads above the surface to feed. Wave-like motions of their bodies create currents that draw water through the burrows, bringing in fresh supplies of oxygen and carrying away carbon dioxide and nitrogenous wastes. They are predators, feeding on small marine animals, but they also consume planktonic organisms and detritus from the mud.

Land and sea snails

Molluscs, which number more than 80,000 known species, are a group of extraordinary diversity, encompassing creatures as different as the periwinkle, the clam and the giant squid. Traces of the basic molluscan plan can be detected throughout all members of the group, but this is often greatly modified.

The molluscan body plan

In essence, molluscs [Key] are soft-bodied creatures with a visceral mass containing such organs as those of digestion and reproduction covered by a sheet of tissue called the mantle. A space enclosed by the overlap of the mantle is called the mantle cavity. The tissue layer covering the outer surface of the mantle, the epithelium, secretes the shell, which is usually calcareous and is one of the group members' most immediately obvious characteristics. Beneath the visceral mass lies the foot, a fleshy muscular pad of tissue often important in movement.

The basic body plan is most easily recognized in the rather specialized chitons [1]. The digestive system has a mouth (bearing a "tongue" or radula for feeding), stomach, intestine and anus discharging into the mantle cavity. Chitons have a heart and blood vessels that make for a simple circulatory system, paired excretory organs, gills that project out into the mantle cavity and a nervous system with ganglia (concentrations of nerve cells) and nerve cords. In the course of the chiton life cycle, as in other molluscs, the fertilized egg develops into a so-called trochophore larva that swims by means of the bands of beating, hair-like cilia that surround its body.

Torsion and its effects

At first glance, at least a few of the gastropods (molluscs with a broad, flat foot and undivided mantle) appear also to have stuck fairly closely to the basic mollusc plan. The limpet [2], for example, has a simple conical shell and an apparently straightforward anatomy. On close inspection, however, that anatomy is not so simple as it first seems. During development from the larva the whole of the visceral mass becomes twisted upon the foot. This process, known as torsion, has several predictable consequences.

The gut, instead of being straight, performs a U-turn while the nervous system, its cords no longer parallel, is twisted into a figure-of-eight. Most striking is the repositioning of the gills and the opening of the anus into the mantle cavity: both are now at the front instead of the back.

Torsion is without doubt the most bizarre event in the life of the gastropod. It may have stemmed from the evolutionary advantages of a clean flow of water into the mantle cavity and over the gills and in the larval stages in the provision of protection for the head.

The mechanics of torsion are simple but elegant. The trochophore larva develops into a further form, the veliger, characterized by expansion of its areas of cilia. The rudiments of the adult shell and foot also become apparent. Then, with the head and foot remaining stationary, the viscera rotate through a half circle, bringing the mantle cavity both forwards and upwards as a result.

Shells and their loss

The other factor that makes recognition of the mollusc body plan difficult in many gas-

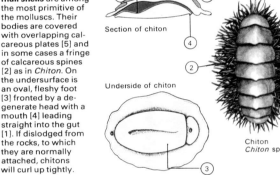

1 Chitons or coat-of-mail shells are among the most primitive of the molluscs. Their bodies are covered with overlapping calcareous plates [5] and in some cases a fringe of calcareous spines [2] as in *Chiton*. On the undersurface is an oval, fleshy foot [3] fronted by a degenerate head with a mouth [4] leading straight into the gut [1]. If dislodged from the rocks, to which they are normally attached, chitons will curl up tightly.

Section of chiton

Underside of chiton

Chiton *Chiton* sp

2 The gastropods have exploited many habitats and adapted accordingly. Limpets, with their conical shells, are some of simplest of them. More widespread are gastropods with coiled shells. Some, such as the whelk, inhabit the sea; others live in fresh water and on land. There are tree snails, marshland snails and even those that live in sand dunes. A few gastropods, such as slugs, have abandoned their shells altogether.

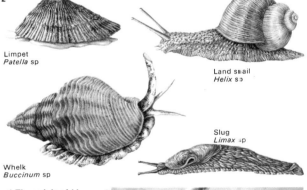

Limpet *Patella* sp

Land snail *Helix* sp

Whelk *Buccinum* sp

Slug *Limax* sp

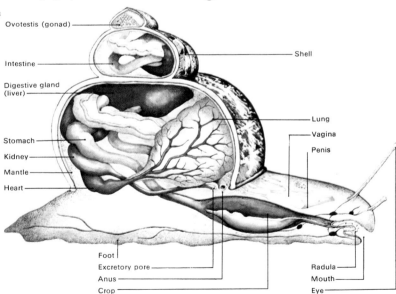

3 Ovotestis (gonad)

Intestine

Digestive gland (liver)

Stomach

Kidney

Mantle

Heart

Shell

Lung

Vagina

Penis

Foot

Excretory pore

Anus

Crop

Radula

Mouth

Eye

3 The anatomy of a snail shows a well-developed head with sensory tentacles and eyes. The foot is flattened and used for creeping. The layout of the internal organs is asymmetrical because of torsion and the coiling of the visceral hump within the shell. The mouth, inside which lies the radula, opens into an oesophagus that leads to the stomach and gut. Because of the changes due to torsion the anus opens towards the front of the animal instead of at the rear as in chitons. Most gastropods respire through a gill located in the mantle cavity, but in the pulmonate group to which land snails belong, this organ has been lost. Instead, the lining of the mantle cavity is liberally supplied with blood vessels, the whole surface acting as a lung. Air moves in and out through a small opening called the pneumostome. The nervous system is made up of paired cerebral ganglia adjacent to the oesophagus, and nerve cords reaching to the foot and the viscera.

4 The radula of *Limnaea*, like that of all gastropods, is a file-like rasping organ rooted within a sac just inside the mouth. It is especially well developed in species such as limpets, which graze on the algae growing on rocks. As the radula moves backwards and forwards out of the mouth, so the food is scraped and macerated. Eventually, the radula and its teeth are worn down. To counteract this, more teeth are constantly formed.

5 A snail creeping across glass shows that it proceeds by a series of wave-like muscular contractions in the undersurface of the foot. These waves can be seen to originate at the rear of the animal and to pass forward along the length. Copious secretions of mucus thoroughly lubricate the surface of the foot, aiding its progress and giving it improved adhesion with the base. Most other gastropods also follow this characteristic means of locomotion.

tropods is the coiling of the shell. This has been achieved by a loss of symmetry in the visceral mass and the disappearance from some of the paired organs of one member. But odd though the coiling may seem it is an arrangement that is far more satisfactory than mere elongation of the shell to an unwieldy length.

The snails [3] in particular have proved remarkably adept at exploring new habitats, both aquatic (marine and freshwater) and terrestrial. Yet the move to land is always a difficult one, and in the case of the gastropods has demanded sacrifice of the gills in favour of a kind of air-breathing lung.

A shell undoubtedly confers favours upon its owner in the way of protection and support. But it also restricts mobility. As a result, some groups – such as the land and sea slugs – evolved a reduced shell, or lost it altogether.

Sea slugs belong to the opisthobranch division of the gastropods and include some of the more colourful and unusual species such as the Spanish dancer (*Hexabranchus* sp) [11]. Associated with the reduction or loss of the shell in opisthobranchs are two

further changes: the uncoiling of the visceral mass and unwinding or detorsion. The change restores some measure of the lost body symmetry.

The gills, again as a result of shell loss, are visible in many members of this group. But in one division the "gills" covering the upper surface do much more than respire. Indeed, by origin they are not really gills at all, but outgrowths of the body surface – called cerata – each of which enclose an outgrowth, a branch or diverticulum of the gut.

Members of this opisthobranch group that feed on coelenterate polyps such as *Hydra* (relations of sea anemones) are somehow able to remove their stinging "cells" (nematocysts) undamaged and transfer them to the gut diverticula in the cerata. The sea slug ingeniously makes use of its victim's defences to defend itself. When the animal is attacked it exudes nematocysts through small pores in the cerata. A predator taking in water containing these nematocysts is stung. And if the attacker persists, it may provoke the harassed mollusc into detaching whole cerata full of poisonous stinging cells.

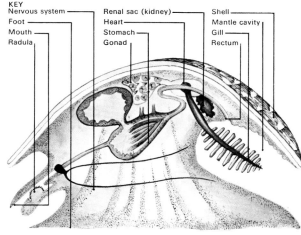

KEY
Nervous system
Foot
Mouth
Radula
Renal sac (kidney)
Heart
Stomach
Gonad
Shell
Mantle cavity
Gill
Rectum

A section through a generalized mollusc shows the body mass, protected by a shell, comprising three components: head, visceral mass and a

large muscular foot. The mouth leads to a stomach and intestine, which discharges through an anus. Respiration is through gills in the

mantle cavity. All molluscs are variations on this plan, although it is more obvious among chitons than in the squid and octopus.

7 The courtship ritual of the snail [A] must be among the strangest of the whole animal kingdom. Two snails will approach each other with the openings to their genital ducts conveniently exposed. Among the glands emptying into the genital atrium is a sac that secretes a small calcerous dart. When the two snails are close enough, each shoots its dart into the other's body, probably as a stimulus to copulation.

8 The slipper limpet (*Crepidula* sp) is so called because of a ledge inside the shell that makes it look like an upturned slipper. *Crepidula* is a static creature that tends to live in chains of up to nine or ten individuals. Younger arrivals settle on the backs of their elders. The sex of each animal depends on its position in the chain, females at the bottom, males at the top. *Crepidula* thus changes its sex with age.

Some gastropods have separate sexes, but snails are hermaphrodite [B], an ovotestis producing both sperm and eggs. At copulation the penis of each partner is inserted into the vagina of the other and sperm is transferred in a packet called a spermatophore. Sperm is then stored in a special chamber, the spermatheca, from which it is released as required. The fertilized eggs are laid in holes in the soil.

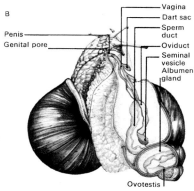

B
Penis
Genital pore
Vagina
Dart sac
Sperm duct
Oviduct
Seminal vesicle
Albumen gland
Ovotestis

9 The great pond snail (*Limnaea stagnalis*) prefers to live in almost stagnant water in ponds and ditches. Its gelatinous egg masses are laid on any solid object that happens to be available, usually some kind of water plant. Many gastropods lay their eggs in this typical kind of mass. One of the more familiar sights on the sea-shore is the empty case of the egg mass of the whelk (*Buccinum* sp).

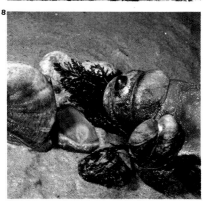

6 The shelled sea butterfly (*Limacina* sp) belongs to an unusual group that swims by means of a flat elongation of the foot. The downward thrust of these "paddles" [1, 2] provides lift. The paddles are then looped upwards in a slower recovery stroke [3–5] and the cycle is repeated.

10 Colour in gastropods is not confined to the shell. The Western Australian species *Amoria grayi* has a slim, translucent, yellowish shell, while its body is covered with rust-coloured striations.

11 The Spanish dancer (*Hexabranchus* sp) is an inhabitant of the Great Barrier Reef of Australia. Shell and mantle cavity have been lost and the animal has reassumed bilateral symmetry and become slug-like in its appearance.

Life between two shells

The shelled animals – the molluscs – are sub-divided into three main groups: the snails or gastropods; the squids and octopuses, or cephalopods; and the bivalves. The bivalves include representatives that are interesting in their ways of life (the rock-boring piddocks), their appearance (the razor shell) and gastronomically (the mussel, the oyster).

Bivalves, as their name implies, have hardened, calcium-containing shells like those of snails, but instead of being in one piece and coiled these are divided into right and left halves. The two shells are joined along their "back" or dorsal margin by a horny hinge. When the animal is relaxed and feeding there is a gap between the two halves; when it is disturbed a pair of powerful muscles attached to the shells pulls them tightly together.

Structure of bivalves
The bulk of the bivalve body lies towards the hinge of the shell and is covered by a structure called the mantle. This grows outwards on either side of the body and forms a lining to the inside of both shell halves. Indeed, it is the tissue lining or epithelium of the mantle

that is responsible for secreting the shell. A muscular foot – flattened from side to side, extensible, and often used in burrowing – projects from the body into the cavity separating the halves of the mantle.

If one shell and half the mantle of a bivalve are removed the most obvious structures exposed are the gills which, besides serving as organs of respiration, are used in feeding. Bivalves are filter feeders [3]. Tubular extensions of the mantle form a pair of siphons, one for the entry of a current of water, the other for its exit. This current is maintained by the beating action of hair-like cilia with which the gills are liberally covered. Water is drawn in through tiny pores in the gills and eventually ejected by way of the discharge siphon. Particles of food are strained out and trapped in the layer of sticky mucus, a substance similar in consistency to human saliva, that covers the gill surfaces.

From there trapped food passes to the edge of the gills farthest away from the hinge and larger particles such as silt fall off into the mantle cavity. The remainder, still entangled in a string of mucus, move forward to the

palps surrounding the mouth. Further sorting then takes place and selected particles enter the mouth and pass into the stomach and gut for digestion.

Digestion and reproduction
A gland surrounding the stomach is the chief organ of digestion in bivalves, but projecting into the stomach, from an intestinal pouch at one end of it is the crystalline style, a gelatinous rod containing carbohydrate-splitting enzymes. Cilia in the pouch make this rod rotate continually and its free end rubs against an area of the stomach known as the gastric shield. As the crystalline style is worn away so the enzymes it contains are released into the stomach contents. Digestion is completed in the intestine, which follows a tortuous course to the anus. This opens into the mantle cavity near the discharge siphon, thus ensuring removal of faeces in the outgoing water.

Paired excretory organs lie on either side of the heart and a "kidney" in each extracts waste products and discharges them, via a bladder, into the gill passages and thence to

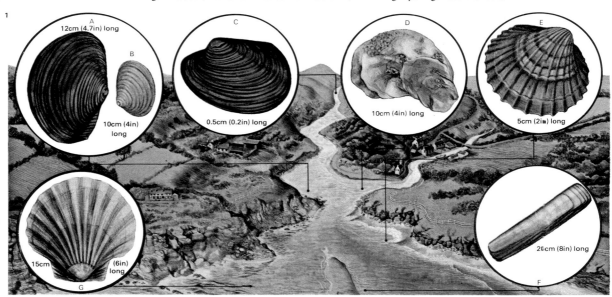

1 The distribution of bivalves can depend on such factors as the force of wave action, tidal movement and the nature of the substrate. The pearl mussel (*Margaritifera* sp) [A] is a freshwater species. The swan mussel (*Anodonta* sp) [B] and the river pea mussel (*Pisidium* sp) [C] are both freshwater species but can also be found in the lower reaches of tidal rivers. The oyster (*Ostrea* sp) [D] lives in estuaries and creeks while the common cockle (*Cardium edule*) [E] can be found on the middle and lower seashore. The pod razor (*Ensis siliqua*) [F] burrows in the sand of the lower shore. The scallop (*Pecten maximus*) [G] lives in offshore waters.

A 12cm (4.7in) long
B
10cm (4in) long
C 0.5cm (0.2in) long
D 10cm (4in) long
E 5cm (2in) long
F 20cm (8in) long
G 15cm (6in) long

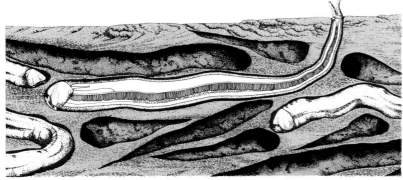

2 Many bivalves burrow in sand or mud but a few species spend their lives in submerged wood or even rocks. The shipworm (*Teredo navalis*) [A] measures up to 30cm (12in) in length. The *Teredo* shell is much reduced, the halves having been modified to act as a drill by which its worm-like body is able to penetrate wooden pilings and ships' hulls – doing much damage in the process. As the bivalve grows it lengthens its burrow, lining it with a calcareous layer

secreted by the mantle. Even more remarkable is the piddock (*Pholas dactylus*) [B] which bores its way into rock; using its foot as a lever it moves its shell backwards and forwards to scrape the rock and slowly hollow out a burrow.

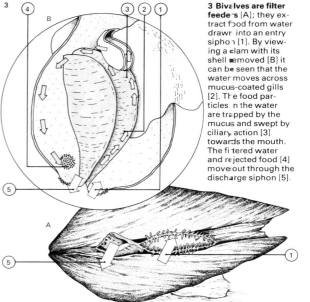

3 Bivalves are filter feeders [A]; they extract food from water drawn into an entry siphon [1]. By viewing a clam with its shell removed [B] it can be seen that the water moves across mucus-coated gills [2]. The food particles in the water are trapped by the mucus and swept by ciliary action [3] towards the mouth. The filtered water and rejected food [4] move out through the discharge siphon [5].

the exterior. Bivalves have a three-chambered heart whose single muscular ventricle pumps blood into arteries that run to the foot, mantle and the body organs. Many of the tissues of the body are permeated not by capillaries, as in higher animals, but by blood spaces or sinuses. Sinuses in the foot enable sand-burrowing – the foot swells and elongates as extra blood is pumped into them. Blood then returns via a series of veins to the paired atrial chambers of the heart.

Bivalves have simple reproductive systems with paired gonads and no glands. The sexes are separate and most allow the spermatozoa and ova to escape into the water where, in marine species, they form part of the plankton. Some freshwater bivalves brood their young, while others release a larval stage called a glochidium that passes through a brief parasitic phase before assuming an independent adult existence.

Methods of movement

Sedentary bivalves, such as mussels, spend their lives attached to a firm base such as a rock by a bundle of threads called the byssus.

These threads are secreted by a gland in the foot and anchor their owner with surprising strength – a necessity if the animal is to avoid being swept away by the force of the waves. A more common mode of life, found in such bivalves as the cockles and razor shells, is sand-burrowing [5] by means of a mobile foot. The depth of burrowing varies among species but the deeper penetrating forms often have long siphons for obtaining surface water for respiration and feeding.

In contrast to the sand-burrowing species the shipworm burrows into wood by back and forth rotation of its shells – it literally drills its way forwards. Some piddocks can even burrow into rock. The piddock must periodically cease boring and project the edge of its mantle on to the shell surface to deposit a fresh layer of calcium carbonate.

In most bivalves the head is reduced or absent and sensory organs are generally insignificant. But the animals do react to water-borne chemicals and to light. Some of the razor shells in particular are extremely sensitive to vibration and will vanish rapidly down their burrows if they are disturbed.

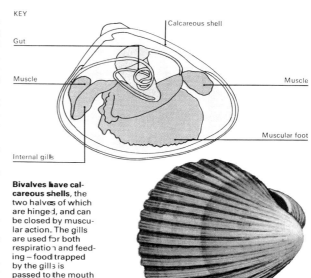

KEY

Bivalves have calcareous shells, the two halves of which are hinged, and can be closed by muscular action. The gills are used for both respiration and feeding – food trapped by the gills is passed to the mouth and gut by hair-like cilia. The muscular foot is used during burrowing.

Cockle
Cardium sp

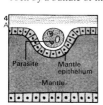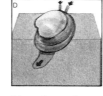

4 Pearl formation is the result of a foreign body – often a parasite or a grain of sand – lodging between the shell and the epithelium of the oyster's mantle [A]. The mantle first surrounds the embryo pearl with its own covering or epithelium [B]; this proceeds to secrete concentric layers of pearly material – actually thin sheets of calcium carbonate comparable with those lining the whole shell. These layers serve to isolate the foreign body, so preventing it from harming the bivalve [C]. Pearls can sometimes be found in European oysters but they are seldom big enough to be of much value.

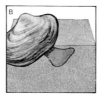

5 As a bivalve burrows, its foot probes downwards [A], expands [B] to anchor the creature and then contracts [C] to pull it into the sand. Diagrams [D] to [F] show how this is accomplished. The foot is forced down by pressure in the blood sinus [D], the closure of the siphons [E] driving water out of the mantle cavity; the foot expands with blood. Retractor muscles in the foot then contract [F] and the cycle is complete.

7 In oysters, as in most bivalves, the sexes are separate. Eggs and sperm are shed into the water and after fertilization the embryo develops into a free-swimming trochophore larva with vestiges of a shell and a bundle of cilia for propulsion and feeding. As the shell enlarges so the second or veliger larval form develops. This has a well-muscled foot, a shell and internal organs. The spat or young oyster remains free-swimming for some two weeks before settling down and attaching itself to a suitable surface. Some bivalves go through a brief parasitic phase that aids dispersal. The larva of these species – a glochidium – must attach itself to a fish for development to take place. Afterwards the young bivalve abandons its host to take up an independent existence.

7

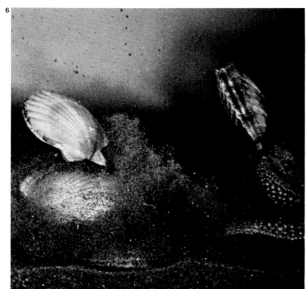

6 Scallops are notable among bivalves in being able to escape predators by "swimming". The great scallop *Pecten maximus* starts life firmly attached to a substrate but later detaches to take up a free life on the sea bottom. It can sense the presence of predators – notably starfish – chemically. The shell at once snaps shut, the closure of its two halves forcibly expelling a water jet, and the creature escapes across the bottom hinge first. If water is expelled on either side of the hinge the scallop "swims" in the other direction. The queen scallop (*Chlamys opercularis*) can even position a flap of tissue in its mantle cavity to choose the direction of its water jet – and thus the direction of its movement.

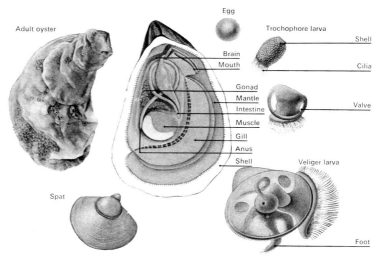
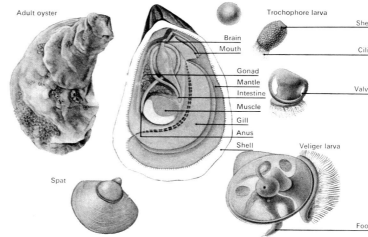

Adult oyster

Egg
Trochophore larva
Shell
Cilia
Brain
Mouth
Gonad
Mantle
Intestine
Muscle
Gill
Anus
Shell
Valve
Spat
Veliger larva
Foot

Squids and octopuses

Cephalopods, the animal group that includes squids, octopuses, cuttlefish and nautiloids, have always held a peculiar fascination for man. Some of these creatures – particularly the squids – grow to an enormous size and may have lent credence to the legend of the "sea serpent". As molluscs the cephalopods are related to such shelled animals as snails, mussels and clams, but are far more advanced and include the largest and most intelligent of the invertebrates. Cephalopods are "jet-propelled" molluscs that live in the sea and have their foot divided to form a number of tentacles or arms surrounding the head.

Support, movement and respiration

Because most cephalopods have internal shells or no shell at all, their relationship to other molluscs is not immediately apparent. The shell itself, where it exists, is markedly different from that of many molluscs. In *Nautilus* it is still in evidence; in the octopus it is completely absent; in the squid it survives as a thin, horny plate (the pen), which serves as a skeleton. In the cuttlefish the shell forms the cuttle-bone, which serves both as a sup-porting structure and an organ of buoyancy. The bone is divided (like the shell of the nautilus) into a series of chambers, the youngest (most recently formed) of which contain gas. The amount of gas is just sufficient to offset the difference between the overall density of the animal and the density of the seawater.

Cephalopods are far more active than other molluscs and therefore require an efficient respiratory system. This is provided by the mantle, which pumps a constant current of water in and out of the mantle cavity and over the gills [Key]. Deoxygenated blood is forced through each gill by a separate gill heart. Oxygenated blood passes from the gills to a single main or systemic heart that distributes it to the vessels and capillaries of the rest of the body. Contractions of the mantle also enable most cephalopods to move with great speed when necessary [4].

Well-developed sense organs

The feature that most distinguishes cephalopods from their lower mollusc relatives is the extraordinary development of their sense organs and nervous system – a feature that reflects their active life-style.

The large cephalopod brain, which encircles the oesophagus directly between the eyes, is made up of several masses of nervous tissue (ganglia) that in most lower invertebrates are spread throughout the body. The nervous system of a squid incorporates a number of exceptionally large nerve cells called the giant fibres and, as a result, scientists have made great use of this animal when studying the physiology of nerves.

Cephalopods have chemical sense and balancing organs but their most striking features are the eyes [8]. Even the most cursory glance at a section of a cephalopod eye is sufficient to reveal its resemblance to the eyes of vertebrates. Within the globe can be seen an outer window or cornea, a lens, a sensitive layer or retina and an opaque backing layer or choroid. Yet there is no evolutionary relationship between molluscs and vertebrates. The arrangement of the cells in the retinas of the two groups is quite different, as is the focusing mechanism. The overall resemblance is a result of evolutio-

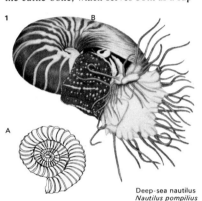

Deep-sea nautilus
Nautilus pompilius

1 Nautilus is one of the few surviving animals resembling the primitive or original cephalopods. The fossilized shells of these extinct forms (ammonites) [A] are quite common. About 100 million years ago there were probably some 2,500 species. Today the hard-shelled cephalopods number about three. One of them, *Nautilus pompilius* [B], is a deep-sea species that lives in tropical waters.

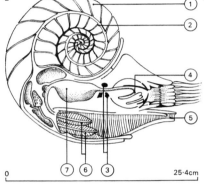

2 A section through Nautilus shows the shell [1] and siphuncle [2] wound in a spiral. Immediately behind the tentacles lies the mouth [4] leading to the intestine [7]. *Nautilus* has an advanced nervous system with a brain [3] and respires by means of gills [6] that are located in the mantle cavity. It swims by forcing a jet of water out of its mantle cavity and through the siphon [5].

3 The cuttlefish *Sepia officinalis* lives on the sea bottom where it feeds on shrimps. It is able to uncover these by blowing jets of water at the sand of the sea-bed. In a similar way the animal can bury itself by blowing the sand from below and allowing the particles to fall back on to its upper surface. The cuttlefish is a master of camouflage: as the environment changes so its surface displays rippling patterns of colour change – the product of many thousands of expandable bags of pigment, each under nervous control.

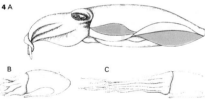

4 The most elegant of movers among the cephalopods is the cuttlefish. When moving slowly [A], as in hunting, for example, it swims by means of a series of undulations in its lateral fins. When it accelerates [B, C], it closes the opening of its mantle cavity, constricts the powerful muscles in the wall of the mantle and forces a jet of water through its siphon. The siphon may be turned through any angle, so the cuttlefish can use this mechanism to steer itself in any direction it chooses. All cephalopods share this system of jet propulsion, which allows them to make a rapid escape from possible danger.

Octopus 9m

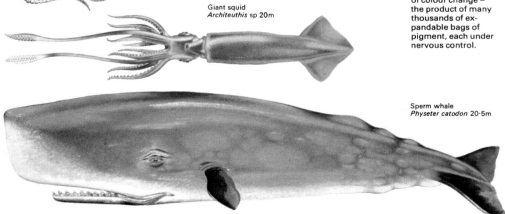

Giant squid
Architeuthis sp 20m

Sperm whale
Physeter catodon 20·5m

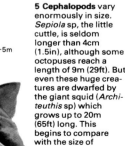

5 Cephalopods vary enormously in size. *Sepiola* sp, the little cuttle, is seldom longer than 4cm (1.5in), although some octopuses reach a length of 9m (29ft). But even these huge creatures are dwarfed by the giant squid (*Architeuthis* sp) which grows up to 20m (65ft) long. This begins to compare with the size of that other giant of the oceans, the sperm whale (*Physeter catodon*). The comparison is significant because the squid forms a major part of the sperm whale's diet. Evidence of conflict between these two species is provided by the scars of huge sucker marks occasionally found on the whale's tough outer skin.

nary convergence according to which two widely differing groups have solved the same problem – that of vision – in a similar manner.

Patterns of behaviour

Their powerful sense organs and large brains allow cephalopods to achieve complex behaviour patterns and account for their pronounced ability to learn. These attributes are clearly valuable assets to predatory animals that feed on such prey as crabs and lobsters, which have the strength and equipment to fight back. All cephalopods are swift and powerful hunters. The tentacles that surround the mouth, armed with suckers on muscular stalks, help to seize the victim and thrust it between the horny jaws. Squids and cuttlefish have two tentacles that are longer than the rest. Normally at least partly retracted, these can be shot out at great speed to capture moving prey. The octopus uses a less graceful but equally effective technique: it jumps on its prey, using its tentacles to enfold the victim before eating it.

Many cephalopods, especially the deep-sea forms, are capable of producing light – a phenomenon known as bioluminescence. This may be useful to illuminate the way, or as an attraction or a warning signal, and can be generated either in special tissues or by bacteria that live symbiotically in various parts of the cephalopod body. One cuttlefish (*Hetereteuthis* sp) can even release a luminous cloud of bacteria when disturbed.

The reproductive behaviour of cephalopods is complex: approaches between cuttlefish, for example, may involve sophisticated colour changes. During the breeding season a male approached by another cuttlefish deepens its colour. If the newcomer fails to change likewise, it is assumed to be female. Both male and female have a single gonad (organ producing sex cells) that discharges into the mantle cavity. The male collects its sperm into a small packet, the spermatophore, and transfers this to the female using a specialized arm called a hectocotylus. Various glands associated with the female gonad produce yolk and shell for the eggs. These eggs are laid and hatch into juveniles that spend the first part of their lives swimming in the plankton.

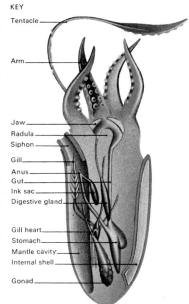

KEY

Tentacle
Arm
Jaw
Radula
Siphon
Gill
Anus
Gut
Ink sac
Digestive gland
Gill heart
Stomach
Mantle cavity
Internal shell
Gonad

The cephalopods or "head-footed" animals, such as this cuttlefish, are the most advanced of the molluscs. The mouth is surrounded by sucker-bearing tentacles and equipped with horny jaws. A muscular mantle covers the cavity that houses the gills. Through these, a pair of gill hearts pump blood. All cephalopods can move by forcing a jet of water out of the mantle cavity and through a tube, the siphon. A dark fluid is discharged from the ink sac when the animals are excited or frightened. The shell is large and external in nautiloids, smaller and internal in squids and cuttlefish and completely absent in octopuses.

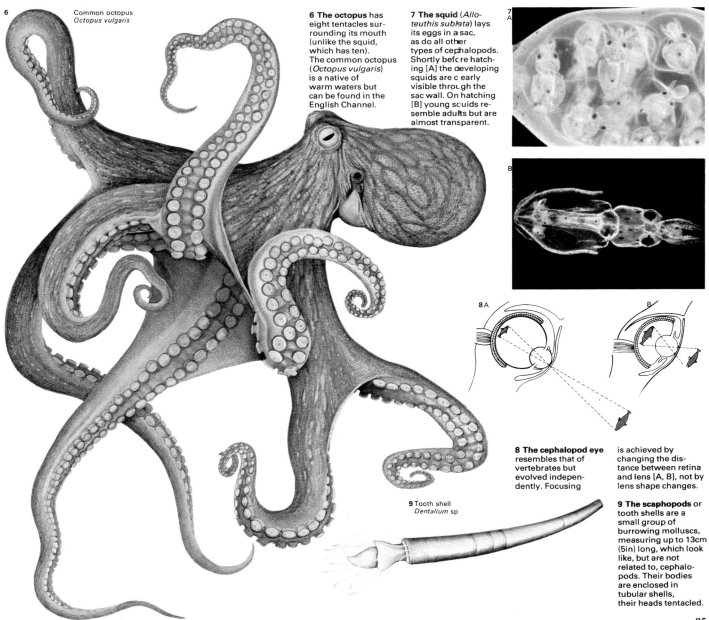

6 Common octopus
Octopus vulgaris

6 The octopus has eight tentacles surrounding its mouth (unlike the squid, which has ten). The common octopus (*Octopus vulgaris*) is a native of warm waters but can be found in the English Channel.

7 The squid (*Alloteuthis sublata*) lays its eggs in a sac, as do all other types of cephalopods. Shortly before hatching [A] the developing squids are clearly visible through the sac wall. On hatching [B] young squids resemble adults but are almost transparent.

8 The cephalopod eye resembles that of vertebrates but evolved independently. Focusing is achieved by changing the distance between retina and lens [A, B], not by lens shape changes.

9 Tooth shell
Dentalium sp

9 The scaphopods or tooth shells are a small group of burrowing molluscs, measuring up to 13cm (5in) long, which look like, but are not related to, cephalopods. Their bodies are enclosed in tubular shells, their heads tentacled.

Shells of the world

More than 80,000 species of molluscs have been identified and named. These animals, many of which have shells, are second only to the arthropods in abundance.

Molluscs are grouped zoologically into six or seven classes, depending on whether the group of deep-sea burrowing forms is placed with the chitons in the molluscan class Amphineura or given their own class Aplacophora. The most curious molluscs belong to the class Monoplacophora, which consists of only one genus, namely *Neopilina*. Another small group, the tusk shells (class Schaphopoda), has 200 species.

The three remaining mollusc classes are the largest and best known. The two-shelled bivalve or molluscs (class Bivalvia) include many familiar shellfish such as cockles, mussels, clams and abalones. The land and sea snails are grouped together in the class Gastropoda and include some of the most interesting and beautiful of all the molluscs. Members of the class Cephalopoda, many of which have lost their shells, include the squid and the octopus. These, the most advanced of all molluscs, show a rudimentary "intelligence".

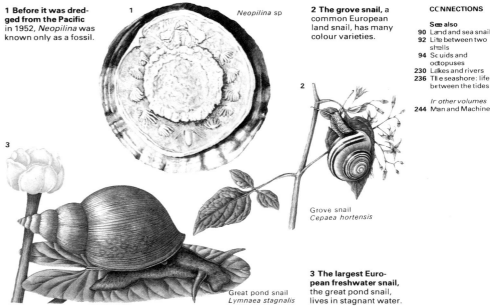

1 Before it was dredged from the Pacific in 1952, *Neopilina* was known only as a fossil.

Neopilina sp

2 The grove snail, a common European land snail, has many colour varieties.

Grove snail
Cepaea hortensis

3 The largest European freshwater snail, the great pond snail, lives in stagnant water.

Great pond snail
Lymnaea stagnalis

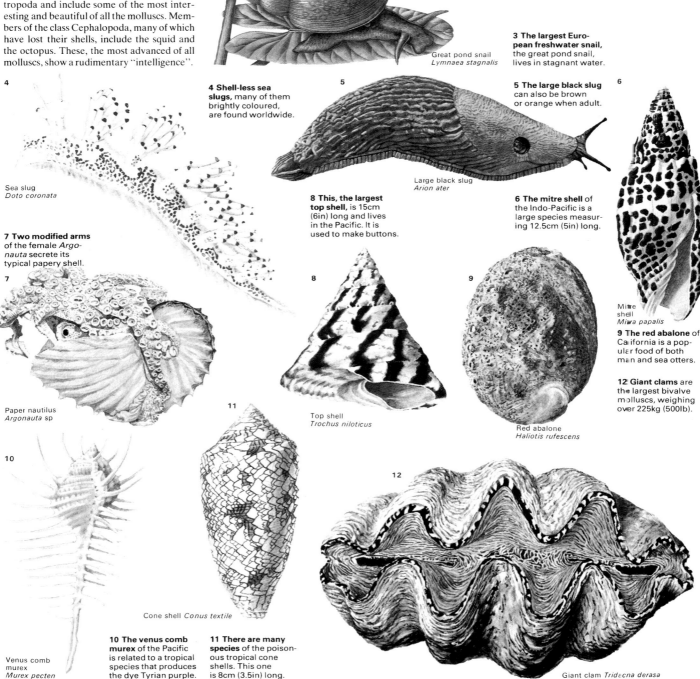

4 Shell-less sea slugs, many of them brightly coloured, are found worldwide.

Sea slug
Doto coronata

5 The large black slug can also be brown or orange when adult.

Large black slug
Arion ater

6 The mitre shell of the Indo-Pacific is a large species measuring 12.5cm (5in) long.

Mitre shell
Mitra papalis

7 Two modified arms of the female *Argonauta* secrete its typical papery shell.

Paper nautilus
Argonauta sp

8 This, the largest top shell, is 15cm (6in) long and lives in the Pacific. It is used to make buttons.

Top shell
Trochus niloticus

9 The red abalone of California is a popular food of both man and sea otters.

Red abalone
Haliotis rufescens

12 Giant clams are the largest bivalve molluscs, weighing over 225kg (500lb).

10 The venus comb murex of the Pacific is related to a tropical species that produces the dye Tyrian purple.

Venus comb murex
Murex pecten

11 There are many species of the poisonous tropical cone shells. This one is 8cm (3.5in) long.

Cone shell *Conus textile*

Giant clam *Tridacna derasa*

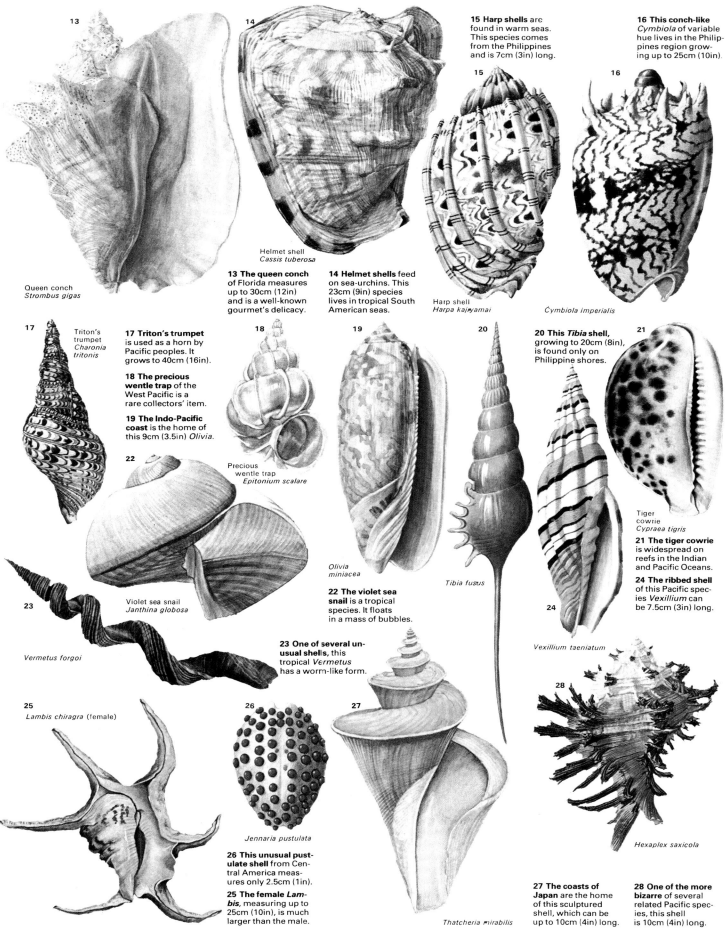

13

Queen conch
Strombus gigas

14

Helmet shell
Cassis tuberosa

15 Harp shells are found in warm seas. This species comes from the Philippines and is 7cm (3in) long.

15

Harp shell
Harpa kajiyamai

16 This conch-like *Cymbiola* of variable hue lives in the Philippines region growing up to 25cm (10in).

16

Cymbiola imperialis

13 The queen conch of Florida measures up to 30cm (12in) and is a well-known gourmet's delicacy.

14 Helmet shells feed on sea-urchins. This 23cm (9in) species lives in tropical South American seas.

17

Triton's trumpet
Charonia tritonis

17 Triton's trumpet is used as a horn by Pacific peoples. It grows to 40cm (16in).

18 The precious wentle trap of the West Pacific is a rare collectors' item.

19 The Indo-Pacific coast is the home of this 9cm (3.5in) *Olivia.*

18

Precious wentle trap
Epitonium scalare

19

Olivia
miniacea

22

Violet sea snail
Janthina globosa

22 The violet sea snail is a tropical species. It floats in a mass of bubbles.

23

Vermetus forgoi

23 One of several unusual shells, this tropical *Vermetus* has a worm-like form.

20

Tibia fusus

20 This *Tibia* shell, growing to 20cm (8in), is found only on Philippine shores.

21

Tiger cowrie
Cypraea tigris

21 The tiger cowrie is widespread on reefs in the Indian and Pacific Oceans.

24 The ribbed shell of this Pacific species *Vexillium* can be 7.5cm (3in) long.

24

Vexillium taeniatum

25

Lambis chiragra (female)

26

Jennaria pustulata

26 This unusual pustulate shell from Central America measures only 2.5cm (1in).

25 The female *Lambis*, measuring up to 25cm (10in), is much larger than the male.

27

Thatcheria mirabilis

27 The coasts of Japan are the home of this sculptured shell, which can be up to 10cm (4in) long.

28

Hexaplex saxicola

28 One of the more bizarre of several related Pacific species, this shell is 10cm (4in) long.

The joint-legged animals

The phylum Arthropoda – literally "joint-legged" animals – is by far the largest and most diversified group of creatures in the animal kingdom and must be regarded as one of the most important developments in animal evolution. Arthropods are of obscure origins, but it is most likely that they arose from primitive annelid (worm-like) stock [Key], possibly from animals rather like modern velvet worms [1], more than 600 million years ago.

Diversity and common features

The phylum Arthropoda contains a great many familiar animals such as insects, spiders, scorpions, centipedes and crustaceans (crabs, shrimps and lobsters) and also more unusual forms such as horseshoe or king crabs, sea spiders and the parasitic ticks and mites. The numbers of individuals in many species are enormous and the number of arthropod species in the animal kingdom far exceeds those of all other animals put together. About 80 per cent of the known species of animals are arthropods.

The fundamental characteristics that all arthropods share are the tough, segmented, external skeleton (exoskeleton) covering the whole of the body (the internal anatomy of which also shows some segmentation) including the jointed limbs and a nerve cord running ventrally along the length of the animal (as in many other invertebrates but not in vertebrates). The heart, if present, is simple and situated dorsally.

There are variations on the basic theme, for arthropods have evolved over a long period of time and through diverse branches and it is not surprising that several species of arthropod are recognizable as belonging to that group only by experts. Some parasitic forms, for example, have lost most of their organs and look more like plant growths in the tissues of other animals. Only their typically arthropod larvae betray them. Others are so small – both winged and wingless species – that they have missed attention for centuries and bear little superficial resemblance to the great marine crustaceans or exotic butterflies and moths. But there must also be many thousands of species waiting to be described for the first time.

All of the arthropod organs have been modified in various ways during evolution and even the heart may be absent. But the dominant feature of the group, the tough exoskeleton [4] upon which so many of the other features depend, is always present, in at least some stage of the life cycle.

The exoskeleton, which has contributed so greatly to the success of the arthropods, also imposes certain restrictions on them. It is excellent as a protective covering, but one-piece armour would prevent any movement. This problem was overcome by the simple solution of dividing the external skeleton into a number of separate plates. The plates are joined together by a thin, flexible membrane. The legs are also segmented and are composed of several tube-like elements also hinged by a flexible membrane.

The rigidity of the exoskeleton

The rigidity of the exoskeleton also complicates growth. The cuticle (outermost skin) can stretch between segments to a certain extent, but for any real overall increase in size the animal has to moult its outer covering and

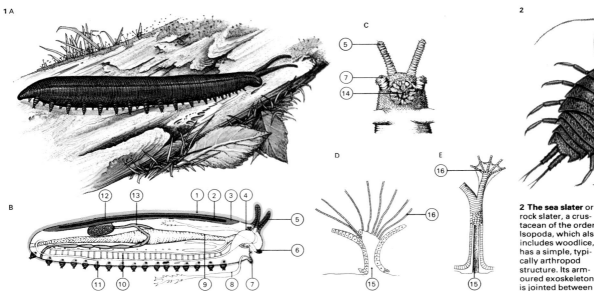

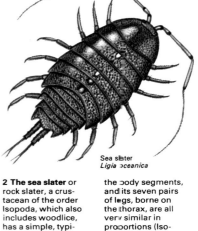

Sea slater
Ligia oceanica

2 The sea slater or rock slater, a crustacean of the order Isopoda, which also includes woodlice, has a simple, typically arthropod structure. Its armoured exoskeleton is jointed between the body segments, and its seven pairs of legs, borne on the thorax, are all very similar in proportions (Isopoda means "equal feet"). Appendages beneath the abdomen function as gills.

1 Velvet worms such as *Peripatopsis* [A] have features [B] typical of both arthropods (the cuticle [1], heart [2] haemocoel [3] and the antennae [5]) and annelid worms (eye [4] and excretory organs [11]). Slime glands [8] discharge through oral papillae [7]. The mouth [6] has jaws [14] and leads to the gut [9]. The underside of the head is shown [C]. Paired nerve cords [10] co-ordinate movement.

Female reproductive organs comprise ovaries [12] and uteri [13]. Spiracles (respiratory openings) [15] and tracheae [16] of *Peripatopsis* [D] and insects [E] are only superficially similar.

Epicuticle

Chitinous endocuticle

Epidermis

Basement membrane

3 The support of muscles for movement is a major function of the exoskeleton. The muscles are attached across a joint [A], as in vertebrates, but inside the skeleton. The contraction of one muscle of a pair and the relaxing of the other results in movement at the flexible joint. A cross-section through the body at a leg insertion [B] shows the part played by muscle attachment points and by the thin flexible cuticle at joints.

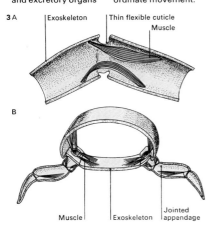

3 A
Exoskeleton | Thin flexible cuticle | Muscle

B
Muscle | Exoskeleton | Jointed appendage

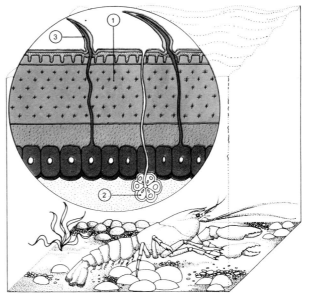

4 The exoskeleton in the Crustacea is hardened by the deposition of calcium salts. As a result most crustaceans, such as the crayfish, have very tough shells consisting mostly of a cuticle [1] secreted by the epidermis. Special glands [2] in the connective tissue secrete the waterproof epicuticle. Epidermal bristles [3] detect various stimuli.

expand before the new skeleton hardens. During that time it is very vulnerable.

The renewal of the exoskeleton places heavy demands upon the animal. The new covering requires a great deal of material for its production and only marine arthropods such as crabs and lobsters with abundant food supplies and available minerals can grow to a large size.

Although excellent as a protective covering the exoskeleton is relatively heavy. With increase in size the exoskeleton becomes proportionately heavier and imposes a size limit on terrestrial arthropods and, more significantly, on flying species.

Success of the group
In spite of the problems inherent in the basic arthropod body plan the group has been remarkably successful within its limitations. One arthropod or another has managed to colonize every habitat invaded separately by all the other groups. The insects, of course, are among the most proficient of flying animals. Indeed, certain flies are, for their size, the fastest of all living things.

Many arthropods, of various groups, can both produce and perceive sounds. Some moths, for example, use ultrasound for confusing the hunting techniques of bats and possibly for other purposes also. Other senses in arthropods are also well developed. They can detect not only airborne vibrations but also those of the surface upon which they are standing. The chemical senses of taste and smell are highly developed and play an important role in communication between individuals of many species. The eyes are extremely well developed in many arthropods and vision can extend beyond man's visible spectrum into either the infrared or ultra-violet zones. The eyes of some insects, spiders and crustaceans can detect the plane of polarization of light from the sky and use it in navigation.

Arthropod muscles are relatively efficient and very powerful. The claw of a large lobster, for example, can crush a man's wrist. The flight muscles of higher insects such as flies possess the remarkable ability of contracting extremely rapidly, allowing the wings to beat at up to 1,000 times a second.

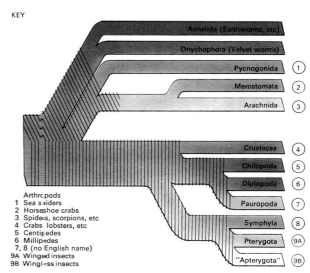

Arthropods
1 Sea spiders
2 Horseshoe crabs
3 Spiders, scorpions, etc
4 Crabs, lobsters, etc
5 Centipedes
6 Millipedes
7, 8 (no English name)
9A Winged insects
9B Wingless insects

The phylum Arthropoda can be divided into ten classes with living representatives and five classes composed of extinct species. The arthropods are the most successful of all invertebrate groups on land, in the water and the air.

5 The jointed limbs of arthropods were originally for locomotion, like the legs on the rock slater's thorax. They have since become modified for different functions, even in the same animal. Examples are a claw [A] and sperm guide [B] of a crustacean; a swimming leg of a water beetle [C]; a walking leg of a ground beetle [D]; an antenna of a cockchafer [E]; and a chewing mandible of a cockroach [F].

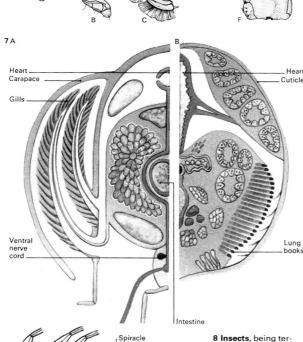

7 The breathing apparatus of arthropods is shown in the cross-sections through a water-dwelling lobster [A] and land-dwelling spider [B]. In lobsters and other crustaceans a carapace (a plate of the skeleton) covers a gill chamber through which water circulates. Blood passing through the gills returns to the heart for pumping to the blood spaces in the body. Most spiders have pairs of lung books made up of "leaves", each containing circulating blood and enclosed in a "vestibule" full of air. This opens to the outside by means of spiracles. As in crustaceans the blood returns to the heart after oxygenation.

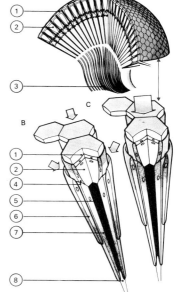

Heart
Carapace
Gills
Ventral nerve cord
Heart
Cuticle
Lung books
Intestine

8 Insects, being terrestrial animals, run the risk of drying out and therefore have a severely limited area open to the outside air. Respiration is by a system of tracheae or tubes opening to the outside by pairs of spiracles along the abdomen.

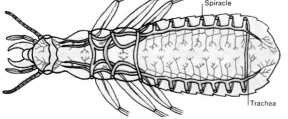

Spiracle
Trachea

6 The compound eyes of arthropods [A] have many units or ommatidia, each covered with a transparent corneal lens [1]. Underneath is the photoreceptor unit [2] that connects with the optic nerve [3]. The ommatidia work together or separately, in the dark [B] or the light [C]. In the dark, pigment in the screening cells [4 and 6] is withdrawn so that the photopigment cells [5] and receptor cells [7] act together to receive the light focused by the crystalline core and corneal lens secreted by the corneagen cells. But in the light the pigment from the screening cells isolates the ommatidia, which send separate messages to the brain via nerve fibres [8].

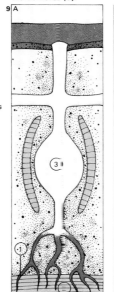

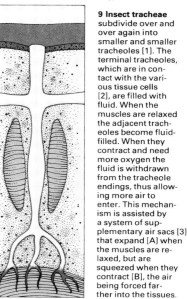

9 Insect tracheae subdivide over and over again into smaller and smaller tracheoles [1]. The terminal tracheoles, which are in contact with the various tissue cells [2], are filled with fluid. When the muscles are relaxed the adjacent tracheoles become fluid-filled. When they contract and need more oxygen the fluid is withdrawn from the tracheole endings, thus allowing more air to enter. This mechanism is assisted by a system of supplementary air sacs [3] that expand [A] when the muscles are relaxed, but are squeezed when they contract [B], the air being forced farther into the tissues.

Crabs and lobsters

Crabs, lobsters, crayfish, prawns and shrimps, together with such creatures as woodlice, water fleas and barnacles, are all members of the zoological class Crustacea. This class is a subdivision of the larger group or phylum Arthropoda (the joint-legged animals) to which the insects also belong.

Diversity of forms

The Crustacea number more than 30,000 species and are one of the most diverse of all animal groups. They vary in form from the familiar lobster to the barnacles, which look more like molluscs than crustaceans, and in size from the giant Japanese spider crab (*Macrocheira kaempferi*), measuring 3–4m (10–13ft) from claw tip to claw tip, to the copepods of the ocean plankton that may be only a millimetre (0.04in) in diameter.

Crustacean life-styles are as diverse as their body forms although most live in water or in damp surroundings. The crabs and lobsters are largely scavengers, while *Sacculina* sp, which looks more like an undifferentiated sac of tissue than a crustacean, lives as a parasite within crabs. The mantis shrimps are

active predators, the gribble (*Limnoria* sp) is a wood-borer and the cleaner shrimp (*Stenopus* sp) feeds on external parasites that live on the bodies and in the mouths of fish. In return for its "cleaning services" the shrimp is granted immunity from potential predators. Many crustaceans, including the sedentary barnacles, are filter feeders, extracting food from the water in which they live by forcing a constant flow of it through their bodies.

Characteristics of crustaceans

The typical water-dwelling crustacean has a body covered with a hard shell or exoskeleton. This external skeleton is composed of protein and chitin and hardened with lime salts. Also typical is the segmented body form, as seen in the Norway lobster [Key], although the middle body section (the thorax) is often concealed by a large plate of skeleton (the carapace).

Essentially, each segment of the crustacean head and body carries a pair of legs or other appendages [6]. These appendages have become modified to perform different functions. On the head, for example, the typ-

ical appendages are the first and second antennae, which are used as sensory receptors, and the mandibles and first and second maxillae, which are used for taking in and crushing food. The thorax bears three pairs of maxillipeds, which are also employed in pushing food towards the mouth, the chelae or pincers for food capture plus four pairs of walking legs. Under the abdomen are five pairs of swimmerets used, as their name suggests, in locomotion, but some of them may be modified for reproductive purposes. The crustacean abdomen ends with a pair of appendages known as uropods. These, with the tip of the abdominal skeleton (the telson), form a tail fan that is used for steering during swimming or, when flicked vigorously down and under, shoots the animal backwards defensively.

Most crustaceans breathe dissolved oxygen from their surrounding water either through gills or through most of the body surface; exceptions to the rule are the woodlice and the extreme parasitic forms. The gills of crabs and lobsters are thin-walled, leaf-like outgrowths that originate at or near the bases

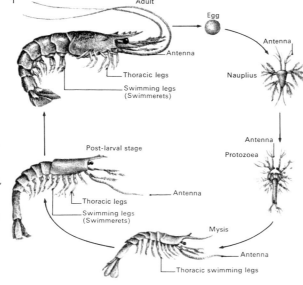

1 Most crustaceans hatch in a form quite different from that of their parents, passing through one or more larval stages before they become adult. The prawn *Penaeus* sp from the Gulf of Mexico shows the type of changes that may occur. The prawn spawns at depths down to 54m (180ft) and from the fertilized egg a nauplius, the basic crustacean larval form, emerges. This metamorphoses into the protozoea stage which, like the nauplius, swims with its antennae, but unlike it has both eyes and thoracic limbs. Thoracic limbs are used for swimming by the mysis larva. The post-larval stage and the adult both swim using abdominal appendages.

Adult
Egg
Antenna
Antenna
Thoracic legs
Naupllus
Swimming legs (Swimmerets)
Post-larval stage
Antenna
Protozoea
Antenna
Thoracic legs
Swimming legs (Swimmerets)
Mysis
Antenna
Thoracic swimming legs

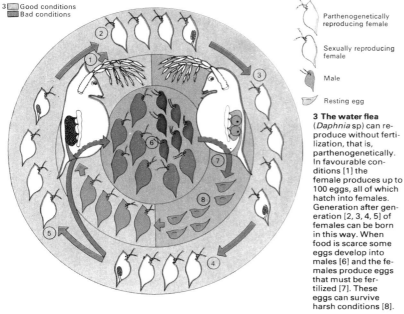

3 ☐ Good conditions
   ◼ Bad conditions

Parthenogenetically reproducing female

Sexually reproducing female

Male

Resting egg

3 The water flea (*Daphnia* sp) can reproduce without fertilization, that is, parthenogenetically. In favourable conditions [1] the female produces up to 100 eggs, all of which hatch into females. Generation after generation [2, 3, 4, 5] of females can be born in this way. When food is scarce some eggs develop into males [6] and the females produce eggs that must be fertilized [7]. These eggs can survive harsh conditions [8].

2 The eggs of the squat lobster (*Galathea squamifera*) are attached to the female on special abdominal appendages. The abdominal swimmerets secrete a substance to which the eggs stick. They are held beneath the abdomen, partly covered by the tail, in a large mass. Females carrying eggs in this way are said to be "in berry". After the eggs have hatched the larvae may also cling to the female's swimmerets for a time.

4 Both fresh and salt waters throughout the world may teem with the microscopic *Cyclops* and its related species which are members of the crustacean sub-class Copepoda. A female is seen here bearing a pair of sacs filled with eggs. *Cyclops* has a single eye – from which it is named – two pairs of thoracic limbs for feeding and four for swimming. There are two pairs of antennae which are sensitive to touch and chemicals.

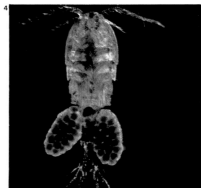

5 The nauplius larva is found in so many different groups of Crustacea that it can justifiably be called the basic larval form. The nauplius usually has a simple triangular or shield-like unsegmented body with a single eye and three pairs of appendages. The first pair are simple and sensory; the second pair, the antennae, are very well developed with two branches covered in special hairs or setae and are used for swimming. A pair of mandibles is employed in feeding.

of the appendages of the thorax. They are enclosed in a chamber covered by a downgrowth of the carapace on each side and a current of water flows over them maintained by the beating of a paddle-like organ at the base of the second maxilla. In the woodlice and other similar crustaceans, however, the gills are modifications of the appendages of the abdomen and in many of the smaller species of Crustacea the body surface itself acts as a gas-exchanger. But whatever their structure all crustacean gills must remain wet and this explains why most crustaceans are confined to watery habitats.

The sense organs of crustaceans, particularly the eyes and antennae, are efficient structures well adapted to the needs of these animals. The paired eyes of crabs and lobsters are borne on stalks and like those of insects they are compound. But in copepods they are simple, single and central – one copepod is in fact named *Cyclops* [4] after the legendary Greek giant with the same characteristic. The two pairs of antennae are used to detect vibrations, maintain balance [7] and also serve as sensors of "taste" and "smell".

Like other creatures that are encased in hard external skeletons, crustacea can only increase in size if they undergo successive moults [8]. Many crustaceans have the ability to regenerate lost parts [9] such as claws and eyes, and often do this following the rejection of an injured part by self-amputation.

Most Crustacea develop from eggs laid by the female [2] following mating and fertilization, although in some species, including water fleas [3], young can be produced by "virgin birth" or parthenogenesis. Larvae development is generally in several stages [1], but most species have a nauplius larva [5].

Food for others

Throughout the world crustaceans play a major role in aquatic food webs, both freshwater and marine, as the food of larger animals. Freshwater ponds teem with microscopic *Daphnia* and *Cyclops*; sea-shores shelter millions of seahoppers beneath seaweed and other flotsam; and in the sea one particular group, known as krill, not only provide food for fish, squids, jellyfish and the like but are the sustenance of some whales.

The Norway lobster, a small, burrowing form up to 20cm (8in) long found off northeast Atlantic coasts, is a typical crustacean. With the crabs, crayfish, prawns and shrimps it is classified in a sub-group or order of the Crustacea called the Decapoda (literally, 10 legs). These legs are borne in pairs [1–5] the first of which is enlarged to form nipping claws or chelae [6]. Of the two pairs of antennae [7], [8] the second may be far longer than the body. With the eyes [9] they are the principal sense organs. The body segments are obvious only on the abdomen [10] for the thorax is covered by a hard carapace [11]. The tail bears a fan-like telson [12].

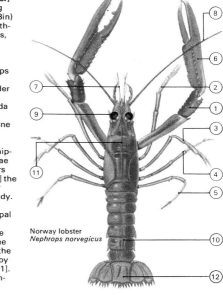

Norway lobster
Nephrops norvegicus

6 The segmented appendages of all crustaceans are built to the same basic plan. Those shown here are from the crayfish, a member of the order Decapoda. Although the appendages are very different along the animal's length, most are essentially two-branched or biramous. This arrangement is most clear in the swimmerets found beneath the abdomen. Variations on the theme are towards expanded or foliacious appendages such as the mandibles and maxillae to create a shape ideal for feeding or respiration. The walking legs are unbranched or uniramous. Structures with common origins like these are said to be homologous.

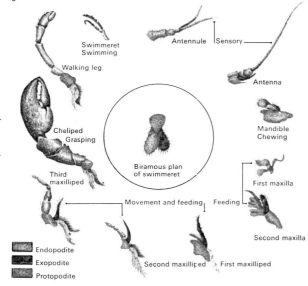

Swimmeret Swimming
Walking leg
Antennule | Sensory
Antenna
Cheliped Grasping
Mandible Chewing
Biramous plan of swimmeret
Third maxilliped
First maxilla
Movement and feeding | Feeding
Second maxilla
Second maxilliped · First maxilliped
Endopodite
Exopodite
Protopodite

7 The balancing organs of Crustacea, as in this crayfish (*Astacus torrentium*) [A], are situated at the bases of the first antennae. Each organ of balance or statocyst [B] lies in a cavity protected by a fringe of bristles or setae [1] and consists of a series of setae [2] on to which sand grains [4] are stuck with a special secretion. As the animal moves the grains also move and as they do so set up stresses on the setae that stimulate sensory cells. These send messages to the brain through the statocyst nerve [3]. New sand grains have to be incorporated each time the crayfish sheds its exoskeleton.

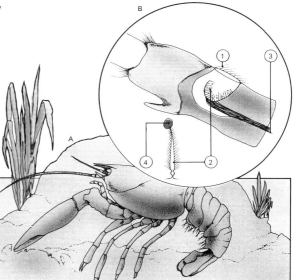

8 Moulting or ecdysis in Crustacea, as in other arthropods, is essential if the animal is to grow. When the crab is ready to change its shell [A] the membrane between carapace and abdomen is ruptured, and the animal hunches up and begins the slow process of withdrawal [B], which may take several hours, starting with the head and ending with the abdomen. Immediately after the moult the animal absorbs water and expands before the new skeleton hardens.

9 The crabs and many related crustaceans are able to regenerate portions of their limbs. This ability is connected with the phenomenon of autotomy whereby a claw or walking leg may be cast off [A] if accidentally trapped. The damaged appendage is snapped off along a built-in line of weakness and then slowly regenerates [B]. It usually takes many moultings for a lost part to be replaced completely.

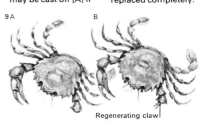

Regenerating claw

Spiders and scorpions

A widespread and successful animal group, spiders, scorpions and their relatives are collectively termed arachnids (class Arachnida) by naturalists. They have a long evolutionary history [Key]. The earliest known fossil arachnids are from the Palaeozoic era. Fossil scorpions occur in Silurian rocks (430 million years old) and fossil spiders are known from the Devonian of Scotland (395 million years old). Their ancestors were aquatic, but most modern and fossil spiders are terrestrial and many are found in extremely arid regions. Their aquatic ancestry is evident in their possession of modified gills (lung books), although the more advanced members of the group possess air tubes (tracheae) that appear externally as spiracles on the abdomen [1].

Spider versus insects
It would seem that the earliest spiders fed upon insects, before the latter evolved wings, and caught them on the ground in the same way as present-day wolf spiders. The development of insect flight opened another niche for the spiders who countered with the evolution of webs and snares to capture flying prey. Insects became masters of the air, but spiders joined the "aerial plankton" (creatures that live hanging in air or blown on its currents), suspended on silk threads.

The best known of the arachnids are the spiders. Many species are found in or near human dwellings and their orb webs or cobwebs are well known [6]. Spiders are generally disliked because a number of tropical species have bites that can be dangerous to humans. But the bite of the large "tarantulas" is not as serious as is generally believed and most can be readily handled.

Spiders as skilful hunters
The main significance of the spiders in the natural order lies in the role they play as predators upon their principal competitors, the insects. There are more than 20 times more insect species than spiders, of which there are about 40,000 species. From sea-level to at least 6,700m (22,000ft) spiders pursue insects with a variety of traps and manoeuvres adapted to fit the habits of their prey [9]. Nocturnal spiders take up the hunt when diurnal ones rest, so that insects are not safe at any time of the day or night. Spiders have devised a great variety of ways of capturing their prey. Some spiders jump upon their prey, others lasso them and many varieties net and snare their victims.

The number of spiders in the world is enormous, a fact first emphasized by the great British expert on spiders, W. S. Bristowe (1901–), in 1939. He found that an English field contained around two million spiders per acre and estimated that British spiders ate a weight of insects each year greater than that of the nation's humans.

The production of silk [7] has been one of the spiders' keys to success. They are able to fashion a number of accessories that aid them enormously in their struggle for existence. Not only are their traps made of silk, but also their shelters and breeding structures, as well as the parachutes that allow the wind to carry them from one place to another. The thinnest spider silk has greater tensile (stretching) strength than the equivalent thickness of steel wire.

Spiders are equipped with poison glands

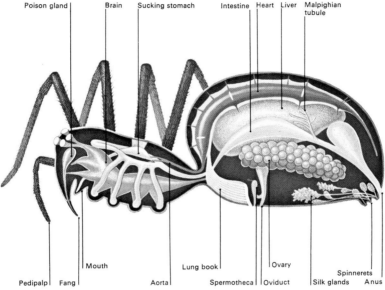

1 Spiders and other arachnids differ from other arthropods in having bodies divided into two sections. Head and thorax are united in a cephalothorax or prosoma, joined to the abdomen by a narrow pedicel. There are four pairs of walking legs and the eyes are simple, not compound as in other arthropods. There are no antennae. Long, straight hairs on the appendages carry out sensory functions. The mouthparts are chelicerae, or fangs, and behind them are limb-like pedipalps with sensory and feeding functions. In male spiders they are modified for the transfer of sperm. Excretory organs (Malpighian tubules) open into the hind gut.

Poison gland — Brain — Sucking stomach — Intestine — Heart — Liver — Malpighian tubule

Pedipalp — Fang — Mouth — Aorta — Lung book — Spermotheca — Oviduct — Ovary — Silk glands — Anus — Spinnerets

2 Garden spider *Araneus diadematus*

2 The European garden spider is a well-known species often found hanging in its complex web, spun where there are insects to be caught. The animal has a typical spider form, with eight legs.

3 A female spider discharges her eggs into a saucer of silk which is then enclosed in a silken sac or cocoon. This may be hidden in the soil, in a tree trunk or attached to the roof of a cave or to plants. In some cases the cocoon will be guarded by the female. Some species of spiders camouflage their cocoons, while others, such as this hunting spider *(Pisaura mirabilis)* protect them by carrying them around with them wherever they go.

4 Newly hatched spiders are vulnerable because they cannot defend themselves nor can they feed or spin silk until after the first moult. These young garden spiders hatch from eggs in cocoons that are hidden and often guarded by the female. Young spiders are vulnerable to predation and sometimes remain together as a ball and if disturbed, disperse in an "explosion" of bodies and colour.

5 Scorpions produce their young alive, not from eggs. Newly born scorpions are carried around on their mother's back, as some spiders carry their young. Scorpion embryos develop in special brood chambers in the mother's body, so when they are born they have only to climb on to her back. They are able to hang on securely by means of small suckers on their feet and will be carried about for days or even weeks – depending on their species and the prevailing conditions – living on the remains of their embryonic yolk supply. As with spiders, scorpions are unable to leave the female until after the first moult. They become adult after six or seven moults, which takes about a year.

that they use to quieten and kill their prey before their silken webs, or they themselves, are badly damaged by the struggles of their victims. Species with the more virulent toxins can deal with prey, or enemies, that are much bigger than themselves.

Small but prolific mites
Less well known than spiders, but an equally successful arachnid group, is the order Acarina, which includes the ticks and mites. These small creatures are easily overlooked, except in the case of the many ectoparasitic species (those living on the surface of their host). Mites and ticks are found on large numbers of vertebrates and invertebrates. All the ticks are blood-suckers and many of them carry diseases in both wild and domesticated animals and man. They are responsible for considerable loss of life. The mites too are of economic importance, many of them as parasites of both plants and animals. Some are beneficial, being involved in the breakdown of plant material in the soil.

The scorpions [5] are a group of arachnids limited to the earth's warmer regions. They

have an ancient fossil history and are a homogeneous group; all 600 or more species are easily recognized as scorpions.

The sting of the scorpion is used for killing prey or defence, the virulence of its poison varying between species. In only a few does the venom prove a threat to human life, for example, species in the family Buthidae. In most it is almost harmless. Scorpions capture their prey with their pincers – technically called the pedipalps. Some species crush their victims without using venom, but those species with less powerful pincers rely wholly on poisoning their prey.

The harvestmen (order Opiliones) are known to most country people. These animals are spider-like but they may be distinguished by the lack of the narrow pedicel between cephalothorax (a union of the head and thorax) and abdomen. They are found throughout the world in low-growing vegetation and in leaf litter, where they feed largely on small insects.

The wide distribution of arachnids and the fact that spiders are carnivorous makes their role in the world's food chains a vital one.

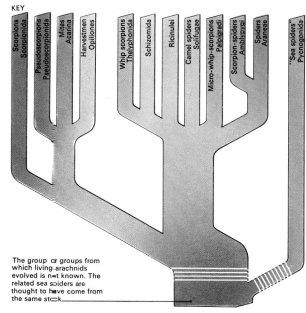

KEY

Scorpions Scorpionida | Pseudoscorpions Pseudoscorpionida | Mites Acarina | Harvestmen Opiliones | Whip scorpions Thelyphonida | Schizomida | Ricinulei | Camel spiders Solifugae | Micro-whip-scorpions Palpigradi | Scorpion-spiders Amblypygi | Spiders Araneae | "Sea spiders" Pycnogonida

The group or groups from which living arachnids evolved is not known. The related sea spiders are thought to have come from the same stock.

6

6 The orb webs of spiders are spun from silk with great accuracy to specific designs of various types. A bridge is first constructed between two supports and then the orb is fashioned beneath.

Radiating spokes support a spiral whose gummy thread is extruded from the silk glands. The gum layer is broken into beads by the spinner's stretching the thread tight and letting go with a snap.

8

7

Cribellum for weaving bands of silk

Spinnerets

7 Spiders' silk is produced in special silk glands and distributed by the spinnerets. A variety of glands produce different kinds of silk for specific parts of the web, for binding prey and for the egg

cocoon. Parts of the glands produce the gum for snare lines. The spinnerets have many minute spinning tubes at their tips. Each thread consists of many filaments fused together.

8 The water spider (*Argyroneta aquatica*) is found in temperate Europe and Asia and is the only spider that spends most of its life beneath the surface of the water. The hairs covering its body

trap a film of air which is taken under water to a bell of silk that is attached to water plants. The spider breathes the air in the bell, emerging only at night to hunt. Eggs are laid in the bell.

9 Spiders show great ingenuity in the ways in which they trap prey. Tropical bolas spiders suspend a sticky gum droplet on the end of a silk thread and swing it to and fro like a pendulum. Passing insects are attracted to it and become ensnared. Trapdoor spiders may make complex burrows from which they dig side chambers where they can retreat in safety from flooding. The Australian funnel-web spider (*Atrax robustus*) has a silken burrow with

9

Bolas spider
Dichrostichus furcatus

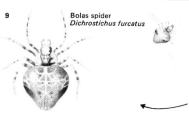

a funnel-shaped entrance. Its prey slips down the funnel to be ensnared below and killed by the spider's venomous bite. *Dinopis bicornis* hangs from a strong thread holding a sticky net

stretched between its legs. When a suitable victim passes near, the spider casts the net over it. As the victim struggles it becomes more enmeshed in the net and the spider catches it with ease.

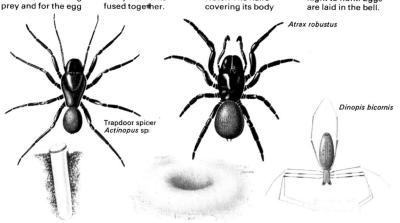

Atrax robustus

Dinopis bicornis

Trapdoor spider
Actinopus sp.

Crustaceans and other arthropods

The many-legged creatures are together classified in the phylum Arthropoda, the group that includes the insects. Those that are not insects, some of which are illustrated here, total more than 90,000 species. These are divided almost equally between the class of the spiders and scorpions, the Arachnida, with more than 50,000 species, and the class Crustacea, which includes the lobsters, shrimps and crabs, with more than 30,000 species. The largest group of the remainder includes the millipedes and centipedes among its numbers and contains more than 11,000 species. The other classes all number less than 450 species each.

All these joint-legged animals have bodies organized in a similar way but show an amazing variety of shapes and sizes. Apart from the spiders, scorpions and their relatives, most are water-dwellers, and planktonic forms are an important link in the food chains of larger marine animals such as fish. Many are more directly important to man either as a source of food (crabs, shrimps, lobsters) or as his enemies (poisonous spiders and scorpions and infesting mites and ticks).

1 The king or horseshoe crab is not a crab at all – it is a primitive arthropod closely related to fossil forms and allied to spiders and scorpions. This marine species measures up to 60cm (2ft) long.

King crab
Limulus polyphemus

2 Sea spiders are members of a small sub-phylum of the arthropods known as the Pycnogonida; their name comes from their spider-like appearance. Unlike spiders, however, they are marine and have from four to seven pairs of legs. The body is so small – 2-3mm (about 0.1in) – that some of the digestive tract is in the legs, which span 2cm (0.75in).

Sea spider
Nymphon rubrum

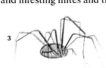

Harvestman
Phalangium africanum

3 The North African harvestman has many relatives throughout the world, including North America where it is known as daddy longlegs. This minute scavenger measures up to 12mm (0.5in).

4 All the false or pseudoscorpions are very small – this large specimen is about 1cm (0.39in) long. The small chelicerae [1] contain silk glands, the large pedipalps [2] have pincers and poison glands.

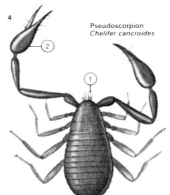

Pseudoscorpion
Chelifer cancroides

5 The mouthparts of the harvest mite, an arachnid parasitic on vertebrate tissue during the early part of its life, are borne on a "false head". The total body length of this species is only 1mm (0.04in).

Harvest mite
Trombicula autumnalis

7 Pill millipedes have the peculiar ability to roll up when disturbed. They differ from centipedes in having two rather than one pair of legs on each body segment. This one is 2cm (0.75in) long.

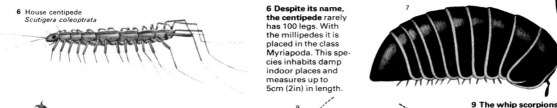

6 House centipede
Scutigera coleoptrata

6 Despite its name, the centipede rarely has 100 legs. With the millipedes it is placed in the class Myriapoda. This species inhabits damp indoor places and measures up to 5cm (2in) in length.

Pill millipede
Glomeris marginata

Bird Spider
Sericopelma communis

8 A humming-bird is the victim of this fiercely predatory bird spider from Panama which measures about 6cm (2.5in) long. Often wrongly called a tarantula, it is relatively harmless – its venom is about as potent as that of a bee.

9 The whip scorpions are members of a small arachnid order comprising about 100 species, most of which live in tropical climates. This one, with the typical whip-like tail, is 6cm (2.5in) long.

Whip scorpion
Mastigoproctus giganteus

10 Little-known arachnids, the 37 species in the order Ricinulei are found in leaf mould and caves in tropical Africa and America. None grows to more than 1cm (0.39in) long. This one is African.

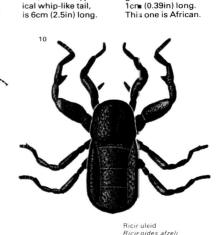

Ricinuleid
Ricinoides afzeli

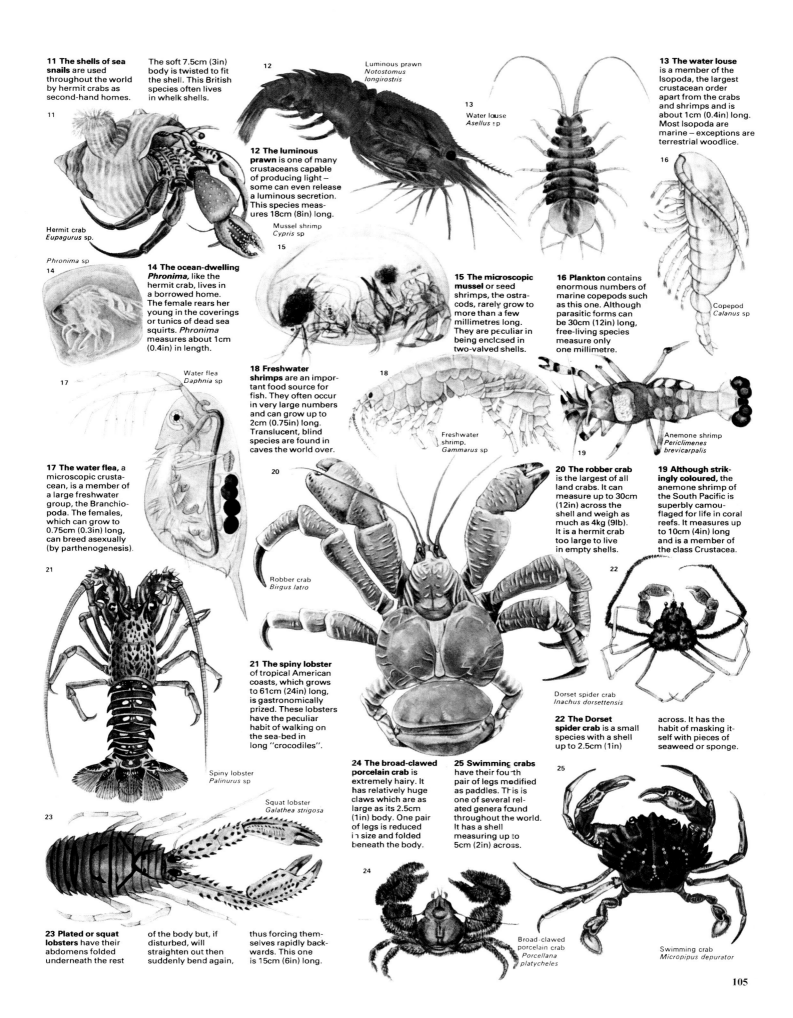

11 The shells of sea snails are used throughout the world by hermit crabs as second-hand homes. The soft 7.5cm (3in) body is twisted to fit the shell. This British species often lives in whelk shells.

Hermit crab
Eupagurus sp.

Phronima sp

14 The ocean-dwelling *Phronima*, like the hermit crab, lives in a borrowed home. The female rears her young in the coverings or tunics of dead sea squirts. *Phronima* measures about 1cm (0.4in) in length.

17 The water flea, a microscopic crustacean, is a member of a large freshwater group, the Branchiopoda. The females, which can grow to 0.75cm (0.3in) long, can breed asexually (by parthenogenesis).

Water flea
Daphnia sp

12 The luminous prawn is one of many crustaceans capable of producing light – some can even release a luminous secretion. This species measures 18cm (8in) long.

Luminous prawn
Notostomus longirostris

Mussel shrimp
Cypris sp

15 The microscopic mussel or seed shrimps, the ostracods, rarely grow to more than a few millimetres long. They are peculiar in being enclosed in two-valved shells.

18 Freshwater shrimps are an important food source for fish. They often occur in very large numbers and can grow up to 2cm (0.75in) long. Translucent, blind species are found in caves the world over.

Freshwater shrimp,
Gammarus sp

13 The water louse is a member of the Isopoda, the largest crustacean order apart from the crabs and shrimps and is about 1cm (0.4in) long. Most Isopoda are marine – exceptions are terrestrial woodlice.

Water louse
Asellus sp

16 Plankton contains enormous numbers of marine copepods such as this one. Although parasitic forms can be 30cm (12in) long, free-living species measure only one millimetre.

Copepod
Calanus sp

20 The robber crab is the largest of all land crabs. It can measure up to 30cm (12in) across the shell and weigh as much as 4kg (9lb). It is a hermit crab too large to live in empty shells.

Robber crab
Birgus latro

21 The spiny lobster of tropical American coasts, which grows to 61cm (24in) long, is gastronomically prized. These lobsters have the peculiar habit of walking on the sea-bed in long "crocodiles".

Spiny lobster
Palinurus sp

19 Although strikingly coloured, the anemone shrimp of the South Pacific is superbly camouflaged for life in coral reefs. It measures up to 10cm (4in) long and is a member of the class Crustacea.

Anemone shrimp
Periclimenes brevicarpalis

Dorset spider crab
Inachus dorsettensis

22 The Dorset spider crab is a small species with a shell up to 2.5cm (1in) across. It has the habit of masking itself with pieces of seaweed or sponge.

23 Plated or squat lobsters have their abdomens folded underneath the rest of the body but, if disturbed, will straighten out then suddenly bend again, thus forcing themselves rapidly backwards. This one is 15cm (6in) long.

Squat lobster
Galathea strigosa

24 The broad-clawed porcelain crab is extremely hairy. It has relatively huge claws which are as large as its 2.5cm (1in) body. One pair of legs is reduced in size and folded beneath the body.

Broad-clawed porcelain crab
Porcellana platycheles

25 Swimming crabs have their fourth pair of legs modified as paddles. This is one of several related genera found throughout the world. It has a shell measuring up to 5cm (2in) across.

Swimming crab
Micropipus depurator

105

The classification of insects

Insects are the most numerous of all living creatures, representing about 80 per cent of all animal species. There are more than 1,000,000 known species and probably as many again are still to be discovered.

There are enough insect fossils and primitive living forms to serve as a guide to the evolution of the 29 orders into which all insects are classified. Most of the evidence dates from the advent of the Carboniferous period some 345 million years ago, when a number of winged insects inhabited the coal-forming swamps of the period.

Primitive insects

Insects are thought to have evolved from a centipede-like ancestor from which they differed principally in having only three pairs of legs. Each pair is attached to one segment of the thorax – the middle part of the body. The most primitive of modern insects are possibly the wingless species belonging to four orders that were once grouped together as the "Apterygota". Of these, the order Thysanura [4] seems to resemble the hypothetical ancestral form most closely. All

other insect orders [5–29] are winged and known as Pterygota.

The Collembola [1] and Protura [2] may have evolved from a creature similar to a dipluran, but the two groups have become modified in different ways. The Collembola have a peculiar forked structure on the abdomen that acts like a spring and enables them to jump considerable distances. In the Protura the antennae are absent and the front legs have taken over some of their functions.

The next major development was the evolution of the wings and the power of flight. Two orders – Ephemeroptera [5] and the Odonata [6] – are combined under the Palaeoptera, or "ancient wings". Their wings cannot be folded or laid over the back at rest. Those insects that can fold wings at rest [orders 7–29] are grouped together in the Neoptera, or "new wings".

The seven Orthopteroid orders [7–13] are considered to be the least advanced of the Neoptera. Most of these have simple mouthparts and are predominantly herbivorous. The order Plecoptera [14] is an evolutionary offshoot with many archaic features. The

hemipteroid orders show a steady progression from the primitive, non-specialized mouthparts of the Psocoptera to those of the Hemiptera [19] which have developed piercing and sucking mouthparts enabling them to feed on sap or blood.

A significant development in the nature of the insect life cycle gave the remaining or Neuropteroid orders [20–29] a great advantage over their more primitive relatives.

Greater flexibility

All the insects in the Palaeoptera and Neoptera are often treated as two groups, depending on their life cycles. In the exopterygotes (orders 5–19) – and in the "apterygotes" – the young insects that hatch from the eggs resemble the adult insect. The young, known as nymphs, undergo a series of moults from which they emerge as fully developed adults. In the endopterygotes (orders 20–29), a larva that does not resemble the adult hatches from the egg. This larva (a caterpillar, maggot or grub) usually eats food which is entirely different from that eaten by the adult. Eventually the larva

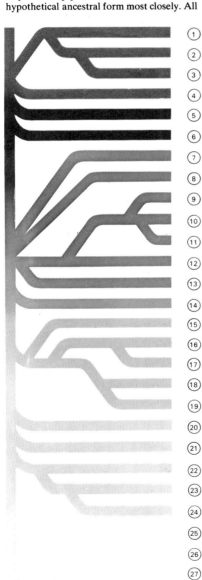

(1)	**Collembola** Springtails Size to 5mm 1,500 species	
(2)	**Protura** Proturans Size 0·5 – 2mm 170 species	
(3)	**Diplura** Japygids and Campodeids Size up to 50mm 660 species	
(4)	**Thysanura** Bristle tails Size up to 20mm 350 species	
(5)	**Ephemeroptera** Mayflies Size 2·5–32mm 1,000 species	
(6)	**Odonata** Dragonflies and Damselflies Size 18—193mm 5,000 species	
(7)	**Dermaptera** Earwigs Size up to 50mm 1,200 species	
(8)	**Grylloblattodea** Grylloblattids Size up to 25mm 12 species	
(9)	**Isoptera** Termites Size 2–110mm 2,000 species	
(10)	**Dictyoptera** Mantids and cockroaches Size 2–120mm 5,300 species	
(11)	**Phasmida** Stick and leaf insects Size up to 320mm 2,000 species	
(12)	**Orthoptera** Crickets, locusts, grasshoppers Size up to 100mm 20,000 species	
(13)	**Embioptera** Web spinners Size up to 20mm 140 species	
(14)	**Plecoptera** Stoneflies Size up to 36mm 1,300 species	
(15)	**Zoraptera** Zorapterans Size up to 3mm 16 species	
(16)	**Psocoptera** Booklice Size up to 5mm 1,700 species	
(17)	**Phthiraptera** Chewing and sucking lice Size 0·5 to 6mm 2,900 species	
(18)	**Thysanoptera** Thrips Size 0·5 to 8mm 5,000 species	
(19)	**Hemiptera** True bugs Size up to 120mm 60,000 species	
(20)	**Megaloptera** Alder and dobson flies Size up to 100mm 500 species	
(21)	**Neuroptera** Lacewings, etc Size up to 70mm 4,000 species	
(22)	**Coleoptera** Beetles and weevils Size up to 150mm 350,000 species	
(23)	**Strepsiptera** Twisted-winged insect parasites Size 1·5—4mm 300 species	
(24)	**Mecoptera** Scorpion flies Size up to 40mm 300 species	
(25)	**Siphonaptera** Fleas Size 1—10mm 1,000 species	
(26)	**Diptera** Two-winged flies Size 1—70mm 70,000 species	
(27)	**Trichoptera** Caddisflies Size 1·5—40mm 3,000 species	
(28)	**Lepidoptera** Moths and butterflies Wing span 4—300mm 165,000 species	
(29)	**Hymenoptera** Ants, bees and wasps Size 0·2—120mm Over 110,000 species	

"APTERYGOT" ORDERS / PALAEOPTERA / ORTHOPTEROID ORDERS / PTERYGOTA / HEMIPTEROID ORDERS / NEOPTERA / NEUROPTEROID ORDERS

The orders of wingless insects. The Thysanura are considered to be the most primitive of the living insect orders. The Collembola, Protura and Diplura have no eyes and are now thought not to be true insects. Most species live in damp places.

The two most primitive orders of winged insects, belonging to widely different lineages. They have non-folding wings.

These orders are grouped together because they are thought to be evolved from a common ancestor. They are considered to be the most primitive of the orders with "modern" wings although the Isoptera has species with a well-organized social system. The orders Isoptera, Dictyoptera and Orthoptera contain some of the most destructive insect pests.

An order not closely related to the adjacent groups.

This group of orders is thought to have a common ancestor. Many members of the Hemiptera are important plant pests. The species of Phthiraptera —the chewing and sucking lice — are mostly parasites feeding externally. The booklice (Psocoptera) can cause minor damage to books and stored food.

The most advanced of the insects are included in this group which contains some of the most numerous and widespread species. Some species of Hymenoptera — bees, wasps, ants — are highly organized social insects. The ovipositor is often long, especially in some parasitic species, and some sawfly species use it as an efficient drill. About a third of all insect species belong to the order Coleoptera — the beetles and weevils. They are found in almost every available habitat. Many species are economically important pests of crops and stored products, some are predators of pests and others, such as the dung beetles, have an important ecological role.

1 North American springtail *Isotoma andrei* 1·2mm

2 European proturan *Acerentomon* sp 1·8mm

3 Campodeid *Campodea folsomi* 4mm

4 Common silverfish *Lepisma saccharina* 8mm

5 North American mayfly *Hexagenia limbata* 5mm

6 Emperor dragonfly *Anax imperator* 75mm

7 Common earwig *Forficula auricularia* 15mm

changes into a pupa, or chrysalis, which may lie dormant for many months while a metamorphosis, or rearrangement of the body tissues, transforms the larva into a mature adult insect.

The difference in life-style of the larva and adult has allowed utilization of extremely diverse habitats. The endopterygote orders contain 84 per cent of insect species and most of those of economic importance.

The Hymenoptera [29] are a large group whose fundamental structure varies little and differs greatly from other endopterygotes. This group shows the culmination of insect social behaviour as exhibited by the social ants, bees and wasps. This is an isolated group but the metamorphosis and larval structure may relate it to the Mecoptera.

Radiative adaptation
The Coleoptera [22], the largest order in the animal kingdom, have distinctly modified hard forewings which cover the rear flying wings. The solidity of the external skeleton and the adaptability of the basic design have been important factors in allowing the adult

to invade many hostile environments.

The remaining endopterygotes form a group centred around the once abundant Mecoptera. The Lepidoptera [28] are recognized by their scale-covered wings and most have the mouthparts specialized into sucking tubes to exploit nectar as a food source. The evolution of this order, and of some of the Diptera [26], coincides closely with that of the flowering plants.

The Trichoptera [27] are an offshoot of the Lepidoptera with hairy wings and chewing mouthparts. The larvae are aquatic.

The Diptera fly with their forewings only, the hind pair being modified into halteres, which act as gyroscopic balancers during flight. The larvae of this order show greater adaptive specialization than any other group of insects. The blood-sucking habit of many of the adult Diptera is associated with their importance as transmitters of diseases.

Closely related to the Diptera are the Siphonaptera [25] which have become wingless and laterally compressed. As with the Phthiraptera [17], this group is entirely externally parasitic on warm-blooded animals.

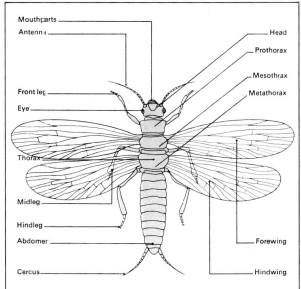

KEY

Mouthparts
Antenna
Front leg
Eye
Thorax
Midleg
Hindleg
Abdomen
Cercus
Head
Prothorax
Mesothrax
Metathorax
Forewing
Hindwing

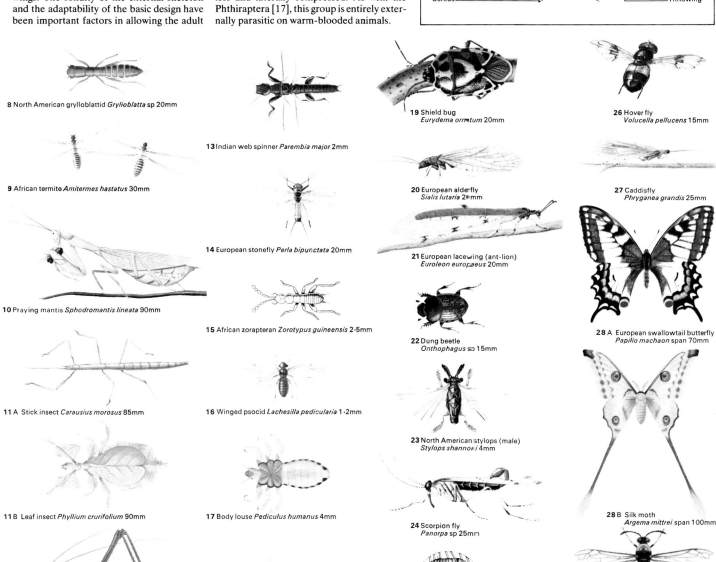

8 North American grylloblattid *Grylloblatta* sp 20mm

9 African termite *Amitermes hastatus* 30mm

10 Praying mantis *Sphodromantis lineata* 90mm

11 A Stick insect *Carausius morosus* 85mm

11 B Leaf insect *Phyllium crurifolium* 90mm

12 Great green bush cricket *Tettigonia viridissima* 50mm

13 Indian web spinner *Parembia major* 2mm

14 European stonefly *Perla bipunctata* 20mm

15 African zorapteran *Zorotypus guineensis* 2·5mm

16 Winged psocid *Lachesilla pedicularia* 1·2mm

17 Body louse *Pediculus humanus* 4mm

18 Onion thrips *Thrips tabaci* 1·4mm

19 Shield bug *Eurydema ornatum* 20mm

20 European alder fly *Sialis lutaria* 20mm

21 European lacewing (ant-lion) *Euroleon europaeus* 20mm

22 Dung beetle *Onthophagus* sp 15mm

23 North American stylops (male) *Stylops shannoni* 4mm

24 Scorpion fly *Panorpa* sp 25mm

25 Rat flea *Xenopsylla cheopis* 3mm

26 Hover fly *Volucella pellucens* 15mm

27 Caddisfly *Phryganea grandis* 25mm

28 A European swallowtail butterfly *Papilio machaon* span 70mm

28 B Silk moth *Argema mittrei* span 100mm

29 Vespid wasp *Eumenes* sp 15mm

The world of insects

The animals that inhabit the earth total about one-and-a-quarter million species. Of these, some 80 per cent are insects, animals classified in the phylum Arthropoda, the group of joint-legged creatures. Numbers alone indicate the success of the class Insecta but they have also colonized the world more widely than any other group.

The ubiquitous insects

There are only a few marine insects; some are surface dwellers, others live between tide marks and one midge even lives on the seabed, but wherever else other animals go, so do the insects – either as free-living forms adapted to an enormous variety of habitats or as parasites living in or on other animals. The insects are a dominant life form from the Arctic to the Equator. Some exist beneath the snow and ice, others in deserts, yet others in salt lakes and hot springs. In southern California there is even a species of small fly (*Psilopa petrolei*) that spends part of its life in pools of crude petroleum.

One of the chief factors in insect success is their ability to fly [2]; apart from the more primitive forms, most species have achieved the freedom of the air, enabling them to colonize new areas and habitats, to escape from predators, to find mates and to prospect for food much more easily than their non-airborne invertebrate relatives. Some of them are even aerial predators.

Although the insects have scored a great evolutionary success through their powers of flight, their weight/wing ratio is such that, theoretically, flight is impossible. In practice, their wing muscles build up energy and then release it rapidly, the speed of the wing-beat compensating for a theoretical lack of lift.

Insect size and shape

Size has also been important in the evolutionary success of insects. When they first appeared more than 350 million years ago, the scale of the environment was similar to that of today and the insects adapted to it, fitting into the many ecological niches that were waiting to be occupied. This explains why insects are comparatively small – although there is a fossil dragonfly with a 76cm (2.5ft) wing span – and can survive and reproduce in niches denied to larger animals.

Another important factor in the success of the insects is their possession of a horny outer covering, the exoskeleton. Although it has to be shed during growth, and thus imposes regular periods of vulnerability, this exoskeleton is both extremely light and virtually unbreakable. Chitin is the basis of the insect skeleton and is tough and flexible. It is made waterproof by a waxy surface layer or by the hardener "sclerotin". The prevention of water loss is essential for land insects. And the material of which the skeleton is composed is so malleable that insects take on a variety of shapes, particularly in the various appendages. Wings, wing covers, legs, ovipositors, mouthparts, bristles, scales, antennae and other appendages all illustrate the ways in which the chitinous exoskeleton may be moulded.

All insect skeletons are moulded round bodies divided into three main parts: the head, thorax and abdomen [1]. The head bears the principal sense organs and mouthparts and encloses the brain, salivary glands and foregut. The thorax bears the wings –

1 **The internal anatomy of all insects,** including this honey bee, is all contained and protected within the confines of the tough, flexible exoskeleton. The typical insect body contains organs of digestion, respiration, circulation, excretion and reproduction. There are muscles through which movement is effected and a nervous system which co-ordinates and controls insect actions on the basis of information received by the sense organs, most important of which are the large compound eyes and the feelers or antennae. The body is in three parts; head, thorax and abdomen.

The antennae or "feelers" of insects are among their most important sense organs. They detect the vibrations of air molecules by means of special structures, situated at their bases, known as Johnston's organs

The heads of insects are very well developed. They carry the sense organs—notably the eyes, antennae and mouthparts. The head houses the brain, feeding glands and various muscles

The cerebral ganglia forming the insect "brain" are the co-ordinating centre for complex activities. A nerve ring connects them with more ganglia from which the ventral nerve cord stems

The honey bee, like most other insects, has two pairs of wings (one has been removed). The veins give added strength

Many-chambered heart

Air sacs in the insect abdomen pump air in and out of an extensive network of tubes (tracheae) by alternate contraction and expansion of the abdomen. At the body surface air moves in and out via holes called spiracles

Tracheal opening

The excretory system is made up of Malpighian tubules which extract waste materials from the blood and convert them into urine which is deposited into the hindgut

Maxilla

Mandible

Foregut

Midgut

Ventral nerve cord

Hindgut

A well-developed pair of compound eyes is a vital part of every insect's sensory equipment. Typically, the eyes are very sensitive to movement, form simple images and can differentiate some colour. The eye of an insect can contain 4,000 separate units or ommatidia each of which acts as a simple eye. The images are then interpreted together in the cerebral ganglia

The mouthparts of the bee are adapted for sucking up food which is liquid—especially nectar from flowers. The mandibles and maxillae have almost lost the ability to "bite" and liquid food is sucked instead through a specialized proboscis

The insect gut is a tube running from mouth to anus with different regions specialized for different functions. It is divided anatomically into foregut, midgut and hindgut of which the first and last are lined with cuticle continuous with the exoskeleton. The foregut is used for storage and grinding, the midgut for digestion and absorption and the hindgut for water resorption

The nerve cord has a ganglion in each segment from which nerves arise

The ovary of this female bee is associated with accessory organs such as shell-producing glands and a vagina

The six legs are jointed and muscled for movement. The claws provide grip

basically two pairs – on the second and third of the three thoracic segments – plus the three pairs of legs. The abdomen consists typically of 11 segments and carries the copulatory organs [3] and, when present, stings which are modified ovipositors.

The jointed legs of insects give them great mobility. This is increased still further by the wings whose structure and movement, like those of the limbs, depend very much on the nature of the insect skeleton. This excellent mobility has been a contributory factor in the interference of insects in the life of man. The locust, for example, combines great fecundity with the power of flight, which means that enormous flocks of countless millions can move to a new area to continue their ravages on man's crop plants when one such food source is depleted.

Patterns of behaviour

The members of the class Insecta show a great variety of behavioural adaptations, particularly in their reproduction. Courtship, mating and care of the young may be remarkably sophisticated, particularly in the social

insects – the ants, bees and wasps of the order Hymenoptera and the termites, which are members of the order Isoptera.

The social organization of one insect species, the honey bee (*Apis mellifera*) is especially efficient involving both different castes – workers, drones and queens – and a division of labour between workers of different ages. This second system is not rigid but can be modified according to the changing needs of the hive as a whole.

Despite the high level of organized behaviour in the beehive, and the success of the insects as a group, insect behaviour is largely instinctive. This involves a genetic "programming" of the animal as a result of which it responds to particular stimuli in a specified way appropriate to the demands of the environment. Intelligence, as applied to human activities, does not enter into this. Thus, although moths of certain species have evolved mechanisms for avoiding bats by unpredictably erratic flight, or by producing bursts of ultrasonics to jam the bats' "sonar", they cannot avoid the attraction of bright lights, which is often fatal.

KEY
A1

Insects develop by metamorphosis – that is they have different forms during their life cycles. There are two types of metamorphosis. In the grasshopper [A] it is gradual. An egg hatches into a nymph [1] which moults several times [2] before adulthood [3]. Butterflies [B] undergo complete metamorphosis with three distinct stages: the larva or caterpillar [4], the pupa or chrysalis [5] which is a quiescent form, and the adult [6].

B4

5   6

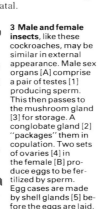

3 Male and female insects, like these cockroaches, may be similar in external appearance. Male sex organs [A] comprise a pair of testes [1] producing sperm. This then passes to the mushroom gland [3] for storage. A conglobate gland [2] "packages" them in copulation. Two sets of ovaries [4] in the female [B] produce eggs to be fertilized by sperm. Egg cases are made by shell glands [5] before the eggs are laid.

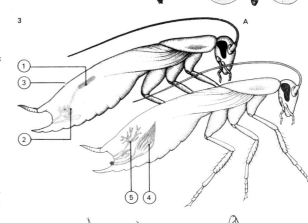

2 Most insect wings provide only enough support for powered flight. In the honey bee [A] the two pairs of wings are connected in flight. The combined wing trajectory follows a figure of eight to

give more lift and thrust on the down stroke [1]. This is made more efficient by wing twisting so that there is little air resistance on the upstroke. The wings move as levers and are indirectly

powered by muscles which change the shape of the thorax [B, C]. The vertical muscles [2] contract to control the up beat while longitudinal muscles [3] contract to power the down beat.

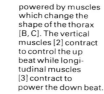

4 Signalling sounds are produced by many insects and in the order Orthoptera, which includes the grasshoppers, locusts and crickets, this is well developed in males and well known. Sounds are made by stridulation in which vibration is set up by the scraping of one hard surface against another. In crickets [A] the forewings or tegmina are specially modified for this purpose [B], for each bears a scraper [1] and a file [2]. When the wings are rubbed together [3] vibrations are set up which produce sounds. Individual variations are created by the "mirrors" [4]. The vibrations of each tooth [C] produces sound of a particular frequency as in the graph [D].

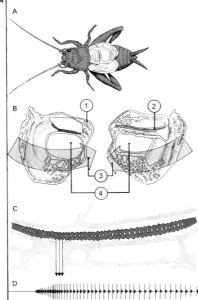

5 Insects have receptors for detecting sound waves. Mosquitoes [A] detect sound velocity with their antennae [B]. At the base of the antennal shaft [1] is Johnston's organ

[C] containing sound receptor cells [3] and their nuclei [2]. These send impulses via nerve fibres [5] to the "brain" where they are interpreted as sounds. Blood vessels and nerves pass through the organ [4]. Each receptor unit [D] con-

tains a sensory cell [6] with a nerve fibre [7] and a sensory hair [8] with a protective cap [9]. The sound receptors on the abdomens of grasshoppers [E] respond to changes in the air pressure. They consist [F] of a rigid frame [10]

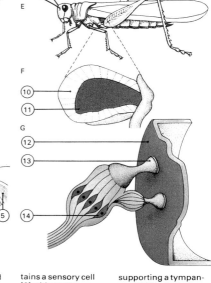

supporting a tympanum [11] like that of a human ear. In detail [G] the tympanic membrane [12] carries a supporting process [13] for sensory units formed as in [D]. The nuclei of the sensory cells [14] give off nerve fibres to the brain.

Locusts, bugs and dragonflies

There are more than a million known living insect species and almost all of them have wings. Truly wingless insects are primitive forms grouped together in the sub-class Apterygota, while winged insects belong to the sub-class Pterygota. Winged insects are further classified into two major groups on the basis of their developmental changes or metamorphosis. The more primitive of these are known as the Exopterygota or "insects with outside wings" because the wings can be seen developing on the exterior of the animal. The other group, called the Endopterygota or "insects with inside wings", includes the more advanced insects such as butterflies, whose wings do not appear on the outside of the pupa stage of the life cycle.

From egg to adult

In the Exopterygota a type of larva called a nymph (basically a miniature adult) emerges from the egg [5]. The proportions are different from those of the adult, as they are in most young animals; there is little indication of wings and the nymph is sexually immature. But in every other respect it is clearly the young form of the adult that it will become.

The nymph grows by gradual progressive stages, or instars, the exoskeleton being moulted between instars so that the insect can rapidly swell and harden off the new exoskeleton that has been prepared beneath the old one. The oldest species – in terms of evolution – tend to have the largest number of instars. Mayflies (order Ephemeroptera), for example, moult 30 times or more, while locusts moult only four or five times. At each moult there is a progressive increase in the relative wing size and gonad development as well as in general proportions. In the Endopterygota the wings develop on the inside of the larva and on the outside of the pupa. Winged insects may also be classified into two groups according to whether the wings are folded along the abdomen or not.

The mode of life of the nymph is basically the same as that of the adult: it lives more or less in the same place and eats the same type of food. There are, however, some notable exceptions. Dragonflies [6] and damselflies (order Odonata) are exceptional: the nymphs are aquatic, the adults free-flying.

A further and remarkable exception to the general rule in exopterygotes that all individuals of a species are similar is seen in the termites or "white ants" (order Isoptera). These insects are fairly closely related to the cockroaches (order Dictyoptera) but have shown an evolution of social organization convergent with that of ants, bees and wasps (order Hymenoptera). Not only are there different castes in one nest, but some of the castes – workers and soldiers – are sterile forms of either sex in an arrested nymphal stage of development. Thus young forms play an adult role, except in reproduction.

The problem of insect pests

Advanced insects of the endopterygote group cause a great deal of human misery and death through the diseases they carry, but only a few exopterygotes, such as the human louse [7], do this. Yet Exopterygota such as locusts can do an almost unbelievable amount of damage to standing crops. One species alone – the desert locust (*Schistocerca gregaria*), which was responsible for the eighth plague of Egypt – may directly affect

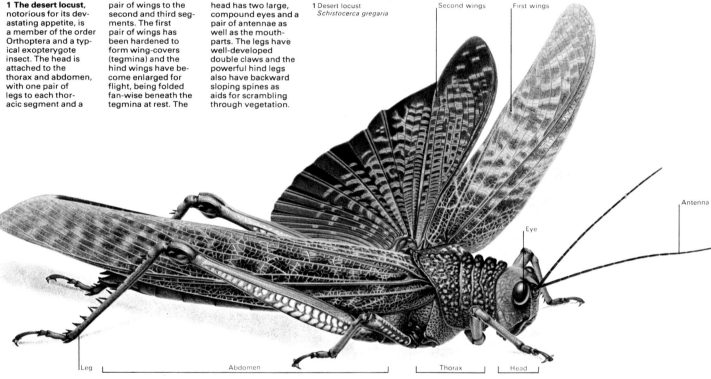

1 The desert locust, notorious for its devastating appetite, is a member of the order Orthoptera and a typical exopterygote insect. The head is attached to the thorax and abdomen, with one pair of legs to each thoracic segment and a pair of wings to the second and third segments. The first pair of wings has been hardened to form wing-covers (tegmina) and the hind wings have become enlarged for flight, being folded fan-wise beneath the tegmina at rest. The head has two large, compound eyes and a pair of antennae as well as the mouthparts. The legs have well-developed double claws and the powerful hind legs also have backward sloping spines as aids for scrambling through vegetation.

1 Desert locust
Schistocerca gregaria

Second wings   First wings   Antenna   Eye

Leg   Abdomen   Thorax   Head

2

Cockroach
Blatta orientalis

2 Cockroaches (order Dictyoptera) are unspecialized insects that live by scavenging. This species is 2.5cm (1in) long and is found in Europe. Cockroaches are not often seen because they are nocturnal, but one may find their egg cases holding 16 eggs.

Egg case

3

Earwig
Forficula auricularia

3 The elegant hind wings of earwigs (order Dermaptera) fold up under short wing cases. These insects have biting mouthparts with which they feed on a wide variety of materials.

4

Greenfly
Aphis sp

4 Greenfly or plant lice of the order Hemiptera and family Aphididae are small insects about 3mm (0.25in) long. They are often serious crop pests as they occur in enormous numbers and throughout the world the 2,000 or more species probably do more damage than any other insect pests. Most aphids have remarkable reproductive powers as the females may produce young parthenogenetically, that is without fertilization by a male.

more than ten per cent of the human population of the world. It is an opportunist, breeding rapidly after unusually heavy rains, and can then suddenly spread out into more than 20 per cent of the world's terrestrial habitat, eating all greenstuff in its path.

The attack on plants

Locust swarms consist of countless millions of insects, blackening the sky across a front of many miles. In a bad year they cause hundreds of millions of pounds' worth of damage and contribute to the death by starvation of many thousands of people. The red locust (*Nomadacris septemfasciata*) and the migratory locust (*Locusta migratoria*) have been effectively controlled by extensive international effort and expense but the desert locust is a much more serious problem.

One of the features that has led to the success of the exopterygote insects is their chewing mouthparts. A number of groups have modified their mouthparts over evolutionary time to produce an apparatus adapted for sucking the juices of plants and animals. The bugs (order Hemiptera) of the sub-orders Homoptera, which include aphids [4] and the cicadas, and the Heteroptera (including the shield bugs), are among the scourges of the horticulturalist, for their habit of imbibing the juices of plants provides a route for the entry into those plants of pathogenic organisms. These bugs and thrips of the order Thysanoptera [8] may thus damage crops indirectly as well as directly.

The same principle, with respect to animal hosts, applies to lice [7]. And when present in large numbers even relatively lowly insects such as earwigs [3] and mole crickets [9] can cause considerable harm to the plants on which they feed. Nobody who grows plants can avoid supporting one or more species of exopterygote insects unless he expends an enormous amount of time, money and energy on their control.

In their way some of the Exopterygota may be said to be as successful as their more advanced and more numerous relatives, the Endopterygota. But there are fewer exopterygotes of benefit to man. One exception is the scale insect (*Dactylopius coccus*) used to produce the red dye cochineal.

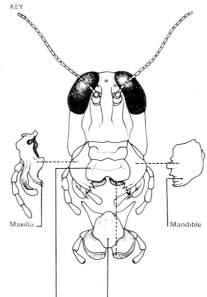

The chewing mouthparts of exopterygote insects are one of the keys to the animals' success. The head-on view of a locust shows the basic pattern. There are three pairs of units: the mandibles, the maxillae and the labium or second maxillae, with an extension of the head skeleton, the labrum, providing frontal protection. The mandibles are very strong and are used for chewing the resistant plant cellulose that forms the staple diet. The first and second maxillae both have sensory and manipulatory parts by means of which the food is tested and subsequently guided between the mandibles and on into the gullet.

KEY

Maxilla

Mandible

Labrum   Labium

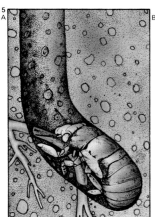

5 Cicadas are true bugs, forming part of the order Hemiptera (Homoptera). They are renowned for their monotonous, high-pitched sound, produced only by males from a pair of drum-like organs at the base of the abdomen. *Magicicada septendecim* of the USA lays eggs in trees. The nymphs drop to the ground and burrow to the roots from which they suck up sap [A]. After 17 years the nymphs return to the surface, climb a tree and finally moult into adults [B].

6 A

Dragonfly
Anax imperator

B Nymph

6 An incomplete metamorphosis, such as occurs in dragonflies (order Odonata), may be an adaptation to take advantage of different habitats. The adult form [A] of *Anax imperator* is a fast-flying predator on other insects while the nymph [B] is aquatic, preying on a variety of life in freshwater ponds.

7 Sucking lice (order Phthiraptera) are small, wingless external parasites of mammals. The human louse exists as two races: the body louse, which lives in the body clothing,

and the head louse which lives on the hair of the head on which it lays its eggs or "nits". Lice are dangerous because they transmit typhus, trench fever and relapsing fever.

9 A   House cricket
Acheta domesticus

Nymph

Adult

B Stick insect
Euryacantha horrida

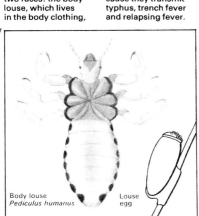

Body louse
Pediculus humanus   Louse egg

8 Thrips are tiny pests of the order Thysanoptera. They are significant for the damage they do to crops and for carrying disease. They have simple, fringed wings – or none at all – and unusual

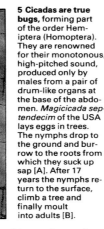

(1)

(2)   Onion thrip
Thrips tabaci

mouthparts with which they suck up plant juices. The onion thrips in both adult [1] and nymphal [2] stages infest a number of hosts to which they may transmit the tomato spotted wilt virus.

C Mole cricket
Gryllotalpa gryllotalpa

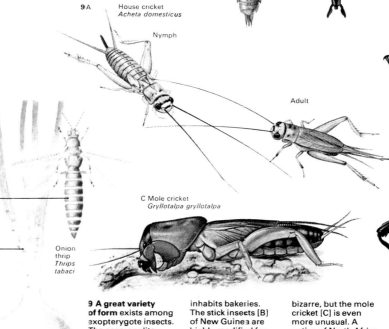

9 A great variety of form exists among exopterygote insects. The cosmopolitan house cricket [A] is particularly common in warm places where food is prepared and cooked and it often

inhabits bakeries. The stick insects [B] of New Guinea are highly modified for the purposes of camouflage, mimicking the plants on which they live. These adaptations may seem

bizarre, but the mole cricket [C] is even more unusual. A native of North Africa and Eurasia, it has greatly enlarged front legs for digging the burrows in which it lives.

Advanced insects

Of all insects, those that have a four-stage life cycle – egg, larva, pupa, adult – are the ones that are both most advanced and most successful. They total more than two-thirds of the 1,000,000 different insect species and include such familiar groups as the moths and butterflies (more than 160,000 species), beetles (more than 350,000 species), bees, wasps and ants (about 110,000 species) and true flies (about 75,000 species). Because of their numbers and worldwide distribution they are of great biological importance. To man they are both a help and a hindrance.

The butterfly life cycle

The butterfly has a life cycle [1] typical of advanced insects. After mating the female lays her eggs on a selected food source which may be completely different from that of the adult. The eggs hatch in a matter of a few days (or even a few hours) into larvae (caterpillars). It is during this larval stage that many insects do great damage to crops, and caterpillars such as those of cabbage white butterflies may strip leaves down to bare "ribs" in a matter of only a few days.

The complete metamorphosis of the larva to the adult stage is via a "resting" phase, the pupa (chrysalis). Within the pupa larval tissues are transformed into those of the adult (imago). The most dramatic changes include the development of wings and the muscles to power them and also, in many insects, a complete change in the feeding apparatus – in the butterfly from chewing larval mouthparts to sucking adult ones.

The life of the adult insect is frequently short and serves only as a dispersal and reproductive phase. Many adult insects do not even feed but rely solely on the energy derived from fat stores laid down in the voracious larvae. However, some adult insects will, if they emerge late in the year, hibernate over the winter and delay egg laying until the spring. These include a number of butterflies, among them the peacock (*Inachis io*).

Keys to success

The advantages conferred on those insects that undergo complete metamorphosis (the endopterygotes) are manifold. Differences between the food eaten by the adults and larvae allows the larvae to exploit food resources that are not available to the adults. The larvae, because they are not involved in reproduction, can be highly camouflaged or live inside their food source – in plant stems or dung, for example – and are thus well protected against predators, such as birds. The immobile pupa, although it is helpless and needs protection from enemies, is necessary for the great change to the adult. The pupa stage also endows the advanced insects with an effective, enforced dormancy period during adverse seasonal weather conditions.

The evolution of advanced insects has favoured those species in which the end of the pupal phase coincides with the onset of conditions conducive to adult survival, and thus the survival of future generations. The flying adult stage increases the chances of successful mating and allows dispersal of the fertilized eggs over a wide area.

The endopterygote insects play a major role in the maintenance of land ecosystems. They form the principal food of many birds and most bats; they act as the chief plant pollinators and contribute enormously to the

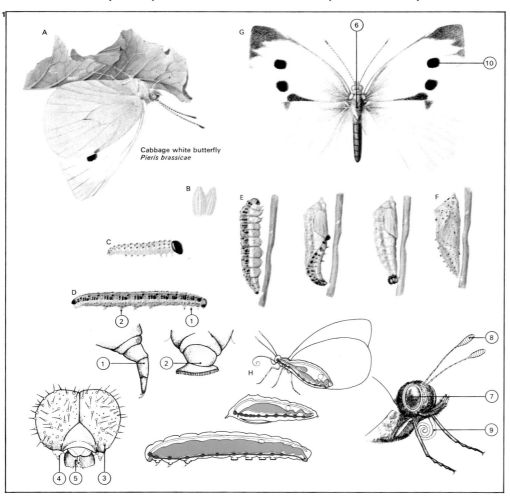

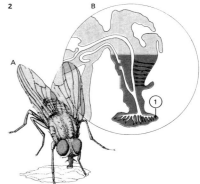

2 Houseflies [A] eat any liquefiable organic matter. The mouthparts [B] consist of a proboscis folded beneath the head when not in use. This extends and its expanded apex spreads over the food [1] Digestive juices are pumped on to the food to liquefy and partially digest it. Left behind on the food are digestive juices as well as bacteria carried by the fly. Flies may thus carry diseases.

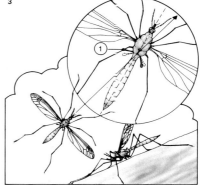

3 Like houseflies, crane-flies or daddy-long-legs are true flies of the order Diptera. The hindwings are reduced to special balancing organs, the halteres [1]. In flight these vibrate with the wings. Any deviation from the stable flight path is detected by them and corrections made by the fly. The operation is similar to that of an auto-pilot on an aircraft. The inset shows a fly deviating from its flight path and the halteres correcting for this.

1 The cabbage white butterfly has a complete metamorphosis. The female [A] lays her eggs [B] on the underside of leaves in batches of 100 or more. The eggs hatch into the first stage larva or caterpillar [C]. The larva moults successively and after four moults is fully grown [D]. It has three pairs of "true" legs [1] which represent the butterfly's legs and four pairs of prolegs [2]; claspers at the end of the abdomen and mandibles [5] for chewing. The head bears rudimentary eyes [3] and antennae [4]. After the fifth moult [E] the caterpillar's skin hardens to form the case of the pupa or chrysalis [F]. Inside this case the tissues of the caterpillar reorganize to form the butterfly [G], which shows the typical insect head, thorax and abdomen. The thorax [6] bears two pairs of wings and three pairs of legs. The head has large, compound eyes [7], clubbed antennae [8] and also a coiled proboscis [9]. The internal systems [H], including those of the nerves [red], blood [yellow] and digestion [blue], have become more complex at each stage. Soon after emergence from the chrysalis, the adult butterflies mate. The male is recognizable by its lack of forewing spots [10].

disposal of dead organisms and waste matter. Many feed on the green plants essential to all animal life, and also form fundamental links in freshwater food chains. They are, however, especially in tropical regions, parasites and carriers of disease [2].

Helpful and harmful insects
The benefits insects confer on man are many. Pollination by insects is essential for many crops, including most fruits. Bees [8] are the most important pollinators and also make honey, one of the oldest crops taken by man. Other products derived directly from insects include bees-wax, used in polishes, and silk from the cocoons of the silkworm moth, spun by the caterpillar before pupation.

The advanced insects are used by man in the field of biological pest control. This may be done by introducing an insect as a predator or parasite. In California, for example, the cottony cushion scale (*Icerya purchasi*) was successfully controlled by the introduction of the Australian ladybird (*Rodolia cardinalis*). Such insects exercise this type of control constantly and help to prevent many

potential pests from becoming actual ones. Scavenging insects, such as dung beetles [4], are also vital. When cattle were first introduced to Australia the accumulated dung rendered grazing lands useless until dung beetles that could utilize it were introduced. On the debit side insects can cause man great economic damage. Insect pests may damage crops and stored products and also carry diseases fatal to man and livestock. Malaria is carried by mosquitoes, while the tsetse fly (*Glossina palpalis*) carries nagana and sleeping sickness, diseases afflicting cattle and man respectively.

Crop damage can be severe. The boll weevil (*Anthonomus grandis*) causes annual losses of $200 million to the American cotton crop, while in the absence of chemical or biological control the codling moth (*Cydia pomonella*) can cause losses in apple yields of up to 50 per cent. The Colorado beetle (*Leptinotarsa decemlineata*), which originally fed on the buffalo burr plant but developed a taste for the cultivated potato, has spread rapidly throughout the world wherever potato plants have been grown.

Bubonic plague is transmitted to man via the rat flea (*Xenopsylla cheopsis*) of the insect order Siphonaptera. The adult rat flea [1] feeds on rat blood before laying eggs. Most of the eggs [2] drop to the ground; there they develop in dirt and litter. Hatched larvae [3] feed on this before pupating [4] and then emerging as new adults which hop on to passing hosts. This type of life cycle is very adaptable because it includes a resting, resistant pupal phase.

4 Typical of many insects that benefit from mammals' faeces are the dung beetles (order Coleoptera). The adult beetles of the genus *Onthophagus* construct brood chambers for their young [A] under cow pats. The male excavates the dung from below [1] and passes it to the female, who fills the brood chambers with it [2], lays an egg in each and seals them [3]. Spoil is taken to the surface [4]. The eggs hatch into larvae which feed on the dung until fully grown when they pupate [B]. On emerging, the adults burrow to the surface.

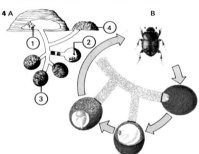
4 A

5 House-fly
Musca domestica

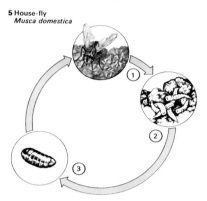

5 The housefly life cycle takes 8–40 days. Eggs are laid [1] in batches of 100 or more and, depending on the temperature, hatch in 1 to 5 days. The larva [2] pupates [3] after a minimum of 5 days.

6 The potter wasps build clay pots attached to plants. They paralyse caterpillars with their stings and place them inside. The female lays an egg in each pot and the emerging larvae feed on the comatose but living caterpillars.

6 Potter wasp
Eumenes coartica

7 Colonies of the wood ant (*Formica rufa*) may contain half a million ants, of three castes – queens [A], female workers [B] and males [C]. The queens and males are winged for their mating flights. Nests of pine needles create mounds 1m (39in) or more high. The queen discards her wings, and lays eggs in brood chambers. The eggs are tended by workers, who also feed the larvae when they hatch. The ant pupae, often called "ants' eggs", are also attended with great care by the workers.

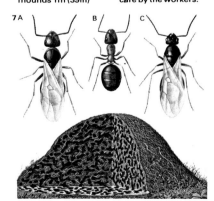
7 A   B   C

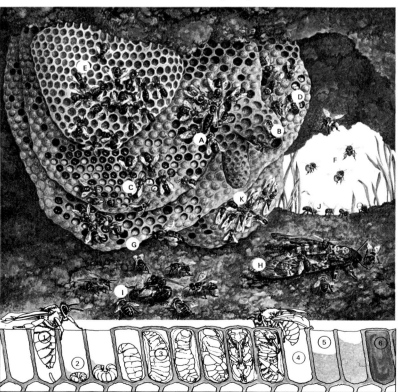
8

8 A strong colony of the honey bee (*Apis mellifera*) may have 80,000 sterile female workers, a fertilized queen [A] and a few hundred males or drones. The colony is organized to maintain a constant internal environment. The queen [1] lays up to 1,500 eggs a day; fertile eggs in queen cells [B] and worker cells, unfertilized eggs in drone cells. The larvae [2] are fed by the workers and when the larvae pupate [3] workers cap the cells. Emerging bees [4] do domestic tasks for ten days, including helping workers [C] and drones [D] to emerge. Then they start comb-building [E]. They take food from foraging workers [F] and store it [G] in pollen [5] or honey [6] cells. They also remove debris or intruders [H], including other queens [I]. The hive is cooled by wing fanning [J] or kept warm by huddling [K]. After three weeks all the workers become foragers.

113

Colourful bugs and beetles

The orders of bugs (Hemiptera) and beetles (Coleoptera) are often regarded with distaste and distrust and lumped together in popular language as "beetles". They are, however, completely distinct groups that are not even closely related. Between them they represent nearly half the known number of insects, the beetles numbering more than 350,000 species, the bugs more than 50,000 species. From this great diversity some of the most spectacular have been chosen for illustration.

Bugs and beetles are found in almost every available habitat on earth, exhibit a wide variety of shape and size and include some of the largest, heaviest and most colourful insect species. A common characteristic of all bugs is their piercing and sucking mouthparts with which they extract their liquid diet from plants or animals. The beetles, in contrast, usually have chewing mouthparts employed equally efficiently on plants and animals. Most bugs and beetles have one pair of wings that is used for flight. The other pair is modified and hardened as a protective covering in all beetles and partially or completely in bugs.

1 Stag beetle
Lucanus cervus

1 The oak forests of Europe and Asia are the home of the fearsome-looking stag beetle (family Lucanidae). Only the male has the "antlers" and can grow to 8cm (3in). The larvae feed in rotten wood and take about three years to mature. Adults emerge after three or four years.

2 Great silver water beetle
Hydrophilus piceus

2 The great silver water beetle (family Hydrophilidae) was popular in Victorian aquaria but is now rare through overcollection. It is one of the largest European water beetles reaching 5cm (2in).

3 European cockchafer
Melolontha melolontha

3 The common cockchafer or may-bug (family Melolonthidae) is a familiar beetle in Europe. It appears in vast numbers every three or four years. The adults grow to 3cm (1.25in) and feed on leaves.

5 The ground beetle is one of several species of the genus *Carabus* (family Carabidae) found in Europe, Asia and North America. It measures about 2.5cm (1in) long and is an active predator.

4 Only the male carpenter's longhorn beetle or timberman (family Cerambycidae) has the enormously elongated antennae. Those of the female are much shorter. This European beetle grows to 2cm (0.75in).

4

Carpenter's longhorn beetle
Acanthocinus aedilis

6 The bold yellow stripes of the European wasp beetle are thought to warn off potential predators. The adults grow to 1.5cm (0.75in) and are classified in the same family as the timberman.

6

Wasp beetle
Clytus arietis

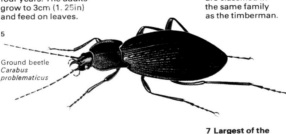

5

Ground beetle
Carabus problematicus

8 Hercules beetle
Dynastes hercules

7 Largest of the European ladybirds, the 7-spot (family Coccinellidae) grows to 8mm (0.4in). Both the adults and the larvae feed on aphids. During the winter large numbers hibernate together.

7

Seven-spot ladybird
Coccinella septempunctata

8 The beetle with the longest known top to tail measurement is the Hercules beetle of tropical Central America (family Scarabaeidae). Including the "horn" it grows to about 15 cm (6in) long. As with the stag beetle only the male possesses the elongated horns, the function of which is unknown. Most members of the group to which this species belongs are found in the South American tropics.

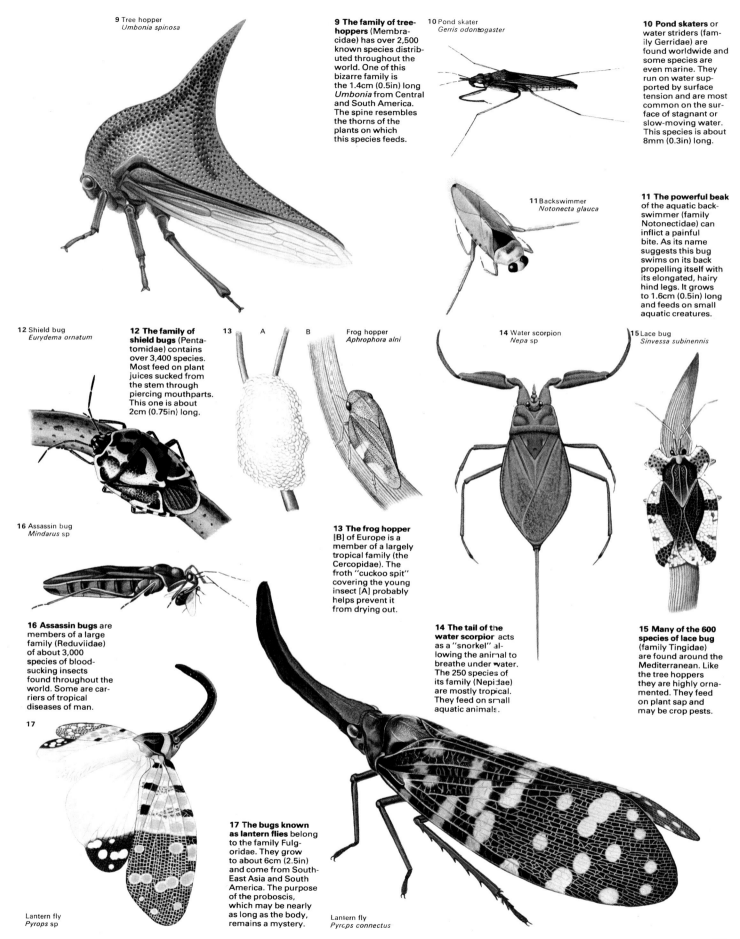

9 Tree hopper
Umbonia spinosa

9 The family of tree-hoppers (Membracidae) has over 2,500 known species distributed throughout the world. One of this bizarre family is the 1.4cm (0.5in) long *Umbonia* from Central and South America. The spine resembles the thorns of the plants on which this species feeds.

10 Pond skater
Gerris odontogaster

10 Pond skaters or water striders (family Gerridae) are found worldwide and some species are even marine. They run on water supported by surface tension and are most common on the surface of stagnant or slow-moving water. This species is about 8mm (0.3in) long.

11 Backswimmer
Notonecta glauca

11 The powerful beak of the aquatic backswimmer (family Notonectidae) can inflict a painful bite. As its name suggests this bug swims on its back propelling itself with its elongated, hairy hind legs. It grows to 1.6cm (0.5in) long and feeds on small aquatic creatures.

12 Shield bug
Eurydema ornatum

12 The family of shield bugs (Pentatomidae) contains over 3,400 species. Most feed on plant juices sucked from the stem through piercing mouthparts. This one is about 2cm (0.75in) long.

13   A   B   Frog hopper
Aphrophora alni

13 The frog hopper [B] of Europe is a member of a largely tropical family (the Cercopidae). The froth "cuckoo spit" covering the young insect [A] probably helps prevent it from drying out.

14 Water scorpion
Nepa sp

14 The tail of the water scorpion acts as a "snorkel" allowing the animal to breathe under water. The 250 species of its family (Nepidae) are mostly tropical. They feed on small aquatic animals.

15 Lace bug
Sinvessa subinennis

15 Many of the 600 species of lace bug (family Tingidae) are found around the Mediterranean. Like the tree hoppers they are highly ornamented. They feed on plant sap and may be crop pests.

16 Assassin bug
Mindarus sp

16 Assassin bugs are members of a large family (Reduviidae) of about 3,000 species of blood-sucking insects found throughout the world. Some are carriers of tropical diseases of man.

17

17 The bugs known as lantern flies belong to the family Fulgoridae. They grow to about 6cm (2.5in) and come from South-East Asia and South America. The purpose of the proboscis, which may be nearly as long as the body, remains a mystery.

Lantern fly
Pyrops sp

Lantern fly
Pyrops connectus

A variety of advanced insects

Butterflies and moths, wasps, bees and ants are some of the best-known insects and a small selection of the many interesting and spectacular species are illustrated on these pages. These insects are members of two of the most advanced of all insect orders. The butterflies and moths are classified in the order Lepidoptera, the bees, ants and wasps in the order Hymenoptera. The families of moths and butterflies include some of the most beautiful of all insects, while the bees, wasps and ants are remarkable for the high degree of social organization shown by some of their number, particularly the honey bees that produce food for man in many different parts of the world. The most economically important species of moths are those from which silk is obtained.

The known species of butterflies and moths total 165,000; of bees, wasps, ants and their allies 110,000. Both groups are distributed worldwide and many species of Lepidoptera conflict with man's interests, being ranked as some of the most noxious of agricultural and forest pests. Their caterpillars can be particularly voracious.

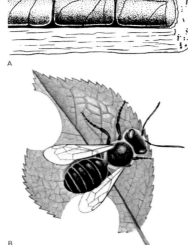

A

B
Leaf-cutter bee
Megachile centuncularis

1 The heavily built, solitary *Megachile* bees (family Megachilidae) are known as leaf-cutters. This name comes from their habit of cutting large discs out of leaves with their powerful jaws [B]. The leaf sections are used to line the egg cells [A]. Each is filled with pollen, then one egg is laid on top and the cell is sealed. Pollen is collected on the hairy underside of the abdomen and forms a thick layer. This bee grows to about 1.2cm (0.5in).

2 The solitary, hairy-legged mining bee belongs to the family Melittidae. Of all bees this group has the largest "pollen baskets" on the hind legs. The bee excavates a nest shaft with cells at the end, each containing pollen and one egg.

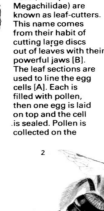

Mining bee
Dasypoda hirtipes

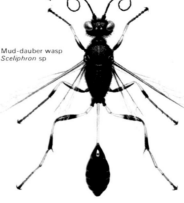

Ichneumon wasp
Rhyssa persuasoria

3 The ichneumon wasp (family Ichneumonidae) has a long ovipositor (here separated from its coverings). It is used to lay eggs under the skins of other insects and larvae. This one is 3.2cm (1.25in) long.

4 The mud-dauber wasps (family Sphecidae) are found worldwide but abound in the tropics. The nest cells of these solitary wasps are stocked with insects and a single egg is laid in each one.

Mud-dauber wasp
Sceliphron sp

Dryinid wasp
Megadryinus magnificus

5 Female parasitic wasps of the family Dryinidae have modified forelegs for holding bug nymphs in which eggs are laid.

6 The velvet ant (family Mutillidae) is in fact a wasp that lays its eggs in the nests of other solitary wasps and bees.

Velvet ant
Mutilla europaea

7 Ants of the sub-family Dorylinae are commonly known as army, driver or legionary ants. Nomadic tropical species, they roam forested areas in long columns feeding on small animals.

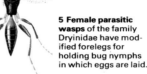

Army ant
Cheliomyrmex andicolus

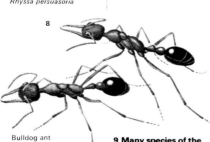

Bulldog ant
Myrmecia forficata

8 The Australian bull-dog ants (sub-family Ponerinae) are large, aggressive and have pincer-like mandibles. They grow to 2.5cm (1in) and make subterranean nests, each containing a few dozen individuals.

9 Many species of the family of gall wasps (Cynipidae) cause cancer-like growths – galls – on plants.

Gall wasp
Andricus kollari

10 All true social wasps belong to the family Vespidae. Many species, such as this one, are solitary and even social ones may form only small colonies. The egg chambers are stocked with moth larvae.

These provide food and shelter for the developing larvae. This species produces galls on oak trees.

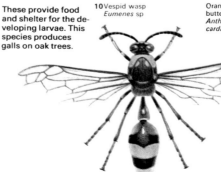

10 Vespid wasp
Eumenes sp

Orange-tip butterfly
Anthocharis cardamines

11 The orange tip (family Pieridae) lays its eggs on plants of the cabbage family (Cruciferae), especially the lady's smock (*Cardamine pratensis*). The chrysalis is attached with a silken girdle.

12 Mimicry is strikingly shown by the leaf butterfly of southern Asia, which even has "veins".

Leaf butterfly
Kallimacha inachus

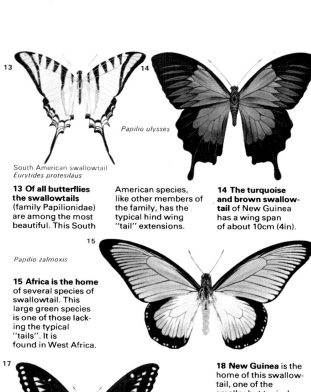

South American swallowtail
Eurytides protesilaus

13 Of all butterflies the swallowtails (family Papilionidae) are among the most beautiful. This South American species, like other members of the family, has the typical hind wing "tail" extensions.

Papilio ulysses

14 The turquoise and brown swallowtail of New Guinea has a wing span of about 10cm (4in).

16 The brilliant colours, "tail streamers" and a slow flight are characteristic of one of the best-known groups of European butterflies – the swallowtails of the family Papilionidae. The caterpillars feed on various plants of the carrot family (Umbelliferae). The chrysalis is attached to the food plant and stays dormant until the following spring. Swallowtails are distributed throughout the world, but in Britain this species is confined to the Norfolk Broads and a few acres of natural fenland in Cambridgeshire. This butterfly, like other European and North American species, is becoming rare because its food plants are being destroyed with herbicides and by fen drainage.

Papilio zalmoxis

15 Africa is the home of several species of swallowtail. This large green species is one of those lacking the typical "tails". It is found in West Africa.

European swallowtail
Papilio machaon

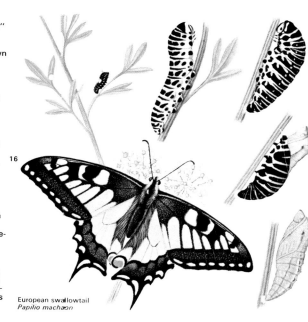

17 This is one of the 15 species of swallowtail found in North and Central America and also in Cuba. Its measurement from wing-tip to wing-tip is about 7cm (3in).

Papilio polyxenes

18 New Guinea is the home of this swallowtail, one of the smaller but typical species. Its wing span is 6cm (2.5in).

Graphium weiskei

19 The scarce or sail swallowtail lives in the warmer parts of Europe but is becoming increasingly rare. The caterpillars feed on the leaves of blackthorn and rowan.

Iphiclides podalirius

Monarch butterfly
Danaus plexippus

European swallowtailed moth
Ouropteryx sambucaria

Viceroy
Limenitis archippus

A

B

20 A form of protection that has been evolved by some animals, particularly butterflies, is mimicry. A harmless species evolves the external appearance of a harmful form that predators have learnt to avoid. The American monarch butterfly, for example, is poisonous to birds due to the presence of cardenolide, a heart poison, and is rapidly regurgitated if devoured. The monarch is mimicked by the viceroy. The bright coloration on the wings is repeated on the undersides of the wings of both butterflies so that they are conspicuous. The butterflies are members of two different families.

Hawk moth
Herse convolvuli

Lackey moth
Malacosoma neustria

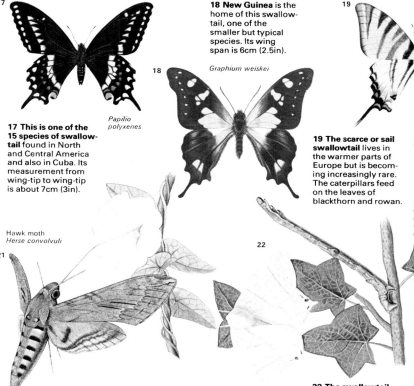

21 Hawk moths (family Sphingidae) such as the convolvulus hawk are found worldwide. Many have a long "tongue", used to extract nectar from tubular flowers. These moths often feed on the wing.

Skipper butterfly
Ochlodes venata

22 The swallowtail moth is a common British species also found from central and southern Europe to Siberia. Its caterpillars eat ivy, sloe and hawthorn. It belongs to the family Geometridae.

23 There are about 3,000 species of the skipper butterflies (family Hesperiidae) found throughout the world. The large skipper of Europe has a wing span of 3.5cm (1.25in). The name is derived from its erratic flight.

24 The caterpillars of the lackey moth (family Lasiocampidae) live communally on hawthorn and similar bushes and trees which they may strip of their leaves. Eggs are laid in a collar around a twig.

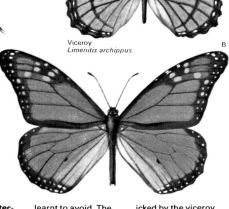

Cinnabar moth
Callimorpha jacobaeae

25 The cinnabar moth (family Arctiidae), seen here alongside its caterpillar, is found in Europe. Its bright colour warns predators that it has an unpleasant taste. The caterpillars feed mainly on ragwort.

Starfish and sea-urchins

Echinoderms, the group of "spiny-skinned" animals [1] that includes the starfish and sea-urchins, are found throughout the world's oceans, on the sea-bed, in rock cavities and shallow coastal waters, and buried deep in sand. There are more than 5,500 species grouped zoologically into five classes.

Echinoderm classes
Sea-urchins (class Echinoidea) are often more familiar as beautifully symmetrical ornaments than as living animals bristling with sharp, often poisonous spines [Key]. The largest species is *Sperosoma giganteum* of Japan with a 30cm (12in) diameter, but most are only about 7.5cm (3in) across. Starfish and cushion stars (class Asteroidea), often brilliantly coloured, are some of the most beautiful creatures in the sea. The 20-rayed star (*Pycnopodia*) of the Puget Sound, USA, is one metre (39in) across, but some cushion stars, with their short arms, are only 1.2cm (0.5in) in diameter. Less familiar are the peculiar limp sea-cucumbers (class Holothuroidea) that crawl along the sand; the tiny, delicate brittle stars (class Ophiuroidea), often found massed together in isolated patches on the sea-bed; and the primitive sea-lilies (class Crinoidea), mostly found in deep water.

All these echinoderms, although it is not obvious in the sea-cucumbers, have pentamerous symmetry – that is, the body can be divided into five parts round a central axis. They may have evolved from a mobile bilateral ancestor that became sedentary, took on a more adaptive radial symmetry and then resumed a free-moving existence.

The starfish and sea-urchin seem unlikely relatives, but if the five arms of the starfish were drawn above the centre of the animal and sewn together, the result would be similar to the body form of the urchin. Another common feature is the external skeleton of bony plates just under the living skin. In sea-urchins, the plates are fused to form a rigid box, but in sea-cucumbers are reduced to microscopic spines (spicules).

How echinoderms move and feed
Most echinoderms move, even if slowly. Starfish, and to a greater extent sea-urchins, have inflexible, rigid skins and for movement have to rely largely on peculiar small appendages called tube-feet, which are blind-ended sacs arising from an internal system of water-filled tubes in the body cavity or coelom. The feet are extended and retracted hydraulically to produce movement. The sea-urchin is clothed in bands of tube-feet armed with suckers [3] that can be extended beyond the long spines when a vertical rock is to be climbed.

Starfish and cushion stars use the rows of feet on the undersides of their arms to cling on to rocks in heavy seas [2]. Brittle stars and basket stars move by wriggling their long arms. Some sea-lilies and feather stars are sedentary. Their flexible arms are joined in a stalk that is anchored to rocks or sand. Tube-feet are used only for respiration and feeding.

Sea-lilies feed on organic particles that fall through the water round them. The particles are caught in grooves on the arms and are moved by hair-like cilia. The sea-urchin shuffles over rocks, rasping off tiny plants and animals using an elaborate chewing apparatus. Waste is ejected through a hole in the top of the shell. With nerve endings

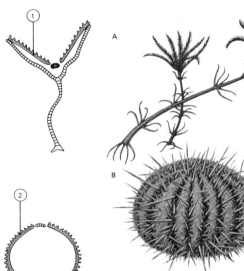
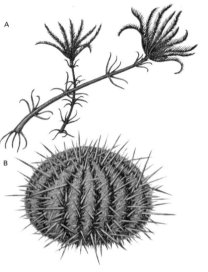
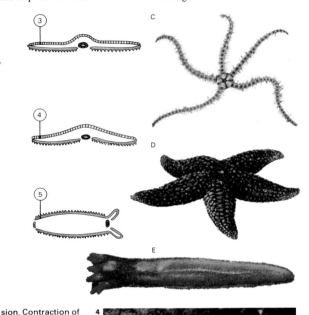

1 **Although diverse in body shape,** these creatures are all "spiny-skinned" animals or echinoderms. The five groups – sea-lilies [A], sea-urchins [B], brittle stars [C], starfish [D] and sea-cucumbers [E] – all have the same basic body plan [1 – 5], with a body pattern that has structures present in fives. They all possess locomotory tube-feet [blue] which are part of an internal system of canals filled with fluid, an external skeleton of calcareous plates embedded under the skin [yellow] and a mouth [red] to rake in food and sometimes to give out waste. They are all marine and mostly live on sand or on rocks.

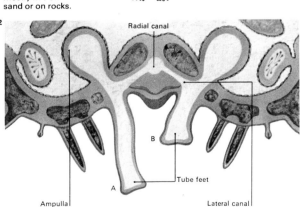

2 **Water powers the movement** of the starfish's tube feet. It is drawn through a series of tubes into a radial canal supplying each arm. The radial canal divides into lateral canals, each with a valve and ending in a bulb-like ampulla and a foot. During movement the ampulla contracts, the lateral canal valve closes and water is forced into the foot which elongates [A], swings forwards and adheres to the substrate. Longitudinal muscles then contract, shortening the foot [B] and forcing fluid back into the ampulla. Each foot is controlled by an intricate system of nerve fibres and works independently, but during forward movement all the feet in the leading arm or arms move in the same direction.

Radial canal

Ampulla

Tube feet

Lateral canal

B

A

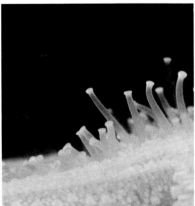

3 **The tube-feet of a sea-urchin** are seen, in close-up, to be capped by suction pads. When the feet contact a solid surface, the centre of the sucker is withdrawn, producing a vacuum and adhesion. Contraction of muscles and removal of water from the feet lifts them from the surface once more. In this way the sea-urchin can move rapidly over rocks and even climb vertical surfaces.

4 **To prise open a shellfish,** the starfish makes use of the adhesive force of its tube-feet, firmly attached to the two shell valves. By applying a strong, steady pressure the starfish opens the two shells far enough to push part of its stomach into the body of its prey and start to digest it, outside its body. Stomach juices reduce the soft parts of the shellfish to a semiliquid mass which can then be drawn into the stomach. The empty shells are finally discarded.

highly sensitive to touch [5], most of the starfish seek out shellfish. They open bivalve shellfish [4] using adhesive tubular feet. Others swallow crustaceans, molluscs and small fish, and the crown-of-thorns starfish (*Acanthaster planci*), lives on coral polyps and has greatly damaged the coral reefs in the Pacific Ocean. Sea-cucumbers extract nutrients from ingested sand.

Means of protection

With no rapid means of escape, most echinoderms have some means of protection against enemies. Most familiar, especially to unwary bathers, are the sharp spines of sea-urchins, such as those of the hat-pin sea-urchin (*Diadema setosum*), which rests on the sand in the clear, shallow water of the Great Barrier Reef and can inflict painful wounds. The spines can also be rotated like a drill and used to burrow into rocks and sand for shelter. Sea-cucumbers of the genera *Holothuria* and *Actinopyga* shoot out sticky white threads through the anus to ensnare enemies. They can also eject viscera through their mouths and later regenerate a new set.

The powers of tissue regeneration of echinoderms are remarkable. The common Pacific sea-star *Linckia* sp can lose all its arms but providing a small piece of one remains attached to the central disc, it will regenerate five new arms. The brittle star *Ophiothrix fragilis* can break into pieces if it is handled roughly, each piece regenerating into a new individual. As well as providing an escape from predators, regeneration is also a means of reproduction. A single male brittle star (*Ophiactis savignyi*) colonized a reef in the West Indies and by self-dividing formed the whole reef population.

Sexual reproduction involves separate males and females. Females shed eggs into the seawater, where they are fertilized by sperm from the male. In the spring spawning each female sea-star may release up to two million eggs. Some cold-water echinoderms brood their eggs. In a Californian sea-cucumber (*Thyone rubra*), development takes place in the coelom and the young leave through the anus. Others brood eggs on the body surface. In most echinoderms fertilized eggs develop into mobile larvae [7].

KEY

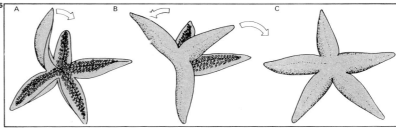

Sea-urchins are the only echinoderms in which the calcareous plates under the skin are fused to form a rigid box or test. Sharp, sometimes poisonous spines are used for protection. for bur-

rowing into rocks and, in conjunction with five columns of tube-feet, for locomotion. Between the spines, pincer-like pedicellariae prevent anything settling on the test. The mouth is equipped with five

movable teeth which rasp encrusting organisms from the rocks. Surrounding the mouth, five pairs of gills are the chief centres of respiration, exchanging oxygen for carbon dioxide.

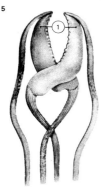

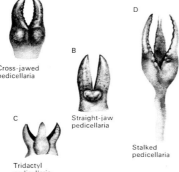

Cross-jawed pedicellaria

Straight-jaw pedicellaria

Tridactyl pedicellaria

Stalked pedicellaria

5 Jaw-like pedicellariae are found between the spines and tube-feet of sea-urchins and starfish. The stalked pedicellariae of starfish have two small bones [1] which articulate with a basal bone

in a scissor- [A] or forcep- [B] like arrangement. Sessile [C] and stalked [D] pedicellariae of sea-urchins have three jaws using a pincer-like movement. The jaws are operated by muscles; if touched

on the outside they open; touched on the inside, they snap shut. Some are poisonous and used for defence; most capture small prey and prevent organisms "lodging" on the rigid body surface.

6 If accidentally turned on its back, the common sea-star (*Asterias rubens*) is capable of righting itself. The tip of

one arm is twisted [A] so that the rows of tube-feet can grip a hard or rocky surface. With this initial foot-

hold, the rest of the arm gradually turns and moves backwards [B] so that the body folds in half. The three gripping arms

now pull the body right over, so that, in a slow somersault, the starfish has regained its normal position [C].

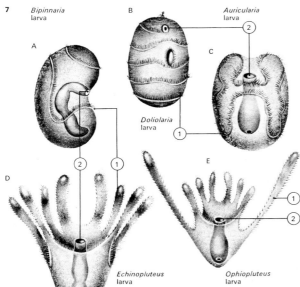

Purple sun star
Solaster endeca

Feather star
Antedon bifida

Goose star
Anseropoda placenta

Basket star
Gorgonocephalus caryi

Purple heart urchin
Spatangus purpureus

7 Larval echinoderms are the free-swimming stage between the fertilized egg and the adult. Larvae vary in appearance, but most have bands of cilia [1] to waft food into the gut [2] and to provide a means of larval dispersal. Two-week-old

starfish larvae [A] have two ciliated bands; three-day-old sea-cucumber larvae [C] have a single band, which later breaks into rings [B]. Sea-urchin larvae [D], and brittle star and basket star larvae [E] have extended larval arms.

8 The sun star and the goose star are both sea-bed dwellers and they illustrate the diversity of shapes within the class Asteroidea. The feather star lives on the European continental shelf and tempor-

arily attaches itself to rocks. The purple heart urchin, adapted to burrowing in sand, mud and gravel, is found offshore from Norway to the Mediterranean. The basket star is a deep-water inhabitant of the North Atlantic.

7 *Bipinnaria* larva

Auricularia larva

Doliolaria larva

Echinopluteus larva

Ophiopluteus larva

Invertebrate oddities

The animals without backbones, the invertebrates, are a group of creatures comprising about 95 per cent of animal species and found in every available habitat on earth. The main groups of invertebrates have been described on the previous pages, but there are, in addition, many small, bizarre groups, some of which are illustrated here. Most of these are aquatic and those that are not have often assumed parasitic life-styles inside other animals.

Although they may appear insignificant many of these "odd" invertebrates play an important part in the food chains of other animals – the marine species, for example, may be eaten by shrimps, the shrimps by fish and the fish by man. And to the zoologist the description and investigation of these animals, many of them too small to be seen with the naked eye, provides an endless source of fascination and discovery.

The invertebrates were the first animals to appear on earth and it may be that these rare groups will provide more clues about the early steps in the evolution of the higher animals, including man himself.

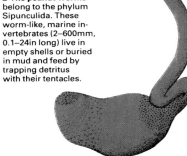

1 Tongue worm
Porocephalus annulatus

2 Peanut worm
Phascolion strombi

1 The tongue worms or pentastomids are small parasitic invertebrates that live in the lungs and nasal passages of vertebrates. They are all less than 15cm (6in) long and are grouped in a small sub-phylum of the arthropods (joint-legged animals) known as the Pentastomida. The classification of this group is based on the appearance of the adults, which have two pairs of claws.

2 The peanut worms belong to the phylum Sipunculida. These worm-like, marine invertebrates (2–600mm, 0.1–24in long) live in empty shells or buried in mud and feed by trapping detritus with their tentacles.

Flustra foliacea

4 The "bear animalcules" or tardigrades are very small invertebrate relations of the arthropods. Most are less than 0.5mm long and inhabit the film of water covering mosses and lichens.

"Bear animalcule" *Echiniscus spinulosus*

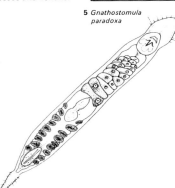

Echiurid
Echiurus echiurus

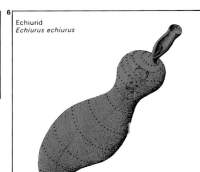

5 *Gnathostomula paradoxa*

5 Marine muddy sands are the home of minute, transparent worm-like animals of the phylum Gnathostomulida, which are related to the free-living flatworms. They are all less than a millimetre long and have heads bearing long, hair-like cilia. The body is covered with short cilia. Both male and female sex organs are found in each individual, but reproduction is by cross-fertilization. About 50 species are known.

3 The colonial, sedentary *Flustra* appears at first sight more like a plant than an animal. In fact it is made up of numerous small creatures called zooids which secrete protective skeletal shells around themselves. *Flustra* is a member of the phylum Ectoprocta (Bryozoa). Starting life as a single zooid it forms a colony by budding. The zooids remain connected internally.

6 The echiurids are a small phylum of worm-like invertebrates that live buried in sand or mud in the sea bed or inhabit rock crevices. They feed on detritus which is spread over the substratum. They trap it using a large tube-like mouth or proboscis. These worms, ranging from 2 to 600mm (0.1 to 24in) long have a similar anatomy to earthworms (Annelida) and may be related.

Nemertean worm
Lineus ruber

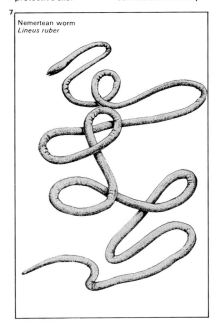

8 Priapulid
Priapulus bicaudatus

9 Arrow worm
Sagitta elegans

7 Most nemertean worms are inhabitants of the seas of the Northern Hemisphere. These unsegmented creatures, which may be up to 27m (90ft) long, live in shallow waters where they feed on both live prey and detritus. Under certain conditions the body of the worm may spontaneously break up into fragments that can stay alive and even regenerate into new individuals. A few nemertean worms live as parasites in other invertebrates.

8 The priapulids are a group of cylindrical worms that inhabit the muddy bottoms of coastal waters in the colder parts of the oceans to depths of about 8,000m (26,000ft) in both hemispheres. Apart from *Priapulus* only one other genus,

Halicryptus, is known. The relationship of these animals to other invertebrates remains a mystery, but their anatomy is somewhat similar to that of the nematodes (threadworms). The species illustrated here measures 5cm (2in) in length.

9 The arrow worm is a member of a small phylum known as the Chaetognatha and is common in the sea's plankton in all the oceans of the world. Although they usually move only in ocean currents, these animals are capable of swimming. They vary in length from 2.5–10cm (1–4in) and each worm contains both ovaries and testes. Reproduction is by self-fertilization and the first stages in the development of the embryo take place inside the body of the adult worm.

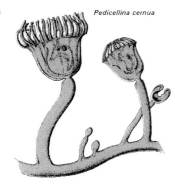

10   *Pedicellina cernua*

11 Rotifer *Trichotria tetractis*

11 The rotifers or "wheel animalcules" are microscopic creatures found chiefly in freshwater lakes and ponds but also in the sea. They are less than 2mm (0.1in) long and belong to the phylum Aschelminthes. They feed on other small floating organisms.

13 *Moniliformis moniliformis*

12

Phoronid
Phoronis architecta

10 Pedicellina is a small invertebrate that lives in the sea. It is only 2–3mm in height and lives attached by a short stalk to rocks, shells and pieces of wood, or to other animals such as crabs and sponges. It is a member of a small phylum, the Endo-procta, and forms colonies that look rather like those of the coelenterate *Obelia*. A ring of up to 24 tentacles surrounding the "body" or calyx is used for catching floating food particles. The calyx contains the digestive and reproductive organs.

12 The phoronids are a group of sea-dwelling, detritus-feeding worms of which only about 15 species are known. Most are less than 200cm (8in) long and live in tough tubes of chitin attached to rocks or buried in sand. Their food consists mostly of detritus.

13 Moniliformis is a parasitic invertebrate worm belonging to the phylum Acanthocephala, a small group of animals consisting of about 90 genera and 600 species. All are parasites and all have simple bodies but complex life cycles which involve several hosts, one of which is usually an insect. The primary hosts are vertebrates and may be freshwater or marine fish, birds, snakes or rodents. Some of these parasites grow to 50cm (20in) but most average 1–2cm.

14

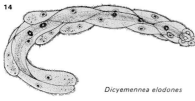

Dicyemennea elodones

14 Dicyemennea is a tiny parasitic invertebrate found in flatworms, molluscs, earthworms, starfish and squids. It belongs to the phylum Mesozoa, whose origins are still uncertain.

16

Pogonophoran
Lamellisabella johanssoni

15

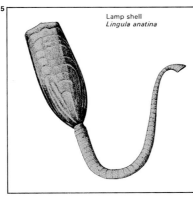
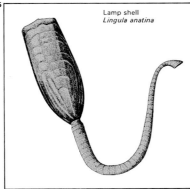

Lamp shell
Lingula anatina

15 The lamp shells, marine invertebrates that look much like molluscs, are grouped in the phylum Brachiopoda. They live attached to rocks or, like *Lingula*, buried in mud or sand. Modern brachiopods are mostly less than 5cm (2in) across and number about 260 species but the group had its heyday in the Devonian, some 370 million years ago. More than 30,000 fossil species have been found and described to date.

16 The pogonophorans inhabit chitinous tubes on the deep ocean bed. Their bodies are divided into three regions; the protosome [2], bearing tentacles [1], a short mesosome and a long "trunk" or metasome. The first specimen of a pogonophoran was discovered in 1900 but the group was not accorded its own phylum until 1955. The 80 known species are from 5cm (2in) to 35cm (14in) long.

17

Kinorhynch
Echinoderes dujardini

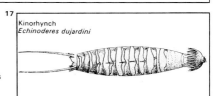

20

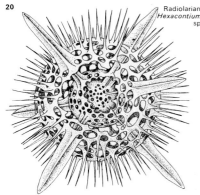

Radiolarian
Hexacontium
sp

18

Foraminifera

18 The shells of Foraminifera, when enlarged, look like snail shells. In fact they are made by minute single-celled creatures that spend their lives among, and feed upon, diatoms.

17 The kinorhynchs are microscopic aquatic animals of the phylum Aschelminthes. The cylindrical kinorhynch body is divided superficially into 13 joints or zonites. The head is retractable and bears long spines. Biologists think that these creatures are close relations of the rotifers for, like them, they possess adhesive glands. Most are less than 1mm (0.04in) long.

19 Venus's flower basket is a deep-sea sponge (phylum Porifera) which grows to 25cm (10in) in length. The skeleton is of silica. A type of shrimp often lives trapped within the body of the sponge.

19

Venus's flower basket
Euplectelle aspergillum

22 Calcareous sponge
Leucosolenia botryoides

20 The radiolarian *Hexacontium* has an ornate skeleton composed of three concentric spheres that are pierced by hundreds of radiating spicules. The skeleton is very hard and made of silica. Radiolarians divide asexually simply by splitting in half. These microscopic animals have existed for millions of years. Their fossil skeletons have been found in rocks of the Tertiary period, about 65 million years ago, and are used as abrasive, siliceous powders.

21 Bath sponge
Spongia mollissima

21 The bath sponge is an inhabitant of warm seas such as the Caribbean and the Mediterranean. After harvesting, the sponge is dried, beaten and washed to remove hard debris so that the only part remaining is the fibrous spongin "skeleton".

22 Shallow coastal water of the Atlantic is the home of this calcareous sponge. Its name is derived from its supporting skeleton, which is composed of calcium carbonate spicules. Each vase-shaped part is about 5cm (2in) in height.

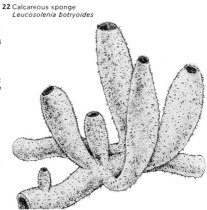

Threshold of the vertebrates

All animals with backbones, including man himself, are chordates. The essential feature that all chordates share, and after which they are named, is the notochord, a stiffening rod running the length of the body [Key]. Animals with backbones are known as vertebrates and are descended from a line of creatures that appear small and insignificant. These animals, the early chordates, possibly arose in the early Cambrian some 570 million years ago. Their exact ancestry is still a mystery, but they are probably related to echinoderms – the starfish.

Chordate characteristics

The chordate notochord is the forerunner of the backbone and is the basis for the attachment of regularly arranged muscles, the myotomes. Above it lies a tubular nerve cord, the anterior or "head" end of which is enlarged and folded to form a brain. In addition all true chordates show some evidence of paired gill openings and a tail.

The modern remnants of the early chordates, or protochordates, from which all the animals with backbones arose, are a few highly specialized "left overs" of a group that millennia ago was probably numerous and highly successful. These comparative rareties are classified in three sub-phyla known as the Hemichordata (acorn worms), the Urochordata (the sea squirts) and the Cephalochordata, whose sole representative is the lancelet or amphioxus. The most primitive representives of the "vertebrates proper" are the hagfish [12] and lampreys [10, 11].

Acorn worms and sea squirts

The acorn worms, marine mud burrowers, are creatures whose physical make-up has some features of the echinoderms and some of the true chordates. Some acorn worms have the tubular nerve cord and most have the gill slits typical of chordates but the slits are used for feeding rather than breathing; and although acorn worms possess an internal structure that looks like a notochord this is formed in the embryo in quite a different fashion from a true notochord. The adult worm may resemble a simple chordate but the larval stages of its development [6] are almost identical to those of starfish and sea urchins and this is evidence for citing the echinoderms as vertebrate ancestors.

Sea squirts [4, 13], peculiar sac-like animals most of which spend their sedentary lives attached to the sea-bed, have few typically chordate features except for gill slits. It is the free-swimming sea squirt "tadpole larva" that reveals the sea squirt's place in the chordate line, for some zoologists argue that vertebrates may have arisen by a process (known as neoteny) in which the tadpole larva [Key] did not mature into an adult [7] but developed sex organs of its own.

The sea squirts are so called because many of them push out a jet of water when they are disturbed. Water is constantly drawn into and pushed out of the sea squirt body, and food and oxygen are removed in the process. The other popular name of "tunicate" comes from their inert cellulose sac or tunic.

Of all the protochordates the lancelets are the most fascinating, because these small creatures, most of which are sand burrowers on the beaches of the warmer seas, have simple bodies bearing all the hallmarks of the chordates. Running nearly the whole dis-

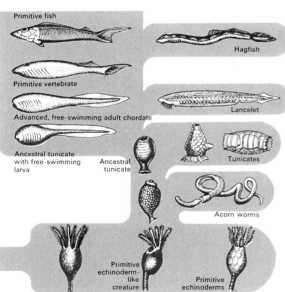

1 **The exact course** of early vertebrate evolution is impossible to trace but protochordates provide what clues there are. Similarities between larval acorn worms and echinoderms suggests that they descended from echinoderm-like creatures. The tunicates may have been the first creatures to have gill slits but a more significant step seems to have been the development of sex organs in tunicate "tadpole" larvae and of a notochord. Free-swimming adult chordates then appeared and the lancelet is most probably a side branch from the mainstream of evolution which resulted in the fish, the first of the true vertebrates.

2 Acorn worm *Ptychodera flava*

2 **The acorn worm** lives in mud and sand on the bottom of inshore marine waters, trapping detritus and plankton in its acorn-shaped proboscis. It is classified as a hemichordate.

3 Water current   Worm cast

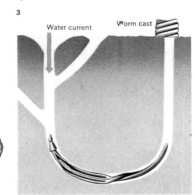

3 **Worm casts** on the beach can be made by burrowing acorn worms such as *Balanoglossus*. The worms live in tubes in the sand, the walls of which the worm secretes. As a result of its feeding habits the worm's alimentary tract becomes filled with sand and it is this that is discharged at low tide to form the familiar curled worm cast. Acorn worms vary from 2cm (0.75in) to 2.5m (7.5ft) long.

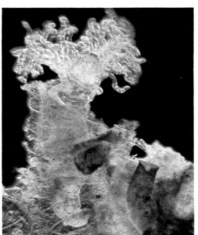

4 Sea squirt *Tunicata* sp

Anus
Gut

4 **The sedentary sea squirt** feeds by a filter mechanism. Water containing food and oxygen is drawn in a steady stream through one siphon [1]. It then enters the pharynx [2] whose aperture is protected by a ring of tentacles [3]. The internal water current is created by hair-like cilia lining the gill slits [4] and passes through them into the atrium [5] and out via the atrial siphon [7]. Mucus, secreted by the endostyle [6], traps food particles. This mucus is then rolled into a rope and passed into the gut for digestion. The anus discharges waste into the atrial siphon.

5 **A branched tube is the home** of each individual of *Rhabdopleura*, a colonial relative of the acorn worm. New individuals are produced by asexual budding from the creeping base of the animal, the stolon. *Rhabdopleura* is about 5mm (0.20in) long.

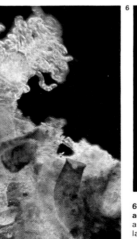

6 **The larva of the acorn worm** is known as a tornaria. This larva is one of the mainstays of the evidence that the early chordates, the vertebrate ancestors, evolved from echinoderms, for the tornaria is so like the larvae of some starfish that for many years it was mistaken for one of them. The significant difference is that the tornaria has the gill clefts typical of chordates. The tornaria is part of the zooplankton and is seen here magnified over 30 times.

tance from head to tail is a notochord below the typically hollow chordate nerve cord. Behind the lancelet mouth on each side of the body lie more than 100 pairs of gill slits. These are used partly for filtering food and partly for extracting oxygen from seawater.

Like other protochordates the adult lancelet does not develop from the egg directly but is the result of metamorphosis from a larva. These larvae, which live in the plankton, may sometimes develop signs of sexual organs – further evidence that true vertebrates may have evolved from larval forms that became mature without the intervention of metamorphosis.

Hagfish and lampreys

The earliest true vertebrates to appear in the Ordovician period were fish-like creatures that were possibly ancestral to the hagfish [12] and lampreys [10] – the most primitive vertebrates known today. Neither the hagfish nor the lampreys have jaws but both have a skeleton made of cartilage and have large notochords as well as gill slits. All lampreys and hagfish have tubular nerve cords lying

above the "spine". The head of the lamprey bears organs of taste, smell and hearing and well-developed eyes while the hagfish head has a cluster of sensory tentacles round the mouth but poorly developed eyes.

Hagfish and lampreys are far from the most attractive of fish, in looks and in habits. The hagfish are ocean scavengers, feeding on any dead or dying fish, crustaceans or molluscs they can find. On locating a dead fish the hagfish actually enters it through the gills or anus and devours the body contents leaving behind only skin and bones.

The world's species of lamprey can be divided into two groups according to their feeding habits. One group is parasitic and individuals attach themselves to their hosts with huge sucker-like mouths, break through the flesh with rasping teeth [11] and suck out the blood. When the lamprey has eaten its fill it releases its oral grip and in doing so may inflict a fatal wound. Free-living lampreys live only a few months. They feed normally as larvae but after metamorphosis the gut degenerates. The adult cannot feed and lives only long enough (a few months) to spawn.

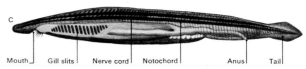

KEY

All chordates conform, at some stage in their life cycle, to a generalized body plan [A]. This has a stiffening rod, the notochord, above

which is situated a tubular nerve cord. Gill slits behind the mouth are used in respiration and there is a post-anal tail. In the sea squirt

chordate features are obvious [B] in the "tadpole" larva's tail. The lancelet or amphioxus [C] has all the chordate characteristics.

---

7 The adult sea squirt [C] is formed through metamorphosis of the "tadpole" larva which attaches itself [A] to a firm substrate then undergoes gradual maturation [B] using the yolk [1] for food.

7 A

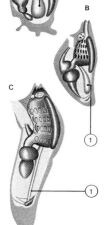

B

C

8 The lancelet, a resident of sandy shores in temperate and tropical seas, is placed in the sub-phylum Cephalochordata. It has a fish-like body but a notochord rather than a true backbone.

8 Lancelet or amphioxus
Branchiostoma lanceolatum

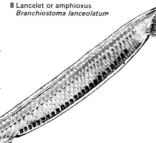

9 A

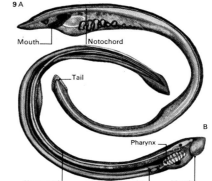

Mouth — Notochord

— Tail

Nerve cord — Pharynx — Gut — Mouth — B

9 The ammocoete larva, the immature lamprey [B], is especially interesting because it resembles embryos of higher vertebrates. Unlike the larva of amphioxus [A] it has a heart, eyes and ears as in vertebrate embryos, as well as such typical chordate features as the notochord. Until metamorphosis was observed this was thought to be a species quite distinct from the lamprey.

10

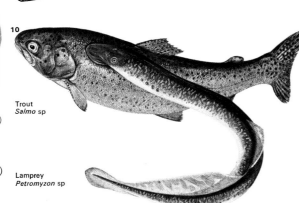

Trout
Salmo sp

Lamprey
Petromyzon sp

10 The parasitic lamprey feeds by blood-sucking from its fish host until gorged. To enhance the blood-flow from its victim the lamprey injects a chemical that prevents blood clotting.

11 Rows of rasping teeth arm the inside of the lamprey's mouth. These are used to penetrate a fish's blood circulation while the perimeter of the mouth acts as a suction device.

11

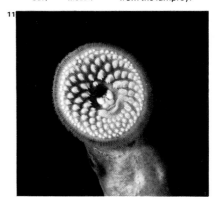

12

Slime hag
Bdellostoma sp

12 Hagfish scavenge the flesh of dead and dying fish using sucking mouths and rasping tongues. At the same time they produce vast quantities of mucus from pairs of slime glands and this is thought to protect the hag and to make its prey die more quickly. All the 21 known species of hagfish are marine and have retained, over millions of years, the basic body organization of the early vertebrates, the first animals to have backbones.

13

13 The star sea squirt (*Botryllus schlosseri*) is seen here encrusting the brown seaweed *Fucus*. This tunicate is a colonial species and classified in the sub-phylum Urochordata and the class Ascidiacea. The colony of tunicates may share the same body covering – the test or tunic that gives them their name. Several *Botryllus* individuals share one water exit or exhalant siphon. Each exit is surrounded by 3 to 12 "petals", the inhalant siphons.

The classification of fish

Apart from the primitive jawless fish or Agnatha – hagfish and lampreys – all fish belong to one of two great classes: the cartilaginous fish or Chondrichthyes, and the bony fish or Osteichthyes.

Cartilaginous fish
On the evolutionary scale [Key] the cartilaginous fish are the most primitive of the two classes. They include the sharks, skates and rays, and their skeletons are composed not of bone but of gristly cartilage. Some dwell on the sea-bed while others swim in mid and open waters.

There are about 620 species of cartilaginous fish, divided zoologically into three groups: the typical sharks (Elasmobranchii), the skates and rays (Batoidea) and a rather odd-looking group of uncertain ancestry, the chimaeras (Holocephali). A characteristic of most cartilaginous fish is the heterocercal tail whose top half is longer than the bottom [1]. To balance the effect of this on swimming ability the shark has fixed pectoral fins and a flattened head.

Most sharks are fast-moving hunters and are by nature fish-eaters, but they will, in exceptional circumstances such as severe hunger, attack and savage mammals, including humans. Skates and rays are found patrolling the ocean floor feeding on sedentary shellfish.

The largest of the cartilaginous fish are ironically the least ferocious and represent no threat to most of their undersea neighbours. The whale shark, which reaches a length of 18m (60ft), the basking shark 14m (46ft) and the awesome manta ray, which has a "wing span" of 6m (20ft), feed only on plankton and other minute marine animals. For all their immense size the whale and basking sharks feed on small creatures and have comb-like structures on their gills through which they strain their food.

Most cartilaginous fish, such as the blue shark and smooth dogfish, give birth to live young, but some lay large, yolky eggs which, before being laid, are individually encased in a tough, leathery cover. After the eggs have hatched, the empty cases can often be found on the sea-shore and are popularly known as "mermaids' purses". Fertilization is internal

and the male has modified pelvic fins, the claspers, with which he holds on to the female during mating.

All cartilaginous fish, with the exception of sawfish and some species of ray, live in the sea. By contrast the 20,000 species of bony fish are found in both sea and fresh water throughout the world.

Classification of bony fish
Bony fish belong to two sub-classes, the Crossopterygii and the Actinopterygii. Very few of the former have survived to modern times, for they are at the base of the evolutionary branch that led to land animals. The best known of them is the coelacanth (*Latimeria* sp) [2], which was believed to be extinct until one was fished up from 67.6m (222ft) off the African coast in 1938. Its lobed fins have fleshy bases that actually look like the beginning of a limb. The other surviving crossopterygians are the lungfish [3], which can breathe air into a primitive version of lungs. They belong to the order Dipnoi and are found in tropical conditions in Australia, Africa and South America.

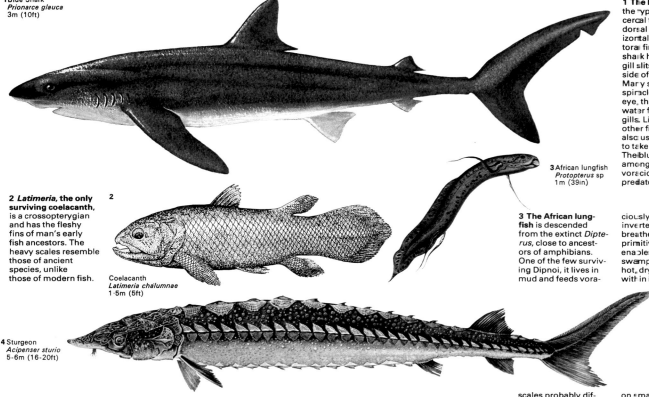

1 Blue Shark
Prionarce glauca
3m (10ft)

3 African lungfish
Protopterus sp
1m (39in)

2 *Latimeria*, **the only surviving coelacanth**, is a crossopterygian and has the fleshy fins of man's early fish ancestors. The heavy scales resemble those of ancient species, unlike those of modern fish.

Coelacanth
Latimeria chalumnae
1·5m (5ft)

1 The blue shark has the typical heterocercal tail, pointed dorsal and horizontally held pectoral fins. This shark has five gill slits in each side of the body. Many sharks have a spiracle behind each eye, through which water flows to the gills. Like most other fish the shark also uses its mouth to take in water. The blue shark is among the most voracious of all predatory fish.

3 The African lungfish is descended from the extinct *Dipterus*, close to ancestors of amphibians. One of the few surviving Dipnoi, it lives in mud and feeds voraciously on fish and invertebrates. It breathes through a primitive lung. This enables it to live in swamps and survive hot, dry summers with in its cocoon.

4 Sturgeon
Acipenser sturio
5-6m (16-20ft)

4 The sturgeon belongs to the Chondrostei, the most primitive group of actinopterygians or ray-finned fish. Its heterocercal tail and cartilaginous skeleton are reminiscent of the shark's. Some species also have a spiracle. Its scales are shiny and large (known as ganoid) and resemble those of the early bony fish.

The sturgeon is not a fierce fish; it swims at the bottom of the sea, snuffling out invertebrate food with its long, sensitive snout or rostrum. The adult fish swims up the river to lay its much-prized eggs. Caviare is processed from the ovary before the fish can spawn.

5 The alligator gar belongs to the Holostei, a group of actinopterygians that lived in the Triassic 225 million years ago. Its thick scales probably differ little from those of its ancestors. Like them, it has a short, symmetrical tail that is a forerunner of the homocercal tail of teleosts. The gar feeds on small fish, which it catches with its well-developed lower jaw. The alligator gar travels along the surface of the water using its tail like an outboard motor propeller.

5 Alligator gar
Lepisosteus spatula
3m (10ft)

The Actinopterygii are the typical modern ray-finned fish. The most primitive, which are also the most shark-like of the bony fish, belong to the infra-class Chondrostei. They share some characteristics with the cartilaginous fish, such as the heterocercal shape of the tail, and they are bottom-dwelling scavengers. But their eggs, unlike those of the cartilaginous fish, are small and fertilized externally. This group includes the highly prized sturgeon [4].

The infra-class Holostei, including bowfin and garfish [5], now contains only the freshwater remnants of a once large seawater group. Its members are fast swimmers, usually with truncated heterocercal tails.

The largest group of ray-finned fish are the members of the infra-class Teleostei. They are the culmination of the evolutionary line and seem to be prefectly adapted to life in water. Their tails are completely symmetrical and with the buoyancy imparted by their swim bladders they do not need rigid paired fins. Many species have developed a streamlined form for maximum speed and minimum friction while swimming, but the locomotion of each type of teleost is adapted to its mode of life. Thus the flatfish "creep" along the sea-bed while the freshwater pike is built for speed and manoeuvrability.

Diverse animals

Teleosts have probably evolved along three main lines to give rise to eight super-orders of living fish. The first line includes the eels (Elopomorpha) [9] and the prolific herring (Clupeomorpha). The second line consists of peculiar tropical freshwater fish (super-order Osteoglossomorpha). The salmon and trout (Protacanthopterygii) [6] belong to the most primitive group of the third line while most freshwater fish, including carp and roach, belong to a more advanced group, the Ostariophysi [7]. The cod and angler fish (Paracanthopterygii) and strange creatures such as flying fish (Atherinomorpha) are also advanced groups but the final super-order, the Acanthopterygii, whose members are typically fish with spiny fins, is much the largest and most diverse. It includes the stickleback [8] and the seahorses, the perch, mackerel, flatfish and puffer fish.

Living fish are classified into two major groups or classes and one minor one. The smallest group is the Agnatha, the primitive jawless fish which include the lampreys and hagfish. From these evolved the cartilaginous fish, the Chondrichthyes, which are further divided into three sub-classes. About 400 million years ago the second large group, the bony fish or Osteichthyes, branched from the cartilaginous fish. Most modern fish are classified in the sub-class Actinopterygii, the main subdivision of the Osteichthyes. This is made up of three infra-classes, the largest of them being the Teleostei.

6 The grayling is a teleost of the Northern Hemisphere and belongs to the salmon group, the super-order Protacanthopterygii. It has an unusually tall and long dorsal fin and its colouring is very variable. During spawning the dorsal, caudal and anal fins become deep purple in colour. The adults are solitary, but juveniles do form shoals. The Latin name comes from the fish's smell.

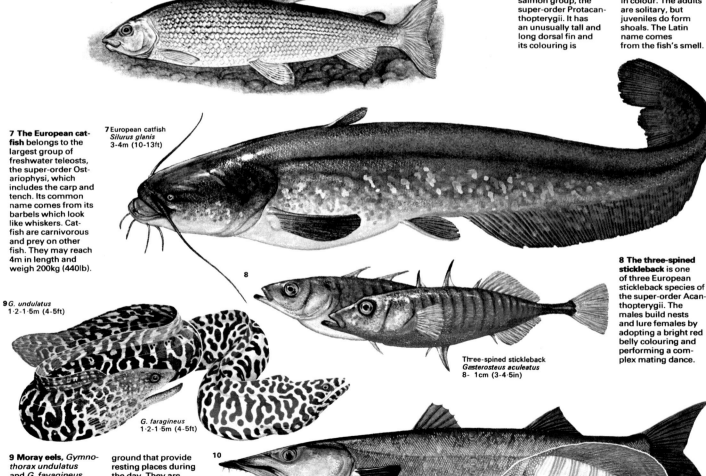

6 Grayling
Thymallus thymallus
60cm (24in)

7 European catfish
Silurus glanis
3-4m (10-13ft)

7 The European catfish belongs to the largest group of freshwater teleosts, the super-order Ostariophysi, which includes the carp and tench. Its common name comes from its barbels which look like whiskers. Catfish are carnivorous and prey on other fish. They may reach 4m in length and weigh 200kg (440lb).

9 *G. undulatus*
1·2-1·5m (4-5ft)

G. faragineus
1·2-1·5m (4-5ft)

9 Moray eels, *Gymnothorax undulatus* and *G. favagineus*, belong to the superorder Elopomorpha which includes the congers, the largest of the eels. Moray eels are found in all tropical seas and favour rocks and areas of broken ground that provide resting places during the day. They are largely nocturnal in their habits and seldom move during the day except to poke their heads out of their hiding places and snap at passing prey. They can inflict severe bites.

8 The three-spined stickleback is one of three European stickleback species of the super-order Acanthopterygii. The males build nests and lure females by adopting a bright red belly colouring and performing a complex mating dance.

Three-spined stickleback
Gasterosteus aculeatus
8- 1cm (3-4·5in)

Northern barracuda *Sphyraena borealis* 45cm (18in)

10 The northern barracuda and a sail-finned surgeon fish appear to have little in common, but both are members of the order Perciformes, the largest order of spiny-finned fish.

Sail-finned surgeon fish
Zebrasoma velifexum 30cm (12in)

The life of fish

Fish can be thought of as the most successful of the vertebrates, the animals with backbones. They are not only more numerous than all other vertebrates, but there are also more species of fish – probably not less than 23,000 of them. Fish vary widely in shape and habits. Some live in sea water and some in fresh; some lurk in the depths while others swim just below the surface; some feed peacefully on seaweed or plankton and some on marine invertebrates and many are aggressive predators that feed on other fish or even on amphibians or land animals.

The bodies of fish

All fish breathe by pumping water past their gossamer-thin gills [4], whose numerous folds offer a large surface area for the intake of oxygen in exchange for waste carbon dioxide. Water is pumped by movements of the mouth and pharynx and, in bony fish, by the opercula (gill covers), in the opposite direction to the flow of blood. As a result the blood is highest in oxygen and lowest in carbon dioxide when it meets the freshest water, with the highest oxygen content.

The typical fish shape has evolved over millions of years to allow for maximum speed and agility in the water. The most predatory of the bony fish are the best swimmers: they can cruise at speeds of between three and six times their body length per second and can turn within one body length. The evolutionary breakthrough for the bony fish came with the development of air bladders to keep them afloat. The cartilaginous fish such as sharks, whose skeletons are composed of "gristle" or cartilage instead of bone, do not have air bladders and sink if they stop swimming. Their "shoulder" or pectoral fins give them lift, but many cartilaginous fish have become bottom-dwellers. The bony fish, however, released from the constant need for lift, can use their pectorals as brakes or paddles for swimming backwards; this adds to their flexibility in movement and allows them to feed in a much greater diversity of niches.

Fish vary greatly in speed and staying power and their different abilities are reflected in their muscle proteins. As a result, red meat comes from fast swimmers such as tunny, or powerful fish with endurance such as migrating salmon, and white meat from slow-moving flatfish such as sole.

Modern fish also have the advantage of having shed the protective heavy armour of their ancestors. Cartilaginous fish, being predators, have no need of such armour and have developed, instead, a tough, abrasive skin. In bony fish the armour has been refined into the familiar light, delicate coat of overlapping scales that protects the fish without hampering its movements.

The senses of fish

For co-ordination of movement when hunting, fleeing or shoaling (for mutual protection), a highly developed set of receptors has evolved which keeps fish informed of their environment. Sharks, for example, have an acute sense of smell for locating prey. Most fish have keen eyesight and react readily to the yellows and greens of their watery world. Many have good hearing, which is used socially to pick up mating or shoaling noises, or sometimes as part of a kind of echo-location system in which the fish's own sounds help in the detection of

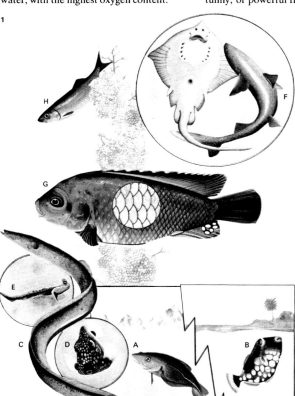

1 **Fish share common features,** but body temperatures vary with water temperature. Thus the cod (*Gadus morhua*) [A] has a lower temperature than the tropical triggerfish (*Balistoides conspicillum*) [B]. Fish swim by means of muscular bodies and tails such as those of the American eel (*Anguilla rostrata*) [C], and by fins as can be observed in the trunkfish (*Lactophrys triqueter*) [D]. The mudskipper (*Periophthalmus chrysospilos*) [E] has specialized pectoral fins for moving across the mud. Gills, that appear as five paired openings in the ray (*Raja clavata*) and the dogfish (*Scyliorhinus caniculus*) [F], are used for breathing. Typical bony fish (teleosts) such as the cichlid (*Labeotropheus fulleborni*) [G], have skin covered with bony scales. Most fish lay eggs: the herring (*Clupea harengus*) [H] probably lays about 50,000 at a time.

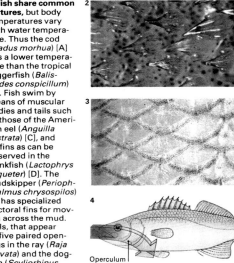

2 **Fish with gristly skeletons,** such as sharks, have tough, flexible skins. The scattered, thorn-like scales (denticles) that grow on their skin are similar in structure to the teeth lining their jaws.

3 **Bony fish have** thin, overlapping cycloid or ctenoid scales that protect them from predators but do not hamper movement. Ctenoid scales have spines on the rear edge for extra protection.

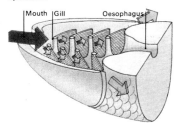

Operculum

Mouth | Gill | Oesophagus

4 **Gills** are the breathing organs of most fish. When a fish breathes it opens its mouth, draws in water, then shuts its mouth again. This forces a continuous stream of water [arrowed] through the gill slits, over the gills and out into the surrounding water. Oxygen from the water is absorbed into blood vessels in the gills while carbon dioxide is carried out by the expelled water.

5 **Members of the shark group** (Elasmobranchii), like most fish, thrust through the water by means of a wave of muscular contraction that spreads down the body. Up to 40 per cent of this thrust may be supplied by the tail, which also helps to keep the fish swimming in a straight line.

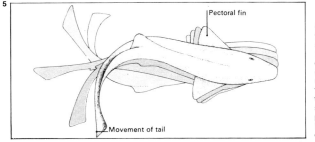

Pectoral fin

Movement of tail

Pectoral fin

6 **Bottom-dwelling fish** are much poorer swimmers than their more streamlined cousins. Most are sedentary bottom-feeders relying on camouflage for protection. If threatened, rays [A] swim by flapping their pectoral fins and flatfish [B] by undulating their dorsal and anal fins.

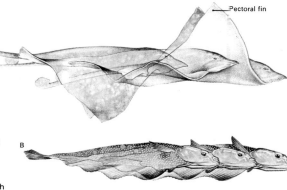

objects in the water. The hearing mechanism forms part of the labyrinth, an organ essential to all fish, that signals position in space and angular acceleration and is crucial to the fish's balance when swimming. Fish also possess a unique organ that puzzled land-dwelling man for a long time. This is the so-called lateral line [10], that works on a similar principle to the vertebrate ear, but instead of detecting sound waves in air it picks up pressure waves due to movement in water. This organ gives the fish a kind of "distant touch sense" for remote objects.

All this information is pooled in the central nervous system where special centres are built on to the basic brain regions [11] that deal with automatic functions such as respiration and heartbeat. The ears, labyrinth and lateral line are linked to the hindbrain by the cranial nerves. The large olfactory bulb, the organ of "smell" or chemical reception, is joined to the cerebrum in the forebrain. Chemoreception is thought to be very important to fish for successful navigation, feeding and mating. The most advanced part of a fish's brain, where

behaviour that has been learned is controlled, is the optic lobe, which is connected to the eyes. The cerebellum has the task of co-ordinating sensory information for the fine control of movement.

How fish reproduce

Fish use various methods of reproduction [12]. Some reproduce by fertilizing eggs within the body; in some fish the female first lays the eggs and these are then fertilized by the male outside the body; and a few fish are even hermaphrodite. But whatever their method of reproduction, fish are enormously prolific. A cod may produce eight million eggs at a time, and most fish produce tens of thousands. The young are usually microscopic and exist at first in the form of animal plankton. Most of them perish before reaching adulthood, but nevertheless many survive. Scientists have estimated, for example, that there are about a million million herrings in the Atlantic. The teeming seas not only signify the tremendous success of the fish as an animal, but also provide a rich and often vital source of protein for man.

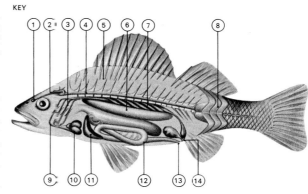

KEY

Fish, the first of the vertebrates, have a backbone [4] that keeps the body rigid against the powerful contractions of the muscles (myotomes [8]) during movement. Fins are based on supporting rods [5] that may be made of cartilage or formed from modified scales. A bony fish has an air bladder [6] above the gut [12]. The ventral aorta [9] takes blood to the branchial arteries [3] which are protected by the opercula. The brain [2] is quite well developed and the olfactory bulb [1] is particularly prominent in sharks. The kidneys [7] lie paired under the vertebral column. The liver [11] is located behind the heart [10]. The gut empties at the anus [13] just in front of the urogenital opening [14].

7

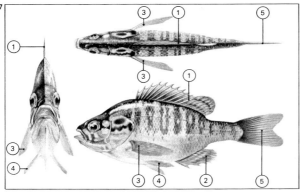

7 The fins of bony fish serve as fine controls for movement. The dorsal [1] and anal [2] fins prevent rolling. The pectoral fins [3] often serve as brakes and the pelvic fins [4] control a tendency to pitch upwards as the fish slows down. The paired fins also control rising and diving. They are used to produce rolling movements. The caudal [5] or tail fin serves as an extremely efficient rudder.

8

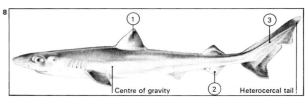

Centre of gravity   Heterocercal tail

8 The shark's fins serve to stabilize its body as it swims. The dorsal [1], anal [2] and caudal [3] fins prevent yawing (deviating off course). Most important for the shark-like fish, which have no air bladder to endow buoyancy, are the tail and the pectoral fins which give lift and keep the nose level.

9 The ear of a fish serves for hearing and for positional sense. The fluid in three semicircular canals [1] shifts in response to changes in movement and transmits this to three ampullae [2]. There sensory cells that transmit the message to the central nervous system are stimulated. For hearing, otoliths [3] are moved by sound waves [4] from the air bladder [5], transmitted in some fish by a chain of ossicles [6].

9

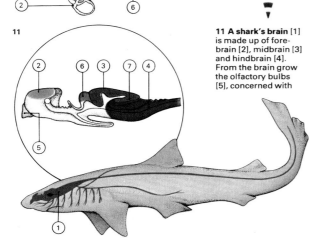

10 The lateral line organs [1] transmit information about the movement of water. The lateral line itself (shown on a red mullet) runs from head to tail on either side of the fish. It consists of a fluid-filled canal [2] with pores [arrowed] opening to the water through scales [3]. Behind each pore is a sensory organ, the neuromast [4]. This is made up of a gelatinous mass, the cupula [5], with a cluster of sensory hair cells [6] whose fibres combine in a nerve [7] running to the brain.

10

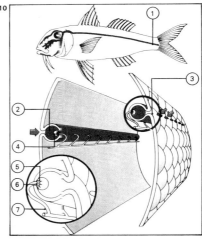

11

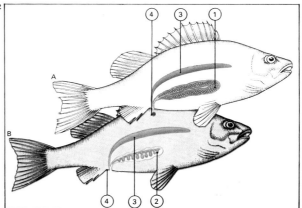

11 A shark's brain [1] is made up of forebrain [2], midbrain [3] and hindbrain [4]. From the brain grow the olfactory bulbs [5], concerned with smell, the optic lobes [6], concerned with sight and the cerebellum [7], which co-ordinates both incoming sensory data and movement.

12 The reproductive organs of the perch are typical of bony fish. The ovary [1] of the female [A] and the testis [2] of the male [B] are entirely separate from the kidney [3] but may expel their products through the same opening, the cloaca [4]. The eggs are fertilized outside the female's body.

12

Unusual fish

Fish have evolved, over a period of 400 million years, an amazing variety of special adaptations to the different environments in which they live. Some of the deep-sea fish – those that lurk at least 3,000m (9,850ft) down in the viscous abyssal depths – display the most bizarre shapes. In contrast the tropical fish of the coral reefs [1] have some of the most brilliant colouring to be found anywhere in the animal kingdom. Each peculiarity, ugly or beautiful to human eyes, has its own vital significance to the fish.

Defence and attack
Deep-sea fish have two special problems: the high density of water more than 180m (600ft) down, which they overcome by having a lightweight skeleton; and the gloom, which, for many, is adequately coped with by the possession of luminous organs. Some fish provide their own luminosity through photophores, which are modified mucus glands. But others use light from luminous bacteria that colonize certain organs where the fish can expose or reveal them, for example, by a fold of skin. Luminous organs are exploited

by abyssal fish to attract and lure not only members of the same species and opposite sex but also other species on which they feed.

Many deep-sea fish have weak jaws with large teeth, in contrast to some of the coral reef fish that have powerful armoured jaws. One group of these, the parrot fish, bite off chunks of coral which are ground up and expelled, after passing through the gut, as a cloud of chalk-dust. The vivid iridescent colours of these fish are ingeniously made up from the interplay of a few pigments – black, red, orange or yellow. This remarkable range of colours is involved in sexual display but poisonous species may also use distinctive coloration as a warning to predators.

A notoriously voracious predator is the small piranha [9], which has formidable teeth and an aggressive temperament so that it actually poses a threat to species as large as man. The puffer fish [3] protects itself, as its name suggests, by puffing itself up in order to appear twice its real size and perhaps to deter attempts by other species to swallow it.

The puffer manages a temporary change in its shape by a special adaptation of its

gullet, but many fish, such as the seahorse [6], have evolved more profound and permanent adaptive changes in form. A fairly typical feature of bottom-dwelling deep-sea fish is the possession of a long thin tail, which, by increasing the length of the lateral line helps the fish detect prey more easily. At the other end of the scale is the flying fish [2], the spectacular adaptation of whose pectoral fins allows it to glide for several metres through the air when propelled from the water by its tail. Usually the fish can rise no higher than 1.5m (5ft) above the surface of the water but in favourable conditions can increase this height to more than 7m (23ft).

Adaptations of the sexes
Sometimes through evolution of form, the two sexes of one fish species become very different. The most grotesque example of such sexual dimorphism is that of the deep-sea fish *Photocorynus* in which the male never grows larger than about 10cm (4in) and leads a parasitic life permanently attached to the female [5]. The virtue of this arrangement is that the female does not have to scour the

1 The fish of tropical coral reefs are often brightly coloured. Many, such as the moorish idol (*Zanclus canescens*) [A] and the longnose butterfly fish (*Chelmon rostratus*) [D], are striped for camouflage. The forceps fish (*Forcipiger longirostris*) [E] has false, rear "eyes" to confuse predators such as the triggerfish (*Balistipus* sp) [B]. Safety from predators is, for the clown anemone fish (*Amphiprion percula*) [F], provided by the shelter of a sea anemone's arms and, for the cleaner wrasse (*Labroides* sp) [G], by its eating of the parasites that infest the predators. The parrot fish (*Scarus* sp) [C] feeds on coral which it bites off in large pieces with its hard, bony jaws.

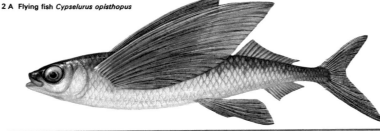

2 A Flying fish *Cypselurus opisthopus*

2 The flying fish [A] has developed wing-like pectoral fins that enable it to glide through the air. To emerge from the sea [B] it holds its fins close to its body, takes off from the crest of a wave then spreads its fins to climb and glide.

3 Blue trunk fish *Ostracion sp*

3 The blue trunk fish and its relative the globe or puffer fish have unique defence mechanisms. The body of the trunk fish is encased in solid bony armour while the puffer fish can inflate itself into a ball shape when danger threatens.

Puffer fish *Spheroides spengleri*

4 The female of the African cichlid fish carries her fertilized eggs in her mouth. After about 12 days the eggs hatch and the young fish form a school around her head. If danger threatens the female "calls" her offspring back into her mouth. After about five days the young depart.

4 African cichlid fish *Haplochromis burtoni*

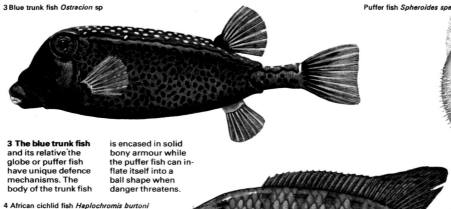

dark ocean depths in a sparse population to search out a mate every time she has eggs that are ready for fertilization.

Some fish have reproductive habits as bizarre as the differences between the appearance of the sexes. In the case of the seahorse, for example, the male has developed a pouch and it is he who nurtures the developing young, instead of leaving them unguarded as is normally the case. Many cichlid fish [4] retain the conventional division of reproductive labour between the sexes but some females appear to spit their young into the world, as a result of their peculiar practice of hatching the eggs in their mouths. Immediately the eggs are laid the female takes them into her mouth. The male has egg-like specks located round the anal opening so that the female, while attempting to pick up the extra "eggs", takes in his sperm and fertilizes the real eggs.

For fish such as eels [7], salmon and steelhead trout spawning can be the end of a very long journey. Salmon return to their native streams to breed after spending a year or more at sea. Pacific species are known to

arrive back after travelling between 1,600 and 3,200km (1,000 to 2,000 miles) and the evidence suggests that they are able to recognize their general course by means of a sun-compass sense of direction and the details of the last parts of their journey by using their well-developed sense of smell.

Electric fish

Electric organs [8] have developed independently in no less than four families of bony fish and also in the cartilaginous torpedoes and rays. These organs are all thought to be derived from muscle and are usually arranged in a series of plates forming an organic battery. Some fish are capable of generating large voltages. One of them, the electric eel (*Electrophorus* sp), which lives in the Amazon, can generate up to 550 volts. These large voltages are probably used for protection or paralyzing prey; most fish that possess electric organs live in muddy or turbid water where visibility is restricted. These fish generate weak electric fields which are thought to enable them to navigate and find prey but not to kill it.

Fish live in almost all aquatic habitats. Some have become adapted to living in extremely inhospitable regions. The grouper inhabits the warm waters of tropical coral reefs while from the freezing Antarctic seas come ice fish. In torrential streams the modified fins of *Sewellia lineolata* serve as suckers. Fish even inhabit caves: the blind cave barb is an African example from Zaïre.

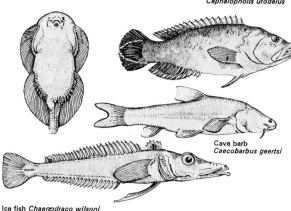

Sewellia lineolata (underside)

Grouper
Cephalopholis urodelus

Cave barb
Caecobarbus geertsi

Ice fish *Chaenodraco wilsoni*

European eel *Anguilla anguilla*

5 *Photocorynus* sp

Male

Female

5 The deep-sea fish *Photocorynus* presents an extreme example of sexual dimorphism in fish. The male, which is a fraction of the size of the adult female, has no independent existence but lives parasitically attached to her. He latches on by his mouth [1] to a special protuberance just above her snout and takes his nourishment from her. In return he supplies her with sperm to fertilize the eggs.

6 The seahorse is related to the pipe-fishes and lives in tropical and temperate seas. It is the only fish with a prehensile tail, which it uses to cling to seaweed. Another distinctive feature is that the male looks after the eggs, carrying them in a pouch in his belly until they hatch. The seahorse swims weakly with an upright stance and is merely carried along by ocean currents.

6 Seahorse *Hippocampus* sp

7 The life history of the freshwater eel starts in the Sargasso Sea where the adults spawn and then die. The tiny larvae, known at this stage as leptocephali, are swept slowly northwards on the Gulf Stream. It then takes them three years to reach the coasts of Europe where they exchange their somewhat fish-like shape for that of a tiny eel. At this stage they are transparent "glass eels" or elvers. During the next stage they migrate inland up the rivers. Eels in the freshwater phase of their life cycle are known as yellow eels. This phase lasts for about a year and then gradually over the next six years the adult eels change from yellow to silver. Not until they are about 10 years old do they start the journey back to the Sargasso Sea to spawn. What induces and guides the migration of eels is still something of a mystery. But it is now thought that very few European eels return to the Sargasso Sea. Most of the adults spawning there seem to have made their way down from the coasts of North America and European eels probably come from these.

8 Gymnotid fish *Gymnarchus niloticus*

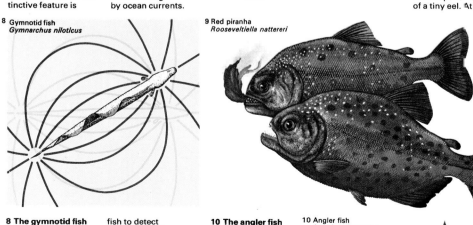

9 Red piranha *Rooseveltiella nattereri*

9 The red piranha of South America is notorious for its ferocity which, allowing for popular exaggeration, is still formidable. It has powerful jaws full of very sharp teeth, and makes up for its small size – 35cm (14in) long at the most – by swimming quite often in large shoals. These represent a threat to larger fish and land animals and even to man. They feed in bouts rather than continuously and locate their prey by means of their sense of smell which leads them, for example, to any flesh that has already been torn and is bleeding. They will eat dead flesh as well.

8 The gymnotid fish is one of the so-called electric fish. It belongs to the weakly electrogenic species that emit regular pulses of current from electric organs modified from muscle or nerve. Their stiff-spined posture enables such fish to detect objects in their environments as disturbances in the electrical field. Special centres in the nervous system control the electric impulses but how the information is interpreted is not yet known.

10 The angler fish is a sluggish predator that lies half concealed in mud waving an appendage developed from the dorsal fin. Smaller fish, attracted by the lure, approach the angler and are sucked into its huge mouth.

10 Angler fish *Lophius piscatorius*

129

Fish of seas and rivers

Fish are found in an extraordinarily wide range of shapes and sizes. This reflects the many different ways in which they have become adapted to the various conditions and habitats that the world's seas and fresh waters have to offer. Many fish that live in weedy habitats, for example, have peculiar outgrowths from their bodies [13] while others are decorated with intricate colour patterns. Both these kinds of adaptations are forms of camouflage, as is the ability of some fish to change colour to merge with their surroundings. A feature of fish that live in muddy waters are the sensory barbels round the mouth [12], which are used in locating food.

Predatory fish [1] are generally sleek and streamlined with large mouths and sharp teeth. Flatfish [8], in contrast, feed on the sea-bed and are slow-moving. Fish of many kinds are vital to man as a source of food, particularly the cods [9], herring [6] and flatfish. Fish are also the source of fishmeal which is used as fertilizer and as pig and poultry feed. The fish, in turn, often feed upon other fish, their eggs and young.

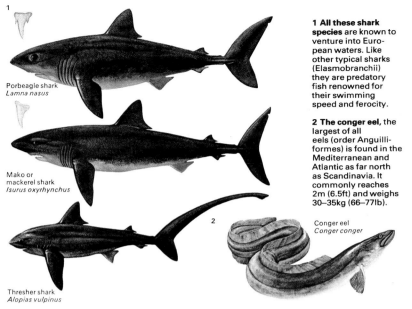

Porbeagle shark
Lamna nasus

Mako or mackerel shark
Isurus oxyrhynchus

Thresher shark
Alopias vulpinus

1 All these shark species are known to venture into European waters. Like other typical sharks (Elasmobranchii) they are predatory fish renowned for their swimming speed and ferocity.

2 The conger eel, the largest of all eels (order Anguilliformes) is found in the Mediterranean and Atlantic as far north as Scandinavia. It commonly reaches 2m (6.5ft) and weighs 30–35kg (66–77lb).

Conger eel
Conger conger

3 Seahorse
Hippocampus sp

Blue-line pipe fish
Doryrhamphus melanopleura

Banded pipe fish
Dunckerocampus dactyliophorus

3 One of the ocean's oddities, the seahorse, is related to the pipefish and to the stickleback. These fish have bodies covered, in varying degree, with bony plates and have tubular snouts.

Sail-fin leaf fish
Taenianotus triacanthus

4 Australian waters are inhabited by about 2,000 species of fish. Sharks and barracoutas patrol the seas and inland waters abound in freshwater types, including the catfish and the murray cod.

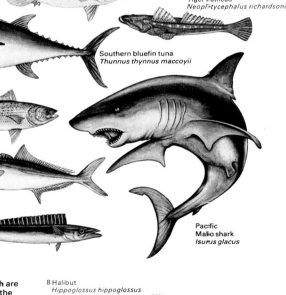

Catfish
Tandanus tandanus

Golden perch
Plectroplites ambiguus

Murray cod
Maccullochella macquariensis

Tiger flathead
Neoplatycephalus richardsoni

Southern bluefin tuna
Thunnus thynnus maccoyii

Australian salmon
Arripis trutta

Kingfish
Seriola grandis

Barracouta
Thyrsites atun

Pacific Mako shark
Isurus glacus

5 The sail fin leaf fish is a kind of scorpion fish. This coral reef fish can change colour to blend with its surroundings and when alarmed may "act dead" by drifting upside down.

6 One of man's most important food fish is the herring. It is classified in the order Clupeiformes and is found on both sides of the North Atlantic. Its average length is 30cm (12in).

6 Herring
Clupea harengus

7 Weedy regions of rivers and backwaters in southern Asia are the home of the knife fish. The fish are named after their vertically compressed shape. Their colour varies widely, even within one species, which makes them hard to identify.

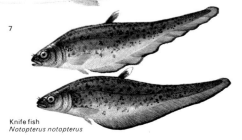

Knife fish
Notopterus notopterus

8 The flatfish are classified in the order Pleuronectiformes. They really have "twisted" bodies, for as these fish develop the bones of the skull twist so that both eyes are on one side of the head. This side becomes the back of the mature flatfish and takes on a heavy pigmentation while the other, the "blind" side, remains nearly white [1]. Some species can change colour as they move from rocky to sandy ocean bottom and vice versa.

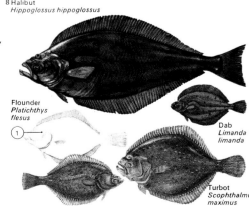

8 Halibut
Hippoglossus hippoglossus

Flounder
Platichthys flesus

Dab
Limanda limanda

Turbot
Scophthalmus maximus

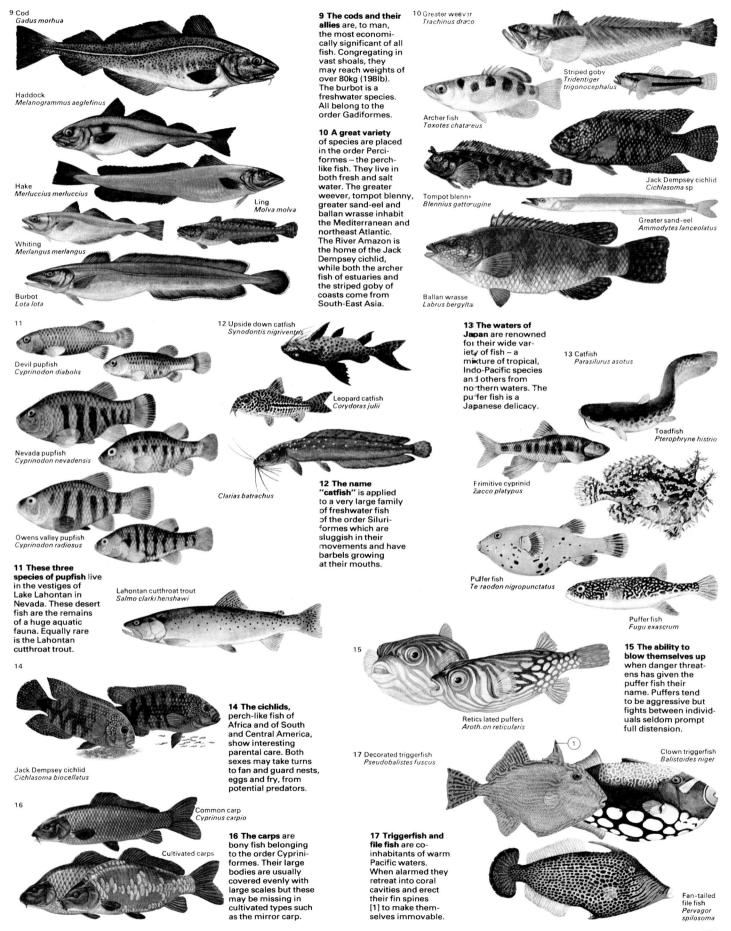

9 Cod
Gadus morhua

Haddock
Melanogrammus aeglefinus

Hake
Merluccius merluccius

Ling
Molva molva

Whiting
Merlangus merlangus

Burbot
Lota lota

9 The cods and their allies are, to man, the most economically significant of all fish. Congregating in vast shoals, they may reach weights of over 80kg (198lb). The burbot is a freshwater species. All belong to the order Gadiformes.

10 A great variety of species are placed in the order Perciformes – the perch-like fish. They live in both fresh and salt water. The greater weever, tompot blenny, greater sand-eel and ballan wrasse inhabit the Mediterranean and northeast Atlantic. The River Amazon is the home of the Jack Dempsey cichlid, while both the archer fish of estuaries and the striped goby of coasts come from South-East Asia.

10 Greater weever
Trachinus draco

Striped goby
Tridentiger trigonocephalus

Archer fish
Toxotes chatareus

Jack Dempsey cichlid
Cichlasoma sp

Tompot blenny
Blennius gattorugine

Greater sand-eel
Ammodytes lanceolatus

Ballan wrasse
Labrus bergylta

11

Devil pupfish
Cyprinodon diabolis

Nevada pupfish
Cyprinodon nevadensis

Owens valley pupfish
Cyprinodon radiosus

11 These three species of pupfish live in the vestiges of Lake Lahontan in Nevada. These desert fish are the remains of a huge aquatic fauna. Equally rare is the Lahontan cutthroat trout.

12 Upside down catfish
Synodontis nigriventris

Leopard catfish
Corydoras julii

Clarias batrachus

12 The name "catfish" is applied to a very large family of freshwater fish of the order Siluriformes which are sluggish in their movements and have barbels growing at their mouths.

13 The waters of Japan are renowned for their wide variety of fish – a mixture of tropical, Indo-Pacific species and others from northern waters. The puffer fish is a Japanese delicacy.

13 Catfish
Parasilurus asotus

Toadfish
Pterophryne histrio

Primitive cyprinid
Zacco platypus

Puffer fish
Tetraodon nigropunctatus

Puffer fish
Fugu exascrum

14

Jack Dempsey cichlid
Cichlasoma biocellatus

14 The cichlids, perch-like fish of Africa and of South and Central America, show interesting parental care. Both sexes may take turns to fan and guard nests, eggs and fry, from potential predators.

Lahontan cutthroat trout
Salmo clarki henshawi

15

Reticulated puffers
Arothron reticularis

15 The ability to blow themselves up when danger threatens has given the puffer fish their name. Puffers tend to be aggressive but fights between individuals seldom prompt full distension.

16

Common carp
Cyprinus carpio

Cultivated carps

16 The carps are bony fish belonging to the order Cypriniformes. Their large bodies are usually covered evenly with large scales but these may be missing in cultivated types such as the mirror carp.

17 Decorated triggerfish
Pseudobalistes fuscus

17 Triggerfish and file fish are co-inhabitants of warm Pacific waters. When alarmed they retreat into coral cavities and erect their fin spines [1] to make themselves immovable.

Clown triggerfish
Balistoides niger

Fan-tailed file fish
Pervagor spilosoma

131

The life of amphibians

Amphibians evolved from fish-like ancestors 350 million years ago in the upper Devonian period. At that time local fern-fringed swamps were vacant and provided ideal humid conditions for the first conquest of the land by animals with no means of conserving their body water.

We know from fossil evidence that the first amphibians resembled giant salamanders with elongated heads and well-developed tails. These animals, often more than 1m (39in) long, moved slowly and clumsily, carrying their heavy bodies from one pool to another. By the Carboniferous period, many different amphibian forms had evolved. These creatures led slow but untroubled lives with little competition from other animals and an abundance of food.

Adaptation difficulties
The change from life in water to life on land posed many problems and it took the amphibians many millions of years to become adapted. In fact amphibians never completely adapted to this harsher environment and still need water to continue breeding.

To move more efficiently, amphibians developed lightweight skeletons and strong muscles to lift their bodies off the ground [8]. The limbs of many of the earlier amphibians were awkward structures with large bones and widely expanded hands and feet, though they showed the typical five-fingered (pentadactyl) limb pattern of higher vertebrates. In order to breathe, the amphibians used a new method of respiration that involved paired air sacs or lungs.

Of the many amphibian groups that once existed, there are only three modern orders: the Anura (frogs and toads), the Urodela (newts and salamanders) and the smallest group, the Apoda or caecilians (elongated, blind and burrowing forms).

There is a great diversity of frogs and toads with over 2,500 species adapted for life in habitats which, apart from wet lands, include tropical forest, grassland and even deserts. A common feature of frogs and toads is that they undergo a complete change in form (known as metamorphosis) during their life history [1].

Male frogs and toads usually call to attract

females for breeding and also in response to danger. Both sexes have vocal organs but only those of the male are fully developed. The typical croaking noise is produced by the vibrations of the vocal cords, a pair of folds of membrane in the larynx. Air is passed backwards and forwards between the lungs and the vocal pouches, formed below the mouth.

Nearly all temperate species of frogs and toads migrate to water in the spring. They find their way with the help of special sensory cells – osmoreceptors – in their mouths. Certain ponds seem to be especially attractive – for reasons unknown – and enormous numbers congregate there in the mating season. The males usually precede the females and then attract them by calling.

The amphibian skin
The larval forms of frogs, toads, newts and salamanders all have external gills for respiration in water but these are lost in most adult forms. Mature frogs can breathe in three ways [Key]. They use their lungs when highly active, the floor of the mouth (buccal cavity) when feeding and their moist skin

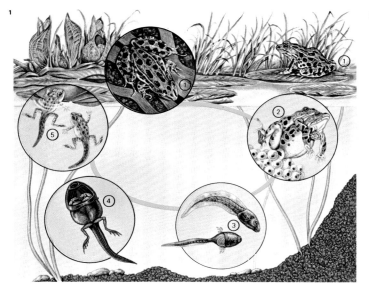

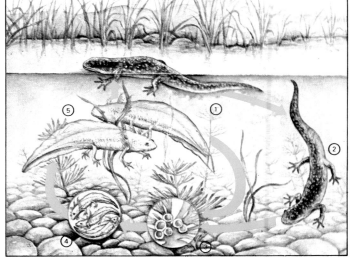

1 The life cycle of the North American leopard frog (*Rana pipiens*) begins when the adult frog [1] leaves the long grass of a wetland meadow and returns to a pond or stream to breed [2]. A few days after the eggs hatch, the tadpoles have external gills [3]. After 8 weeks, hindlimbs are well developed [4]. Young frogs at 3 months have all limbs developed but still have a tail [5].

2 The phenomenon of paedogenesis is seen in the Mexican salamander. Adult male spermatophores[1] are picked up by the female [2] and the fertilized eggs are laid on waterweed [3]. The larva has external gills [4]. The fully-grown sexually immature larva [5] develops reproductive ability. At higher than normal temperatures and iodine concentrations, sexually mature larvae will metamorphose.

3 The fire salamander (*Salamandra salamandra*) is the largest European salamander, attaining a length of 28cm (11in). It lives in hilly, wooded areas and uses striking coloration to warn off predators.

3 Fire salamander
Salamandra salamandra

6 South American caecilian
Siphonops annulatus

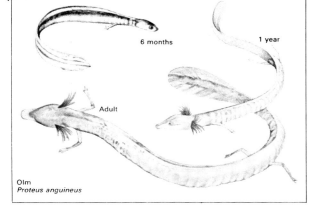

6 months

1 year

Adult

Olm
Proteus anguineus

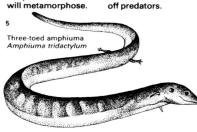

Three-toed amphiuma
Amphiuma tridactylum

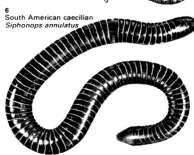

4 The olm (*Proteus anguineus*) is a neotenous amphibian that lives in caves. The young have eyes and are dark-coloured like salamanders but these features soon disappear in the adult forms.

5 The three-toed amphiuma (*Amphiuma tridactylum*) grows to a length of 1m (39in) and is found in the south-eastern USA. It is nocturnal and spends most of its time in water.

6 The worm-like *Siphonops annulatus* is a typical member of the caecilian group. It is blind, lives underground and probably feeds on earthworms. Caecilians live only in subtropical and tropical regions of the world. This species comes from South America. It incubates its eggs and grows to a length of 50cm (20in).

when hibernating. The skin is kept moist by secretions from mucus glands in the outer layer of the skin, the epidermis. The skin may also contain poison glands which are well developed in tropical frogs such as *Dendrobates* and *Phyllobates*. The virulent poison secreted is used on arrows by South American Indians to paralyse prey such as birds and monkeys. Many of the poisonous amphibians are brightly coloured, thus warning predators to leave them well alone. The use of colour for camouflage is highly developed in the amphibians. Three layers of pigment cells in the skin can produce colour changes by expansion and contraction.

Newts and salamanders show less deviation from the generalized amphibians than do the more specialized frogs and toads. The body shape is usually lizard-like with a distinct head. Adult and larval forms are very similar and do not generally show the complete metamorphosis typical of frogs and toads. There are eight families with approximately 225 species. Like frogs and toads, they generally breed in the water. Most of them lay eggs and internal fertilization is common. The male releases a packet of sperm (spermatophore) which is taken up by the female using her cloaca.

Courtship displays
Newts become brightly coloured during the breeding season and there is often an intense courtship display. Some salamanders exhibit neoteny, in which the adult retains features of the young larvae, such as well developed external gills and translucent unpigmented skin. In paedogenesis [2] the creature reaches sexual maturity in the larval stage. An example is the axolotl (*Ambystoma mexicanum*).

The caecilians are the smallest and least well known group of amphibians. Many are burrowing [6] and all are limbless, showing such interesting primitive features as the retention of scales in the skin. They have vestigial eyes whose function is largely replaced by special sensory "tentacles" with which they feel their way through the earth. One of the better known species is the Ceylonese caecilian (*Ichthyophis glutinosus*), first studied at the end of the nineteenth century.

The common frog (*Rana temporaria*), uses its lungs [6], skin [9] and buccal cavity [3] for breathing. By means of the hyoid bone [5], which lowers the cavity floor, air is drawn through the nostrils [2]. The frog's eyes [1] are set high on the head for improved vision. The tongue [4] is used to catch food. The heart [8], typically amphibian, has two atria and one ventricle. The eardrum [7] is visible superficially.

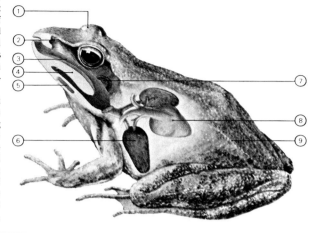

7 Specialized breeding methods are used by many different species of frogs and toads. In tropical areas, eggs may be laid in a nest of leaves or attached to a twig over a stream or pond so that when the tadpoles hatch they fall into the water to continue developing. These methods have the merit of giving protection from predators in the vulnerable egg stage. Another protective step is for the parents to carry the eggs. The male of the European midwife toad (*Alytes obstetricans*) winds the eggs around his hind feet and carries them until they hatch. The ultimate in parental care is shown in tropical species whose young hatch directly from the female's back. Frogs that show this degree of protective care produce fewer young but the chances of survival are much greater than in species that lay a large number of unprotected eggs.

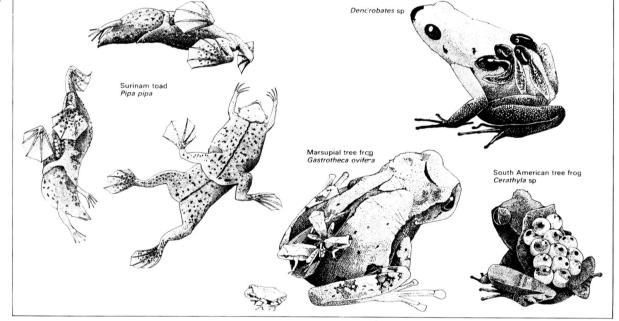

Dendrobates sp
Surinam toad *Pipa pipa*
Marsupial tree frog *Gastrotheca ovifera*
South American tree frog *Cerathyla* sp

The male Surinam toad (*Pipa pipa*) courts the female by uttering metallic calls, then holds her in his arms. The sticky eggs are fertilized as they are laid singly and fall on to the female's back, sinking in so that the skin covers them. The young do not go through a true tadpole stage but hatch out from the female's back 3-4 months later. The South American tree frog (*Cerathyla* sp) carries her eggs in a basket-shaped hollow on her back until they are ready to hatch in a cup of rainwater formed in a leaf. The tadpoles of the South American frog *Dendrobates* are carried on their father's back but their development is completed in the water. In the "zip-bag" birth of the marsupial tree frog (*Gastrotheca ovifera*), about 20 fully formed young frogs hatch out from a pouch on the female's back.

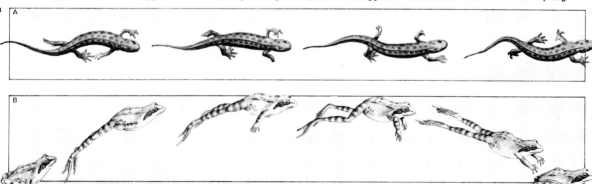

8 Amphibians move on land by various methods. [A] A newt walks by raising its body on to legs which act like levers. The backbone acts like a girder by carrying the weight of the body. When frightened, a newt can wriggle along on its belly.
[B] A frog's long hind legs give very good leverage for jumping. Both legs are thrust out simultaneously to produce a highly effective leap.

133

The life of reptiles

Reptiles, as inhabitants of the earth, are older than man and all other mammals, and older than the birds. Indeed, early reptilians were the ancestors of both the other classes, and their descendants continue to share with birds such characteristics as the laying of shelled eggs, a single knob on the back of the skull fitting it to the backbone, and a single bone in the middle ear.

Reptiles appeared about 300 million years ago, descendants of the early amphibians. In the course of time, some grew to huge proportions, like the dinosaurs, but for the past 70 million years they have mostly been small animals, except for the Crocodilia, relations of the dinosaurs.

Reptilian orders
Four main orders of reptile, totalling some 6,000 species, exist today. Tortoises and turtles, of the order Chelonia, are perhaps the most primitive group, with skulls like the early reptiles, the Saurians. The crocodiles and alligators are members of the Crocodilia while snakes and lizards, most recently evolved and most numerous of the reptiles, belong to the order Squamata. The tuatara, sole surviving member of the order Rhynchocephalia, lives in New Zealand and existed as a species 200 million years ago, before the dinosaurs came into existence.

The turtles and terrapins [4, 7] have shells, typically with a horny layer covering a bony box derived from the backbone and ribs plus bony plates which originate in the skin. It is an effective armour, but heavy. Land tortoises are slow-moving, with legs adapted for weight-bearing; they sometimes have elephantine feet. The most common chelonians are terrapins, amphibious freshwater forms with flatter shells than those of land tortoises. Their toes are webbed. Sea turtles are completely aquatic and eat mainly animal food, while land tortoises are largely vegetarian. No species of chelonian has teeth but deals with its food with horn-rimmed jaws, the front legs assisting in the tearing process.

Crocodiles and alligators are found in lakes and rivers throughout the tropics, using their muscular, flattened tails for driving themselves through the water and their webbed feet for steering. One species, the estuarine crocodile (*Crocodylus porosus*) makes its way to the sea. The sharp, conical teeth with which crocodiles and alligators are armed are good for seizing their prey but less efficient at cutting up a carcase; to overcome this drawback the crocodile will spin over in the water with its victim in its grasp, until part is torn off. It keeps its sharp bite into old age for it continually grows new teeth to replace the old ones. The alligator differs from the crocodile in its dentition. The large fourth tooth of a crocodile's lower jaw sticks up inside the upper jaw whereas, in the alligator, it disappears into a pit inside the jaw [8].

Snakes and lizards
Members of the Squamata, the lizards and snakes, share many similarities despite the obvious difference between them (lizards have five-toed limbs, while snakes are legless). Lizards usually have eyelids, visible external ears and small scales under the belly as well as on the back. Snakes have no true eyelids, no sign of external ears and possess a single row of large scales along their bellies.

Most lizards are carnivorous, feeding on

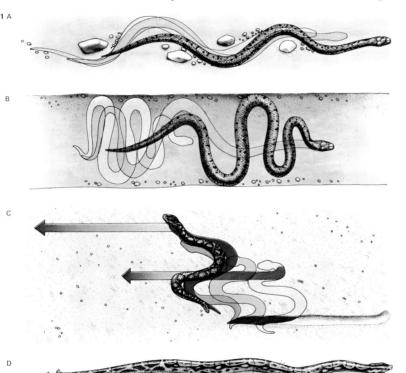

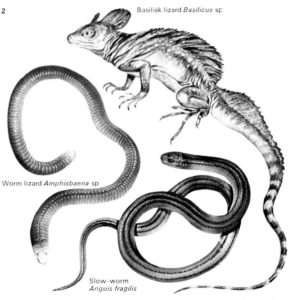

1 Serpentine locomotion [A] is the conventional means of movement for most snakes on land or in water; in a burrow a snake may use a concertina movement [B]. Some desert vipers and rattlesnakes reduce contact with the hot sand by "sidewinding" [C]. A boa constrictor moves in a straight line by contraction of its belly muscles [D].

Normal

In flight

3 The flying snake glides between trees by flattening its undersurface.

Flying snake *Chrysopelea pelias*

Basilisk lizard *Basilicus* sp

Worm lizard *Amphisbaena* sp

Slow-worm *Anguis fragilis*

2 Lizard locomotion shows several variations on the typical four-footed method. Basilisk lizards can run at 11km/h (7mph) on their hind legs alone. At these speeds the long tail acts as a counter-balancer. Worm lizards are legless burrowers that tunnel by ramming the ground with blunt, strong-skulled heads. In existing tunnels they move like earthworms, pulling their rings of scales forwards in groups. The slow-worm belongs to a family of lizards that includes some with normal legs and feet, but like many burrowing or semi-burrowing species this one is limbless.

Freshwater spiny soft-shelled turtle *Trionyx spiniferus*

Sea-dwelling loggerhead turtle *Caretta* sp

4 Some reptiles are well adapted to a life in water, both fresh and salt. Several families of terrapin or turtle are found in lakes and rivers, while all true sea turtles belong to one of two families. The leatherbacked sea turtle is the only member of the family Dermochelidae; all the remaining sea turtles are placed in the family Chelonidae. The sea turtles possess flippers, but the freshwater species usually have only modified webbed feet.

insects and other small prey, but some larger species such as monitors prey upon vertebrates. They tend, like other reptiles, to spend much of the time immobile, waiting for the prey to approach before they grab it, or make a short rush for it [5]. Iguanas and some skinks are among the few vegetarian lizards.

Camouflage and immobility are their main means of defence, but monitors use both claws and teeth to protect themselves and even small lizards may bite if cornered. Others protect themselves by autotomy – if harried they shed part of their tail at a special breaking point, leaving it behind to confuse the predator. A replacement is then grown.

In movement, most lizards have a straddling gait. Many climb and the gecko, which has ridges and microscopic hairs under its toes, can walk on an apparently smooth ceiling. The flying dragon of southeast Asia is a lizard, which glides using flaps of skin.

Big eaters
Most snakes move by bending their bodies from side to side using muscles along the backbone. The sides of their bodies grip the ground. Some species, however, have adapted to modified means of locomotion [1].

All are carnivores, swallowing their prey whole. Jaws loosely connected to the skull and to one another permit an enormous gape [10], and allow prey much wider than the snake's head to be engulfed by "walking" the jaws alternately round it. In this way a 4.8m (16ft) rock python has been known to swallow whole a 59kg (130lb) impala, and an 8m (26ft) anaconda a 45kg (100lb) peccary.

Large snakes such as these suffocate their victims by construction, squeezing the air out of them. Others – a minority – have poisonous bites. The most dangerous of these are the two families of front-fanged snakes, the cobras, whose venom mainly attacks the nervous system, and vipers and rattlesnakes, whose poison affects the blood and tissues.

Venomous snakes, like others, usually flee before a threat, but some use a deterrent display – the cobra spreads its hood, the rattlesnake rattles its tail. The coral snakes have developed garish warning colours, but other non-poisonous species have developed near-identical colour patterns as a defence.

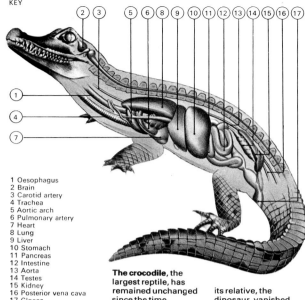

KEY

1 Oesophagus
2 Brain
3 Carotid artery
4 Trachea
5 Aortic arch
6 Pulmonary artery
7 Heart
8 Lung
9 Liver
10 Stomach
11 Pancreas
12 Intestine
13 Aorta
14 Testes
15 Kidney
16 Posterior vena cava
17 Cloaca

The crocodile, the largest reptile, has remained unchanged since the time its relative, the dinosaur, vanished.

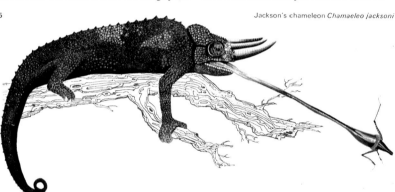

5 Jackson's chameleon *Chamaeleo jacksoni*

5 Flicked out in ¹/₂₅ **sec**, a chameleon's tongue tip is sticky and partly prehensile, giving the insect victim no chance. In locating food, the eyes swivel independently until prey is sighted, then focus together. Another remarkable characteristic of this reptile is its ability to change colour. Colour changes are activated largely by changes in light, a signal from the eye passing via the nervous system to pigment cells in the skin. Most chameleons are well adapted for climbing with toes fused and grouped to form grasping hands and feet. The toes have sharp claws.

6 The egg-eating snake [A] gulps eggs, cracks and rejects the shells. The boa constrictor [B] coils round its prey, suffocating it. The camouflaged fishing snake [C] seizes fish in its back-fanged jaws.

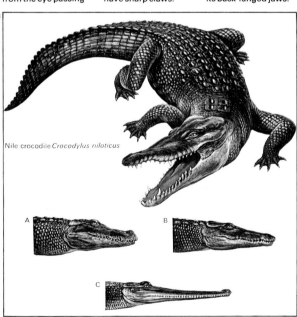

Nile crocodile *Crocodylus niloticus*

7 The alligator snapper, *Macrochelys* sp, has a lure for fish on the back of its tongue – a pink projection that is wriggled like a real worm. Fish are taken in by this bogus bait.

8 Large alligators [A] and crocodiles [B] are powerful enough to kill and feed on animals up to the size of a cow. The specialized gharial [C] has long, thin jaws which it swings from side to side snapping at fish in the vicinity.

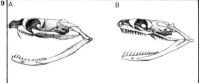

9 The fangs of the Indian cobra [A] are at the front of its jaws, and well placed for delivering the venom. Back-fanged snakes such as the boiga [B] generally have smaller fangs.

10 Snake venom is produced in modified salivary glands, delivered in certain snakes by grooved teeth. In rattlesnakes [A] and their like, venom is squeezed down a duct [1] from the venom gland [2] into a fang like a hypodermic syringe. The fangs are erected and the huge gape [B] becomes apparent as the snake strikes. Three kinds of cobra spray venom at their adversaries' eyes. Their fangs have forward-directed orifices [C], unlike the "hypodermic" fang [D].

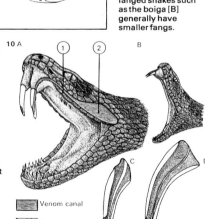

Venom canal

Enamel

Dentine

135

Snakes, lizards and turtles

There are about 6,000 species of reptiles living in the world today. They comprise about 25 species of crocodiles, 250 species of turtles and tortoises, 2,800 of snakes and 3,000 of lizards. Most reptiles live in the tropics [Key], but a few hardy species, such as the common lizard, the adder and the greater snake, are found in the Arctic Circle, and one lizard lives in Tierra del Fuego.

Temperature regulation

Most reptiles cannot tolerate cold because of the way their bodies work. Reptiles are cold-blooded animals. This means that they have no built-in control over their body temperature and take on the temperature of their surroundings, to within a few degrees. Reptiles are most active when they are warm and they are best suited to tropical climates. Cold conditions make them sluggish and eventually kill them. But reptiles can control their temperature to some extent [11] by basking in the sun when cold and so raise their temperature even above that of the surrounding air. However, they are not able to maintain constant temperatures like mam-

mals. Even with body temperatures as high as those of mammals they may produce energy only one-tenth as fast. Behaviour which takes advantage of heat when available and avoids unnecessary cooling is important.

In the tropics, reptiles may avoid large fluctuations in body temperature, but far from the Equator or at high altitudes it is not so easy. In northern Europe, low winter temperatures make life almost impossible, even for hardy reptiles, and they survive only by hibernating during the winter.

Reproduction and growth

All reptiles reproduce by laying eggs. Crocodiles and pythons may lay up to 100 in a clutch, and turtles lay even more, but most lizards and snakes have smaller clutches. The eggshells are hard in tortoises and crocodiles, but in other species are usually leathery. The eggs are fertilized within the female before laying. All male reptiles, with the exception of the tuatara, have an intromittent organ – single in turtles, tortoises and crocodiles, double in lizards and snakes – which is turned outwards from the genital opening, the

cloaca, for mating. Most reptiles have no family life and the eggs are abandoned after being laid, although they are often buried first. Warmth from the sun or decaying vegetation incubates them.

Some reptiles retain their eggs inside the oviducts until hatching, or even afterwards, so that the young are born alive – a condition known as ovoviviparity. This process is particularly common in species that live in hostile surroundings, such as deserts or cold regions, and it has the advantage that the mother, by avoiding extremes, protects the developing embryos from them too.

Newly hatched reptiles are miniatures of their parents, ready to fend for themselves. Poisonous snakes can bite immediately after hatching. The rate at which reptiles grow depends on food and warmth. An alligator in suitable surroundings might grow an average of 2.5cm (1in) a month for the first years of its life and pythons might grow three times as fast as that. Reptiles grow quickly until sexual maturity, which is reached in less than a year by some small lizards but not until ten years or more by some tortoises and crocodiles.

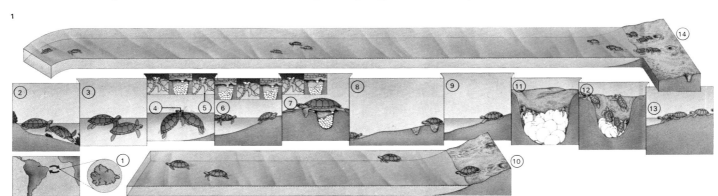

1 The green turtle (*Chelonia mydas*) ranges through tropical seas but the number of major breeding sites is small. In the Atlantic, Ascension Island [1] is used by turtles that feed on eelgrass and algae in warm shallow coastal waters off South America [2]. By unknown means the turtles navigate [3] 2,200km (1,375 miles) to Ascension in December. Courtship and mating [4] take place just off shore. Females may mate several times [5] but sperm can be stored for long periods and several clutches (insets) can be fertilized from one mating. The female leaves the water [6] at night. She lays hundreds of eggs in a season in sandy pits [7]. When the eggs are covered and camouflaged [8] she returns to the sea [9] and after the breeding season [10] returns to Brazil. After incubating for up to 10 weeks, eggs hatch [11] almost simultaneously. The baby turtles emerge together [12] at night and scramble toward the sea [13] guided by the lighter sky over it. They head for open water [14] and an unknown destination.

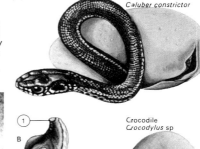

5A

Blue racer
Coluber constrictor

Crocodile
Crocodylus sp

B

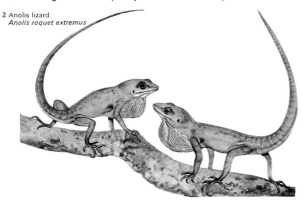

2 Anolis lizards display to each other by raising their bodies and tails and flashing the brightly coloured dewlap. This is an aggressive display between males. Other lizards display with flashing colours and by bowing.

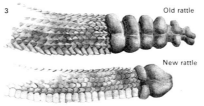

3 Old rattle

New rattle

3 A rattlesnake's rattle begins at birth with a hard button on the tail. Each time the skin is sloughed, a new hard tip forms, but the segments from the old skin remain. In this way a rattle is built up of hard, hollow pieces.

4 The "combat dance" is a display of ritual fighting by pairs of rival male snakes over a female. Venomous snakes, such as these rattlers, do not attempt to bite but merely try to push each other to the ground to establish superiority.

5 Reptile embryos have special devices with which to cut themselves out of the egg. Hatching crocodiles [B] and turtles have a special horny thickening of the skin on the snout [1] called the egg-caruncle. The hatching blue racer snake [A], like other snakes and lizards, employs an egg-tooth, which develops in the mid-line of the upper jaw. The tooth is large, very sharp and projects forwards. After hatching it is no longer of any use and is then discarded.

2 Anolis lizard
Anolis roquet extremus

Unlike mammals, reptiles may still grow even when adult, although more slowly.

The sense organs

The sense organs of reptiles vary according to species. Crocodilians have reasonable eyesight, with the slit pupils characteristic of nocturnal animals, but cannot distinguish colours. Their hearing is good and their sense of smell is probably adequate. Turtles and tortoises have good vision and respond to colours; they also respond to scents, but their hearing is limited. Like many reptiles, they hear low notes best.

Lizards that are active during the day probably have the best reptile vision, seeing sharply and in colour, as would be expected from their use of colour in display. Hearing is less acute, the animals responding, if at all, to the lower end of the sound scale. The exceptions seem to be the geckos, which have good hearing, and these and crocodiles are the only reptiles to produce much sound themselves. The sense of smell varies: it is poorly developed in tree lizards but good in many others.

Snakes have the oddest sense organs of all. They have an organ called Jacobson's organ [9], made up of two small cavities in the roof of the mouth, which is used in conjunction with the tongue. The organ is lined with sensitive cells and the twin cavities may explain why snakes and some lizards have a forked tongue. Jacobson's organ helps a snake's sense of smell and many snakes can follow a trail by using it. Snakes also have unusual eyes. These are covered with a special transparent scale or spectacle, which replaces the two lids, and a sideways-moving nictitating membrane (third eyelid) which is normal reptile equipment. Snakes are colour-blind and their eyesight is not as good as that of lizards. They lack the external eardrum that is present in other reptiles and, although they have internal ears, snakes are deaf to airborne sounds but often sensitive to ground vibrations.

Some snakes possess warmth detectors – sense organs that are unknown in other vertebrates. These take the form of large pits [10] on the face of rattlesnakes and a series of smaller pits lined with sensitive cells round the lips of some pythons and boas.

KEY

Numbers of genera
- Aquatic chelonia
- Terrestrial chelonia
- Lizards
- Snakes
- Crocodilians
- Marine turtles
- Marine lizards
- Marine snakes

Tropical regions are the home of many reptile species. Numbers on the map refer to the genera present in each area. Although they come ashore to breed in particular areas, sea-turtles are the Komodo dragons of Indonesia with lengths of up to 3m (10ft). Largest of the chelonians is the leatherback sea turtle, more than 2m (6.5ft) long and weighing more than half a tonne.

are widely distributed throughout the warmer oceans. The only genus of marine lizard lives on the Galapagos Islands. Australasia, despite its size, has relatively few reptile genera.

Galapagos and Seychelles giant tortoises are the largest on land. The smallest lizards are New World geckos *Sphaerodactylus* sp which, excluding the tail, are only 1.8cm (0.75in) in length.

6 Indian python *Python molurus*

6 Parental care is shown by few reptiles. Some snakes, such as cobras and pythons, coil round their eggs, and female crocodiles and alligators sit over their nests. Because these animals are cold-blooded, they cannot keep the eggs warm, so the main effect of their presence is to deter potential predators. But there is some evidence to show that a brooding Indian python may produce heat by muscular contraction. Some lizards brood eggs too. The skink (*Eumeces obsoletus*) actually cleans and turns the eggs, helps the young to hatch and then grooms them for more than a week after hatching.

7 The longest reptile that was reliably recorded was a reticulated python of 10m (32ft). The anaconda is reputed to grow larger still but while it must be the heaviest species of snake the most acceptable record of its length is only 9m (30ft). The longest poisonous snake is the king cobra, at 5.5m (18ft). Several kinds of crocodiles may occasionally reach 6m (20ft). The giants among lizards

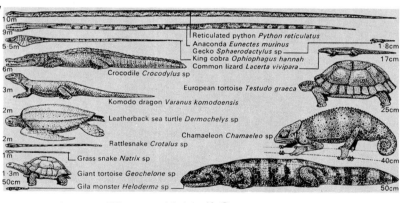

7
- Reticulated python *Python reticulatus*
- Anaconda *Eunectes murinus*
- Gecko *Sphaerodactylus* sp
- King cobra *Ophiophagus hannah*
- Common lizard *Lacerta vivipara*
- Crocodile *Crocodylus* sp
- European tortoise *Testudo graeca*
- Komodo dragon *Varanus komodoensis*
- Leatherback sea turtle *Dermochelys* sp
- Chamaeleon *Chamaeleo* sp
- Rattlesnake *Crotalus* sp
- Grass snake *Natrix* sp
- Giant tortoise *Geochelone* sp
- Gila monster *Heloderma* sp

8 A snake sheds its skin regularly. The old skin undergoes chemical changes as a new layer of cells comes into readiness below. The old skin splits, usually round the lips, the snake wriggles out, and the old skin is often left as a perfect but colourless cast. Colours show up brilliantly on the newly moulted snake, as in this African boomslang (*Dispholidus* sp).

9 The tongues [1] of lizards and snakes convey particles to Jacobson's organ [2] where smells are detected. Originally part of the nose [3], the organ is now separated and well developed.

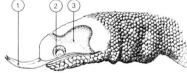

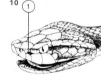

10 A sensory pit [1], sensitive to infra-red radiation, is located on either side of the face of pit vipers and rattlesnakes. Even in darkness they can detect and strike at warm-blooded prey such as mice.

11 Reptiles may control body temperature by variations in behaviour. In the morning, when it needs to absorb heat, an agama lizard lies with flattened body on a sloping surface in order to absorb as much as possible [A]. Later, when less heat is needed, it turns its head to the sun and presents a smaller surface to the rays [B]. Finally, during the hottest part of the day, the lizard may seek shelter from the sun [C] or burrow into cooler soil [D].

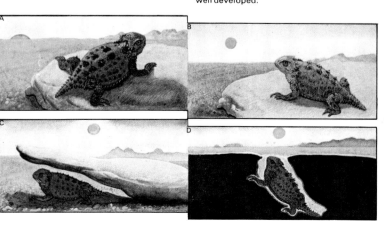

Unusual reptiles and amphibians

The amphibians and reptiles are two groups of cold-blooded vertebrates, most of which lay eggs. The amphibians, numbering more than 2,000 species, include frogs, toads, newts, salamanders and caecilians, and the 6,000 reptile species include the snakes, lizards, crocodiles and turtles. They range in size from 2cm (0.75in) tree frogs that live in bromeliad "vases", to the 8m (25ft) constricting snakes.

Amphibians, because they must keep their skins moist in order to breathe, are found in wet and humid situations. Most species of amphibians live in the tropics where they can take advantage of the hot, wet climate. Reptiles are also more numerous in tropical than in temperate regions, but do not rely on the presence of water in order to survive. They are common desert animals and in this environment avoid becoming overheated by hiding during the day.

No amphibians and relatively few reptiles live in the sea. Most marine reptiles – the sea snakes are an exception – come ashore to breed. The biology of amphibians and reptiles is described on pages 518–23.

1 European green tree frog
Hyla arborea

Box tortoise
Terrapene carolina

Soft-shelled turtle
Amyda ferox

Tropical hawk-billed turtle
Eretmochelys imbricata

1 The European green tree frog has discs at the ends of its toes that enable it to grip slender branches of the trees in which it lives. It is found throughout central Europe, southern Italy and eastwards to Asia. When frightened, or when the sky becomes overcast, it changes colour from bright green to grey. For this reason, some people keep a caged frog to forecast rain.

2 Tortoises and turtles have a massive bony shell [1] made of plates of keratin that are fused to the backbone [2] and ribs [3]. Most can pull back their heads under the shell when danger threatens. The North American box tortoise spends most of its time on land, whereas the soft-shelled turtle and the tropical hawk-billed turtle are both entirely aquatic.

Bell's ceratophrys
Ceratophrys ornata

3 The horned frog, Bell's ceratophrys, lives in Argentina. The horns are outgrowths of the upper eyelids. It uses its large pointed teeth to attack other frogs, which it eats.

4 Boulenger's arrow-poison frog lives at high altitudes in the South American Andes. It is easily caught and local people use a venom secreted from its skin to poison the tips of their hunting arrows.

5 Green turtle
Chelonia mydas

Boulenger's arrow poison frog
Atelopus boulengeri

Two-toned arrow poison frog
Phyllobates bicolor

5 The green turtle lives in tropical seas but has to go ashore on sandy beaches to lay its eggs. This turtle has a flat shell and limbs well adapted for swimming. It is valued as food.

6 The two-toned arrow-poison frog is native to Peru. The poison exuded by the skin is used by Indians to coat their arrows. The rim of the frog's upper jaw is armed with small teeth.

7 The timber rattlesnake is common in the northeastern United States and southern Canada. Highly venomous, it grows up to 2m(6.5ft) long and is found in groups of a hundred or more in winter.

7 Timber rattlesnake
Crotalus horridus

Boa constrictor
Constrictor constrictor

8 The boa constrictor is a large snake that grows up to 3.6m (12ft) long. It lives in underground holes or in trees in many areas of South America. It preys on birds and small mammals such as rats and agoutis, which it kills by entwining them in its coils and crushing them until they suffocate.

Wagler's pit viper
Trimeresurus wagleri

Mangrove snake
Boiga dendrophila

Banded sea snake
Laticauda colubrina

Dog-headed water snake
Cerberus rhynchops

9 Snakes found in mangroves include tree-living species and species that have adopted an aquatic way of life. The two arboreal snakes have become adapted to different feeding methods. The bird-eating *Boiga* moves rapidly in order to catch its prey, while Wagler's pit viper is more likely to lie in wait for prey which it detects with heat-sensitive pits between the nostrils and the eyes. The aquatic snakes hunt fish and molluscs.

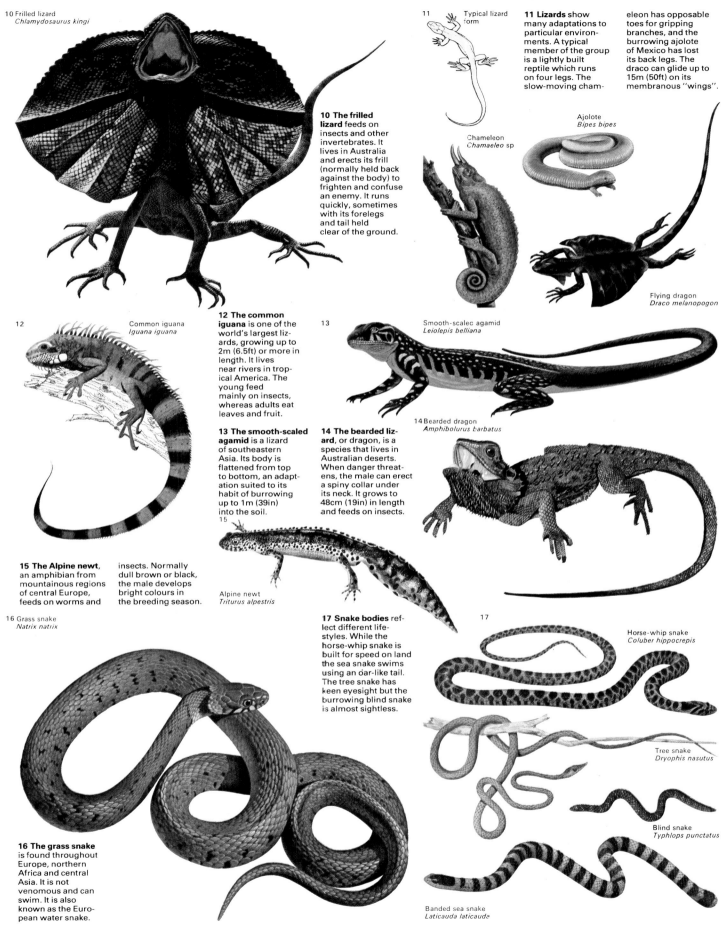

10 Frilled lizard
Chlamydosaurus kingi

10 The frilled lizard feeds on insects and other invertebrates. It lives in Australia and erects its frill (normally held back against the body) to frighten and confuse an enemy. It runs quickly, sometimes with its forelegs and tail held clear of the ground.

11 Typical lizard form

11 Lizards show many adaptations to particular environments. A typical member of the group is a lightly built reptile which runs on four legs. The slow-moving chameleon has opposable toes for gripping branches, and the burrowing ajolote of Mexico has lost its back legs. The draco can glide up to 15m (50ft) on its membranous "wings".

Chameleon
Chamaeleo sp

Ajolote
Bipes bipes

Flying dragon
Draco melanopogon

12 Common iguana
Iguana iguana

12 The common iguana is one of the world's largest lizards, growing up to 2m (6.5ft) or more in length. It lives near rivers in tropical America. The young feed mainly on insects, whereas adults eat leaves and fruit.

13 Smooth-scaled agamid
Leiolepis belliana

13 The smooth-scaled agamid is a lizard of southeastern Asia. Its body is flattened from top to bottom, an adaptation suited to its habit of burrowing up to 1m (39in) into the soil.

14 The bearded lizard, or dragon, is a species that lives in Australian deserts. When danger threatens, the male can erect a spiny collar under its neck. It grows to 48cm (19in) in length and feeds on insects.

14 Bearded dragon
Amphibolurus barbatus

15 The Alpine newt, an amphibian from mountainous regions of central Europe, feeds on worms and insects. Normally dull brown or black, the male develops bright colours in the breeding season.

15

Alpine newt
Triturus alpestris

16 Grass snake
Natrix natrix

17 Snake bodies reflect different lifestyles. While the horse-whip snake is built for speed on land the sea snake swims using an oar-like tail. The tree snake has keen eyesight but the burrowing blind snake is almost sightless.

17

Horse-whip snake
Coluber hippocrepis

Tree snake
Dryophis nasutus

16 The grass snake is found throughout Europe, northern Africa and central Asia. It is not venomous and can swim. It is also known as the European water snake.

Blind snake
Typhlops punctatus

Banded sea snake
Laticauda laticauda

The classification of birds

The study of wildlife is possible only if each specimen can be labelled adequately and unambiguously. The system of labelling birds currently in use was propounded by the Swedish naturalist Carl von Linné (1707–78), whose name is more familiar in its Latinized form of Carolus Linnaeus. He proposed that every living creature should have a "binomial" consisting of a generic and a specific name. The carrion crow, for example, belongs to the genus *Corvus* and is specifically *Corvus corone*. Each of the 8,600 bird species has its own name.

The basis of grouping
Several genera of closely related birds are grouped together in families and when more than one family is considered to have descended from a single ancestral form then these families are grouped in the same order. In the case of the Corvidae, the family to which the crow belongs, the appropriate order is the Passeriformes [28] – the largest bird order, containing 57 families, including finches, family Fringillidae, the starlings, family Sturnidae, the thrushes, family Mus-

cicapidae and the swallows, family Hirundinidae. All the 28 orders of birds are classified in the class Aves and the sub-phylum Vertebrata (the back-boned animals).

Although birds are placed in their respective groups largely on the basis of detailed anatomical and behavioural comparisons, in recent years the analysis of the egg – and especially of the proteins in the egg white – has been used. This procedure has caused a great deal of controversy among taxonomists because affinities of various bird groups have been suggested that are not in accordance with traditional ideas. The taxonomic problems created by several aberrant groups have been solved, however, by egg-white protein analysis. The hoatzin – long given a family of its own – is now, as a result of this technique, considered to be a strange cuckoo and placed in the family Cuculidae which, with the turacos, form the order Cuculiformes [20].

Examination of the anatomy of the flamingos (family Phoenicopteridae), showed similarities to the storks (family Ciconiidae, order Ciconiiformes) but studies of their behaviour and feather lice suggested

relationships with the ducks and geese (family Anatidae, order Anseriformes). On the evidence of egg-white proteins, however, flamingos are now thought to be more closely related to herons (family Ardeidae) in their original order, the Ciconiiformes.

Convergent evolution
The several species of large ground-dwelling birds – the ostrich, rhea, emu and cassowary – all resemble each other quite closely but are thought to have arisen independently and as such are examples of a phenomenon called convergent evolution. They are thus classified in separate orders – the ostrich in the Struthioniformes [1], the rhea in the Rheiformes [2] and both the cassowary and the emu in the order Casuariiformes [3].

The strange New Zealand kiwi also has its own order, the Apterygiformes [4]. Similarly the divers are alone in the order Gaviiformes [5], the grebes in the Podicipediformes [6] and the penguins in the Sphenisciformes [7]. The families of tube-nosed sea birds, the petrels, albatrosses and shearwaters, are united in the order Procellariiformes [8]. All the six

1 **Africa is the home of the ostrich**, the largest living bird. The single species is the only member of the order Struthioniformes.

Ostrich
Struthio camelus

2 **The South American rhea** is ostrich-like. Its order, the Rheiformes, has two species in one family.

Greater rhea
Rhea americana

3 **The cassowary and emu** of Australasia belong to separate families in the Casuariiformes.

Cassowary
Casuarius casuarius

4 **Kiwis come from New Zealand**, where there are 3 species. They belong to one family in the order Apterygiformes.

Brown kiwi
Apteryx australis

5 **The Arctic loon** is one of four species of the order Gaviiformes found in the Arctic and tundra regions.

Arctic loon
Gavia arctica

6 **Slavonian grebes** belong to the order Podicipediformes, which contains 20 species in 1 family.

Slavonian grebe
Podiceps auritus

7 **The largest of all penguins**, the emperor penguin, is found in Antarctic seas. There are 17 species of penguins in a single family in the Sphenisciformes.

Emperor penguin
Aptenodytes forsteri

8 **The great shearwater** is a sea bird of the order Procellariiformes which contains four families.

Great shearwater
Puffinus gravis

9 **The darter**, tropicbird, gannet, cormorant, pelican and frigate bird families are placed in the order Pelecaniformes.

Darter
Anhinga anhinga

10 **The scarlet ibis** is a colourful member of one of the six families in the order Ciconiiformes.

Scarlet ibis
Eudocimus ruber

11 **From Australia comes the black swan**, one of the 147 species of the duck family in the order Anseriformes.

Black swan
Cygnus atratus

12 **Diurnal birds of prey** in the order Falconiformes include the falcon, secretary bird, vulture, osprey and hawk families. The hobby is a typical falcon.

Hobby
Falco subbuteo

13 **The great tinamou** (order Tinamiformes) is found in South America, as are the other fifty species of the order.

Great tinamou
Tinamus major

14 **Game birds** (order Galliformes) of which the guineafowl is one, are classified in six families.

Helmeted guinea-fowl
Numida meleagris

families contained within the order Pelecaniformes [9] are water dwellers.

The six families of large, long-legged wading birds – herons, the shoe-bill, the hammerhead, storks, ibises and flamingos – are in the order Ciconiiformes [10]. Although differing in appearance from the ducks, geese and swan anatomical studies place the three species of screamer (family Anhimidae) in the same order, Anseriformes [11].

Family relationships

Diurnal birds of prey, including the vultures, are placed in five families and combined in the order Falconiformes [12]. The single family of terrestrial game birds of South America, the tinamous, have an order, the Tinamiformes [13], to themselves. Pheasants, grouse, guinea-fowl and turkeys are all ground-dwelling game birds classified with the remarkable megapode family of Australasia and the South American curassow family in the order Galliformes [14].

Many members of the Gruiformes [15] are aquatic birds. Those that are terrestrial, such as the bustards and hemipodes, fly only rarely. Wading behaviour is a feature of several Charadriiformes [16], an order that includes the plovers, avocets and oyster-catchers, the gulls and auks.

The sandgrouse are a one-family order, the Pteroclidiformes [17], as are the pigeons and doves of the Columbiformes [18], and the parrots of the Psittaciformes [19]. The Strigiformes [21], owls and barn owls, is composed of just two families.

Nocturnal life-styles and large mouths are the features shared by most birds in the five families of the order Caprimulgiformes [22], a group that includes the frogmouths and nightjars; and small feet and legs characterize and give the name to the order Apodiformes [23], the group to which the swifts and humming-birds belong. Some rather peculiar birds, the mousebirds or Coliiformes [24] and the trogons or Trogoniformes [25], make up two single-family orders. The spectacular kingfishers, hornbills and cuckoo-rollers represent three of the nine families in the order Coraciiformes [26], while the Piciformes [27] are the woodpecker-like birds of which the toucans comprise one of the six families.

KEY

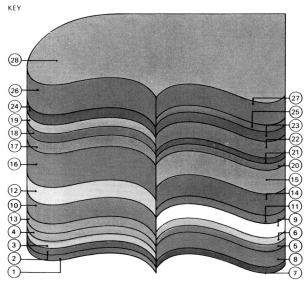

The 28 bird orders are arranged in evolutionary sequence, starting with the most primitive at the bottom. The numbers on the tree refer to the illustrations below.

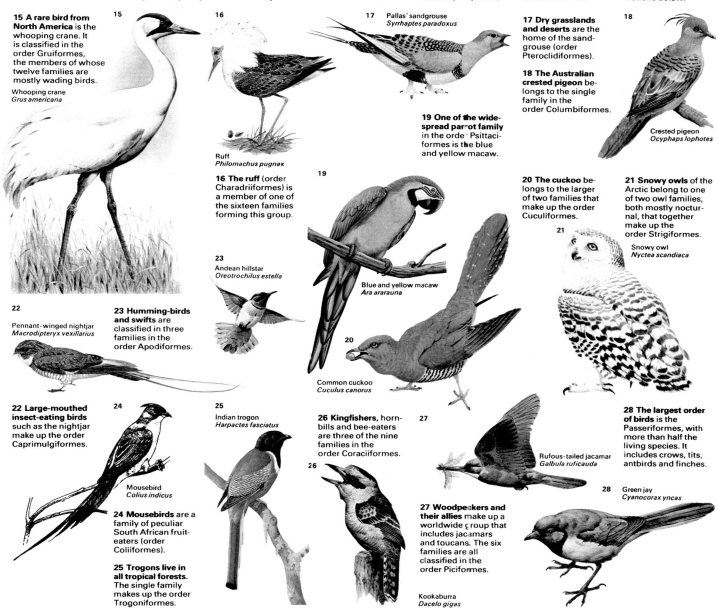

15 A rare bird from North America is the whooping crane. It is classified in the order Gruiformes, the members of whose twelve families are mostly wading birds.
Whooping crane
Grus americana

16 The ruff (order Charadriiformes) is a member of one of the sixteen families forming this group.
Ruff
Philomachus pugnax

17 Pallas' sandgrouse
Syrrhaptes paradoxus

19 One of the widespread parrot family in the order Psittaciformes is the blue and yellow macaw.

17 Dry grasslands and deserts are the home of the sandgrouse (order Pteroclidiformes).

18 The Australian crested pigeon belongs to the single family in the order Columbiformes.
Crested pigeon
Ocyphaps lophotes

23
Andean hillstar
Oreotrochilus estella

Blue and yellow macaw
Ara ararauna

20 The cuckoo belongs to the larger of two families that make up the order Cuculiformes.

21 Snowy owls of the Arctic belong to one of two owl families, both mostly nocturnal, that together make up the order Strigiformes.
Snowy owl
Nyctea scandiaca

23 Humming-birds and swifts are classified in three families in the order Apodiformes.

22
Pennant-winged nightjar
Macrodipteryx vexillarius

22 Large-mouthed insect-eating birds such as the nightjar make up the order Caprimulgiformes.

Common cuckoo
Cuculus canorus

Mousebird
Colius indicus

24 Mousebirds are a family of peculiar South African fruit-eaters (order Coliiformes).

25 Trogons live in all tropical forests. The single family makes up the order Trogoniformes.

25
Indian trogon
Harpactes fasciatus

26 Kingfishers, hornbills and bee-eaters are three of the nine families in the order Coraciiformes.

Kookaburra
Dacelo gigas

27 Woodpeckers and their allies make up a worldwide group that includes jacamars and toucans. The six families are all classified in the order Piciformes.

Rufous-tailed jacamar
Galbula ruficauda

28 The largest order of birds is the Passeriformes, with more than half the living species. It includes crows, tits, antbirds and finches.

Green jay
Cyanocorax yncas

The anatomy of birds

Birds are the only group of vertebrate animals (apart from bats) that are capable of true flight, as distinct from mere gliding, and their bodies are designed to make this possible. Though they are masters of the air they are also quite at home on land or in the water, and some – certain ducks for example – are efficient in all three.

The development of feathers has been crucial to this success. Feathers undoubtedly came before flight. They were evolved, like fur, for insulating a warm-blooded body and were only later developed and used as aerofoils. Birds were probably feathered for millions of years before they flew.

Physical adaptations for flight
The adaptations of a bird for flight involve all of the principal structural and behavioural features of the animal. The structural changes are centred upon the development of strength with lightness. Therefore the bones are hollowed or honeycombed, or moulded into thin, curved plates in such a manner that they are quite strong enough for the jobs they have to perform. The light beak takes the

place of heavy teeth and the feather covering is very light – even though it may weigh more than the skeleton. A series of air sacs in the body cavity, connected with the lungs, assist in respiration.

The rearrangement of bones to deal with the mechanical stresses in an animal that walks on its hind legs and flies with its front ones is seen principally in the pectoral (shoulder) and pelvic (hip) regions. The pectoral girdle is firmly attached to the breastbone so that the body is efficiently suspended from the wings in flight [7]. This is brought about by an extra development of the coracoid bones which are absent in mammals. Similarly, the pelvic girdle is so strengthened and arranged that the hind legs can efficiently carry the bird's weight on the ground – or when it is climbing, perched or in the water – and especially they can act as shock-absorbers on landing. Because bones are so delicate, they have been strengthened by means of fusion. As in mammals, the three bones of the pelvic girdle on each side are fused together and to the backbone. There has also been a fusion of vertebrae in the

spine, from the last thoracic (chest) one through all the lumbars and sacrals to the first few caudal (tail) vertebrae. These fused vertebrae form the synsacrum, which supports the pelvic girdle and lets the legs and wings function efficiently without impairing the other body functions.

Limbs and feathers
The limb skeleton of birds is also much modified from the basic vertebrate arrangement. The bones of the lower leg and the tarsal bones of the foot have become elongated and fused to provide an extra joint in the limb [11]. The thigh bone of birds is usually concealed in the body wall and feathers. A most unusual feature of this limb is the mechanism for ensuring a good perching grip. The toe flexor muscles originate above the knee. Their tendons pass in front of the knee, behind the ankle and beneath the toes. When a bird's leg is flexed (bent), as when perched, this arrangement ensures that the toes are pulled in and grasp tightly even during sleep.

In the forelimb the "hand" is much reduced, the few remaining bones being

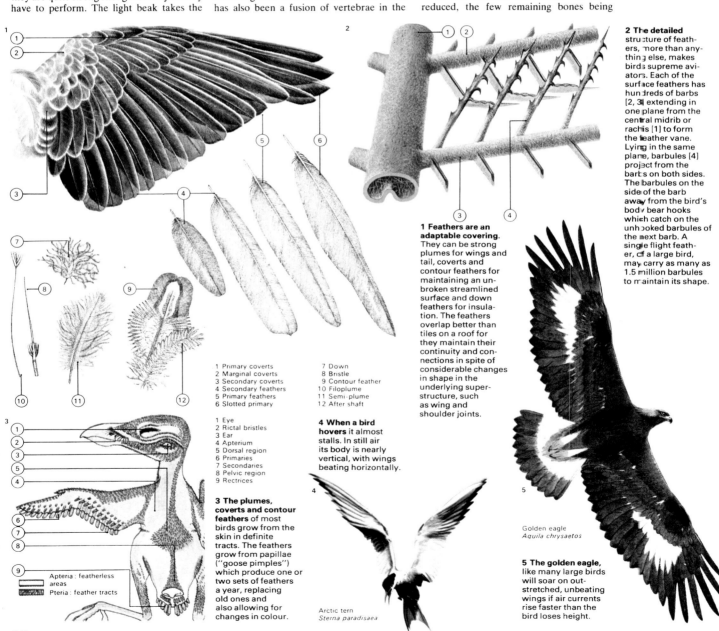

2 The detailed structure of feathers, more than anything else, makes birds supreme aviators. Each of the surface feathers has hundreds of barbs [2, 3] extending in one plane from the central midrib or rachis [1] to form the feather vane. Lying in the same plane, barbules [4] project from the barbs on both sides. The barbules on the side of the barb away from the bird's body bear hooks which catch on the unhooked barbules of the next barb. A single flight feather, of a large bird, may carry as many as 1.5 million barbules to maintain its shape.

1 Feathers are an adaptable covering. They can be strong plumes for wings and tail, coverts and contour feathers for maintaining an unbroken streamlined surface and down feathers for insulation. The feathers overlap better than tiles on a roof for they maintain their continuity and connections in spite of considerable changes in shape in the underlying superstructure, such as wing and shoulder joints.

1 Primary coverts
2 Marginal coverts
3 Secondary coverts
4 Secondary feathers
5 Primary feathers
6 Slotted primary
7 Down
8 Bristle
9 Contour feather
10 Filoplume
11 Semi-plume
12 After shaft

1 Eye
2 Rictal bristles
3 Ear
4 Apterium
5 Dorsal region
6 Primaries
7 Secondaries
8 Pelvic region
9 Rectrices

4 When a bird hovers it almost stalls. In still air its body is nearly vertical, with wings beating horizontally.

3 The plumes, coverts and contour feathers of most birds grow from the skin in definite tracts. The feathers grow from papillae ("goose pimples") which produce one or two sets of feathers a year, replacing old ones and also allowing for changes in colour.

Apteria : featherless areas
Pteria : feather tracts

Arctic tern
Sterna paradisaea

Golden eagle
Aquila chrysaetos

5 The golden eagle, like many large birds will soar on outstretched, unbeating wings if air currents rise faster than the bird loses height.

largely fused to form an attachment for the primary flight feathers. The first remaining digit forms a support for the bastard wing which acts as a slot to prevent stalling at low flight speeds. The secondary flight feathers are attached to the ulna in the forearm. Together with the remarkable structure of feathers [1, 2], this results in a wing of extreme efficiency and adaptability.

Feathers in flight
The flight feathers of the wing (remiges) and those of the tail (rectrices) provide lift and guidance in flight, but their aerodynamic properties are not fully understood. In normal flapping flight [8] the wing is beaten strongly downwards and forwards and then more quickly upwards and backwards. On the down stroke the wing has such a high angle of attack that it would stall if the primary feathers did not act as individual adaptable aerofoils to prevent this. Each feather twists up and back along its length so that the total resultant thrust is strongly forwards, the separation of the feather tips assisting this. Also, at a certain angle of attack, the bastard wing or alula lifts forwards from the front of the wing creating a "slot" to cut down turbulence over the aerofoil and reduce the risk of the wing stalling. Birds capable of flying slowly have particularly well-developed wing slots (large spaces between the primary feathers); for example those of the golden eagle (*Aquila chrysaëtos*) [5] may take up as much as 40 per cent of the total wing area when in use. In vultures the tail area is particularly large and helps provide lift during gliding.

At the other extreme from eagles and vultures, seabirds such as the albatrosses have long, thin wings. These birds rarely flap their wings but soar in the wind, principally by gliding and accelerating downwind and then turning and climbing sharply upwind until they almost stall. Their flight is so specialized that on still days albatrosses are "grounded".

The specialized wings of humming-birds, in which flight feathers are primaries only, may beat up to 50 times a second or more while the bird is hovering, the stroke being horizontal and powered in both forward and backward directions.

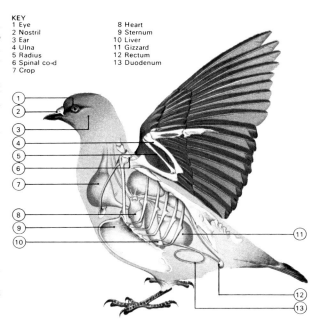

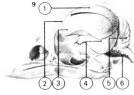

Whinchat
Saxicola rubetra

6 On landing, a bird reduces its speed. This is usually achieved by swinging the body into an upright position with its tail spread and its wings beating against the direction of flight.

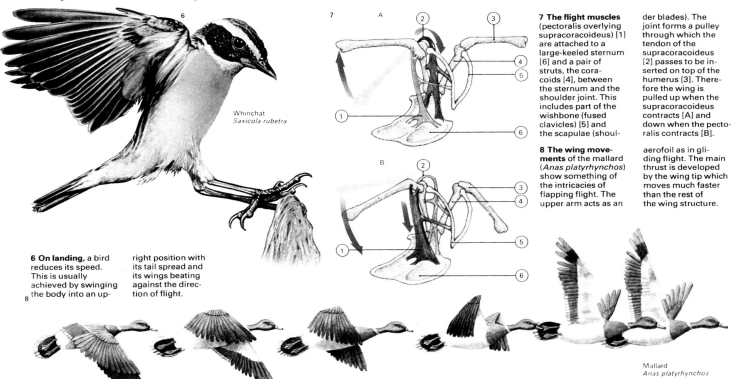

7 The flight muscles (pectoralis overlying supracoracoideus) [1] are attached to a large-keeled sternum [6] and a pair of struts, the coracoids [4], between the sternum and the shoulder joint. This includes part of the wishbone (fused clavicles) [5] and the scapulae (shoulder blades). The joint forms a pulley through which the tendon of the supracoracoideus [2] passes to be inserted on top of the humerus [3]. Therefore the wing is pulled up when the supracoracoideus contracts [A] and down when the pectoralis contracts [B].

8 The wing movements of the mallard (*Anas platyrhynchos*) show something of the intricacies of flapping flight. The upper arm acts as an aerofoil as in gliding flight. The main thrust is developed by the wing tip which moves much faster than the rest of the wing structure.

Mallard
Anas platyrhynchos

9 Birds have "instinct" brains for the co-ordination of complex inborn behaviour patterns and a well-developed corpus striatum which controls these activities. Mammals possess intelligence brains with a highly developed cerebral cortex. The bird's cortex is well enough developed for advanced types of learning to occur and this, with instinctive behaviour, the power of flight and good eyesight, enables birds to show complex behaviour.

1 Cerebral cortex
2 Olfactory bulb—small birds have poor sense of smell
3 Corpus striatum
4 Large optic lobes
5 Cerebellum—muscular co-ordination centre
6 Medulla oblongata— origin of most of cranial nerves

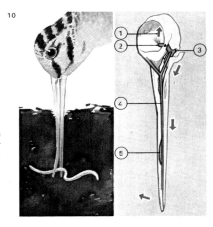

10 Most manipulations are performed by the bill. The woodcock (*Scolopax rusticola*) shows how complex such procedures can be when it plunges its bill deep into mud to take a worm. When it has reached its prey, the bill tip is opened by the quadrate bones [2] rocking forwards in their seating [3] when muscles [1] contract. This pushes forwards on the jugal bones [4], these in turn push the tip of the upper beak open beyond a thin, hinge-like area [5].

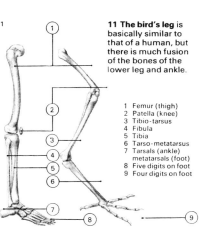

11 The bird's leg is basically similar to that of a human, but there is much fusion of the bones of the lower leg and ankle.

1 Femur (thigh)
2 Patella (knee)
3 Tibio-tarsus
4 Fibula
5 Tibia
6 Tarso-metatarsus
7 Tarsals (ankle) metatarsals (foot)
8 Five digits on foot
9 Four digits on foot

How birds reproduce

The laying of a clutch of bird's eggs is the result of mating, which in turn is the result of courtship. The courtship displays [1] of birds are as varied as the birds that employ them and are essential to successful procreation.

Courtship displays

Visually – to human eyes at least – some bird displays may seem to be almost non-existent, as in the robin (*Erithacus rubecula*), but this is a species with a highly developed song. Songbirds have evolved their characteristic songs for the establishment and maintenance of the pair-bond, as well as for territorial advertisement and warning off intruders. In species of similar appearance living in the same area, such as the willow warbler (*Phylloscopus trochilus*) and the chiffchaff (*P. collybita*), the differences in song seem to be the principal means of both specific and individual recognition, to the birds as well as to humans.

The most complex bird courtship displays are those of the bowerbirds and other members of the family Ptilonorhynchidae, widespread in Australasia. Not only are the males of many species very brightly coloured but many also have outstanding powers of mimicry – probably the best in the animal kingdom – and build elaborate display grounds (bowers) in which they conduct their courtship. Some species build mounds, huts or avenues; the ground and their buildings may be decorated with fruits, flowers, shells or bones; some species even paint their bowers with a mixture of saliva and dried grass, charcoal or fruit pulp.

Nests and eggs

The nests of birds [4, 5, 7], where the eggs are laid, vary from mere scrapes in the ground as made by nesters like the waders, to large intricately woven and even communal structures like those of the weaver birds (Ploceidae). Some birds, such as the auks (Alcidae), make no nest at all, and lay their eggs on the bare rock. A most extreme case is the emperor penguin (*Aptenodytes forsteri*), which incubates its egg on its feet, covered with a flap of feathered belly, through 64 days and nights of darkness and bitter cold in the Antarctic winter.

The bird's egg [6] is an adaptation for embryonic development on dry land. It is a closed system: within the shell when it is laid, there is provision for the nourishment of the growing embryo, room for it to develop, mechanical protection and a means of dealing with waste material. Birds' eggs are uricotelic; that is, the excretory product (waste) of the embryo is largely uric acid. This is a highly insoluble substance and therefore does not dissolve in the embryo's body fluids. It remains inert in the egg and is left behind when the fully developed embryo hatches and is removed by the adult bird along with the discarded egg shell.

The formation of an egg in the oviduct [Key] follows a carefully controlled sequence of events. Following the courtship, the act of mating can take place either on the ground or in the air. The male has no penis, so to transfer his sperm to the female he must press his genital opening, or cloaca, close to her cloaca. This "cloaca kiss" takes only a few seconds to complete. The male's sperm then swims up the female's oviduct and fertilizes the egg. After fertilization, the egg proper – ovum plus nutrient yolk – passes

Ruffed grouse
Bonasa umbellus

Blackcock
Lyrurus tetrix

Hawfinch
Coccothraustes coccothraustes

Yellow-thighed manakin
Pipra mentalis

Greater bird of paradise
Paradisaea apoda

Ruff
Philomachus pugnax

Kingfisher
Alcedo atthis

Great frigatebird
Fregata minor

Adélie penguin
Pygoscelis adeliae

Shag
Phalacrocorax aristotelis

Gannet
Morus bassanus

Black-headed gull
Larus ridibundus

Robin
Erithacus rubecula

Jackdaw
Corvus monedula

Pochard
Aythya ferina

Pintail
Anas acuta

1 Courtship leads birds into a wide variety of displays, some of which are shown here. The most obvious is male adornment as with the plumage of the bird of paradise and the grouse. Bizarre movements often accompany plumage changes, as in manakins and penguins. Simple feeding gestures can also be significant, as in robins and kingfishers. Hawfinches tend to be courtly in their approach. If stress builds up, "cut off", shown here by black-headed gulls who are head flagging, often reduces tension.

2 Adélie penguin
Pygoscelis adeliae

2 Between Adélie penguins, courtship includes the presentation of nest stones by the male. This reinforces the attraction of the birds to a particular spot to which they return yearly.

The birds occupy "rookeries" in the Antarctic spring (Sept/Oct) travelling more than 300km (200 miles) over sea, ice and snow to reach them. They usually hatch their two eggs after 35 days.

3 The tailor bird (*Orthotomus sutorius*) is one of the species that has fitted in with human development, occupying gardens and verandas where there are suitable plants for nest-making. This involves sewing two leaves together to make a basket, inside which a nest is built. The bird makes holes in the leaf margins with its bill and "sews" the edges together with cotton, grass or wool.

4 Cave swiftlets (*Collocalia fuciphaga*) make their nests in the walls and roofs of caves. The nests are often in total darkness and the birds use a form of echo-location for navigation. The principal nest material is saliva, sometimes mixed with feathers or other materials, depending on the species. Pure saliva nests are highly prized for the production of bird's-nest soup.

5 The rufous oven-bird (*Furnarius rufus*), of Brazil and Argentina, lays its three to five eggs in a conspicuous domed nest of mud and straw. This gives it a local name of *el hornero* (the baker).

down the oviduct. It is then covered by the protective jelly-like albumen or "white" of the egg, then by two shell membranes and the shell. The last is deposited in layers and is followed by a thin cuticle. The pigments that give the egg its specific appearance are deposited mainly in the outer layers of the shell and in the cuticle. They consist of two basic colours: red-brown and blue-green. These two pigments give a remarkable range of egg patterns.

Some of the most strikingly coloured eggs are those of birds that nest on the ground in the open, such as the waders, gulls, terns and nightjars. The eggs of tinamous are unusual in that they have the appearance of polished porcelain and a range of colours including chocolate, grey, purple and near-black.

Shapes and sizes

The shapes of birds' eggs vary considerably, from the well-known ovoid to the almost spherical eggs of some birds of prey and the pear-shaped ones of plovers and guillemots. There is also a great range of size in birds' eggs from the largest, that of the ostrich, to the smallest humming-bird's egg. The egg of an ostrich (*Struthio camelus*) is, on average, 15–20cm (6–8in) long and weighs about 1.5kg (3.5lb). That of the bee humming-bird (*Mellisuga helenae*) is only 11.4mm (0.45in) long and weighs about 0.5gm (0.18oz), less than one 3,000th of the weight of an ostrich egg. The largest egg known was laid by the now extinct elephant bird (*Aepyornis maximus*), from Madagascar. Its fossilized eggs are about 32.5cm (13in) long and must have weighed about 12.25kg (27lb).

The bird's egg is such an efficient structure for protecting the growing embryo inside that it is necessarily difficult to break out of. The hatching bird therefore has to employ a special combination of "equipment" and behaviour in order to get out of its shelly prison. By means of an egg-tooth at the tip of the upper bill and a series of vigorous upward nods of the head, the chick makes a series of ruptures ("pips") round the blunt end of the egg, turning anticlockwise on its axis as seen from that end, until enough of the shell has been cracked to enable the chick to break the rest with one blow.

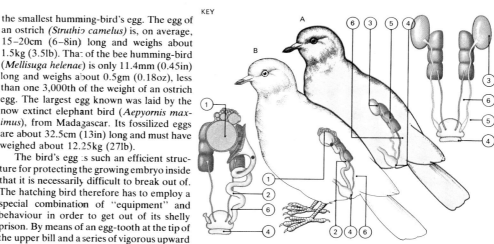

KEY

In adult pigeons, the male [A] and female [B] urinogenital systems have basic similarities. In the female, the reproductive system is made up of the left ovary [1] and the oviduct [2] only, the right side being reduced to leave room for the relatively large egg. Eggs and sperm from ovaries and testes [3] pass to the cloaca [4] via the oviduct and vasa deferentia [5] respectively. The kidney ducts [6] also open into the cloaca.

6 The chicken's egg is surrounded by a strong calcareous shell which has two lining membranes. These enclose the albumen that surrounds the yolk, on top of which is the ovum [A]. A fertilized egg develops quickly if incubated, the heart pumping blood after only three days. After nine days, capillaries (minute blood vessels) extend over the yolk, transporting nutrients to the developing chick [B]. This is ready to hatch after 21 days [C]. Downy chicks [D] such as chickens or ducklings can follow the parent as soon as they are dry after hatching, and are called nidifugous. Nidicolous young are almost helpless [E].

6A B C D E

8 The gaping reflex, one of the few responses of nidicolous baby birds, is at first stimulated by vibration. Later it is directed towards the parent birds.

7

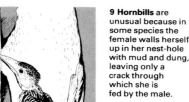
8

7 The lapwing nests in open fields [A] while the little tern [B] and the ringed plover [C] nest on the sea-shore. Eggs and chicks [D] and often the adult birds of these and other ground-nesters match their surroundings almost to perfection and are said to be cryptically coloured. This camouflaging of the young is most effective when they crouch motionless at the approach of a potential predator.

11

Common European cuckoo
Cuculus canorus

9 Red-billed dwarf hornbill
Tockus camurus

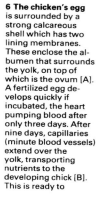

9 Hornbills are unusual because in some species the female walls herself up in her nest-hole with mud and dung, leaving only a crack through which she is fed by the male.

10 Pheasants such as the koklass illustrate the differences between the sexes in many birds. The males generally sport bright attractive colours while the female has cryptically coloured plumage for defence while incubating.

10

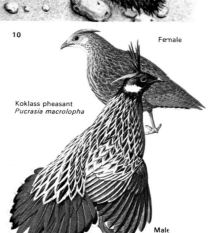

Female

Koklass pheasant
Pucrasia macrolopha

Male

11 "A cuckoo in the nest" is serious if the species is the common European cuckoo. This bird lays its egg in the nests of a small host species. The nestling ejects its nest-fellows, thus receiving all its foster parents' attention.

Birds and migration

Birds owe much of their success as a group to their unusual powers of migration. The phenomenon of migration, for the scientist bewilderingly complex and for the animal extremely exhausting, is deep-rooted in many species and has undoubtedly exerted great influence on their evolution. The change in residence of migrants parallels changes in appearance, diet or even the behaviour of more sedentary bird species.

The longest journeys

True migration involves a regular, seasonal movement between one area, in which the animal breeds, and another, in which the climatic and other conditions are more suitable when the breeding season is over. Thus some bird species, particularly wildfowl and waders, breed in the tundra and migrate to temperate regions for the winter. Others, particularly insectivores of the Northern Hemisphere such as the swallows and warblers, breed in temperate lands and move south to winter in the tropics. In the Southern Hemisphere, the migratory movements are largely in the other direction, although the

greater land area in the Northern Hemisphere means that there are more species moving south than north.

Bird migration varies hugely in its scope [Key]. The greatest travellers, particularly the sea birds, may fly almost from pole to pole and many land birds also make long trans-equatorial migrations. At the other extreme, some species may move only from an inland breeding area to the coast or from mountain breeding areas to winter quarters in the valleys. Sea bird colonies provide a vivid example of local movement, with vast numbers of breeding birds moving from their coastal nest sites to a much wider dispersal area over the open sea.

The misfits: "partial migrants"

Not all bird species show the consistent, cohesive movements that characterize the true migrants. These "partial migrants" are species in which not all individuals behave uniformly; some may go on a post-breeding migration, others may not. In Britain, the song thrush (*Turdus philomelos*) and the lapwing (*Vanellus vanellus*) show this

inconsistent behaviour, which may provide a clue to some of the selection pressures involved in the evolution of migration. These species may be at the edge of their "comfortable" winter range, and some individuals will be poorly adapted to local conditions. These birds will tend to move out for the winter.

Migration does not, strictly speaking, include movements such as irruptions [3], or extensions of range, even though birds often travel greater distances in the process. Extensions of range are largely the result of a preferred food source becoming available over a wider area. A most impressive extension of range has been seen in the collared dove [6] and the fulmar petrel (*Fulmarus glacialis*) of the eastern Atlantic. In recent decades the fulmar has progressively colonized Britain from the northern islands, and in the last 100 years has increased its population five-fold.

The origins and mechanics of migration

The evolutionary origin of migration is extremely complex and as yet incompletely understood. The phenomenon is undoubtedly related to the necessity of

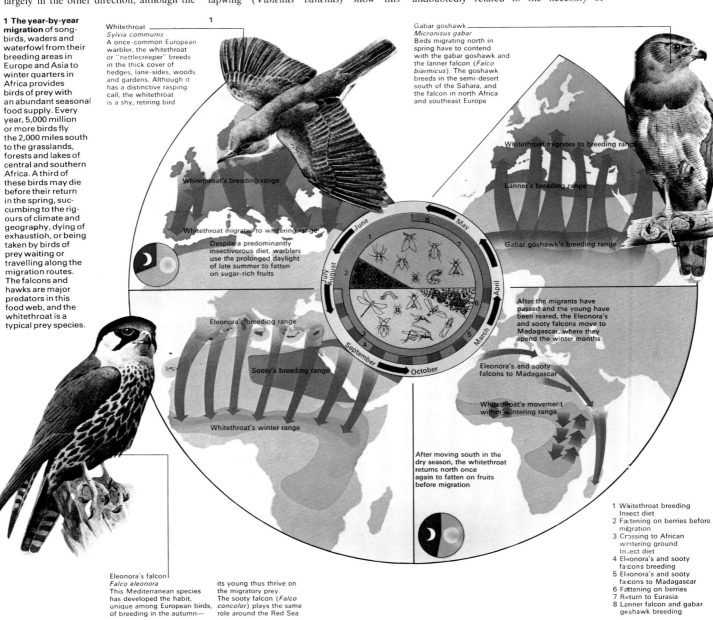

1 The year-by-year migration of songbirds, waders and waterfowl from their breeding areas in Europe and Asia to winter quarters in Africa provides birds of prey with an abundant seasonal food supply. Every year, 5,000 million or more birds fly the 2,000 miles south to the grasslands, forests and lakes of central and southern Africa. A third of these birds may die before their return in the spring, succumbing to the rigours of climate and geography, dying of exhaustion, or being taken by birds of prey waiting and travelling along the migration routes. The falcons and hawks are major predators in this food web, and the whitethroat is a typical prey species.

Whitethroat
Sylvia communis
A once-common European warbler, the whitethroat or "nettlecreeper" breeds in the thick cover of hedges, lane-sides, woods and gardens. Although it has a distinctive rasping call, the whitethroat is a shy, retiring bird

Whitethroat's breeding range

Whitethroat migrates to wintering range

Despite a predominantly insectivorous diet, warblers use the prolonged daylight of late summer to fatten on sugar-rich fruits

Eleonora's breeding range

Sooty's breeding range

Whitethroat's winter range

Gabar goshawk
Micronisus gabar
Birds migrating north in spring have to contend with the gabar goshawk and the lanner falcon (*Falco biarmicus*). The goshawk breeds in the semi-desert south of the Sahara, and the falcon in north Africa and southeast Europe

Whitethroat migrates to breeding range

Lanner's breeding range

Gabar goshawk's breeding range

After the migrants have passed and the young have been reared, the Eleonora's and sooty falcons move to Madagascar, where they spend the winter months

Eleonora's and sooty falcons to Madagascar

Whitethroat's movement within wintering range

After moving south in the dry season, the whitethroat returns north once again to fatten on fruits before migration

Eleonora's falcon
Falco eleonora
This Mediterranean species has developed the habit, unique among European birds, of breeding in the autumn—

its young thus thrive on the migratory prey. The sooty falcon (*Falco concolor*) plays the same role around the Red Sea

1 Whitethroat breeding Insect diet
2 Fattening on berries before migration
3 Crossing to African wintering ground Insect diet
4 Eleonora's and sooty falcons breeding
5 Eleonora's and sooty falcons to Madagascar
6 Fattening on berries
7 Return to Eurasia
8 Lanner falcon and gabar goshawk breeding

finding appropriate food and weather for the raising of a brood. The migratory instinct has probably been developing over a period of 40 to 50 million years and, though greatly influenced by the Pleistocene glaciations, did not originate with them. Migratory movements are governed by inherited factors but initiated and guided by environmental influences. Individuals of a migratory species will thus show "migratory restlessness" at the onset of the migration period, even though there is usually no external indication that the season is changing. However, the weather may then delay or accelerate departure and to some extent modify the route taken. Despite occasional deviations, when birds get lost or blown off course, the direction of migration remains remarkably constant.

The physiological mechanisms behind these great powers of navigation and orientation are still a mystery. If the means by which migrating birds compute their bearings is unclear, it is well established that they use landmarks when convenient and sky-signs, almost certainly both the sun and the stars. Recent studies prove that Adélie penguins navigate by the sun. Nevertheless, the birds apparently follow a magnetic bearing rather than a bearing to a particular point on the earth's surface.

The habits of certain species show that they have an innate knowledge of the migratory route. The young of these species depart in autumn before or after their parents. Individual adult birds displaced laterally by humans while on migration will resume their flight on release on exactly the same magnetic bearing, finishing up with the same lateral displacement from their destination. On the other hand, some birds can navigate back home after an unusual displacement. One famous example of the innate migratory urge is that of a Manx shearwater (*Puffinus puffinus*) that was taken by plane to Boston, Massachusetts, from its burrow on Skokholm Island, South Wales. It was back on its nest 12.5 days after release in America – ten hours before the mail arrived giving details of its release. How the bird managed to find its way over this distance, with no landmarks to guide it, and in such a short time remains a mystery.

KEY

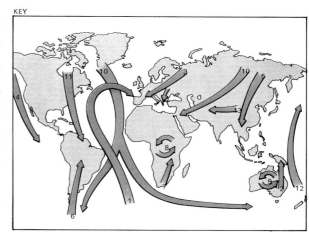

When migrating, many birds travel along fairly well defined flightpaths or fly-ways. These often follow winds or sea currents [1, 2], land contours or coast-lines [3, 4]. Birds from high latitudes move towards the Equator [5, 6, 7] during the winter. Some travel many thousands of miles [10, 11 12]. In Eurasia east-west movements take place [13]. On large land masses such as Australia and Africa, fruit-eaters and insectivores follow seasonal food sources [8, 9].

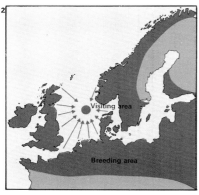

2 Shelduck (*Tadorna tadorna*) undertake mass post-breeding migrations from the northwest European coasts to Heligoland to sit out flightless moulting periods. They return in smaller parties.

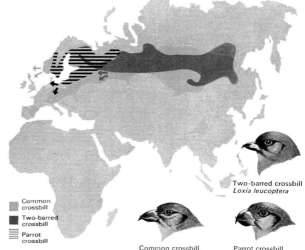

Common crossbill

Two-barred crossbill

Parrot crossbill

Two-barred crossbill *Loxia leucoptera*

Common crossbill *Loxia curvirostra*

Parrot crossbill *Loxia pytyopsittacus*

3 Irruptions are special "migratory" movements which occur irregularly in a few species as a response to unusual food conditions. The crossbills of Eurasia, for example, may leave their normal breeding areas in large numbers, moving south and west to exploit exceptional crops of their main food, conifer seeds. The parrot crossbill, the heaviest-billed, specializes in tough pine cones, while the common crossbill has an intermediate bill, and prefers spruce. The two-barred species, which has the lightest bill, is restricted to larch cones. These species do not usually set up new, long-term viable populations.

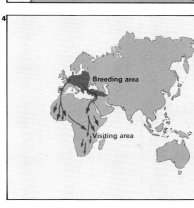

1 Sun obscured
2 Intermittent cloud
3 Clear sky

Adélie penguin *Pygoscelis adeliae*

4 The European populations of the white stork (*Ciconia ciconia*) winter largely in the steppes and deserts of southern Africa. Flying via land masses they avoid long sea crossings with poor soaring conditions.

5 The penguins navigate by sky-signs just as flying birds do. Faced with a return journey to the breeding grounds of up to 200 miles across snow and ice, Adélie penguins (*Pygoscelis adeliae*) use the sun to aid their spot-on homing navigation. This was conclusively proven when birds were taken hundreds of miles from home and released on featureless terrain. With the sun obscured, the navigation of the displaced birds was extremely erratic but with a clear view of the sun they were well able to orientate themselves accurately. Intermittent cloud gave intermediate results. How the birds compensate for the daily movement of the sun is as yet unknown.

6 Extensions of range are not examples of true migration. The collared dove spread slowly from Asia early this century and had reached Hungary by 1928. Then its rate of progress increased and its range was extended by 1,200 miles in 20 years. It reached Britain in 1955, is still moving westwards and may have bred in Iceland in 1973. This success is due largely to its adaptability: it nests in walls and shrubs as well as trees, raising up to five broods a year and feeding on grain wasted by man.

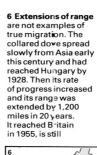

Collared dove *Streptopelia decaocto*

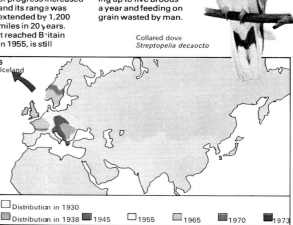

Distribution in 1930
Distribution in 1938
1945
1955
1965
1970
1973

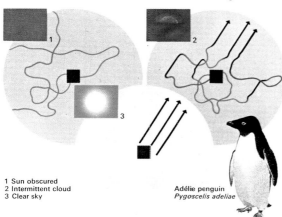

Bird life and variety

Within the limits imposed by their special adaptations for flight, birds show a remarkable range of form and habit. The largest living bird is the ostrich (*Struthio camelus*), which stands up to 2.4m (8ft) tall and weighs up to 136kg (300lb), while the smallest is the bee humming-bird (*Mellisuga helenae*), less than 6.3cm (2.5in) long and under 2.5gm (0.1oz) in weight. In fact, the world population of bee humming-birds – about 100,000 would weigh approximately the same as a pair of large ostriches.

Limitations of flight
The size of birds imposes a limit on their ability to fly, for weight increases proportionally more than lifting area as size increases. A heavier bird therefore needs proportionally more power for its weight. The trumpeter swan (*Cygnus cygnus buccinator*) is probably the heaviest living flying species, with a weight of up to 17.2kg (38lb). Its 3m (10ft) wing span is exceeded only by that of the wandering albatross (*Diomedea exulans*), which may be 3.5m (11.5ft) or more. It would be interesting to know the flying abilities of

the apparently larger, but extinct, giant condor *Teratornis* which lived in the Pleistocene period some two million years ago.

The flying abilities of birds, coupled with their capacity for regulating a high body temperature, have made possible a range of ecological and behavioural variation seen in few other groups. The 8,600 species (approximately) of birds are distributed throughout almost every part of the world.

Breeding and feeding
Birds breed almost anywhere except in the sea. Nests may be on or in the ground, in holes in trees, cliffs or buildings, in low bushes or the tallest trees, even floating on still water. Some species nest in enormous colonies, others make no nest at all, laying their eggs on the ground or on cliff ledges. The "mound builder" birds (Megapodidae) use the heat from the fermentation of rotting vegetation, or hot sand, to incubate their eggs. And the emperor penguin (*Aptenodytes forsteri*) breeds in the depths of the Antarctic winter. The male bird, which incubates the egg, stands on ice at temperatures below

−60°C (−76°F) for 64 days holding a single egg on its feet before the chick hatches.

A remarkable variety may also be seen in feeding habits. Birds have adapted to all the principal food sources, from the smallest planktonic organisms to the largest whale carcasses. Different avian predators specialize in feeding on a great range of prey organisms, from the smallest invertebrates to birds and mammals several times their own weight. Some birds store food and use the stores in winter; others feed on parasites that live on mammal skins, the skin tissue itself and on blood. The Egyptian vulture (*Neophron percnopterus*) casts stones at ostrich eggs to break them open, and one Galapagos finch (*Camarhynchus pallidus*) forces insects out of tree holes and crevices by means of a cactus spine or twig held in its bill.

Some birds are parasites upon other avian species. The European cuckoo (*Cuculus canorus*) is a well-known example. This species, like the brown-headed cowbird (*Molothrus ater*) of North America and some other species, lays its egg in the nest of a "host" bird, leaving the host to rear the

1 Reproductive isolation through different plumage and displays in closely related species can be clearly seen in the ducks of the genus *Anas* and the American wood warblers of the genus *Dendroica*. In North America there are 13 *Anas* species of dabbling ducks – or "puddle" ducks – and 20 *Dendroica* warblers. Several species often live together and must be reproductively isolated in order to avoid producing unhealthy hybrids. This is achieved by a combination of colour, form, movement and vocalization which is distinctive for the male of each species, together with an instinctive preference on the part of the female for the male who gives the "correct" display. Thus half a dozen species of ducks may be found on the same stretch of water, even in the breeding season, without any significant variation occurring. The same applies to the warblers. In both the *Anas* ducks and the *Dendroica* warblers the distinctiveness of the males is enhanced by a striking breeding plumage. The females are very similar in appearance, apparently even to the birds themselves. *Anas* and *Dendroica* males in the post-breeding season are extremely difficult to identify.

Wood warblers *Dendroica* spp	Dabbling ducks *Anas* spp
1 Blackpoll warbler *D. striata*	8 Gadwall *A. strepera*
2 Magnolia warbler *D. magnolia*	9 Pintail *A. acuta*
3 Townsend's warbler *D. townsendi*	10 Cinnamon teal *A. cyanoptera*
4 Blackburnian warbler *D. fusca*	11 Mallard *A. platyrhynchos*
5 Chestnut-sided warbler *D. pensylvanica*	12 Blue-winged teal *A. ciscors*
6 Black-throated blue warbler *D. caerulescens*	13 Common teal *A. crecca*
7 Prairie warbler *D. discolor*	

young. Other birds, such as the skuas (family Stercorariidae) are kleptoparasites, forcing neighbouring sea birds to disgorge their food.

The plumage of birds shows a wide range of form, pattern and colour. Some species are brighter than the most exotic flowers or jewels, others are as sombre as desert sand or the blackest night. Some of the brightest species are the pheasants (family Phasianidae) and the birds of paradise (family Paradisaeidae).

An additional and unusual example of behavioural variety in birds is that of the satin bowerbird (*Ptilonorhynchus violaceus*). Like the other bowerbirds, all found in Australia and New Guinea, the male of this species builds a bower of twigs on the ground in which to display itself to the female. The bower is decorated with objects such as feathers and flowers similar in colour to those of the bird's rivals, and is orientated north-south so that the bird is not dazzled by the sun when displaying. Additionally, this species "paints" its bower with fruit pulp held in a spongy wad of fibre retained in the bill. The male bird may display for several months, posturing with the display objects held in the bill until, with the seasonal appearance of insect food for the young, mating takes place.

Learning by experience

Bird behaviour, although advanced, does not necessarily indicate "intelligence" on the part of the bird. But the variety of bird behaviour increases through "insight learning". This may occur when tits (family Paridae) find there is cream in the tops of milk bottles, quickly learning through their own experience or by watching other birds that they have only to peck through the bottle top to reap the benefit. A similar process may also occur when tits and crows (family Corvidae) learn to pull in a length of string in order to obtain food attached to the end. That this kind of behaviour may involve insight is suggested by the results of experimental work with crows, parrots (family Psittacidae) and finches (family Fringillidae), showing that they can learn to "see into" a situation and modify their behaviour accordingly. Such abilities may increase enormously the variety of bird form and activity.

KEY

Puffin
Fratercula arctica

Great black-backed gull
Larus marinus

Cormorant
Phalacrocorax carbo

Gannet
Morus bassanus

Razorbill
Alca torda

Guillemot
Uria aalge

Fulmar
Fulmarus glacialis

Shag
Phalacrocorax aristotelis

A sea cliff often provides a habitat of such variety that many different bird species may use it. These eight species of five families have each found suitable, though specialized, nesting sites.

Jackass penguin
Spheniscus demersus

2 Penguins are the most aquatic of all birds, the 18 living species showing a mastery of their element that enables them to prey on fast-swimming fish with great success. The jackass is typical in its streamlined form, flipper-like wings and steering-paddle feet placed far back. The feathers have been modified to form a close-fitting scale-like wet suit, and the specific markings are largely on the head.

3 Puffin
Fratercula arctica

3 The puffin is a marine bird that finds its prey — primarily fish — entirely under water. Its aquatic adaptations are less extreme than those of the penguins. It is not so streamlined, the feet are not so far back (hence it does not have to stand bolt upright) and the wings are still used for flight. But it can catch smaller or slower-moving fish for itself or for its unfledged chick.

4 Great hornbill *Buceros bicornis*

4 The 45 species of hornbills are found in tropical Africa and Asia. The enormously developed bill seen in the great hornbill is used for display and nesting purposes rather than for feeding.

5 Toucans are the New World counterparts of the hornbills, some 35 species being found in the forests of tropical America. But they have even bigger and brighter bills.

5 Toco toucan
Ramphastos toco

8

New Holland honeyeater
Phylidonyris novaehollandiae

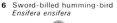

6 Sword-billed humming-bird
Ensifera ensifera

7 Double-collared sunbird
Nectarinia mediocris

6 The humming-birds of the Americas, numbering more than 300 species, live largely on nectar. Some use tubular tongues to suck nectar from flowers and may add to their diet insects trapped in or near blooms. Others have brush-like tongues.

7 The sunbirds, numbering about 100 species, fill the nectar-feeding niche in Asia and Africa. The double-collared sunbird lives high in the Kenya mountains and conserves heat by lowering its body temperature at night.

8 The 167 species of honeyeaters are the main insect- and nectar-feeding birds in Australasia. Long isolation without much competition has resulted in a wide adaptive range of form, of which the New Holland honeyeater is typical.

How birds behave

Studies of bird behaviour were hampered until recently by the idea that birds had little learning ability and were mainly creatures of instinct. This misconception was based largely on birds' lack of brain structures similar to those of the cerebral cortices of mammals (the structures responsible for complex behaviour and consciousness). Recent tests indicate, however, that birds have learning capacities in some areas that are exceeded only by the highest mammals [8, 9, 10] as well as unmatched navigational skills which are demonstrated in their migratory activities. Bird behaviour is a complex mixture of the learned and the instinctive, the flexible and the rigidly pre-programmed.

Instinctive behaviour

Instinct can be seen at its most blind in the habits of birds in the incubation of their eggs. Herring gulls, for example, will sit on any large egg at the expense of their own. Much instinctive behaviour [3] takes the form of intricate sequences known as fixed action patterns. These are inherited and tend to be stereotyped; they can be seen in such

activities as fly-catching by insect-eating birds, or in nest-building. They are also a conspicuous feature of courtship. It is possible to cross closely related but different species to produce hybrids whose courtship and nesting behaviour has elements of the patterns of each parent, as in lovebirds [5]. This rigidity of behaviour is thought to be an essential element in the natural process of keeping species distinct from each other by preventing mating between them.

Signs and rituals are also involved in territorial behaviour [2]. Most birds have well-defined territories within which they are noisier and livelier; and they generally seem to behave as if they are more at home in these territories than they are outside them. Such birds defend their territories from intruders of the same species – or strictly speaking from any animal or inanimate object of roughly the right size that has a similar appearance. In the robin the red breast is the sign stimulus for defensive behaviour [1]. Such behaviour, however, rarely involves a fight; instead the defending bird engages in a stereotyped threat display – a sign that the intruder should

retreat. The intensity of display depends on the distance of the encounter from the territorial centre. Cautious "sparring" occurs at the edge of a bird's territory but fighting may be involved at the centre.

Displacement and imprinting

A bird sometimes finds itself in a situation that elicits two conflicting behaviour patterns. At the boundaries between their territories, for example, males may be torn between fight and flight. And in the mating season many birds have to overcome their natural aversion to physical contact. Once the level of sex hormones reaches its peak the sex drive swamps the aversion, but there is a stage at which the two tendencies more or less balance one another. In such situations the bird resorts to actions known as displacement activities. These activities have nothing to do with the animal's innate drives: the male bird, for example, may preen its feathers or peck at something [7] instead of approaching a female.

In most of these behaviour patterns there is an innate recognition of a specific inducing

1 **The threat posture of a robin** defending its territory is designed to display the red breast [A] and varies according to the position of intruders [I]. A defending robin adopts posture [B] when the intruding bird is above it, but posture [C] if the intruder is below it. If the threat display fails as a deterrent the defending robin may actually launch an attack [D].

2 **Rigid demarcation of territory** can be observed even when birds gather in large social groups. An example is the uniform spacing between nests in a colony of gannets (*Morus* sp).

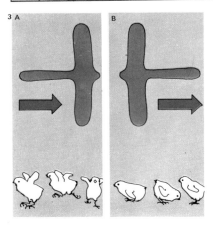

3 **Instinctive fright reaction** to predators flying overhead is shown by chicks. The same cardboard model of a bird that induces fear of short-necked, long-tailed predators [A] is ignored if the direction of movement is reversed so that it looks like a long-necked, short-tailed and therefore harmless species [B].

4 **Goslings became "imprinted"** to the German ethologist, Konrad Lorenz, who studied the behaviour of tame geese.

stimulus. But much bird behaviour is modified by experience and some is entirely learned. An example of pre-programmed behaviour with a single learned component is the phenomenon known as imprinting [4]. Newly hatched ducks, goslings and chicks will imprint on and follow the first moving thing they see. This is usually their mother but behaviour experiments have shown that baby birds will adopt a wide range of objects.

In imprinting the behaviour is pre-programmed but there is no innate recognition of the parents. This knowledge is quickly learned and the young bird tends to adopt as its parent the first large moving object it sees. Not only will the young bird follow this object as a parent; later it will direct sexual advances towards it, as demonstrated by the famous experiments of Konrad Lorenz (1903–).

This kind of following is simple, automatic and innate, but more complex behaviour, such as the production of the bird's characteristic song, sometimes involves a subtle combination of innate and learned influences [6]. Some young song birds, if reared in isolation, produce a song

that has roughly the right length and number of notes for its species but the wrong tune. If a group of chaffinches, for example, is reared together without adults, each develops its own version of the species song, identical within the group but quite distinct from the usual song of the wild male.

Navigational expertise

One of the most striking examples of learning in birds is still shrouded in mystery, namely the ability of many birds to home accurately (as opposed to the migration of young birds, which has no learned component). Recently experiments have begun to reveal the complex combination of signals a pigeon uses to orientate itself. One of these is certainly the sun. But because pigeons can orientate themselves even under overcast skies, it seemed that there must be some other kind of "map". It was found that pigeons bearing magnets could manage to navigate in sunny conditions but became disorientated under heavily overcast skies. However, there are still many questions about the pigeon's navigation system that remain unanswered.

Herring gulls will sit patiently on abnormally large models of their own eggs, ignoring their real eggs. The normal stimulus for incubating behaviour is the real egg and the outsize egg acts as a greater stimulus than normal and thus will distract the gull from incubating the real eggs. The rigidity of instinctive behaviour compels the birds to be "tricked" in this way.

5 Various methods of carrying material and nest-building are shown by Madagascar [A], peach-faced [B] and Fischer's lovebirds [C]. Hybrids [D] show conflicting patterns and are incapable of breeding.

6 The development of bird song is the result of a combination of innate and environmental influences. A bird will usually produce some kind of song however it is reared, but if a song-bird is deafened at birth and cannot hear adult male songs [A], it will produce a totally abnormal sound pattern in its own song [B]. A song-bird learns to sing during the first four months of life and after that the song pattern cannot be changed.

7 Displacement behaviour occurs when conflicting instincts are aroused. Rival male blackbirds [A], torn between fight and flight, may settle for pecking at leaves, whereas herring gulls [B] may react to a threat by pulling at grass. A threatened oystercatcher [C] may resolve the same conflict by taking refuge in "sleep". Sex-shy avocets [D] channel their drives into preening (for the male) or bill-dipping (for the female) before progressing to coition.

8 Blue tit
Parus caeruleus

8 Tits have learned to peck through milk bottle tops to drink the cream. They not only learn where to find bottles but can also distinguish the colours of tops and select bottles richest in cream.

9 Jay
Garrulus glandarius

9 Ability to adapt to a novel situation is demonstrated by jays and other birds that can be taught to obtain food attached to a string by pulling up the string and anchoring it with a foot.

10 Raven
Corvus corax

10 Ravens and other birds have been found capable of counting up to seven. Presented with a series of marks, they can associate these with the same number of marks on a box containing a reward.

Island birds

When islands first emerge from the sea, only nesting sea birds and turtles can find a use for such barren hulks of rock and coral. Land-based birds arriving at this stage must either depart or perish; only after vegetation is well established can these species have any chance of surviving. Purely insect-eating species must wait even longer before the island can provide a life-supporting food supply. Those few birds that do settle tend, in contrast to mainland species, to become more versatile and adaptable than their mainland counterparts, and exploit the relatively few feeding niches that are available to them.

Over a quarter of all island species have become extinct through over-specialization and the fortunes of the remainder turn on the delicate ecological balance of their habitats; the smaller the island the finer the thread on which survival depends. On the smallest islands extinction may follow natural fluctuations in population and the arrival of man, with his introduction of rival bird species and predators, does nothing to improve their survival. Hunting is also a major factor in the decline or extinction of many island birds.

1 The palm chat lives only on Hispaniola in the Caribbean, feeding on fruit and berries. It builds huge communal nests.

Palm chat
Dulus dominicus

2 The rare blue vanga is one member of a unique family found in Madagascar.

Blue vanga
Leptopterus madagascarinus

3 The scaly ground roller is a shy resident of the dense east coast rain forest of Madagascar. It spends much of its time at ground level in sparse undergrowth, feeding on insects.

Scaly ground roller
Brachypteracias squamigera

4 The Bornean bristle-head is so rare that little is known about it. It lives in lowland forests, feeding on insects.

Bornean bristle-head
Pityriasis gymnocephala

5 The world population of the Narcondam hornbill is confined to one island in the Bay of Bengal. Its numbers, 400, are controlled by lack of nest sites.

Narcondam hornbill
Rhyticeros narcondami

Cuckoo roller
Leptosomus discolor

6 The cuckoo roller lives high in the forests of Madagascar feeding on insects and reptiles. Its food includes chameleons, caterpillars and insects.

7 The Papuan hawk owl is a rare nocturnal resident of the forests of New Guinea. Nothing is known of its breeding behaviour. It feeds on insects and rodents.

7 Papuan hawk owl
Uroglaux dimorpha

8 Victoria crowned pigeon
Goura victoria

8 The large (turkey-sized) Victoria crowned pigeon was hunted for the magnificent plumes on its head. Unlike other New Guinea pigeons it flew into the open when disturbed, so was easy to shoot.

9 Black-mantled goshawk
Accipiter melanochlamys

9 The black-mantled goshawk is one of eight species found in New Guinea. It ambushes smaller birds at water holes, flying in to seize its prey on the wing.

10 The buff and green colours of the emerald dove match the background of the evergreen and deciduous forests in south Asia, where it is widespread in shady wooded ravines.

10 Emerald dove
Chalcophaps indica

11 The Luzon bleeding heart is a sedentary bird, one of five species confined to the Philippine Islands.

11 Luzon bleeding heart
Gallicolumba luzonica

12

Pheasant pigeon
Otidiphaps nobilis

12 A shy, crow-sized bird, the pheasant pigeon is capable of only short bursts of flight. It inhabits dense hill forests of New Guinea and adjacent islands.

PHILIPPINES

Key to distribution

Luzon bleeding heart

Pheasant pigeon

Nicobar pigeon

Emerald dove

NEW GUINEA

13 Nicobar pigeon
Caloenas nicobarica

13 The Nicobar pigeon is a ruff-necked species that inhabits densely forested islands from Malaysia to the Solomon Islands. It lives mainly on hard seeds, but also on fruits and insects.

14

14 A quiet, subdued bird, the Princess Stephania bird of paradise is much sought after by New Guinea tribesmen for its tail plumes.

15 The magnificent riflebird is an agile climber in its search for insects. It is sombre-looking until seen in the rays of the New Guinea sun.

13 Kokako
Callaeas cinerea

17

18 The fernbird has been forced to retreat as New Zealand marshlands have been reclaimed.

18

Fernbird
Bowdleria punctata

15

Princess Stephania bird of paradise
Astrapia stephaniae

Tui
Prosthemadera novaeseelandiae

16 The tui of New Zealand mimics other birds.

17 The kokako of New Zealand is considered to have a fine song.

Magnificent riflebird
Ptiloris magnificus

20

Rifleman
Acanthisitta chloris

21

21 The forest dwelling New Zealand pigeon was saved from extinction by protective measures.

19 A New Zealand mountain parrot, the kea, feeds mainly on carrion. It was the victim of an extermination campaign when falsely branded as a sheep-killer.

19

New Zealand pigeon
Hemiphaga novaeseelandiae

22 The takahe, a giant rail, was thought extinct until rediscovered in 1948 in New Zealand.

Kea
Nestor notabilis

20 The rifleman is the commonest of the three living species of a unique New Zealand bird family.

24 The iiwi is a long-billed, nectar-feeding honeycreeper fairly common in Hawaii.

22

24

23

Brown kiwi
Apteryx australis

Takahe
Notornis mantelli

23 The brown kiwi is the commonest of the three New Zealand kiwi species. It seems less vulnerable to predators than the others. It lives in forest areas where it feeds on worms and insects.

Iiwi
Vestiaria coccinea

153

The classification of mammals

The mammals, the animal group of which man is a member, are the most highly organized and among the most successful of all the creatures on earth. From humble origins as contemporaries of the dinosaurs they have evolved to become the dominant animal form of our planet, despite the fact that compared with other groups their numbers are relatively few. The mammals total some 4,500 species, but there are more than 1,000,000 insect species, more than 20,000 fish species and about 8,600 bird species.

Common characteristics

All mammals share several common characteristics. They are all warmblooded, are usually hairy, have relatively large brains and suckle their young. Monotremes, although mammals, lay eggs, but all other mammals give birth to live young. Female mammals, except monotremes, have a placenta by which the young are nourished before birth.

Zoologists classify mammals on the basis of their anatomy and their behaviour and as a result they recognize 19 groups or orders of mammals. Each order contains mammals alike in essential features – all the primates, for example, have brains with extremely large and dominant cerebral hemispheres. The animals in each order are further classified into families and each family consists of creatures that share closer relationships to each other than to those of other families. Thus all the great apes in the family Pongidae resemble each other more closely than they do the Old World monkeys of the family Cercopithecidae. The number of families in each mammal order varies widely. An indication of the relative size of each order is given on these pages by the relative sizes of the illustrations of representatives of each order.

The classification of mammals also reflects their age in evolutionary terms, the monotremes being the most ancient. The most advanced mammal order, by this reckoning, is the Pinnipedia because its members appeared most recently; the primates are well down the list.

During the early part of the Eocene period, about 60 million years ago, the mammals found a vast range of habitats available to them due to the extinction of the dinosaurs and evolved along many and varied lines. Several of these early mammal groups themselves became extinct after a few million years and changing conditions left even highly successful orders with only the few representatives that exist today. Thus the elephant, hyrax and aardvark represent orders (Proboscidea, Hyracoidea and Tubulidentata, respectively) that have only one modern family remaining.

The orders of mammals

The two most primitive mammal orders, the Monotremata [1] and the Marsupialia [2], are very different from the remainder. The monotremes lay eggs, whereas marsupials such as the kangaroos possess pouches to house and protect immature young.

The Insectivora [3] are the least specialized of all the higher mammals. The insectivores resemble most closely the ancestors from which all mammals are derived. They have simple teeth and body plans and relatively small bodies and brains.

The bats, grouped together in the order

1 Long-nosed spiny anteater
Zaglossus bruijni

1 The long-nosed spiny anteater of New Guinea is a monotreme of the family Tachyglossidae. It can grow from 45 to 77.5cm (1.5 to 2.5ft) long. The order contains one other family (Ornithorhynchidae).

Northern native 'cat'
Satanellus hallucatus

2 Rocky areas and wooded plains of northern Australia are the home of the northern native "cat". This marsupial is one of the 45 species in the family Dasyuridae whose members are found in Australia, Tasmania, New Guinea, the Aru Islands and Normanby Island. It is a nocturnal creature that feeds on small vertebrates, insects and molluscs. It grows up to 35cm (14in) long, excluding the tail.

3 The common European mole belongs to the family Talpidae, a group of Insectivora which all spend most of their lives underground. The mole uses its wide, clawed forefeet to dig tunnels which may be as deep as a metre (39in). Its minute eyes are sensitive only to light and dark but it preys on a variety of animals including insects, worms, mice, snakes and small birds. It grows up to 18cm (7in) long.

3 European mole
Talpa europaea

4 The colugo, the gliding lemur of the Philippines, is one of the only two species in the order Dermoptera and the family Cynocephalidae. This nocturnal vegetarian is about 40cm (16in) long.

Philippine colugo
Cynocephalus volans

6 Sacred langur
Presbytis entellus

7 Hoffman's sloth is an edentate and a relative of the anteaters. It is placed in the family Bradypodidae, along with other sloths, on the basis of its sluggish habits and its skeletal structure. And like other members of the genus Choloepus it has two clawed toes on its forefeet. All sloths spend the greater part of their lives upside down. Hoffman's sloth grows to 65cm (2ft). It has no tail.

5 The largest bat in the New World, with a wing span of up to 91cm (36in) is the Linnaeus false vampire bat. It is a member of one of the largest mammal orders, the Chiroptera, and is further classified in the family Phyllostomidae. The bat's name indicates its feeding habits for it does not suck blood like a true vampire but feeds on rodents, birds, insects and fruit.

5 Linnaeus false vampire bat
Vampyrum spectrum

6 The sacred langur, found in forests of South-East Asia, is an Old World monkey, a primate, and belongs to the family Cercopithecidae. Like other primates the sacred langur is a social creature and lives in a group comprising up to 40 individuals dominated by an adult male. These monkeys are known as leaf-monkeys and they have a vegetable diet which includes fruits and flowers as well as foliage. They are arboreal animals, and grow to about 1m (39in) long, with tail.

Hoffman's sloth
Choloepus hoffmanni

Chiroptera [5], are closely related to the insectivores but their bodies are entirely modified for flying. They are the only true flying mammals. The "flying" lemur only glides from tree to tree and is separately classified in the order Dermoptera [4].

Primates [6], the mammal order to which man belongs, are all very similar in appearance and show various degrees of adaptation to life in trees. Only man among the primates is exclusively ground dwelling and he is the most "intelligent" of all mammals. The two orders Edentata [7], the anteaters, sloths and armadillos, and Pholidota [8], the pangolins, contain species that are toothless and highly modified for living on a diet of ants and termites. Only the sloths among the edentates possess very simple teeth and feed on leaves.

The Lagomorpha [9], the rabbits and hares, and the Rodentia [10] – although they both have teeth and intestines specifically adapted to a vegetarian diet – are not closely related to each other. And while there are only two families of lagomorphs, the rodent families number about 33 and are the most numerous and widespread of all mammals.

Mammals that are adapted for eating flesh are all grouped together in the order Carnivora [12], a name that reflects their diet.

The aquatic mammals
Three mammal orders have returned to the water, the environment which their amphibian ancestors left many millions of years before. The Cetacea [11], the whales and dolphins, are wholly adapted to this life-style and never return to land, but the seals, order Pinnipedia [13], must breed on shore.

The other aquatic group, the sea cows or Sirenia [16], are a small group closely related to two other small but land-dwelling orders, the elephants or Proboscidea [14] and the hyraxes, the Hyracoidea [15]. The sole surviving member of a group that was once more widespread and numerous is the aardvark, whose order is the Tubulidentata [17].

The two remaining mammal orders both consist of hoofed beasts or ungulates. The Perissodactyla [18] have an odd number of hoofed toes whereas the Artiodactyla [19], which includes the cloven-hoofed animals, have an even number.

KEY

Monotremes	1
Marsupials	2
Insectivores	3
Primates	6
Bats	5
Flying lemurs	4
Edentates	7
Pangolins	8
Rodents	10
Lagomorphs	9
Whales	11
Carnivores	12
Seals	13
Artiodactyls	19
Perissodactyls	18
Aardvark	17
Elephants	14
Hyraxes	15
Sea cows	16

Primitive mammal stock

The mammals, which all developed from a shrew-like ancestor, underwent a period of rapid development of species in the Eocene. Although many groups have become extinct, all the 19 modern mammal orders originate from that period. The numbers refer to the illustration numbers below.

8 The giant pangolin is a member of one of the smallest mammal orders, the Pholidota, which consists of just one family, the Manidae, and one genus, *Manis*. Lengths range from 65 to 176cm (2–5.5ft).

Giant pangolin
Manis gigantea

9 Cotton-tail
Sylvilagus sp

9 The cotton-tail is a New World lagomorph of the family Leporidae, all of which have typically short, furry tails and long ears. Cotton-tail species range in length from 27.5 to 50cm (11 to 20in).

10 Bandicoot rat
Bandicota indica

12 Ocelot
Felis pardalis

11 The porpoise is grouped with the whales and dolphins in the order Cetacea. Both porpoises and dolphins are in the family Delphinidae. Porpoises can grow from 1.2 to 1.5m (4–5ft) in length.

12 The ocelot of the Americas is a member of the flesh-eating mammal order (Carnivora). It measures from 80 to 147cm (31–58in) long and is grouped with all other cats in the family Felidae.

11 Common porpoise
Phocaena phocoena

10 The largest order of mammals, to which this bandicoot rat belongs, is the Rodentia. Its family, the Muridae, includes all the other Old World rats and mice. The bandicoot rat is found in India and the Far East and is up to 62cm (24in) long.

14 Indian elephant
Elephas maximus

14 Only one family of elephants, the Elephantidae, exists today – once there were six. The Indian elephant is one of the two surviving genera of the order Proboscidea. It grows to 6.4m (20ft) long.

13 Grey seal
Halichoerus grypus

13 Turbulent waters round North Atlantic rocks are the preferred habitat of the grey seal of the order Pinnipedia and the family Phocidae. Males grow to 3m (9ft) long and females to 2.25m (7.5ft) long.

15

Rock hyrax
Procavia capensis

15 The rock hyrax is one of nine species extant in the order Hyracoidea and the family Procaviidae. This African animal grows to about 38cm (15in) long and lives in colonies of over a hundred.

16

Dugong
Dugong dugon

16 Sea cows of the order Sirenia inhabit coastal waters and estuaries. They can grow from 2.5 to 4m (8–13ft) in length. Each family, Dugongidae (dugongs) and Trichechidae (manatees), has but one genus.

Hippopotamus *Hippopotamus amphibius*

17 Aardvark
Orycteropus afer

18 The woolly tapir belongs to the Tapiridae, one of three families in the Perissodactyla, the order of hoofed, odd-toed animals. This tapir is an Andean species and grazes on grass and other low-growing vegetation. Like other tapirs it is a docile creature and a good runner, swimmer and diver. It has similar dimensions to a donkey, measuring up to 2.5m (7.5ft) long.

19 Hoofed mammals with an even number of toes are grouped together in the order Artiodactyla. The hippopotamus has a family of its own, the Hippopotamidae. It is a water-loving mammal that was once common in deep water habitats all over Africa but is now, due to man's ravages, severely restricted in range. The heavyweight hippopotamus tips the scales at between 3 and 4.5 tonnes. The largest animals grow to about 4.6m (15ft) in length.

17 The termite-eating aardvark is the sole survivor of the mammal order Tubulidentata and the family Orycteropodidae. Its name comes from the Afrikaans word meaning "earth pig" and this refers both to its habit of digging deep burrows and to its appearance. It may measure up to 1.5m (5ft) in length and weigh up to 70kg (154lb). Apart from the head region it is covered in bristle-like hair. Aardvarks are found where ant and termite food abounds, south of the Sahara.

18

Woolly tapir
Tapirus roulini

19

The life of mammals

The first mammals emerged at some time in the Triassic period, about 200 million years ago, and they were probably similar to living shrews and opossums. The story of mammal evolution is one of growing independence from the immediate pressures and constraints of the environment. They have become able to regulate their body temperature automatically and to maintain it at a constant level in heat or cold, usually above that of their surroundings. The infant mammal is protected with warmth and a guaranteed food supply during the earliest and most vulnerable part of its life. And the mammalian brain has developed to the point where the animal is capable of controlling its environment instead of being subject to it.

How temperature is controlled

Temperature control in mammals is effected largely by means of skin glands and by blood vessels located just below the skin. A part of the brain, called the hypothalamus, incorporates a mechanism for detecting changes in blood temperature. If the temperature is too high, activity in the hypothalamus causes the blood vessels of the skin to dilate so that blood heat can be more readily lost and, to encourage this, the sweat glands secrete liquid on to the skin so that it is cooled by evaporation. When the temperature is too low, the skin blood vessels, conversely, are constricted, the sweat glands dry up, and another reflex comes into play to erect the fur (or in the case of man the meagre remnants of it). In a respectably furry animal, the fluffed up hairs make an effective insulating layer. When the skin temperature is uncomfortably low, reflex shivering is induced so that heat is generated by the work of the muscles. Many mammals, especially marine species, have layers of insulating fat beneath their skins.

Within the warm cocoon of its mother's body, the unborn mammal is insulated from the environment for anything from the first few days to the first year or so of its life and most are provided with nourishment by a specially developed organ, the placenta. Nutritional care of the young continues after birth, with suckling.

Mammals have many features in common, but one of the most outstanding characteristics of the group is their extreme diversity [1], particularly in body form. A vast array of animal types has radiated from the small insectivore-like ancestors and these vary enormously in habit and life-style.

Diversity of mammals

There are herbivores with hoofs, carnivores (flesh-eaters) with claws, omnivores of all kinds; animals that burrow, nest or build dams; animals with and without teeth, tails or toes. By strange adaptations of body form, the mammal has gained the mastery of land, sea and air. Kangaroos, by the enormous development of the hindlimbs, bound over the huge areas of the Australian outback. The fleetest sprinters come from the ungulates, deer and antelopes, and from the carnivores – especially the big cats. Squirrels and primates move along tree branches with ease and grace; the sloth is so completely adapted to an arboreal existence that it is virtually unable to walk on the ground.

Otters, coypus, beavers and many other carnivores and rodents are proficient swimmers and lead a semi-aquatic life; but some

1 The diversity of mammals is shown in their modes of locomotion, which range from the swimming of seals [B] to the flying of bats [E]. They have important features in common. All mammalian young (except for monotremes) spend the first part of their lives inside the mother, who supplies their food and oxygen, for example, the reindeer [D]. Young are suckled like those of wild boars [F]. A special centre in the brain regulates temperature, and usually a fur coat like that of lemmings [C] helps. The brain is highly developed in mammals enabling some, like chimpanzees [A], to use tools.

3 The spiny anteater (*Tachyglossus aculeatus*) displays its primitive cloaca and mammary glands. It lacks separate urinogenital and gut openings and has no proper nipples.

6 The long-nosed bandicoot (*Perameles nasuta*) has a pouch that faces backwards. This reversal stems from the animal's vigorous burrowing habits which would otherwise fill the pouch with earth and harm the young.

4 The young of the spiny anteater lap milk from a primitive mammary gland that turns inwards instead of protruding. They lick the milk off the mother's fur.

Mammary gland

Ovary
Fallopian tube
Uterus
Kidney
Bladder
Rectum
Vagina
Urinogenital sinus

5 The marsupial mouse (*Dasyuroides byrnei*) has no pouch. As a result, the helpless young are obliged to cling to their mother's fur.

7 Kangaroos and wallabies have roomy pouches in which the young can grow to a considerable size. Even when they are old enough to leave it, young kangaroos return to the pouch in order to feed.

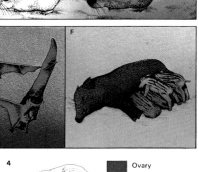

Temporary birth canal

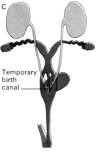

2 The reproductive arrangements of mammals divide them into three groups (gestation periods in blue). Monotremes, the Prototheria, have female reproductive systems [A] resembling those of reptiles and pro-

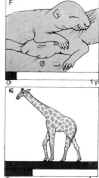

duce eggs [B]. Thus the development of the young takes place outside the mother's body. Marsupials, the Metatheria, have a two-part vagina [C] (but no placenta) and give birth to live young [D] which crawl into the mother's pouch and live there for the remainder of foetal development. Placental animals, the Eutheria [E], have a uterus that can carry the young to an advanced stage of development ranging from the immature rat [F] to the fully formed giraffe [G].

mammals have really "taken the plunge", exchanging fingers for flippers and down the ages losing their legs in the sea. Seals, sea lions, sea cows and, most completely of all, the whales and dolphins [11] have abandoned the land and their four-footed pretensions to return to the primeval ocean from which they painfully emerged so many millions of years ago.

The bats are the only flying mammals. Others, such as the so-called flying squirrel, may prolong what is really no more than a leap by passively holding out sheets of loose skin and gliding. But only the bat has wings that flap – as well as its personal version of a sonar system, using a continuous emission of ultrasonic bleeps to locate obstacles and potential food sources that it cannot see.

Diversity from genetic development
The remarkable diversity of form among mammals has developed during a relatively short period in evolutionary terms. Reptiles and amphibians, over a much longer time, have diversified much less dramatically. The embryos of vertebrates resemble each other during the early stages of development and there is not much difference between the appearance of a fish and a human embryo to begin with. But as development progresses mammalian embryos begin to show characteristic mammalian features and reach a stage at which they all look very similar to one another and quite different from the "lower" orders of vertebrates.

This is presumably because the genes that regulate the early development of fish have changed little in the course of events leading to the emergence of mammals. Those that control later development, however, have clearly evolved very fast indeed. It is now believed that the diversity of the mammals is due to an explosive acceleration in the rate of change of the genes regulating critical developmental events – genes whose activities are still not understood, but which are responsible not only for features such as the elephant's trunk, the neck of the giraffe and the camel's hump, but also for the development of the cerebral hemispheres of the brain, an organ that in turn is able to initiate complex behaviour patterns.

Male
Female

Eutherian, or placental mammals have reproductive systems typified by that of the rabbit. The erectile penis [1] is inserted into the female vagina [2], and sperm is emitted from the testes [3]. This enters the uterus [4] to fertilize ripe ova descended from the ovaries [5]. Urinogenital tracts [2, 6], separate from the rectum [7], replace the primitive cloaca.

8

8 The spiny anteater has two problems, and has found a solution to both. The first problem arises from the presence of spines on its back instead of hair. This makes copulation in the normal belly-to-back position painful. The second is the absence in the male of an erectile penis, which means that the genital openings must be brought as close together as possible for fertilization. So the spiny anteater mates belly-to-belly or just tail-to-tail.

9 Foxes control their body temperatures differently. The Arctic fox [A] is protected from cold by a thick fur coat that turns white in winter. Its ears are small and almost submerged in hair to prevent heat loss. The fennec fox [B], on the other hand, lives in the Sahara and has enormous ears whose copious blood supply and large surface area make them a very efficient device for losing heat. The temperature range over which it has to survive starts somewhat above where that of the Arctic fox leaves off.

9 A Arctic fox
Alopex lagopus

50 40 30 20 10 0 10 20 30 40 50 60
degrees centigrade

B Fennec fox
Fennecus zerda

50 40 30 20 10 0 10 20 30 40 50 60
degrees centigrade

10

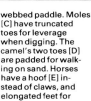

10 The feet of mammals have evolved in many different ways from the basic mammalian foot [A] possessed by the earliest shrew-like mammals. Seals [B] have developed evenly graduated toes for a webbed paddle. Moles [C] have truncated toes for leverage when digging. The camel's two toes [D] are padded for walking on sand. Horses have a hoof [E] instead of claws, and elongated feet for speed, as has the cheetah [F]. Bats [G] have enormously elongated digits to support wings. Kangaroos' toes [H] are for hopping. Lemurs [I] and sloths [J] have forelimbs for grasping trees.

11 Bottle-nosed dolphin
Tursiops truncatus

Dorsal nostril (blowhole) | Streamlined body

Dorsal fin for stability | Tail flukes for propulsion

Toothed jaws

Sunken eye aids streamlining | No external ear

Flippers are modified forelimbs | Young are born tail-first | No hindlimbs

11 The dolphin has come full circle in evolutionary history and has returned to the sea. Typically mammalian features are not very easily recognizable because of the dolphin's streamlined, fish-like body. After birth of the young, which feed from teats close to the genital opening, female dolphins separate themselves from the school with one other female. Dolphins have well-developed social lives and are considered to be highly intelligent.

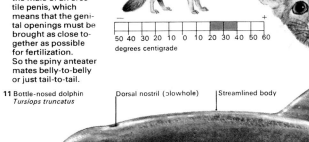

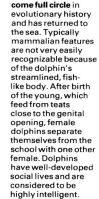

157

Monotremes and marsupials

Monotremes and marsupials are the two most primitive groups of true mammals. The monotremes bear the marks of their reptilian ancestry in their bony structure as well as in the feature that gave them their name, for monotreme means literally "single hole" and refers to the fact that excretory and reproductive functions are both served by one passage, the cloaca. Higher mammals have separate excretory and genital tracts.

Monotremes: the egg-laying mammals

The three surviving genera of monotremes, the duck-billed platypus (*Ornithorhynchus* sp) [1] and the Australian and New Guinea anteaters (*Tachyglossus* spp and *Zaglossus* spp), all lay eggs that have a typically reptilian soft leathery shell and the latter take the hatched young into a temporary pouch. The young feed on milk secreted from modified sweat glands which, however, are not combined into a single milk duct ending in a nipple. Although the monotremes have fur that evolved to help maintain body temperature, the temperature-regulating mechanism in the brain is imperfect; as a result they tend

to be cooler and have a more variable body temperature than higher mammals.

Both kinds of monotremes have, however, developed marked specializations peculiar to their different ways of life. The platypus has webbed feet as well as its duck-like bill to equip it for an aquatic existence [1]. The spiny anteater has long claws and a pointed nose for digging ants out of their nests. Neither has teeth but the platypus bill has a horny ridge which is used to demolish the shells of molluscs.

Marsupial characteristics

Although the pouch is the distinguishing characteristic that comes to most people's minds when they think of marsupials, strictly speaking the important zoological point is not the presence of the pouch but the absence of the placenta, an internal organ designed to nourish the fetus within the mother. (Many American opossums, for example, generally lack a pouch.) The egg yolk plus uterine secretions nourish the embryonic marsupial for the first few days and in some cases a primitive placenta forms. The young animal

is then obliged to crawl, still in an unformed fetal state, into its mother's pouch, where it latches on to one of her nipples for sustenance during the rest of its development.

Both the marsupial mother and the fetus are highly specialized for this arrangement. The fetus, for example, develops its forelimbs far in advance of its hindlimbs for the long climb between the vagina and pouch. The mother's teats are not the passive objects of the placentals, but long, muscular organs that actively inject milk into the young.

Only in the Americas and Australasia have marsupials survived. In South America they are represented principally by the opossums and the rat opossum of the Andes. Marsupials are classified into two main groups, the polyprotodonts and the diprotodonts. A polyprotodont, which literally means "many front teeth", has more than three incisors in each jaw. A diprotodont, "two front teeth", is more specialized, and has only two projecting teeth in the lower jaw for grazing.

The South American marsupials, with the single exception of the rat opossum, are all

1 The duck-billed platypus uses its distinctive bill to "paddle" for the molluscs, chiefly mussels and snails, on which it feeds. It lives by the water, building tortuous burrows in the river banks where it nests. The female remains in the nest (which may be lined with gum leaves) until the young hatch, a process that takes about two weeks. The platypus usually lays two eggs at a time which are 1.6–1.8cm (0.5in) in length and 1.4–1.5cm in diameter.

2 The New Guinea anteater is also an egg layer. While the platypus is entirely covered in short, brown mammalian fur, the anteater has spines over its upper body and only the soft underside is furry. Anteater eggs hatch about 10 days after they are laid, after which the young transfer to the mothers pouch for 6 to 7 weeks.

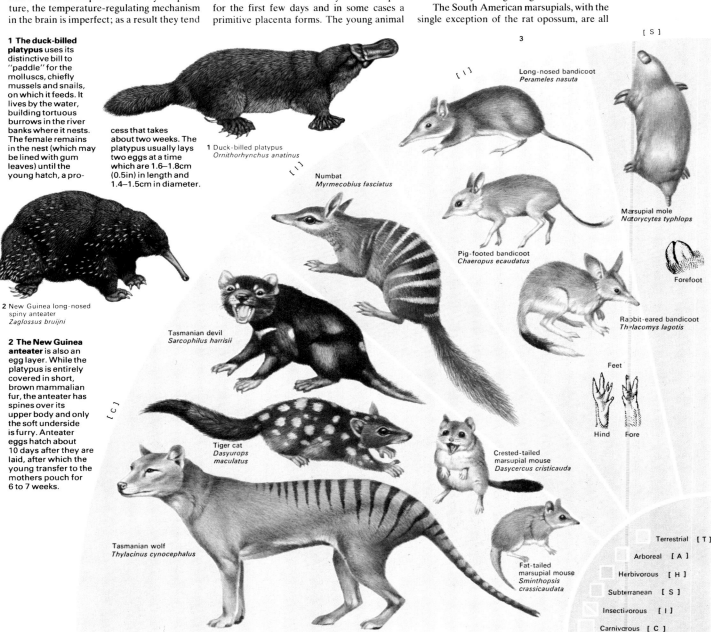

1 Duck-billed platypus
Ornithorhynchus anatinus

2 New Guinea long-nosed spiny anteater
Zaglossus bruijni

Long-nosed bandicoot
Perameles nasuta

Numbat
Myrmecobius fasciatus

Marsupial mole
Notoryctes typhlops

Pig-footed bandicoot
Chaeropus ecaudatus

Forefoot

Tasmanian devil
Sarcophilus harrisii

Rabbit-eared bandicoot
Thylacomys lagotis

Feet

Tiger cat
Dasyurops maculatus

Hind   Fore

Crested-tailed marsupial mouse
Dasycercus cristicauda

Tasmanian wolf
Thylacinus cynocephalus

Fat-tailed marsupial mouse
Sminthopsis crassicaudata

Terrestrial	[T]
Arboreal	[A]
Herbivorous	[H]
Subterranean	[S]
Insectivorous	[I]
Carnivorous	[C]

polyprotodonts. They have failed to develop a typical diprotodont feature – the fusion of the two digits (syndactyly) for purposes of hair-combing, which is found in all diprotodonts except the bandicoot.

Australasian marsupials

Australasian marsupials [3], which have had 130 million years to evolve without competition from placental mammals, are much more varied than their South American relatives. They fall into four rough groupings, defined by their feeding habits – omnivorous, insectivorous, carnivorous and herbivorous.

The South American opossums (family Didelphidae), which are generally pouchless, are insectivorous or omnivorous [4, 5]. The bandicoots, also omnivorous or insectivorous, occupy a family of their own and have developed backward-facing pouches. This is a special trait of marsupials that dig or burrow and serves to protect the young from flying earth. The wombat shares this specialization.

Carnivorous marsupials are rare in South America, but numerous in Australia, and include parallels to the placental cats and wolves, although all are quite small. The insectivores include the marsupial mole, numerous pouched mice (which occupy the niche taken up by the shrews of the placental group) and the rat opossum.

The most specialized of the marsupials are the Australian diprotodonts. All have developed rodent-like teeth for gnawing vegetation or, in kangaroos and wallabies, for shearing grass. This group contains three genera of "flying" mammals, including the flying squirrel (*Petaurus*). Apart from the squirrel-like phalangers, there is the wombat of similar size and habits to large placental rodents such as marmots and woodchucks, and the famous Australian "bear", the koala.

The kangaroos and wallabies invite no easy comparison with placental equivalents. They alone have developed the peculiarly powerful hind legs and muscular tails that have given the group its name – the Macropodidae, meaning "big feet". But they are unique in other respects, combining gnawing types of teeth with grazing habits and often a very large body size.

KEY

Order Monotremata — Families Tachyglossidae / Ornithorhynchidae

Sub-order Polyprotodonta — Families Didelphidae / Dasyuridae / Thylacinidae / Notoryctidae / Peramelidae

Order Marsupialia

Sub-order Paucituberculata — Family Caenolestidae

Sub-order Diprotodonta — Families Vombatidae / Phascolarctidae / Phalangeridae / Petauridae / Burramyidae / Macropodidae / Tarsipedidae

Placental mammals

Only two branches of monotremes, the egg-laying mammals, exist today: the spiny anteaters of Australia and New Guinea and the Australian duck-billed platypus. They belong to the sub-class Prototheria and although specialized are the most primitive mammals. Spiny anteaters develop a temporary pouch. Monotremes and marsupials – Metatheria – lack the true placenta of placental mammals – Eutheria. Marsupials live only in Australasia and the Americas; about 250 species are still extant.

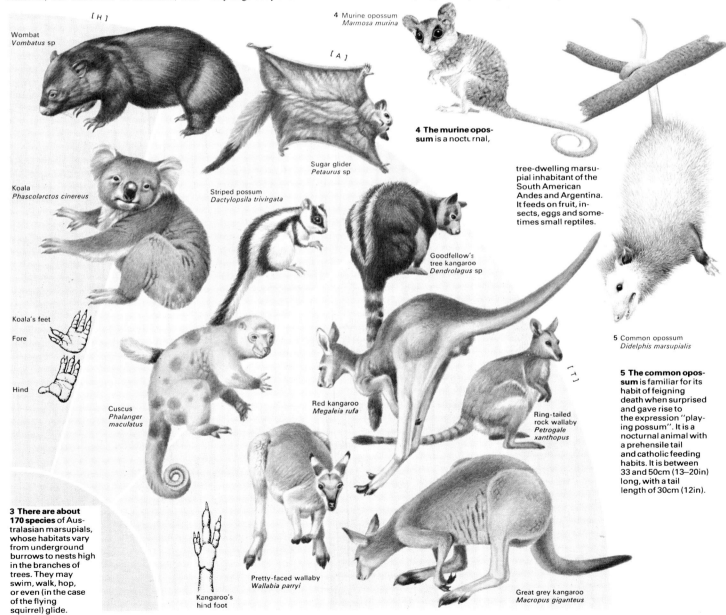

[H]

Wombat
Vombatus sp

[A]

Koala
Phascolarctos cinereus

Sugar glider
Petaurus sp

Striped possum
Dactylopsila trivirgata

Koala's feet

Fore

Hind

Cuscus
Phalanger maculatus

4 Murine opossum
Marmosa murina

4 The murine opossum is a nocturnal,

tree-dwelling marsupial inhabitant of the South American Andes and Argentina. It feeds on fruit, insects, eggs and sometimes small reptiles.

Goodfellow's tree kangaroo
Dendrolagus sp

Red kangaroo
Megaleia rufa

Ring-tailed rock wallaby
Petrogale xanthopus

[T]

5 Common opossum
Didelphis marsupialis

5 The common opossum is familiar for its habit of feigning death when surprised and gave rise to the expression "playing possum". It is a nocturnal animal with a prehensile tail and catholic feeding habits. It is between 33 and 50cm (13–20in) long, with a tail length of 30cm (12in).

3 There are about 170 species of Australasian marsupials, whose habitats vary from underground burrows to nests high in the branches of trees. They may swim, walk, hop, or even (in the case of the flying squirrel) glide.

Kangaroo's hind foot

Pretty-faced wallaby
Wallabia parryi

Great grey kangaroo
Macropus giganteus

Rodents, insect-eaters and bats

Two out of every five species of mammals in the world are rodents (order Rodentia), the chisel-teethed gnawing animals. The insectivores fall into two groups, first those classified in the order Insectivora and second other insect-eating animals, namely the bats (order Chiroptera), the anteaters (order Edentata), the pangolins (order Pholidota) and the aardvark (order Tubulidentata).

Despite the differences between them, which are mainly connected with feeding, the Rodentia and the Insectivora share a number of common features and these are, in turn, similar to those of early mammals. For example, they are small, built to a simple plan, and most have five toes on each foot.

Rodents range in size from the capybara [1] of South America, which is as large as a small pig, to the African pygmy mouse (*Mus minutoides*) which averages about 7.5cm (3in) in length, only slightly smaller than the shy European harvest mouse (*Micromys* sp).

Rodents and rabbits
Rodents are found in almost every available habitat the earth has to offer, apart from the

sea. Most live on the ground, but niches from treetops (squirrels) to underground burrows (mole rats) have been exploited; beavers and some others have even become adapted, though not totally, to life in water, having webbed feet and heavy waterproof fur. Rodents move in various ways. On the ground, progression is usually by running, as with the Patagonian hare, or by leaping, as with the jerboas. Above the ground, rodents may climb or glide, as can be seen in the tree squirrels and flying squirrels [2].

Rabbits and hares, and their relatives the pikas, belong to the order Lagomorpha, a word that means "hare-shaped". Although they resemble rodents, lagomorphs have different kinds of gnawing teeth [Key B5]. Compared with rodents, there are few species in the rabbit group, but it is nevertheless a highly successful one with vast numbers of individuals. Hares differ from rabbits in having longer hind legs and ears that are longer than their heads [10]. Rabbits and hares can walk, but they usually hop over the ground instead; however they can move at great speed when danger threatens. Pikas are

the smallest of the group. They are easily distinguished by their small ears, almost complete lack of a tail and the possession of all four legs of equal length.

The insect-eaters
The group of mammals belonging to the order Insectivora has a long evolutionary history. Fossil insectivores are known from the rocks of the late Mesozoic era. These creatures shared their world with the dinosaurs, but they and their descendants survived when the great reptiles died out. Today, members of the order Insectivora are found in all parts of the world except for the polar regions and Australasia (where the insect-eaters are monotremes and marsupials) and live in a wide range of habitats from mountains to lowland rivers, and from tundra to tropical forest. They include a number of widely differing forms, but all feed on insects or other small invertebrates. Most widespread are the ever-hungry shrews, the smallest of all mammals. Somewhat larger, though no less voracious, are the moles. Moles are all burrowing animals that loosen the soil with their

1 Capybara
Hydrochoerus hydrocharis

2 European flying squirrel
Pteromys volans

3 Patagonian hare
Dolichotis patagona

4 Indian porcupine
Hystrix indica

4 Protection for the porcupine is provided by a coat of spines. It can erect these when danger threatens. The spines are loosely attached and may become embedded in predators.

5 Black rat
Rattus rattus

6 Naked mole rat
Heterocephalus glaber

7 Muskrat
Ondatra zibethicus

8 A native of the cold scrublands of North America, the jumping mouse is related to the desert jerboas.

Like them, it hops on its long and powerful hind legs.

6 The naked mole rat is almost blind but is nevertheless perfectly adapted to its burrowing life in the deserts of eastern Africa.

8 Jumping mouse
Zapus hudsonius

9

Relative sizes

1 **The largest of all rodents** is the capybara. An excellent swimmer, it lives in small colonies in swamps and near rivers of tropical South America.

2 **The European flying squirrel** lives in forests in the far north of the continent. Similar squirrels are found in both northern and tropical forests.

3 **Another South American rodent** is the mara or Patagonian hare. It is closely related to the guinea-pig and is called a "hare" because it hops away when it is frightened.

5 **Black rats** lived in Asian forests originally, but over the centuries they have accompanied man to many parts of the world. Because they need warmth, they are often found in buildings. Fleas that live on them sometimes spread disease to man.

7 **The scent of the muskrat,** from which it gets its name, comes from special glands. The animal was originally a native of North America but it has since been introduced into other parts of the world because of its fur.

9 **Rodents and insectivores** vary in size from the bulky capybara down to the tiny pygmy shrew, one of the world's smallest mammals.

short, powerful forelimbs and shoulder their way through to consolidate a tunnel round their bodies. Still larger are the hedgehogs [12], which are native only to the Old World. They are protected by a coat of spines that has the advantage, rare in animal armour, of being fairly lightweight. The porcupine [4], among the rodents, has the same kind of protection, but there is no close relationship between it and the hedgehog.

The bats
Most bats [15, 16], the only flying mammals, eat insects. Their wings are attached to their forelimbs, which are long and slender and have elongated finger bones that support the fragile membrane. The rear of the wing is attached to the bat's hind legs and is sometimes extended to include the tail. Bats make up almost a quarter of all placental species. They owe their success possibly to the fact that they occupy a niche left vacant by the diurnal birds, because bats are nocturnal insect-eaters. They feed mainly on night-flying creatures such as beetles and moths, which they detect with a highly sophisticated

sonar system [17]. There are far fewer bats in temperate regions than in the tropics, where their diet comprises a much wider range of food, including blood, fish and fruit.

Many other groups of animals eat insects. They include the pangolins [13] of the tropical Old World, which are armoured with overlapping horny plates, giving them the appearance of animated pine cones. Some species can climb trees. In the tropical New World there are no pangolins but their place is taken by the anteaters. These belong to a curious South American order called the Edentata ("animals with no teeth"). This is a misleading name because some members of the group – the tree sloths and armadillos – are equipped with many teeth, although these are degenerate structures that lack enamel and are restricted to the back of the mouth. The anteater's extremely weak jaws [Key D] are long and tubular, and its mouth forms a tiny slit at the end. Its long, sticky tongue flicks out like a whiplash to scoop up the termites on which it feeds. The aardvark of South Africa is in no way related to the anteater but has a similar life-style.

KEY

The teeth of animals betray their feeding habits. Rodents such as rats [A] and lagomorphs such as rabbits [B] have gnawing incisors [1] that are open-rooted and continue growing for life. Then comes a toothless gap, the diastema [2], followed by grinding teeth [3]. Rodents have a single pair of incisors in both upper and lower jaws, and only the front surface is enamelled [4]. Lagomorphs have a second, smaller pair of incisors [5] that lie behind the main upper teeth, which are fully enamelled. The teeth of insectivores such as shrews [C] have sharp shearing edges. Pangolins and anteaters [D] feed on ants and have long, tubular, toothless jaws [6].

10 Jack rabbit
Lepus californicus

12 Long-eared hedgehog
Hemiechinus sp.

14 North American pygmy shrew
Microsorex hoyi

14 One of the smallest of all mammals is the North American pygmy shrew. It measures only about 9cm (3.5in) overall and weighs less than 3gm (0.1oz). Like all shrews, it is highly active but has only a short life-span.

10 The American jack rabbit, so called, is actually a hare, and its huge ears are a clue to its correct classification. It squats in a scrape in the ground, whereas the true rabbit lives in a burrow. Young jack rabbits are born fully furred.

11 Grant's desert mole (*Eremitalpa granti*), from South Africa, is a golden mole. These resemble true moles but are not closely related.

11

Grant's desert mole

12 A coat of spines protects the long-eared hedgehog, which lives in dry regions ranging from the Near East to Mongolia. The hedgehog hunts by night and rests in a burrow during the day.

15 The long-eared bat is found in the temperate regions of Europe, Asia and North Africa. It holds its long ears erect in flight and folds them down when it is at rest.

15 Long-eared bat
Plecotus auritus

16

16 Bats are flying insectivores. The Mexican big-eared bat (*Macrotus mexicanus*) [1] is typical. The fisherman bat (*Noctilio leporinus*) [2] takes fish from surface water. The vampire bat (*Desmodus rotundus*) [3] slices the skin of a tapir and laps up the blood. The long-tongued bat (*Glossophaga soricina*) [4] feeds on nectar and pollen, whereas the wrinkle-faced bat (*Centurio senex*) [5] prefers pulpy fruit.

17 Almost all bats use echo-location with which to navigate and find their food. As they fly, they emit a series of high-pitched squeaks, each lasting about 1/500 of a second, at a rate of about 50 a second. These signals bounce off any object in their paths and the resulting echo is detected by the bats' sensitive ears. Acting on this information the bats can then take the necessary action.

13

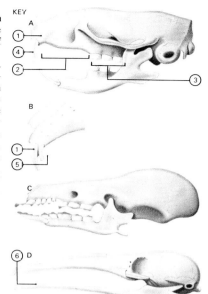

African tree pangolin
Manis tricuspis

13 The African tree pangolin is an unusual animal. Its short, powerful forelegs are armed with sharp claws that help it to climb and tear open the nests of tree ants on which it feeds. Its tail also helps its progress through the trees by holding on to branches or by pressing its overlapping scales against the trunks.

17

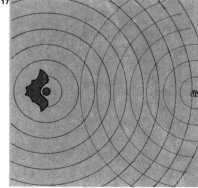
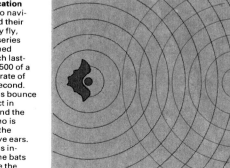

Hoofed mammals

There are more than 200 different species of hoofed animals or ungulates, of which the horse and cow are the most familiar. All hoofed animals are mammals and all are herbivores, most feeding on tough vegetation which they masticate with specialized and complex teeth. Hoofs are large, flat toenails which developed in the course of evolution as a result of the tendency of some animals to stand and walk on their toes rather than using the full length of the foot in the way that man and his close relatives do. This has the advantage of increasing the length of the leg by the length of the foot, and sometimes by that of the toes as well, resulting in a longer stride and faster movement.

Odd-toed and even-toed animals

Most hoofed animals belong to two great orders: the even-toed, or Artiodactyla, which have two or four toes on each foot; and the odd-toed, or Perissodactyla, which have a variable, but almost always an odd number of toes. There are approximately 190 species of even-toed animals and roughly 15 species of odd-toed animals. There is, in addition,

another group of hoofed mammals, the subungulates, which includes animals as diverse as the elephant, the rock hyrax and the sea cow which has forelimbs developed as flippers and no hindlimbs. It is the relationship between the teeth that has confirmed the relationship between group members.

The herbivores

The ungulates arose from early mammalian stock at the start of the Cenozoic era some 60 million years ago and by the Eocene, some 20 million years later, they had become large, heavy-bodied herbivores many of which were destined to be replaced, in the Miocene, by fleet-footed grazers. Even in the early days of the ungulates three distinct groups began to emerge. Thus although the cow group (even-toed) and the horse group (odd-toed) may seem linked they have a long history of separate evolution.

Of the two, the even-toed species have proved to be the better survivors, for almost all the medium to large plant-eaters in the world belong to this group. The more primitive even-toed species include pigs and the

peccaries (found in South America), which have four well-developed toes on each limb. They are omnivorous in their choice of food and their dentition is less specialized than in many other forms. Hippopotamuses, which retreat to water during the daytime, come ashore to feed at night, and in some areas of Africa are major crop thieves. Camels, which are highly adapted for desert life, have only two toes on each foot, as do their South American relatives the closely similar llama, guanaco and alpaca. All of these animals have some upper teeth.

The upper front teeth of 'the other even-toed, cud-chewing ungulates are missing and are replaced by a horny pad, but they can take in food tremendously fast, often using the tongue to tear at vegetation. The food is passed to a holding compartment in the stomach and regurgitated later to be masticated thoroughly before being swallowed a second time, after which digestion proper starts. One advantage of chewing the cud is that food which may have been gathered in dangerous areas and rapidly eaten can be digested later in a place of comparative

African elephant
Loxodonta africana

1 The African elephant is the largest land mammal; a big male may stand at nearly 4m (12ft) and weigh over 7 tonnes. The tusks are upper incisor teeth that continue to grow throughout the long life of the animal.

Rock hyrax
Procavia capensis

3 Several species of rock hyrax live in parts of Africa and the Near East. They are small, superficially rodent-like animals, living in rock crevices or in borrowed burrows.

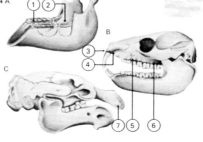

2 The aardvark, which measures up to 1.5m (5ft) long, is a subungulate found in much of Africa. It is rarely seen for it is nocturnal and extremely shy. Its presence may be detected by the large burrows it digs, using the hoof-like claws on its forefeet. It feeds chiefly on termites and ants, which it catches with its long tongue, coated with sticky saliva. It also eats vegetation.

2 Aardvark
Orycteropus afer

5

Florida manatee
Trichechus manatus

4 The dentition of subungulates varies widely. In the elephant's lower jaw [A] the larger molar tooth [1] is replaced as it wears out by another tooth growing forwards [2]. In the rock hyrax [B] the upper incisors [3] are open rooted. The lower incisors point forwards like small tusks. There is a big gap [4] between these and the grinding premolars [5] and molars [6]. In the manatee [C] the incisors are replaced by horny pads [7] behind fleshy lips.

5 The manatees are subungulates of the tropical coasts of the Atlantic, although one species travels as far north as Georgia. The fleshy lobes on the upper lip are sufficiently mobile to seize food and pass it to the mouth.

safety. Deer, giraffes, antelopes, cattle, sheep and goats all chew the cud.

Odd-toed animals were at one time more numerous than at present; today only horses, rhinoceroses and tapirs survive. The rarity of the first two is largely due to persecution by man, and the true wild horses and asses and the three species of Asiatic rhinoceroses are among the rarest mammals in the world, although the zebra is relatively common in parts of Africa. No odd-toed animal chews the cud, and none has any true horns, those of the rhinoceros being made of compressed hair. The rhinos have three toes on each foot, the tapirs have four on the front feet and three on the hind feet, while horses (including donkeys and zebras) have lost all but the large central toes of each foot. The balance of the leg is therefore completely different from that of the even-toed animals.

Elephants, hyraxes and sea cows
The subungulates are the smallest group of hoofed animals. Two species of elephant survive to the present day in Africa and southern Asia, the sole remnants of an order of giant

animals which once inhabited most of the world. Their size alone distinguishes them from all other land mammals, but the possession of a trunk and the peculiarities of the teeth, which include two tusks in the upper jaw and only four other teeth, also set them apart. Elephants eat much food and may travel many miles a day in search of it.

A group of animals found in Africa and the Near East and thought to be closely related to the elephants is that of the hyraxes, small, dumpy yet agile creatures which live socially in forested or rocky areas. All have hoof-like nails on the four toes of the front feet and a large claw on the innermost of the three toes of the hind feet.

Fossil remains have shown us that the early relatives of the sea cow were similar to those of the elephant. Now, however, the sea cows are entirely aquatic, living in tropical rivers and offshore areas where they feed on vascular plants akin to eelgrass. They are large, placid, sluggish animals, which swim by means of a tail fluke. The forelimb is transformed into a paddle and in some species this carries small flattened fingernails.

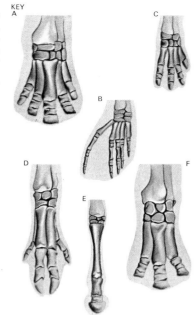

KEY

	Forearm
	Wrist (carpals)
	Metacarpals
	Phalanges

The front limbs of a selection of hoofed animals show that the same sequence of joints occurs however many the toes. The elephant's foot [A] has 5 toes slightly spread and backed by tissue pad. The manatee's flipper [B] and the rock hyrax's trotter [C] both have a hand-like skeleton. The pig [D] has 4 toes but walks on only 2. The horse's "hand" [E] contains only one toe, with the last of the finger bones carrying the hoof, while the rhinoceros [F] has 3 toes to each foot.

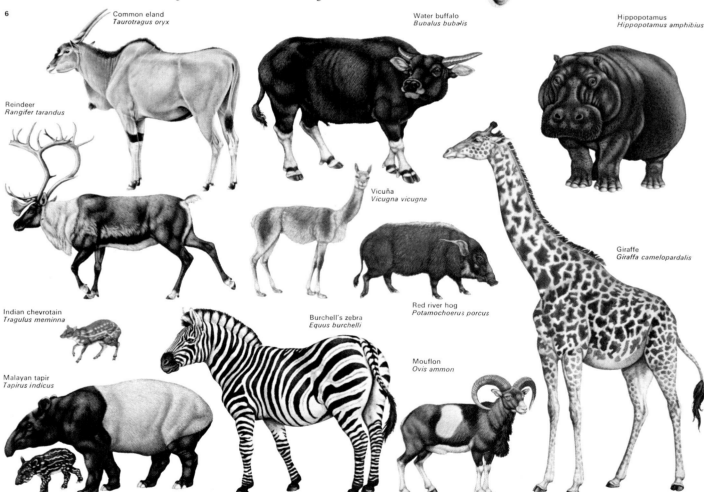

6   Common eland
     Taurotragus oryx

Reindeer
Rangifer tarandus

Water buffalo
Bubalus bubalis

Hippopotamus
Hippopotamus amphibius

Vicuña
Vicugna vicugna

Red river hog
Potamochoerus porcus

Giraffe
Giraffa camelopardalis

Indian chevrotain
Tragulus meminna

Burchell's zebra
Equus burchelli

Mouflon
Ovis ammon

Malayan tapir
Tapirus indicus

6 Hoofed animals include many species economically important to man. The eland is the largest of the antelopes. Recently attempts have been made to domesticate it for its excellent milk, meat and hides. The water buffalo is

found as a wild animal in some parts of India but has been domesticated in much of the tropical world. The hippopotamus, an African animal, is found in rivers, lakes and swamps, in social groups. The reindeer migrates vast dis-

stances between summer and winter feeding grounds. Both male and female carry antlers. The vicuña is a South American relative of the camel and lives in small herds in arid or mountainous areas. Its numbers have been greatly

reduced by hunters seeking their fine silky fleeces. The red river hog, so-called because of its reddish bristles, is found in Africa. It lives in family groups and forages at night. Chevrotains live in India and Asia. They are small,

shy, nocturnal animals that inhabit dense bush. The males develop long tusks. Burchell's zebra is the commonest of the striped horses of Africa. Why it is striped is a puzzle, for it lives in family groups in open country where

the stripes seem to make it more obvious. The Malayan tapir is a shy, nocturnal inhabitant of dense forests in SE Asia. The mouflon is a mountain dweller found only in the remoter parts of Sardinia and Corsica. It is related

to the ancestor of the domestic sheep, whose fleece has been developed from the wool layer beneath the mouflon's top coat of coarse fur. The long neck of the giraffe enables it to reach trees up to a height of 6m (18ft).

Flesh-eating mammals

Many kinds of animals are flesh-eaters. Land mammals of this type are classified zoologically into the order Carnivora (which includes cats, hyaenas, dogs, weasels, bears, racoons and civets), a group in which most of the members are adapted to a life of preying on other creatures.

One group of mammals (the seals, sea lions and walrus) became adapted – although not completely – to existence in water. These are treated by most authorities as the order Pinnipedia, but are sometimes grouped as a sub-order within the Carnivora because both groups are descended from a common mammalian ancestor that lived in the Eocene period more than 50 million years ago.

Some members of these two groups have diverged from a strictly carnivorous way of life. Thus the hyaena [5], although a ferocious hunter, feeds on carrion, bears [6] eat a variety of food including fruit leaves and berries, the aardwolf [10] feeds almost exclusively on termites and the walrus [12] feeds on marine shellfish. The panda and kinkajou are Carnivora that eat mostly plants.

The carnivores are native to almost all parts of the world except Australia, New Zealand and many Pacific islands, while the seal group has members widely distributed on the shores of temperate and polar seas.

The carnivorous life-style

Successful carnivores are well equipped for preying on other animals and all are built on a relatively primitive plan, with a long, fairly flexible body and tail. The terrestrial species have well-developed and often long legs. Among the dogs are several long-distance runners, and among cats is the cheetah, the world's fastest sprinter. Even the short-legged carnivores can move rapidly over short distances. The proportions of the head differ in the various groups of flesh-eaters, but certain features are common to all of them. Most have well-developed canine teeth with which they hold their prey, comparatively small incisors (cutting teeth) and two pairs of molars or carnassial teeth adapted to flesh-eating needs. The carnassials have special cutting edges that act like scissor blades.

Carnivores have fairly good eyesight; in many the eyes are sufficiently far down the face to give some degree of stereoscopic vision and this helps them to judge distances when pouncing on prey. Their sense of smell is usually well developed and their sense of hearing is acute. Most carnivores are intelligent animals – this is because they must be versatile opportunists if they are not to be outwitted by their prey.

Social, marine and subterranean carnivores

Most carnivores are solitary creatures, although the young often depend on their mothers – or occasionally both parents – for a long time after birth. Two exceptions are the dogs, which usually live and hunt in packs, and lions which form prides of males, females and juveniles. Lions may also hunt singly or in pairs. Another "great cat", the tiger, is a lone hunter living in jungle – unlike the lion, which lives in open grassland. Most of the big cats gorge their kill in one meal and follow this with a drink and a long period of rest, usually in a shady spot.

In the colder parts of the world, members of the weasel family have occupied many

3 Leopard
Panthera pardus

2 A carnivore's skull bears small, nipping incisor teeth in the front of the mouth [1], large, tearing canine teeth [2] and a series of premolar and molar teeth for slicing and crushing food [3]. The fourth upper premolar tooth and the first lower molar have cutting edges that slice against each other. These are known as the carnassial teeth [4] and are characteristic of carnivores. The orbit [5] or bony cavity that contains the eyeball is large, open and directed forwards.

1 The jackal (*Canis* sp) feeds in typical carnivore fashion, using the large, slicing carnassial teeth at the back of its mouth to cut the meat into chunks small enough to swallow. Most carni- vores have toes con- nected with a web of skin that endows flexibility yet gives enough strength for digging. Carnivores sometimes dig for their prey; at other times they excavate or enlarge a hole for a den. The jackal lives in family groups but many canids travel in larger groups. As a result the task of hunting becomes less onerous for each individual and the prey is shared by the pack.

3 The leopard and the "black panther", a colour phase of the leopard, are found in tropical rain forests of Africa and Asia and feed on any animal that they can over- power. The leopard is one member of the cat family (Felidae) of which the Siber- ian tiger is the largest and the South African black- footed cat the smal- lest. Most cats, including the leo- pard, are solitary, secretive animals that live in forests or dense scrub, where they are cam- ouflaged by beauti- fully spotted or striped coats. Many cats can climb and some, in- cluding jaguars and tigers, can swim well.

4 A formidable snake- killer of Africa and India is the banded mongoose. Lithe and swift, it relies on speed and agility to evade the poisonous fangs of its prey. It also feeds on small ground game.

4 Banded mongoose
Mungos mungo

6 The brown bear (*Ursus arctos*) is found in Europe, North America and Asia. Many forms have developed, dif- fering widely in colour and size. An island race, the Kodiak bears from northern Canada may weigh up to 725kg (1,600lb) and are the largest land carni- vores. Bears will eat almost anything, including fish, carrion, plant food and even honey when they can get it. They doze through the winter months but do not hibernate fully.

5 The spotted hyaena lives in the savan- nas of Africa. It feeds as a scavenger, taking the remains of a lion's kill, but it is also an efficient hun- ter and competes with lions. Its powerful jaws can crush bones that even a lion could not crack.

5 Spotted hyaena
Crocuta crocuta

niches. Some are found in water (the otters), some in trees (martens) and others (weasels and stoats) are small enough to follow their prey into underground burrows. The badgers dig vast underground tunnel systems from which they emerge at night to forage. No carnivore, however, has become completely adapted to a subterranean way of life. In the tropical Old World, mongooses [4], genets [11] and their relatives hunt mostly small, ground-living prey, although there are some that eat insects or fruit. In the New World and parts of the Old, mongooses have their counterparts in the racoon group [8], which embraces a wide range of animals including fruit-eaters. The marine carnivores, the seals [13], sea lions [14] and walruses [12], feed mainly on fish and molluscs.

Because of their way of life, the seals have evolved a highly streamlined form. They have, unlike the whales, retained their body fur but like them have a thick layer of insulating blubber beneath the skin. The seals are skilful swimmers; their limbs are modified into flippers, the forelimbs being used for propulsion in the eared seals and

walrus and the hindlimbs in the true seals. On land, seals are clumsy creatures, although the eared seals can turn their hindlimbs forward, which enables them to walk and even run. The true seals cannot do this and, as a result, have to drag themselves by their flippers.

Seals come ashore to breed and at this time they are highly gregarious, although the males are also intensely competitive. Breeding colonies of seals can contain up to a million animals in an area of only 50 square kilometres (19 square miles).

Scent glands and skin
Most carnivores except the seals possess a pair of anal scent glands that produce an odoriferous fluid for defining territories, attracting a mate, or in some cases as a means of defence, the most notorious example being the skunk [7]. Most of these scents are offensive to human beings, but some have a sweet component and are used as bases for perfumes. Many flesh-eating animals, especially the cats, weasels and seals, have pelts of great softness and beauty and have been hunted almost to extinction for their skins.

KEY

Flesh-eating mammals (carnivores) were among the earliest of placental mammals to appear. Primitive carnivores (creodonts) lived alongside the last of the dinosaurs. Early in the Tertiary, during the Eocene, cat-like and dog-like forms evolved from and replaced the creodonts. All modern members of the order Carnivora are descended from these two groups. The bear and racoon families are the least specialized for flesh-eating.

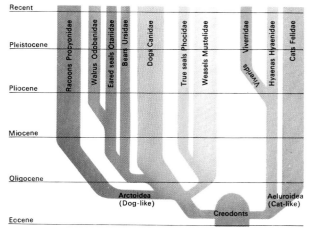

7 The skunk's fur is boldly striped in black and white and serves as an unmistakable warning to would-be predators, for it is well able to look after itself. If molested, the skunk ejects a foul-smelling liquid from its anal scent glands.

7 Skunk
Mephitis mephitis

8 Racoon
Procyon lotor

8 An all-rounder among carnivores is the racoon, which is found in many parts of North America. It eats almost any kind of food, climbs and swims well and often lives close to man.

9 The kinkajou is an oddity among carnivores because it has a prehensile tail. It uses this as it climbs among the tree tops of tropical America in search of the soft fruit that is its main food.

9 Kinkajou
Potos flavus

11 Genet
Genetta genetta

11 There are six genet species but only one, *Genetta genetta*, lives in Europe. The rest inhabit forest and dense brush in Africa. These nocturnal animals spend the day in rock crevices, hollow trees or any similar shelter. They feed on the ground, eating rodents, birds and any other animals that they can catch. The civet cat (*Civettictis civetta*), a close relative, is one of the sources of commercial musk.

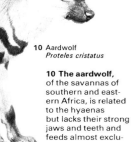

10 Aardwolf
Proteles cristatus

10 The aardwolf, of the savannas of southern and eastern Africa, is related to the hyaenas but lacks their strong jaws and teeth and feeds almost exclusively on termites and insect larvae, varying its diet with mice and eggs of ground-nesting birds. It may gain protection by mimicking the striped-coat colouring of the hyaena.

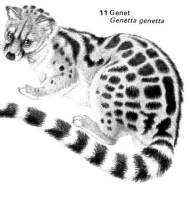

12 A unique relative of the seal, the walrus is assigned a family of its own. It feeds on molluscs which it rakes up from the sea-bed with its large tusks. A large male (a bull) may be more than 3m (7ft) long and weigh a tonne. The tusks may reach 1m (39in) in length.

13 Bearded seal
Erignathus barbatus

12 Walrus
Odobenus rosmarus

13 The bearded seal (*Erignathus barbatus*) is a maritime carnivore that lives in small groups in Arctic waters. Like all seals it has a streamlined body that is heavily insulated with fat beneath the skin.

14 South American sea lions breed in large, widely distributed colonies. One bull guards a harem of up to nine females and may be very aggressive.

14 S. American sea lion
Otaria byronia

Whales, porpoises and dolphins

The whales comprise a group that includes, as well as the large marine animals of that name, all the smaller species such as porpoises and dolphins. All are members of the mammal order Cetacea. The whales are not only the largest but also among the most intelligent of animals that have ever lived. But as long as he has been a seafarer, man has hunted whales for food and today modern methods used to capture whales threaten the continued existence of many species.

The origins of whales

The whales, with the carnivores, are descended from a common mammalian ancestor that lived in the Cretaceous period more than 65 million years ago. While most of the carnivores developed on land, the ancestors of the whales took to the water and became completely adapted to an aquatic way of life. As they evolved, tail flukes developed for locomotion, the hindlimbs disappeared and the forelimbs became modified to form balancing and steering paddles.

There are two main groups of whales: the baleen whales, of which the relatively few species vary from 6 to over 30m (20 to 100ft) in length and the largest weigh more than 100 tonnes, and the toothed whales, which are far more numerous and mostly much smaller [Key]. The baleen whales have no teeth but carry instead triangular plates of horny material in the roofs of their mouths. These "whalebones" [10] are used to strain krill – small shrimp-like organisms that are the baleen whale's only food – from the seawater. Most toothed whales have large numbers of simply shaped teeth in their mouths and feed almost entirely on fish and squids.

Whales are the only mammals that live the whole of their lives in water. Unlike seals they do not even return to the land to breed. Although they are aquatic creatures they must surface at regular intervals to take in oxygen and release waste carbon dioxide. The nostrils, situated at the top of the head, are tightly closed during submergence.

Whale life and development

All whales can hold their breath for a long time – some of them for up to two hours. But if water enters a whale's lungs it drowns as any other air-breathing creature would. The distinctive spout that displays the presence of a whale, thrown up just before it surfaces, is a mixture of water and condensed vapour from the animal's breath.

Whales have been known to dive to depths of 1,500m (4,800ft) and they are able to do this by means of various physiological adaptations. Before diving the air in the whale's lungs is changed completely; to do this the whale breathes several times in quick succession. During the dive the heartbeat slows and blood is shunted from the muscles to the brain. The centres in the brain that control breathing, unlike those in man, are relatively insensitive to the increase in carbon dioxide in the blood and as a result the whale is not "forced", as a man would be, to take another breath when the carbon dioxide concentration rises.

Whales can grow so huge because their bodies are supported by water. The blue whale (*Sibbaldus musculus*) [11] is the largest animal ever to have lived; sizes of up to 34m (113ft) have been recorded for the female blue whale. Almost all traces of hair and fur

1 La Plata dolphin
Stenodelphis blainvillii

1 The La Plata dolphin is representative of the family of freshwater dolphins and is found in the River Plate of South America.

The Ganges dolphin (*Platanista gangetica*) and Amazon dolphin (*Inia geoffrensis*) occur in the respective rivers and yet another (*Lipotes*) vexillifera) is found in Lake Tungting Hu on the River Yangtze. The long, beaked snout is characteristic and contains large numbers of teeth – up to 222 in the La Plata dolphin. Except for the Amazon dolphin, the river dolphins are virtually blind and orientation and food location are achieved by means of sound and touch.

2 The bottle-nosed whales belong to the groups called beaked whales. They have only one or two pairs of teeth and feed on squid and fish. They range in length from 4.5 to 13m (14 to 40ft).

2 Bottle-nosed whale
Hyperoodon ampullatus

3 Narwhale
Monodon monoceros

4 Killer whale
Orcinus orca

3 The narwhale is found in Arctic seas. It reaches a length of 3.6 to 5m (11 to 16ft). The distinctive tusk can grow to 2.8m (9ft) in length. It is generally found only in the male and develops from the left tooth of a pair in the upper jaw. The function of the tusk is not known. Narwhales feed on fish and squid.

5

Beluga
Delphinapterus leucas

4 The large triangular dorsal fin and the black and white body are the two obvious features of the killer whale (*Orcinus orca*). It is found mostly in polar seas and is generally considered to be the most fero- cious of the whales. Any creatures in the sea are considered as food by the 9m (30ft) killers.

7 Pilot whale *Globicephala* sp

5 Another Arctic species of whale is the beluga. It usually travels in a school which may number hundreds. Its food consists of many species of fish and squid. The beluga grows to a size of 4.25m (17ft).

6 The Atlantic bottle-nosed dolphin is the one commonly kept in captivity. It grows to a length of 3.6m (10ft) and feeds on a variety of fish. It is a highly intelligent animal and can be taught to fetch and carry objects.

6 Bottle-nosed dolphin
Tursiops truncatus

7 The pilot whale or black fish is often seen in schools containing several hundreds. It grows to 8.5m (26ft) in length and can weigh over 1,000kg (2,200lb). It feeds mostly on squid. If frightened near a sloping shore a whale school may become stranded and die. The three species of pilot whale are found in most oceans except the polar seas. They migrate between warm and cold waters depending on the season. Breeding takes place in warm water. They have a life-span of about 50 years. The Newfoundland whaling industry is based on this species.

have been lost, but these insulating devices are replaced by thick layers of fat or blubber beneath the skin. The efficiency of the blubber as an insulator is so great that the heat produced in the decomposition of a dead whale can even cook and char the flesh.

Whales lack a sense of smell and cannot see very well. As a result they have to rely greatly on their senses of hearing and touch. Recent studies have also shown that whales make a wide range of sounds and some of these have been shown to involve communication between individuals. This may be so elaborate that it comprises a primitive language. Sound is used for both echolocation of prey and orientation – the strandings of some species on sandy beaches may be due to an absence of echo reflection.

Whales and dolphins are found in all the oceans of the world and some dolphins live in freshwater rivers [1] and lakes. Many whales, especially those from the colder seas, migrate to warmer waters to breed. One of the best known is the Californian grey whale (*Eschrichtius glaucus*) [8], which makes its annual migrations down the west coast of North America to the lagoons and estuaries of Baja (Lower) California.

Young whales may be up to a third of the size of the female. They are born tail first to ensure that they do not drown during birth. The female pushes the young to the surface to take its first breath and suckles it from teats lying in slits on either side of the genital opening. To allow the young whale to suckle and breathe simultaneously the female floats on her side and ejects the milk forcibly.

The endangered species
The species of whales most threatened by man are those of the open oceans whose carcases are used to produce oil. These include, in particular, the sperm whale (*Physeter catodon*), the blue whale (*Sibbaldus musculus*) [11], the humpback (*Megaptera novaenagliae*) and the fin whale (*Balaenoptera physalus*). Properly managed, however, the world's whale stocks could still provide man with a large, steady source of food as well as allowing him to study these fascinating and unique creatures. Instead, he seems determined to exterminate them.

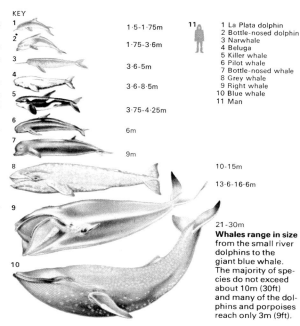

KEY

1 1·5-1·75m
2 1·75-3·6m
3 3·6-5m
4 3·6-8·5m
5 3·75-4·25m
6 6m
7 9m
8 10-15m
13·6-16·6m
9 21-30m
10

11 Man

1 La Plata dolphin
2 Bottle-nosed dolphin
3 Narwhale
4 Beluga
5 Killer whale
6 Pilot whale
7 Bottle-nosed whale
8 Grey whale
9 Right whale
10 Blue whale
11 Man

Whales range in size from the small river dolphins to the giant blue whale. The majority of species do not exceed about 10m (30ft) and many of the dolphins and porpoises reach only 3m (9ft).

8 The grey whales are confined to the North Pacific. There are two populations, one living on the eastern side and one on the western side. From the northern seas they migrate south in winter to breed in the shallow, warmer seas off Baja California and South Korea. They feed on plankton, which they strain from the water by means of baleen plates in their mouths.

8 California grey whale *Eschrichtius glaucus*

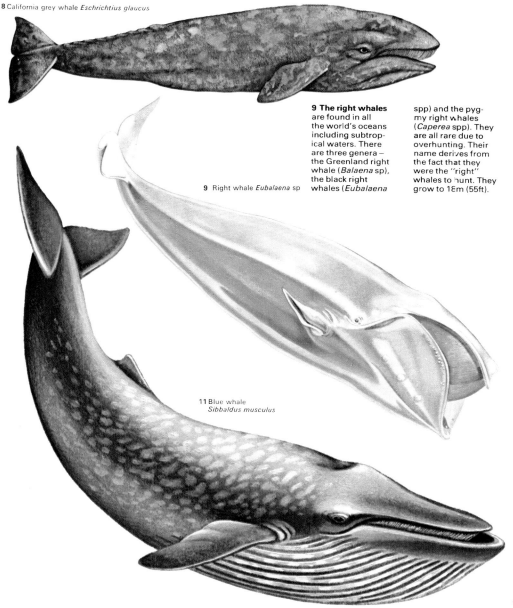

9 Right whale *Eubalaena* sp

9 The right whales are found in all the world's oceans including subtropical waters. There are three genera – the Greenland right whale (*Balaena* sp), the black right whales (*Eubalaena* spp) and the pygmy right whales (*Caperea* spp). They are all rare due to overhunting. Their name derives from the fact that they were the "right" whales to hunt. They grow to 18m (55ft).

10

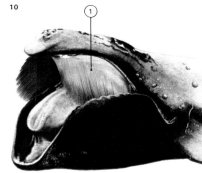

10 The head of the baleen whale is enormous and is about a third or a quarter of the animal's total length. Fringing each side of the mouth is a series of baleen plates [1] which act as food strainers. When feeding, the whale swims through swarms of krill (small, shrimp-like marine animals) and when sufficient have been trapped it dives, closes its mouth and swallows.

11 Blue whale
Sibbaldus musculus

11 The largest animal in the history of the earth is the blue whale. It grows to a length of 30m (90ft) and can weigh 112.5 tonnes. It usually travels at speeds of 10 to 12 knots. Its other name of sulphur-bottom whale is derived from the film of yellowish microscopic algae that sometimes forms on its undersurface. It inhabits polar seas and only rarely enters tropical oceans, but it does move to warmer waters to breed. It, like the right and grey whales, feeds on krill, which it strains from the water with plates of baleen. Its stomach can hold about two tonnes of food. It is rarely encountered in schools and seems to be a rather solitary animal. It is now protected, but possibly too late to save it from extinction, following over-hunting by whalers. Two related and endangered species are the fin whales (*Balaenoptera* spp) and the humpback (*Magaptera novaeangliae*).

Primates: relatives of man

Man, as a mammal, is a member of a group of mainly tropical creatures called primates. Like other mammals the early primates [Key] were contemporaries of the last of the dinosaurs. Today, the non-human primates, which include the monkeys and apes, abound in warmer parts of the world. They are in many respects a conservative group and retain, as does man, a number of primitive skeletal features such as the collarbone long since discarded by most other mammals. They have also kept, as few others have, the primitive number of five toes on each foot.

In evolutionary terms, primates have undergone few physical changes. Such specializations as they do show are almost all towards perfecting them to a tree-dwelling life and even man still retains many of these features. The earliest primates to take refuge in trees differed little from their insect-eating ancestors on the ground.

Variety of lower primates
The tree shrews (Tupaiidae) of South-East Asia [1] are thought to represent an early stage of primate development. They can be regarded as living fossils. Indeed, some authorities prefer to exclude them from the primates proper and group them with the insectivores or to place them in their own separate order. They are active, squirrel-like animals with long tails, pointed faces, large eyes and small, human-like ears. As they climb in search of food – mainly insects – they grasp the branches with their clawed fingers and toes. This grasping ability [8] is amplified in all other primates. It enables them to make more use of small branches than the purely clawed species can and, when at rest, to hold and handle food and other objects.

The prosimians of Africa and Asia are an array of species including the lemurs (Lemuridae) from Madagascar [7]. Many have hands and feet with opposable thumbs and great toes (that is, the first digit can be rotated and crossed over the palm of the hand or ball of the foot) [8]. These creatures can hold firmly on to whatever they grasp and their handling power is enhanced by the development, on most toes, of flat nails rather than claws. The bush babies and lorises (Lorisidae), although retaining the dog-like naked nose of the lower primates, have large eyes and a short snout which makes them look more human than their close relatives, at least facially.

The tarsier (Tarsiidae) [2], similar in many respects to the other prosimians, has an even more human-looking face. Its huge eyes look directly forward and the face has shrunk beneath them, enabling the animal to achieve a high degree of stereoscopic vision. This is a necessity (as is its foot structure) if the tarsier is to leap about in its forest home. It needs to be able to judge distances, like other primates, with accuracy. This comes with eyes that look forward rather than to the side. With the shortening of the face there is a tendency for the sense of smell to become less acute. The face is similar to that of the higher primates rather than the dog-like face of the lemurs. The tarsier's brain is larger and more complex than that of its near relatives.

Man's closest relatives
Animals belonging to the highest group of primates are called anthropoids (Anthropoidea). They include monkeys and apes, as

1 The tree shrew, like some of the earliest primates, hunts insects.

1 Tree shrew
Tupaia glis

2 The tarsier is so named because of its elongated tarsals or foot bones. These enable it to leap great distances.

2 Philippine tarsier
Tarsius syrichta

3 Lower and higher forms of primates are found in tropical Africa. Bosman's potto is a primitive primate, slow-moving and nocturnal, that clings tightly to tree branches. The related dwarf galago spends much of its time on the ground but can leap out of danger with great agility. Most female primates establish close, long-term bonds with their young, as shown by mona monkeys. The colobus monkey, like other African forest species, has decorative fur, for which it has been hunted to near extinction.

Bosman's potto
Perodicticus potto

Dwarf galago
Galagoides demidovii

Mona monkey
Cercopithecus mona

3

4 The uakari is a species of monkey that inhabits the tropical forests of South America. Soon after its ancestors reached the area it became isolated by a sea barrier, and the monkeys developed along slightly different lines from their Old World relatives. All are very agile forest dwellers and many, but not the uakari, are helped in climbing by the prehensile tail, which can be used as an extra hand. Some carry obvious tufts of hair; others have bald and often brightly coloured skin on their hands.

4 Uakari
Cacajao rubicundus

5 Gorilla
Gorilla gorilla

Colobus monkey
Colobus abyssinicus

6 Cotton-top marmoset
Saguinus oedipus

5 The gorilla, one of the rarest of the man-like apes, lives in the forests of equatorial Africa and is the biggest of the great apes. It is a peaceable animal, keeping to its ways in the forest, where it feeds almost entirely on vegetable matter.

6 The cotton-top marmoset is found in wooded areas of South America.

7 The ring-tailed lemur is a prosimian found only in Madagascar.

7 Ring-tailed lemur
Lemur catta

well as man, and many lead complex social lives in which individuals live as families and the young are cared for over a long period. Communication among these primates is also highly developed.

Two groups of monkeys are known. One is found in the tropical New World (South and Central America) [4], while the other inhabits the Old World tropics (Asia and Africa). Some species are found in cooler areas, such as the foothills of the Himalayas or in Japan. The main difference between the two is their nose shape. Old World monkeys have their nostrils closer set which gives them a pronounced nose. The nostrils of New World monkeys are set wide apart giving them a flattish nose.

Grasping or prehensile tails are found only in species of New World monkeys but all monkeys, with one exception (the night monkeys or douroucoulis) are diurnal; they sleep at night and are awake during the day. Both types of monkeys are social animals, living in groups with a well-developed hierarchy. Some Old World forms such as the baboons, have largely deserted the trees to take up life on the ground. But they are still good climbers and when threatened can scamper to safety among branches or rocks.

The lesser and great apes
The apes are distinguished from monkeys by their greater size, their bigger and more complex brains and their complete lack of a visible tail. The siamangs and the gibbons of South-East Asia, referred to as the lesser apes, are spindly-limbed forest acrobats [9] that can travel at high speed through the tree tops, leaping and swinging by their arms. The great apes, which include the orang-utans, gorillas [5] and chimpanzees, are all larger and heavier creatures. They spend much of their time on the ground, although they retreat to the branches to sleep.

Man's evolutionary relationship with the great apes is particularly close. Apart from man's much larger brain, the main bodily differences are associated with the ape's specialization to a forest life, and man's to open country living – walking on hind legs, rather than on all fours or swinging from the branches of trees by his forelimbs.

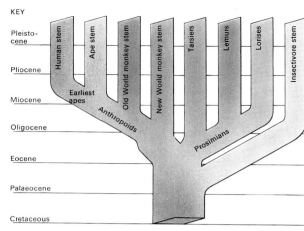

KEY

A family tree of primates shows that the earliest members of the group were closely related to the insectivores, as are today's tree shrews shown in illustration 1). In the Eocene period, lemurs flourished in many parts of the world although today they are found only in Madagascar and the nearby Comoro Islands. Both the New World and the Old World monkeys probably evolved from tarsier-like ancestors during the Oligocene period. The first true apes appear later in the Miocene. These forest-dwelling apes gave rise to man's immediate ancestors.

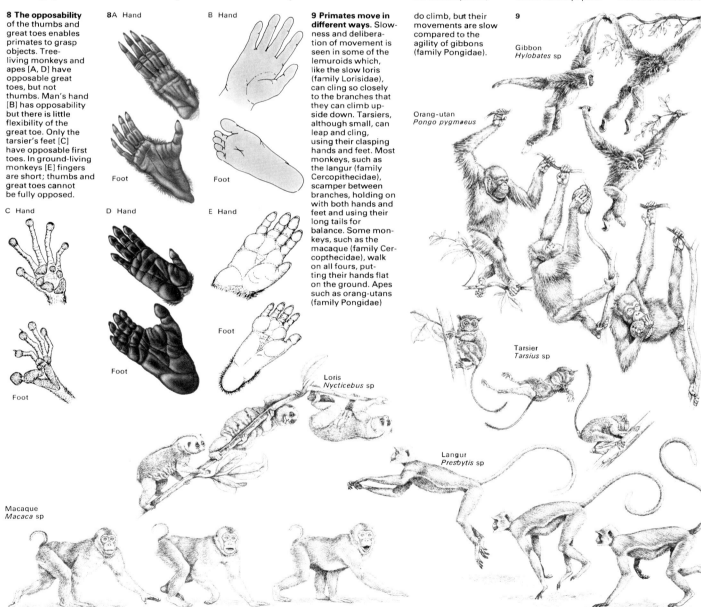

8 The opposability of the thumbs and great toes enables primates to grasp objects. Tree-living monkeys and apes [A, D] have opposable great toes, but not thumbs. Man's hand [B] has opposability but there is little flexibility of the great toe. Only the tarsier's feet [C] have opposable first toes. In ground-living monkeys [E] fingers are short; thumbs and great toes cannot be fully opposed.

8A Hand
Foot

B Hand
Foot

C Hand
Foot

D Hand
Foot

E Hand
Foot

9 Primates move in different ways. Slowness and deliberation of movement is seen in some of the lemuroids which, like the slow loris (family Lorisidae), can cling so closely to the branches that they can climb upside down. Tarsiers, although small, can leap and cling, using their clasping hands and feet. Most monkeys, such as the langur (family Cercopithecidae), scamper between branches, holding on with both hands and feet and using their long tails for balance. Some monkeys, such as the macaque (family Cercopthecidae), walk on all fours, putting their hands flat on the ground. Apes such as orang-utans (family Pongidae) do climb, but their movements are slow compared to the agility of gibbons (family Pongidae).

9

Gibbon
Hylobates sp

Orang-utan
Pongo pygmaeus

Tarsier
Tarsius sp

Loris
Nycticebus sp

Langur
Presbytis sp

Macaque
Macaca sp

How mammals behave

Mating and feeding, the primary needs for survival, provide the motives for much animal behaviour. Whatever the motives that stimulate the animal, its encounters – whether sexual or aggressive – always tend to be surrounded by rituals. These rituals are probably most varied among the mammals.

Rituals associated with copulation and those associated with aggression may be hard to distinguish within one species although both commonly involve genital sniffing. In some rodents the female is larger and fiercer than the male and sex-play may take the form of a battle from which the male may not escape without injury.

Group survival patterns
Much of the social behaviour of mammals is directed towards the survival of the species as a whole. Wolves hunt in packs to enhance their chances of cornering prey and the prey in their turn have developed ways of thwarting bands of predators. In the musk ox [1] this involves the formation of phalanxes, a tactic that is extremely successful.

Co-operation of this sort involves communication between the members of the species, and in mammals this is both widespread and diverse, reaching its most sophisticated level in whales and dolphins and in man and other primates. One of the most familiar non-verbal phenomena is the warning system of rabbits, which thump with their hind feet and flash their white tails as they retreat from danger [3]. Mammalian signals are not just visible and audible, for both touch and smell [4] are also important.

The importance of smell
The importance of smell as a stimulus is easy to underestimate from the human point of view, because the sense of smell is relatively poorly developed in man, but is crucial to the behaviour of many mammals. Many male mammals are equipped with scent glands that attract females and from the female's smell the male can tell if she is in heat – that is, in the fertile phase of her sexual cycle. Responses to these substances, known as pheromones, can be quite subtle and unexpected. Pregnant female mice, for example, will abort their young if they are put in a cage

with a male of another strain. This does not occur if they are put into an empty cage, so the phenomenon has been attributed to a pheromone. Effects of a similar kind may underlie the natural regulation of population density which has been noticed in many colonial species of mammals.

The sense of smell may be vital in establishing the bond between a mother and her young [2]. A kind of olfactory "imprinting" seems to occur among some ungulates, hoofed mammals, that give birth to a single young and then have to keep track of their offspring in a moving flock or herd. The imprinting takes place immediately at birth and enables the mother to recognize her own offspring.

Exclusive bonds between a mother animal and her own young are not always the rule among mammals. Prairie dogs, which are actually large rodents, live in populous colonies that for part of the year are divided into smaller territories containing groups of one or two adult males with females and young. The young are suckled indiscriminately by any female and groomed in a friendly way by

1 A herd of musk oxen forms a powerful phalanx in a bare landscape to make a defensive ring of bulls around the more vulnerable cows and calves, which are protected in the safety of the centre. Living in the harsh Arctic tundra of northern Greenland and Canada the musk ox *Ovibos moschatus* finds natural, predatory enemies in the wolf packs of that inhospitable region. Instead of scattering as the pack attacks the oxen muster together with their great horns pointing outwards. This tactic shows how the basic drive to flee from danger by running away has been superseded by the development of a social instinct requiring communication, co-ordination, discipline and courage. In their defensive formation the oxen can defy the wolf pack but if discipline broke down the pack would quickly attack and pull them down.

2 Imprinting is responsible for the bond that is formed between a female animal and her offspring within hours, possibly even minutes, after the birth. The female goat soon learns to distinguish her own young from others in the herd. The immediate post-natal nuzzling and licking is probably necessary to imprinting. If immediately after the birth [1] the kid is removed [2, 3] for 3 hours the mother will reject it, butting or even biting it [4]. If immediately after the birth [5] the mother is allowed a mere 5 minutes of licking and nuzzling her offspring [6] before a similar 3-hour separation [7] she will usually accept and suckle the kid when it is returned to her [8]. The rapidly formed bonding between the female and her offspring is probably essential to the survival of those animals living in herds that are constantly on the move, like the reindeer (*Rangifer tarandus*).

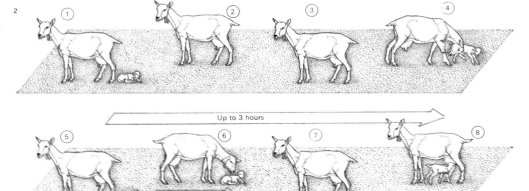

Up to 3 hours

5 minutes

any male. The same kind of system is found in small communes of African hunting dogs. Here, however, conflict can arise between females who compete to suckle the young.

The most common source of conflict between mammals is competition for females during the breeding season. This is the underlying reason for the ritual clash of antlers among deer and is one of the defining features of dominance hierarchies in animal groups: the more dominant males have access to larger numbers of females.

The weak and the strong

Much ritualized animal behaviour is concerned with preventing competition from ending in harm to either competitor within a species. Dogs and wolves, for example, have recognized signals of threat and submission. A dominant wolf stands stiffly, hair raised, growling through bared teeth. The subordinate animal deters attack by crouching with ears flattened in a posture of submission.

Dominance and subordination signals also occur in rats, both between members of one colony and between the members of a colony and intruders. It is believed that members of the same colony recognize one another by smell, and that intruders are immediately liable to attack. The attacker adopts an arch-backed posture and minces round the intruder, flank-on, chattering his teeth. The intruder may adopt a submissive posture – by lying down – and thus prevent actual attack. If he fails the next move will be leaping and biting which is sometimes fatal.

The most curious social phenomenon observed in rat colonies is that of the fatal effect of social stress [5]. An intruder, or a rat low in the dominance hierarchy of a colony, may be subjected to so much threat that he literally gives up and dies although he is unwounded.

Mammalian behaviour is not always directly controlled by the primary drives of mating and feeding but has come to reflect the complex demands of social life. Limited freedom from the immediate demands of hunger, thirst and sex can be seen in an everyday context in the willingness of dogs to run errands for their masters and of both cats and dogs to play with humans.

KEY

Two male rats in competition for a female threaten [1], attack [2] and bite each other [3]. A female not in heat [4] rejects a male but submits [5] when she is in heat (oestrus).

3 A

B

C
21 March—11 April   24 May—13 June

Blue—Adult
Green—Young

3 A colony of wild rabbits [A] occupies a warren made up of a system of burrows that perform various functions. In spring [1] rabbits graze the pasture around their warren, stripping it of grass and leaving nettles, ragwort and ground ivy to flourish. The rabbit population is swollen by young twice a year [C] for each female. In the summer and the autumn [2] rabbits destroy pastures and crops. In winter [3] they may eat such crops as winter wheat. The cross-section cut through a typical rabbit warren [B] shows a bolt-hole [4] offering a quick refuge, [5] a breeding burrow lined with leaves and grass and [6] young venturing above ground where they encounter predators [7, 8] that deplete the population.

4 Black-tailed deer [A] have scent glands on their ankles. They use the glands, called tarsal organs [1], for recognition. Members of a herd [B] sniff one another's tarsal organs [2]. Fawns recognize their mothers by similar ankle sniffing [3]. When a stranger [4] enters the herd's territory its smell betrays it at once.

5 A strange rat [A] entering a colony may be subjected to social stress [B] severe enough to prove fatal. It may not have suffered any physical injury.

4 A   B

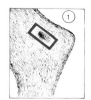

5 A

B

171

How primates behave

Primate behaviour is a source of fascination for humans as the evolutionary predecessor of their own. But the behaviour of many primates is radically different from that of man. Most species are social, but a few, particularly the nocturnal prosimians, are solitary [1]. In addition to the endlessly fascinating question of the mental abilities of subhuman primates [Key] the social organization of these mammals is also significant.

Interpreting social signals
Most of the more advanced primate species live in troops with a clearly defined role structure, the dominant males (usually the older and larger) being allowed first choice of both food and females. By comparison with other groups primates make elaborate use of facial expression, in addition to posture, in order to signal dominance or submission [3]. Aggressive facial expressions often involve bared teeth and other signs that would be recognized as threatening by a human. But the interpretation of specific signals is not always obvious in human terms; for instance, curling of the lip is believed to signal pleasure in

gorillas whereas in humans it portrays anger. Roaring and breast-beating, however, are more easily deciphered as hostile although they may also express exuberance and fear. The correct submissive response is to shake the head slowly from side to side, avoiding the gorilla's eyes.

Within a troop, and subject to the constraints of the dominance hierarchy, primates are in general sexually promiscuous. A female chimpanzee has been seen to accept as many as seven males in succession. The degree to which sexual activity is limited to the oestrus period (fertile period) of the female is variable. In rhesus monkeys, copulation seems to be stimulated by a pheromone (a chemical that is smelt) produced by the female only during oestrus; but chimpanzees seem to copulate at more or less any stage of the oestrous cycle. This dissociation of sexual behaviour from its immediate reproductive function is generally seen as a reflection of the advanced thinking abilities and need for social cohesion of primates. Their play, through which they gain much experience, is more elaborate than that of

other advanced mammals. The reasoning capacity of the more intelligent primates is certainly well developed but it is only recently that detailed observations of animals in the wild have supplemented those in a laboratory environment in confirming this.

Weaponry and skills
Tests conducted in the wild have proved decisively that chimpanzees habitually use sticks and stones as tools [5] or weapons. The immediate reaction of a troop of savanna chimpanzees confronted with a stuffed leopard was to retreat in fear to a nearby group of trees. The adults made much noise and then cautiously advanced and began throwing sticks in the direction of the leopard. When it did not move the boldest chimpanzees came closer and began striking it with sticks. After the stuffed head became detached from the body the chimpanzees realized that the leopard was harmless and went over warily to sniff it. Finally they ignored it altogether and wandered off. Baboons [2] show similar behaviour in the wild and have been seen trying to frighten a

2 The orang-utan (*Pongo pygmaeus*), one of the intelligent great apes, lives in the tropical rain forests of Sumatra and Borneo. Orang-utans live in small groups of up to six individuals. Mother and infant [A] comprise the basic family unit and all feed mainly on fruit [B]. Adult males [C] are larger and heavier, usually live apart and have pronounced cheek pads which have evolved for display purposes.

1 Lemurs, relatively primitive primates from Madagascar, are solitary animals. The sportive lemur (*Lepilemur mustelinus*) never forms groups larger than a mother-child pair. By day [A] this lemur sleeps in a hole in a tree or a nest of leaves and twigs. At dusk it wakes to feed on bark and leaves [B]. It is active at night within its home range [C] – a radius of some 50m (160ft) from its nest site to which it returns at dawn [D].

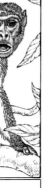

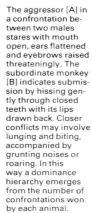

The aggressor [A] in a confrontation between two males stares with mouth open, ears flattened and eyebrows raised threateningly. The subordinate monkey [B] indicates submission by hissing gently through closed teeth with its lips drawn back. Closer conflicts may involve lunging and biting, accompanied by grunting noises or roaring. In this way a dominance hierarchy emerges from the number of confrontations won by each animal.

4 Baboons live in troops with well-defined dominance hierarchies and engage in co-operative hunting. The troops may be relatively small and, to avoid inbreeding, there seems to be a system whereby young males periodically transfer from one troop to another when they have reached their full adult size at about six to eight years of age. The females do not transfer and their troops have traditional home ranges into which others rarely intrude.

3 Facial expressions are important in the social interactions of rhesus monkeys.

Primates on the brink

Conservationists are in despair over the plight of the lemur. An international appeal for £5m to help save the Northern sportive lemur – the world's most endangered – has fallen on deaf ears, despite being popularised in the cartoon film, Madagascar.

60 left

Northern sportive lemur
MADAGASCAR
They take their name from the boxing pose they assume when they feel threatened. One of the most endangered animals in the world

250 left

Northern brown howler
BRAZIL
The monkey located in the southern Bahia region is a fruit-eater, vital for dispersing seeds in the surrounding area.

2,000-10,000 left

Grauer's gorilla
DEM. REP. OF CONGO
Gorillas are the largest primates in the world and Grauer's are the biggest of the lot. A single animal may eat as much as 40 pounds of food daily.

50 left

Cat Ba langur
VIETNAM
Crowned by a large tuft of blond fur, hence its other name, golden-headed langur. It inhabits the moist tropical forests of the Cat Ba archipelago.

300-2,000 left

Silky sifaka
MADAGASCAR
Wears a bright white coat. The species suffers from a vitiligo-like skin condition which leads their skin to slowly turn pink.

NOTE: POPULATIONS ESTIMATED

SOURCE: PRIMATES IN PERIL

27.02.2015

briefing

FROM

The ⬤ INDEPENDENT

**FRIDAY
27 FEBRUARY 2015**
Number 1,328

i@independent.co.uk
Twitter: @theipaper
facebook.com/theipaper

Sport
EUROPA LEAGUE

Italian red
*Celtic crash
out to Inter Milan
after first half
sending off*

P61

» British fanatic behind
IT graduate known to sec
» Questions for MI5 about
» Family and neighbou

COMMENT
Isabel Hardman
Who'd be an MP?
Chris Blackhurst **The**
republic of Manchester

team of researchers by dropping stones on them from a safe distance up their home cliff.

Despite the artificiality of the conditions, some of the most revealing observations and rigorous tests on primates have been made in the laboratory. Studies have been made of mother chimpanzees, for example, and have demonstrated how they encourage the independent climbing skills and movements of their infants. And a famous series of experiments has shown that the social development of rhesus monkeys depends heavily on the mother-infant bond initially and later on interactions with other infants.

Laboratory tests of rhesus monkeys have shown they can grasp abstract concepts such as "oddness". If given training in a special test apparatus the monkey can learn to select an object because it has a different shape or colour from others offered [6].

Language experiments
A rich source of controversy concerning primate behaviour is whether apes are capable of language, the crucial human achievement. American researchers have bypassed the problem of apes' inability to produce human sounds by teaching them sign language as used by the deaf and dumb. Apes taught this language by people fluent in it who were responsible for their day-to-day care as infants (much as human infants are taught to speak by their parents) learned words fast and could use them in an appropriate context. What they have not yet been able to do, however, is string them into sentences with a clear syntax; their communication lacks the factor that distinguishes language from social signalling of other kinds.

Ethologists (students of animal behaviour) point out that this kind of experiment is only meaningful in the context of the apes' ecology and highly developed communication system. Researchers have counted as many as 36 distinguishable sounds made by certain monkey species and found that different sounds are used in complex combinations. Nobody has yet learned an ape language, but enough is known to make it clear that the sound signals of these animals are used exclusively for conveying social information – not abstract concepts.

The problem-solving abilities of chimpanzees enable them to work out novel solutions to practical dilemmas by the use of tools. Confronted with a bunch of bananas that are hanging too high for it to reach, a chimpanzee is capable of reasoning that, by putting a box beneath the bananas and using this as a step, it will raise itself to a level at which the fruit can be reached. While many animals make use of various materials for building purposes, the cognitive process involved in solving complex problems by the regular use of tools has evolved in only a few species besides man and the other primates.

5

5 The high intelligence of chimpanzees is demonstrated in their practice of "fishing" for termites. This involves a sequence of behaviour that is quite complex in logical thought. The first step is for the chimpanzee to find a suitable twig. This is thrust into a hole in a termite mound where the termites will attack the intruding object. The chimpanzee then withdraws the twig, to which a number of termites remain clinging, picks off the insects with its mobile lips and eats them. This sequence of behaviour with its regular use of an accessory tool (the twig) has been observed in wild chimpanzees in many parts of Africa. It is a remarkable example of the way in which the curiosity and puzzle-solving abilities that have been seen in laboratory tests on chimpanzees is actually applied to life in the wild.

6 A monkey's intelligence enables it to grasp the concept of "oddness" in a laboratory test using covered food wells, one of which contains a reward. In [A] the objects covering the wells are the same colour but two are cubes and one is a pyramid. The monkey quickly learns to choose the pyramid. In [B], given two blue objects and a red one all of different shapes, it then selects the pyramid, proving that it has learned to recognize shape and colour.

6 A

B

Fossils: the life of the past

Fossils, traces of dead organisms found in the rocks of the earth's crust, tell us what life was like at the time the rocks were formed [1]. It is through the study of fossils (the name comes from the Latin meaning "something dug up") that the evolution of life on earth has been traced and affinities revealed between groups of animals and plants, both living and extinct.

Fossilized organisms may be found in many different states of preservation, but only rarely totally intact. More usually only the hard parts are preserved or only the shape impressed on some other material. In trace fossils [2] only the footprints or some other signs of an animal's passing are left.

Other more unusual types of fossils include gastroliths and coprolites. Gastroliths are smooth, rounded pebbles often found within the rib-cages of dinosaurs and swimming reptiles. These were swallowed by the animal possibly for the same reason that a modern fowl swallows gravel; to help grind up food in the stomach. Coprolites are fossilized dung, most commonly associated with Tertiary mammals or Carboniferous fish. Such fossils give valuable clues to the diets and habits of extinct animals.

Myths and mystery

Fossils have not always been recognized for what they are. Although ancient Greeks such as Herodotus (*c.* 485–425 BC) noted their similarity to living animals and plants, later civilizations did not have the philosophies to account for them. Whole mythologies were built around fossils: mammoth skulls found on the Greek islands may have given rise to the Cyclops legends because the fused nostril cavity suggests a single eye socket. The American Indians regarded dinosaur skeletons as the remains of great serpents that lived beneath the surface of the earth and died when they came too close to the light. Christian chronology, based as it was on biblical events and dating the Creation at about 4000 BC, made no attempt to account for the time involved in turning sediments into rock and raising mountains, so the true nature of fossils was obscured for many hundreds of years. For centuries they were regarded as sports of nature or as tricks of the devil conjured up to test the faith of the people. Those who were convinced that they represented the remains of once living organisms could account for them as the remains of creatures destroyed in the Great Flood. Not until the beginning of the nineteenth century did palaeontology (the scientific study of fossils) become a recognized science.

Fossil investigation

The detailed study of fossils, rather like a crime investigation, involves the piecing together of many diverse fragments of evidence and a vast number of techniques are used [6]. The fossil could be a 25m (82ft) long dinosaur skeleton, which must be attended to by a large number of people on the spot where it is found, or it may be a microscopic structure found in the depths of a piece of rock revealed in laboratory studies.

Initially, in the case of the dinosaur, the bones have to be removed from the matrix [6], encased in some protective substance and transported to a museum or laboratory, careful note and measurements first having been taken of the position in which the bones

1 **Fossils** are the traces of animals and plants of past times, found preserved to some extent in rock. The preservation can take place in a number of different ways. In some cases the organism can be preserved in its entirety [A] when it becomes embedded in an antiseptic medium. Examples of these are insects entombed in amber and mammoths buried whole in frozen mud. The hard parts may be preserved unchanged [B] when an antiseptic medium encases them after the soft anatomy has decomposed. Mammal bones are preserved in tar pits like this. More often only very little of the original material is left [C] as in the fossils of fern leaves preserved in Carboniferous shale as a thin film of the carbon constituents. Sometimes the original tissues are replaced, molecule by molecule, by another substance to give an exact replica of the original. Fossil wood replaced by silica is an example of this [D]. The organism may rot away completely after its burial [E] leaving a hole called a mould in the shape of the original. Moulds of Tertiary water snails are often found. After a mould has formed it may fill with minerals deposited by percolating ground water [F] giving a solid with the external shape of the original, called a cast. Calcite casts of ammonites are common.

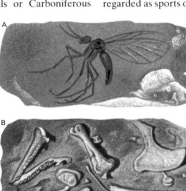
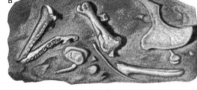

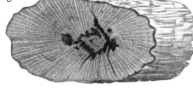

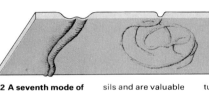

2 **A seventh mode of fossilization** is one in which no part or shape of an organism is preserved – only the footprints, burrows and feeding trails. These are known as trace fossils and are valuable as indicators of the life-styles of extinct creatures. Sometimes an animal is known only from its fossil traces. For some animals these consist of tunnels or tunnel systems which are given group names according to their shapes and complexity. *Repichnia* traces [A] are long and fairly straight; the *Pascichnia* types [B] are meandering and cover a wide area. The *Cubichnia* burrow types [C] are usually short and straight and those of *Fodinichnia* [D] radiate from a single spot to cover a wide area.

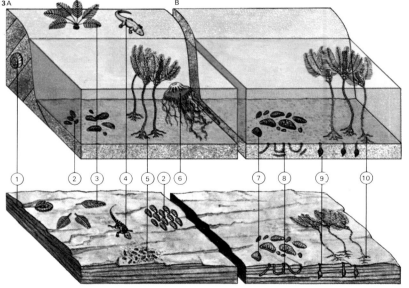

3 **Fossils lie in two kinds of assemblies;** one that does not reflect the life-style of the organisms (called a death assemblage or thanatocoenosis), or one that does (a life assemblage or biocoenosis). Features of a thanatocoenosis [A] include derived fossils [1], fossils of an earlier age removed from a rock eroded at the time of deposition of the assemblage, shells disarticulated and aligned with the current [2], plant and animal remains from another environment brought in after death [3, 4], and fragile creatures such as crinoids, broken and scattered [5]. Some soft-bodied creatures such as jellyfish [6] are never fossilized. A biocoenosis [B] has shellfish in their living positions [7], burrows and burrowing organisms undisturbed [8, 9] and crinoids still in one piece [10]. A thanatocoenosis is the result of currents and erosion disturbing remains and in a biocoenosis there is little or no disturbance of them.

are found. The nature of the surrounding rock is also noted because this gives valuable information about the environment in which the creature lived. In the laboratory the bones are cleaned and treated to make them less fragile. Usually the individual bones are cast in plaster or a plastic material so that a facsimile of the skeleton can be assembled and put on display while the actual specimens are stored for research purposes.

The dinosaur remains are then compared to the hard parts of modern animals and a reconstruction is made, usually by means of a painting or a model, showing the appearance of the animal when alive. The colour of extinct animals is never preserved and is a matter of informed guesswork. When microfossils are studied, a likely looking rock is taken into the laboratory, sliced up and treated with solvents that either corrode rock and fossil at different rates, so throwing the fossil into relief, or corrode only the rock.

The fossil hunters

Anyone with sufficient interest can find and collect fossils, but to identify them and study them in depth an interest in and knowledge of animal and plant anatomy is essential. To locate fossils with accuracy an understanding of earth movements and processes that give rise to the preservation of organisms is also necessary. This will enable the fossil hunter to interpret the clues that they give about past conditions. Once they are removed from their matrix all fossils must be treated with respect and preserved in the best way.

Although fossils may be collected as a hobby, they are used by the palaeontologist as the identifying "fingerprints" of a particular rock bed. An otherwise featureless bed of limestone containing fossils of a particular short-ranged species found at one locality can be identified at another locality by the presence of that fossil species. Or knowledge of the rates of animal and plant evolution can be used to give an age to the sedimentary rocks in which identifiable fossils are found. This branch of geology is known as stratigraphical palaeontology; techniques involved are extremely valuable in the constant search for oil. Fossils also provide clues on the evolution of life.

KEY

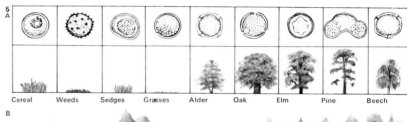
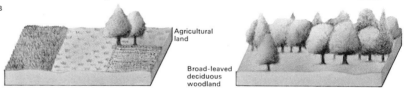

Cereal Weeds Sedges Grasses Alder Oak Elm Pine Beech

Agricultural land

Broad-leaved deciduous woodland

Coniferous woodland

Tundra

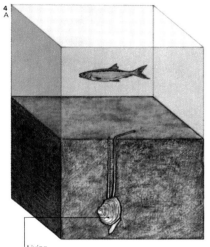

Living Scrobicularia

Fossil Scrobicularia

4 A facies fossil is indicative of a particular environment of deposition. The shellfish *Scrobicularia* lived only in oxygen-free mud [A]. It shows that such muds formed the rock in which it is found fossilized [B]. The prawn and fish were mobile, not dependent on the mud, and thus are not facies fossils.

5 The study of microscopic fossils, micropalaeontology, includes the study of pollen grains [A] found at different depths in recent soils. They indicate changing vegetation – and hence changing climates – over the past few million years [B]. This new science is known as palynology.

6 A fossil in a rock such as this belemnite is often a most unprepossessing item [A] and sometimes it takes a palaeontologist to recognize it. The find could be made on a building site or it could be at the top of a mountain in Peru; the fossil hunter must work with it wherever the specimen is found. The removal of the fossil from the rock [B] can take place on the spot but more usually the fossil and matrix are transported back to the laboratory where the right equipment, preserving fluids and technicians are on hand. Sometimes, if the fossil and the rock are of two very different materials – such as a calcite shell embedded in shale – removal is easy. More often the task involves months of laborious work scraping away the embedding material from round the valuable specimen. The end product, after all this, is a prepared and mounted specimen [C], a solid reminder of life that may have flourished millions of years before man.

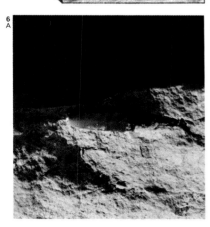

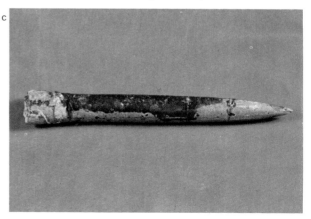

Plants of the past

The first traces of life on earth are those of plants; strange, primitive growths spawned in the primeval ocean, the rich "broth" of materials absorbed from the earth and atmosphere. These first traces – humped or branching masses of silica – occur in Precambrian rocks, 2–3,000 million years old, in North America and South Africa.

Fossil evidence of plant life is hard to come by, for plants do not fossilize easily; they tend to lack durable hard parts, with the result that plant fragments are usually destroyed by weather or chemical processes before they have had time to be preserved in the rocks. Such plants that have become fossilized, however, are exhaustively studied by palaeobotanists and from them much has been deduced about the history of plant life.

Importance of plants
Of the two biological kingdoms, plants and animals, the plant kingdom is by far the most important and arose first – the earliest traces of animals do not appear until the early Cambrian, about 570 million years ago, in Australia. Plants take in energy directly from sunlight and use it to synthesize foodstuffs. Without these foodstuffs the food chains that involve the animal kingdom would have no starting-points: with no plants there would be nothing to feed the herbivores, with no herbivores, nothing to feed the carnivores, and so on. But at first the importance of plants was even greater. The early atmosphere of the earth was probably so rich in carbon dioxide that no animal could have breathed it. Plants changed all that and made animal life possible, since the process of food production involves the removal of carbon dioxide from the atmosphere and the production of free oxygen. Thus once evolution began to transform the primeval living protoplasm into different forms of life, the plants came first.

The first plants that would have been recognizable as such would have been one-celled algae [1], in which all vital processes would have taken place in a single cell. As time passed more specialized algae developed. They consisted of more than one cell and various processes such as reproduction were carried out by different parts of the plant. These would have resembled present-day seaweeds. By this time animals, too, would have developed; herbivores would have fed on these seaweeds and carnivores would have fed on them in turn.

Life out of the water
All this early activity took place in the sea – the land and its atmosphere was too alien and hostile for much to happen there. Some algae became adapted to fresh water and it is from there that land plants are thought to have evolved. In time physiological changes came about in forms that enabled the plants to spend more and more time exposed to the atmosphere. The most important of these changes was the evolution of a vascular system – a form of plumbing that could carry water up from the base where it was available and synthesized foodstuffs down from above [2]. With the formation of light-trapping organs held towards the sun [7] and a system of reproduction that would work in the air the true land plant had evolved. The atmosphere on land probably resembled that of today although it may have contained a higher proportion of carbon dioxide. With the

1A

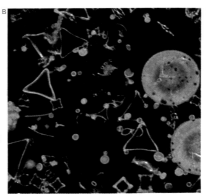

1 The earliest known plants, the primitive algae, are known from the rock structures (stromatolites) they produced [A]. These were created by the fossilization of concentric layers of mud and algae. Other fossil algal remains commonly found are the "shells" of diatoms [B]. These small, brown algae have been abundant since the Cretaceous period, in all aquatic habitats. The diatomaceous earths are composed entirely of their shells.

B

2 A

B Nematophyton

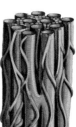

2 To live on land plants needed "plumbing" equipment – a means of transporting water up from the roots and foodstuffs down from the leaves. This is done by means of the vascular system, a network of tubes throughout the plant. In modern plants this is well organized, but extinct plants such as *Nematophyton* had a trunk [A] consisting entirely of tubes of two widths [B].

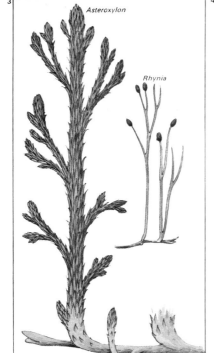

3

Asteroxylon

Rhynia

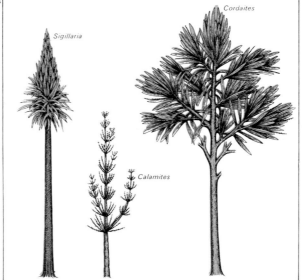

4

Sigillaria

Cordaites

Calamites

5

3 The earliest land plants known from detailed fossils are psilophytes preserved by a freak volcanic eruption during the Devonian in Scotland. *Asteroxylon* grew to a height of 1m (39in) and *Rhynia*, named after the site at Rhynie, to 50cm (20in).

4 The vegetation of the great coal forests consisted of lycopods (club-moss relatives), horsetails and primitive gymnosperms. *Sigillaria* was typical of the lycopods, growing to a height of 30m (100ft) in very marshy ground. *Calamites* was a 9m (30ft) horsetail that grew as "reed beds" in the waters of the swamps. Drier areas supported gymnosperms such as *Cordaites*, a primitive relative of the conifers. The undergrowth of these forests consisted mainly of ferns.

5 The great size of the coal forest trees is evident in this picture of a fossil *Sigillaria* stump. Although most trees of the Carboniferous forests lost all their recognizable structures on being decomposed and compressed into coal, sometimes the shapes of roots ("stigmaria") and fallen trunks are preserved in the sands and muds on which they grew when the ground was clear of other plant remains. These remains are most often found in "seat earths" underneath coal seams.

emergence of terrestrial plant life there was food available for animals to follow them.

During the Devonian and Carboniferous, over 300 million years ago, plants had developed into a great variety of complex forms. These were still primitive by today's standards, being close relatives of the horsetails, club-mosses and ferns [Key], but nevertheless they formed vast forests [4, 5] and adopted all the growing habits of modern plants – trees, bushes, creepers, undergrowth and the like. They grew in such numbers that the thick layers of their rotting remains became solidified into the present-day coal beds [6]. One feature that characterized this flora was its lack of a self-contained reproductive body – the seed.

The evolution of the seed

The seed evolved during the Carboniferous (and certain ferns that bore a seed-like structure were common at this time) but the seed-bearing plants did not come into their own until the Permian and in the Triassic (225 million years ago) the dominant plants were gymnosperms [8] such as the conifers.

During the Cretaceous, which began about 135 million years ago, the angiosperms [9] or flowering plants came to the fore and a flora was established that was closely related to today's. For the past 130 million years no major plant group has arisen but there has been an enormous proliferation of grasses and herbs with a correlated influence on the animals associated with them.

Because the division of the geological column is based on the stages of evolution of animal life, the story of plant evolution does not fit easily into the time scale. In essence, the earth's flora consisted entirely of algae up to Silurian times – half way through the Palaeozoic era, the age of invertebrates. Then land plants – consisting of mosses and liverworts (bryophytes) and ferns, horsetails and club-mosses (pteridophytes) – evolved and dominated the flora until the Permian, the end of the Palaeozoic. Gymnosperms were the dominant plant forms from the Triassic until the beginning of the Cretaceous, half way through the Mesozoic, the age of reptiles. At this time the angiosperms developed and continued until the present.

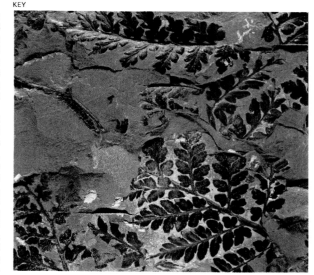

The fronds of this fossilized fern have been preserved as a thin film of residual carbon pressed between beds of Carboniferous shale, the most common method of plant fossilization.

6 A creeping form that made up part of the undergrowth of the coal forest was the horsetail *Sphenophyllostachys*. Whorls of six leaves 2cm (0.75in) long sprouted at intervals from the stem and spores were formed in a strobilus or cone on a side branch. *Lepidodendron* was of the Carboniferous lycopods, growing to a height of 30m (100ft). The trunk had a characteristic diagonal pattern of leaf scars and branched dichotomously (repeatedly into two equal parts) to give a crown of branches with strap-shaped leaves. Under the ground the stem branched dichotomously to produce a woody root stock.

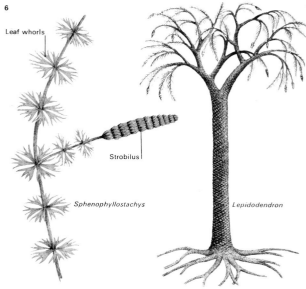

Leaf whorls

Sphenophyllostachys

Strobilus

Lepidodendron

7 For a leaf to be an efficient organ for trapping light with which to produce food and energy, it must present as great an area as possible to the sun. In primitive plants this was achieved by a larger number of branches. It is thought that outgrowths of these branches coalesced, forming single plates of a large area, so that the leaf was evolved [A, B, C]. The primitive position [D] of reproductive bodies [1] is at the ends of dichotomously branched stalks. More advanced forms show a reduction in the branching [E] which is carried still further in some species of fossil fern [F].

8 From the Triassic to the end of the Jurassic, the gymnosperms were the dominant type of plant life. These resembled the gymnosperms found today – the conifers and the cycads – and were characterized by their properly developed seeds. These seeds were open to the air rather than borne in enclosed fruits. The gymnosperms of the Mesozoic were mostly conifers and Bennettitales, the latter being related to the cycads and consisting of a bulbous trunk which may have carried "flowers", surmounted by palm-like fronds. Examples include *Williemsonia* [1] and *Bennettites* [2]. Contemporary conifers include *Voltzia* [3] and *Araucarites* [4].

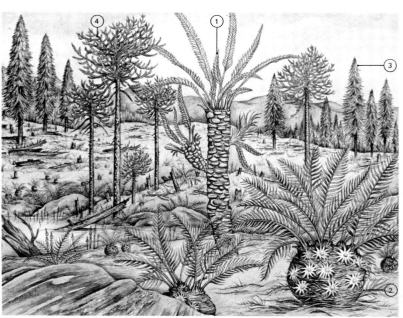

9 Angiosperms, higher plants with enclosed seeds, took over from gymnosperms when the dinosaurs roamed the earth at the beginning of the Cretaceous. Since then the flora, both herbs and trees, has remained very similar to that of the present day. The leaf shown is that of a plane (*Platanus* sp) [A] preserved in sediments of the Mesozoic, and is about 100 million years old. The structure of this leaf is almost identical to that of a modern specimen of the same genus [B].

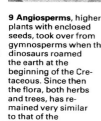
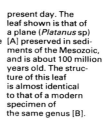

177

Fossils without backbones

The most abundant and significant fossils in the rocks of the earth's surface are those of invertebrates. These range in size from microscopic creatures to cephalopods with coiled shells 2m (6ft) in diameter and in age from the early Cambrian, 570 million years ago, to historical times.

Joint-legged creatures and shells

Some of the earliest fossils were those of the arthropods, which are found from the Cambrian onwards. These are the joint-legged animals that have shells of chitin – today's insects, spiders, crabs and lobsters. In Cambrian times they were represented by the trilobites [1] and other strange creatures. The trilobites were generally unspecialized marine animals with bodies divided into a large number of similar segments. They appeared in the Ordovician, flourished during the lower Palaeozoic, but started to wane during Devonian times and by the end of the Permian were extinct. The closely related eurypterids [3] had a shorter range – Silurian to Devonian – but while they lived they were the terror of the seas. Ostracods

are another important group of arthropods. They were encased in a pair of tiny, almost microscopic, shells hinged along the back. These shells had characteristic decorations and are valuable as index fossils because the distribution of each species through time is well known, when a species is found in a rock the age of the rock can be determined.

Arthropods were among the first animals to colonize land, the millipede *Archedesmus* being found in Devonian rocks. Insects evolved soon afterwards and the Carboniferous forests sheltered such creatures as *Meganeura* [4] – a dragonfly the size of a parrot. Unfortunately, fossil land arthropods are rarely found and consequently little is known about their evolution. It is known, however, that the arachnids and simpler insects were well established by the upper Palaeozoic while the flies and social insects did not develop until the upper Mesozoic.

Molluscs are another major group of organisms with a long ancestry and the tentacled molluscs – the cephalopods – have been found as far back as the Cambrian. These first cephalopods had straight, conical shells and

this primitive shape, the orthocone, lasted until the Carboniferous. Meanwhile other branches of the cephalopods developed, evolving shells that were curved and, eventually, shells coiled in a tight spiral. From these nautilus-like creatures of the upper Palaeozoic evolved the ubiquitous ammonites of the Mesozoic [10], the fossil shells of which are a familiar sight in Jurassic rocks all over the world. These also make excellent index fossils because of their great variety in shell shapes and ornamentation [11, 12] and the short time range of each species. Ammonites died out during the general change of fauna at the end of the Mesozoic and since then the only cephalopods have been the squid, cuttlefish, octopus and nautilus.

The snails and the bivalves, the other chief members of the mollusc group, have also been present since the Cambrian but it was not until the Tertiary that the bivalves became the important "sea shells" that they are today. During the Palaeozoic and Mesozoic their ecological niche was occupied by a primitive group, the brachiopods, which superficially resembled them – being sessile

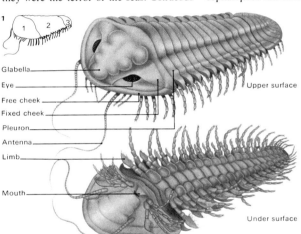

Glabella
Eye
Free cheek
Fixed cheek
Pleuron
Antenna
Limb
Mouth
Upper surface
Under surface

1 A trilobite looked rather like today's woodlouse, being covered by a chitinous skeleton. This was divided into the cephalon [1], or headshield, which carried sensory organs and the glabella, a bump that housed the stomach; [2] the thorax, a region of articulated segments below each of which was a pair of legs; and [3] the pygidium or tail shield. Each limb consisted of a jointed organ for walking, a swimming and breathing organ and a paddle that swept food particles towards the mouth.

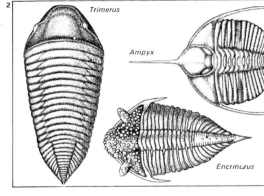

Trimerus
Ampyx
Encrinurus

2 Trilobites adapted to their environment in a variety of ways. *Trimerus* was a burrower, *Ampyx* was a lightweight swimmer and *Encrinurus* was a slow-moving bottom dweller.

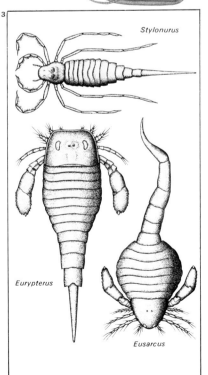

Stylonurus
Eurypterus
Eusarcus

3 The eurypterids, probably related to the trilobites, included the great sea scorpions of the Silurian and the Devonian. Although some carried claws and grew to a length of 3m (10ft), those shown here were smaller and more like today's scorpions. *Stylonurus* was a long-legged form found in Silurian and Devonian rocks. *Eurypterus* was the first of the group to be discovered and lived in the Devonian of Scotland. *Eusarcus* was scorpion-like. These lived only in the sea in the Silurian but later moved into fresh water.

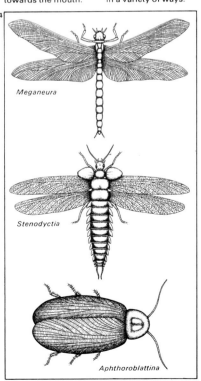

Meganeura
Stenodyctia
Aphthoroblattina

4 Insects appeared in the Devonian and flourished in the Carboniferous. These included a dragonfly called *Meganeura*, with a wing span of about 80cm (31in); *Stenodyctia*, which was a generalized type; and *Aphthoroblattina*, a cockroach.

5 This trilobite is *Ogygiocarella* and was preserved in Ordovician rocks 500 million years old.

6 The fossil of the extinct lobster *Eryon* was found in 140 million-year-old Jurassic limestone.

(sedentary) and having their soft anatomy enclosed in two shells – but were in no way related to them. Instead of having a left and right shell, as bivalves do, the brachiopods had a top and bottom shell. They have been declining in importance since the end of the Mesozoic and since then the bivalves have largely replaced them.

Development of corals
The corals are an important group of fossil organisms and can be divided into three groups, two of which are extinct. The first group, called the rugose corals because of the wrinkled appearance of the skeleton, were mostly solitary organisms looking like the related sea anemones but encased in cup-shaped shells. Some later rugose corals were colonial, consisting of more than one individual. They appeared in the middle Ordovician, reached their peak in the lower Carboniferous and died out in the Permian. The second group, the tabulate corals, were all compound, consisting of large numbers of small individuals. They appeared in the middle Ordovician, flourished in the Silurian

and Devonian and became extinct in the Permian. The third group, the scleractinian corals, are the reef-building corals of today. They did not appear until the middle Triassic and most have been colonial.

Some vertebrate relatives
The phylum known as the chordates, to which man belongs, includes several groups of small, apparently simple worm-like animals. An early off-shoot from this phylum may have given rise to the graptolites – communal drifting organisms that were abundant in the oceans of Ordovician and Silurian times [7]. Their evolution took place rapidly from the complex types of the lower Ordovician to the straight, simple forms of the Silurian, making them good index fossils for identifying Palaeozoic black shales.

The echinoderms – the starfish, sea-urchins and sea lilies – are probably close to the ancestors of the vertebrates and, because they are mostly covered with armour, they tend to fossilize easily. Certain Carboniferous limestones are made up largely of crinoid (sea lily) plates.

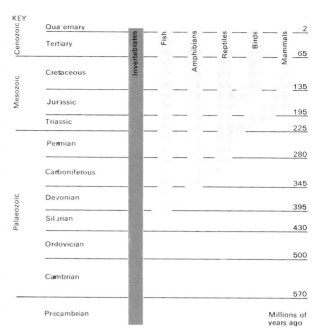

KEY

		Invertebrates	Fish	Amphibians	Reptiles	Birds	Mammals	
Cenozoic	Quaternary							2
	Tertiary							65
Mesozoic	Cretaceous							135
	Jurassic							195
	Triassic							225
Palaeozoic	Permian							280
	Carboniferous							345
	Devonian							395
	Silurian							430
	Ordovician							500
	Cambrian							570
	Precambrian							Millions of years ago

7

Common canal

Theca

Lophophore

7 The graptolites were free-floating colonial organisms that abounded in the seas of Ordovician and Silurian times. They did not grow much above 10cm (4in) long and started off with a single individual (zooid) produced by sexual means. Subsequent zooids grew from the side of the original and budded off from one another producing the characteristic shape of the particular species. Each zooid rested in a cup called a theca and was connected to all the others by a common canal. The chitin skeleton had an inner banded layer and an outer layer secreted by an outer skin. Each zooid carried a feathery feeding organ (lophophore).

8 A B C

8 Graptolite genera can be recognized by the shape and arrangement of their thecae (cases) and of their branches (stipes). *Diplograptus* [A] had two stipes joined together growing upwards back to back.

Didymograptus [B] had two stipes separately growing downwards and *Phyllograptus* [C] had four stipes joined together growing upwards with long and spiny thecae.

9 The rapid evolution of the graptolites and the fact that they are found all over the world makes them ideal index fossils. The presence of large numbers of *Didymograptus* in this shale identify it as being of lower Ordovician age. Slightly younger or older shale would have different graptolites.

9

Skeleton 10

Zooid

10 The ammonite had a soft anatomy similar to that of the modern nautilus, which lives in the open end of its shell. As the animal grew it secreted more shell and moved forward into the new part, walling off the old section with a septum. The walled-off chambers were used for buoyancy, being supplied with air from a tissue filament or siphuncle connecting them all. The septa met the shell wall in suture lines that had identifiable patterns for each species and

A B C

became more complex as the group advanced. [A] shows a primitive *Ceratites* suture, [B] *Hildoceras* and [C]

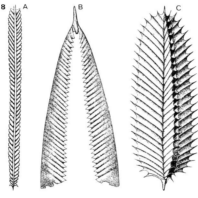

Operculum
Intestine
Tentacles

Septum
Siphuncle
Gills
Syphon
Jaws

Shell 11 A

B

C

D

E

advanced *Baculites*. The arrow shows the position of the siphuncle and the direction in which the head lies.

11 Ammonites can be identified by the ornamentation on the shell. *Promicroceras* [A] had ribs, *Douvilleiceras* [B] had tubercles, *Harpoceras* [C] had a keel, *Hildoceras* [D] had a longitudinal series of ridges and *Kosmoceras* [E] had its aperture guarded by a pair of paddle-shaped little flaps.

12 The shape of an ammonite's shell determined its mode of life. *Scaphites* [A] was a passive drifter. *Baculites* [B], *Amaltheus* [C] and *Dactylioceras* [D] were streamlined active predators; inactive *Cadoceras* [E] moved only up and down by varying its buoyancy; and snail-like *Turrilites* [F] was confined to the bottom of the sea.

12 A
B
C
D
E
F

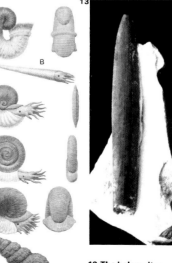

13

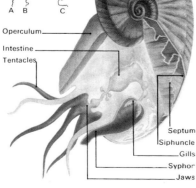

13 The belemnites were related to the ammonites, but were squid-like and had pencil-shaped shells.

179

Fish and amphibians of the past

Vertebrates, the animals with backbones, are evolved from segmented, worm-like animals. This is seen in modern vertebrates such as ourselves by the segmented nature of the backbone and the fairly repetitive patterns of the ribs. The pattern of early vertebrate evolution shows well the development of new features from a simple, basic form.

Towards the age of fish
The first recognizable skeletal features of early vertebrates would have been the jointed support for the main muscular system – the backbone – and a box to contain the brain, namely the skull. This is more or less the layout of the earliest and most primitive fish, the agnathans [1]. These Silurian forms had no jaws, the mouth being merely a sucking organ. They breathed through gills which were paired and supported on bony outgrowths. As time progressed the first pair of gill-supports moved forwards and became hinged to the skull to form the jaws, and bony structures evolved to support swimming organs. The true fish had developed.

The jawed fish are divided into three groups: the Placodermi, armour-plated fish of the Devonian; the Chondrichthyes, cartilage-skeletoned sharks and rays that have existed since the Devonian to the present day; and the Osteichthyes, bony fish that also appeared in the Devonian and represent the major type of fish found in the seas today.

The Devonian period is known as the age of fish because during this time all three groups of fish came to the fore and dominated the seas. Even in those early times the bony fish were divided into the ray-finned and lobe-finned types. The ray-finned fish are what we think of as typical – swimming by means of paired fins supported by a "ray" of bony outgrowths and balancing by means of a "swim-bladder" of air within the body. The lobe-finned fish have their fins supported by muscular lobes and in some instances the swim-bladder is connected to the gullet to form a lung. After Devonian times the lobe fins decreased in importance and today are represented only by four genera, but before that they were to play a very significant role in the history of life on our planet.

The Devonian lobe-finned fish lived in freshwater lakes among the mountains of the Northern Hemisphere. These lakes would occasionally dry out, killing everything that lived in them [5]. The lungfish, however, with their ability to breathe out of the water, and their lobed fins providing a clumsy but serviceable means of locomotion on land [6], could exist, at least for short periods, out of the water. Often this was enough to enable them to find their way to the next pond. As time passed the bones in the lobe were formed into a more efficient support and the tetrapod limb was developed.

Life comes ashore
The early limb, built of two main sections and several digits at the end, became the basic pattern of all land vertebrates to follow. The earliest amphibian, *Ichthyostega* [8] from the upper Devonian of Greenland where forests of sizeable land plants were beginning to grow, had the early tetrapod limb but also retained fish-like characteristics in the bones of the skull and in the tail.

By the time of the Carboniferous amphibians were plentiful and they mostly

1 One of the jawless fish of the Silurian was *Cephalaspis*. It had a triangular head shield bearing the mouth and paired fins beneath, and a pair of vibration-sensitive organs on the upper surface.

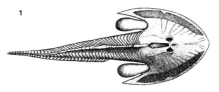

2 *Dinichthys* was an armoured placoderm. With a length of 9m (30ft) it was a fearsome fish of the Devonian seas.

3 *Drepanaspis* was a Devonian agnathan. It was a flattened bottom-dweller with a complicated pattern of armour.

4 The typical habitats of Devonian fish were inland seas, many of which existed among the freshly formed mountain chains of the north continent at that time. Shown here is a selection of the fish types found in such an area. *Bothriolepis* was a typical placoderm reaching a length of 24cm (9in). Its head and body were armoured as were the first pair of fins. *Xenacanthus* was an early member of the Chondrichthyes and had many shark-like features. It swam by means of paired, leaf-shaped fins and had a long, pointed tail. It was 75cm (2.5 ft) long. *Moythomasia* was one of the ray-finned Osteichthyes while *Holoptychius* was one of the lobe-finned type. *Fleurantia* was a lobe-finned lungfish similar to the ancestral stock that gave rise to the first amphibians such as *Ichthyostega*, shown here on the surface.

Ichthyostega

Moythomasia

Holoptychius

Fleurantia

Xenacanthus

Bothriolepis

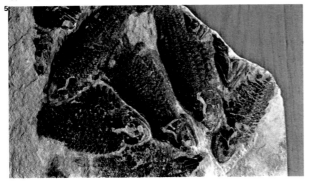

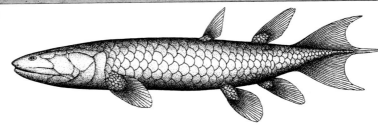

5 As the Devonian desert pools dried out the inhabitants crowded together in the last drop of moisture and died, like these *Holoptychius* from Scotland. The land vertebrates evolved from creatures that survived such conditions.

6 To be able to live out of the water a fish would need a lung system for breathing and at least one pair of muscular leg-like fins to allow it to travel. These features are found in the Devonian 60cm (2ft) long lobe-fin *Eusthenopteron*.

belonged to one extinct group called the labyrinthodonts – so called because of the convoluted nature of the tooth enamel. These had a much more powerful vertebral column than *Ichthyostega*, reflecting the fact that they spent much more time out of the water where their weight would have been buoyed up. They probably had the same general life-style as modern amphibians, spending most of their adult life on land and returning to the water to lay eggs. The eggs would have hatched into fish-like tadpoles with external gills and these would have lived in the water only until they were mature, exactly as frog tadpoles do today.

Adaptations of amphibians

The Carboniferous amphibians adapted swiftly to the different ecological niches offered by the dry-land environments and they were all carnivores, the smaller ones eating the insects that abounded in the coal forests and the larger ones living like crocodiles and eating either fish or members of their own kind in the swamps.

Other upper Palaeozoic amphibians, unrelated to the labyrinthodonts, adopted very different and specialized modes of life. Some spent all their time in the water, losing their legs and adopting a snake-like body. One such creature was Carboniferous *Dolichosoma* [10]. Others became flattened and their skulls grew into broad, flat horns, the purpose of which is obscure. These, like *Diplocaulus* [11], must have lived in the mud that lined the bottoms of pools.

The labyrinthodonts and the other specialized amphibians became extinct during the Triassic period leaving only their descendants; the small groups that we know today. But some time before this, the amphibians gave rise to an intermediate group that was to abandon the water altogether and become reptiles. These, the seymouriamorphs, showed very advanced features, especially in the reptilian nature of the limbs, although the skull and vertebrae remained distinctly amphibian. Typical of this group is the Permian *Seymouria* [13] which, however, could not be the direct ancestor of the reptiles since primitive reptiles are known from the earlier Carboniferous period.

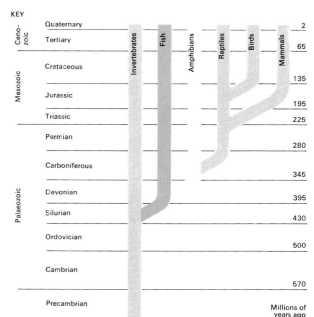

KEY

			Millions of years ago
Cenozoic	Quaternary		2
	Tertiary		65
Mesozoic	Cretaceous		135
	Jurassic		195
	Triassic		225
Palaeozoic	Permian		280
	Carboniferous		345
	Devonian		395
	Silurian		430
	Ordovician		500
	Cambrian		570
	Precambrian		

Invertebrates, Fish, Amphibians, Reptiles, Birds, Mammals

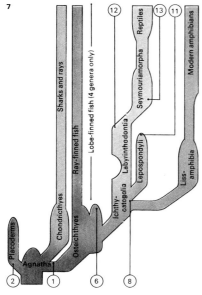

7 Amphibians and advanced fish evolved from primitive Devonian forms. The numbers refer to the illustrations shown here.

Placoderms, Agnatha, Chondrichthyes, Sharks and rays, Osteichthyes, Ray-finned fish, Lobe-finned fish (4 genera only), Ichthyostegalia, Labyrinthodontia, Lepospondyli, Seymouriamorpha, Reptiles, Lissamphibia, Modern amphibians

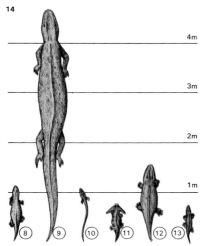

14 The sizes of the upper Palaeozoic amphibians varied as much as the shapes, as the illustrations on these pages show.

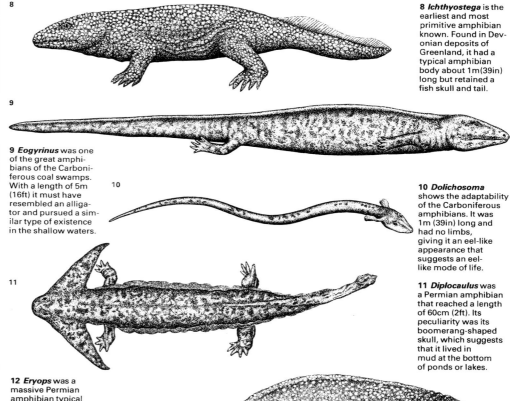

8 *Ichthyostega* is the earliest and most primitive amphibian known. Found in Devonian deposits of Greenland, it had a typical amphibian body about 1m (39in) long but retained a fish skull and tail.

9 *Eogyrinus* was one of the great amphibians of the Carboniferous coal swamps. With a length of 5m (16ft) it must have resembled an alligator and pursued a similar type of existence in the shallow waters.

10 *Dolichosoma* shows the adaptability of the Carboniferous amphibians. It was 1m (39in) long and had no limbs, giving it an eel-like appearance that suggests an eel-like mode of life.

11 *Diplocaulus* was a Permian amphibian that reached a length of 60cm (2ft). Its peculiarity was its boomerang-shaped skull, which suggests that it lived in mud at the bottom of ponds or lakes.

12 *Eryops* was a massive Permian amphibian typical of the grotesque forms that developed just before the age of amphibians passed into that of the reptiles. It was 1.5m (5ft) long and was well adapted for life on land.

13 *Seymouria* was a Permian amphibian so advanced that there is still much discussion as to whether it was an amphibian or a primitive reptile. It was 60cm (2ft) long and the bones of its limbs were distinctly reptilian.

181

Towards life on land

The Palaeozoic era spans the 345 million years between the beginning of the Cambrian period and the end of the Permian [Key]. During the lower Palaeozoic era (the Cambrian, Ordovician and Silurian periods), life was confined to the sea. It came ashore during the upper Palaeozoic era (in the Devonian, Carboniferous and Permian periods), flourished, and developed into the innumerable, interrelated forms of today.

Tell-tale fossils

Evidence of life on earth has been found in rocks 3,000 million years old, but it is only since the beginning of the Cambrian period some 570 million years ago that the record has been clear. Before then all living organisms were soft-bodied and they fossilized only under exceptional conditions [1]. After the beginning of the Cambrian, several groups of animals developed hard parts that were easily fossilized, and since then enough remains have been preserved to provide a detailed picture of ancient environments.

During the Cambrian period [2], a time of low lands and mild climates, the most promi-

nent creatures were arthropods (joint-legged animals) of various types, the most important of which were the trilobites (insect-like creatures that lived in the sea). Because of their great age, Cambrian rocks provide only fragmentary fossil evidence, but enough exists for quite detailed environments to be reconstructed with some accuracy.

The Ordovician deposits [3], 500 million years old, indicate that there was a great submergence of the lands with the spread of shallow seas. Fossil evidence from the seas shows that the arthropods continued to develop and there had arrived on the scene a number of cephalopods (molluscs), some of them very large with shells and tentacles. The most prominent shellfish during those times were the brachiopods, which only superficially resembled the more familiar bivalves. In modern times bivalves have taken over and brachiopods are now quite rare.

The trends of life on the sea-bed during Ordovician times continued into the Silurian period, 430 million years ago. In this period came the development of massive coral reefs, the rise of the first vertebrates and the first

tentative land forays of primitive plants. Much dry land appeared as mountain ranges rose from the sea. Extensive deposits from the Ordovician and Silurian periods have been found in parts of Europe, North and South America and Australasia, and it is because such detailed remains have been found in particular localities that the ecology of entire periods can be reconstructed with a high degree of accuracy.

Life reaches the land

Life began to reach the land in earnest during the Devonian period [4], 395 million years ago. The land plants that flourished at that time released great quantities of oxygen into an atmosphere previously too rich in carbon dioxide to support animal life out of the water. Once a certain level of oxygen had been established, the way was open for the colonization of the land by arthropods such as mites and millipedes. These were followed by vertebrates, starting with lungfish and progressing to amphibians. Most of this activity took place at the edge of shallow inland waters, away from the destructive

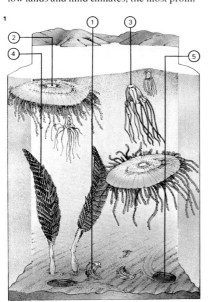

1 Really ancient fossils dating from before the Cambrian period, some 570 million years ago, are extremely rare. This is because organisms before that date were simple and lacked hard parts that fossilized easily, and also because rocks older than this are usually badly deformed, destroying all traces of organic remains. But occasionally individual organisms such as lime-secreting algae are found in these ancient rocks, some of which are 2,500 million years old. The best Precambrian fossil assemblage so far discovered dates from just before the beginning of the Cambrian and was found at Ediacara in South Australia. The rocks indicate that there was a shallow sea environment with a sandy bottom inhabited by soft-bodied creatures such as worm-like *Spriggina* [1]; *Eoporpita* [2] and *Kimberella* [3], jellyfish; *Arborea* [4], a sea pen; and *Dickinsonia* [5], an invertebrate of uncertain affinities.

2 Fossils of the Cambrian period are more widespread because many groups of organisms developed preservable hard parts. The most prominent of these were the arthropods, including the jointed trilobites. Cambrian fossils have been found in all parts of the world but a particularly good fauna of approximately 70

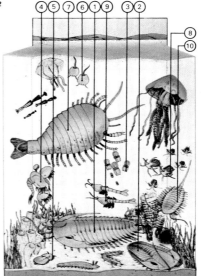

genera is known from the middle Cambrian Burgess shales of British Columbia. This rock sequence indicates a deep-water environment that was suddenly poisoned by a seepage of gas. Among arthropods found there are trilobites such as 20cm (8in) *Olenoides* [1] and *Ogygopsis* [2], and 2 cm (0.75in) *Agnostus* [3]. Other arthropods include malacostracans [4], *Naraoia* [5], *Burgessia* [6], *Sidneya* [7], *Marella* [8], *Waptia* [9] and *Emeraldella* [10]. As well as these there were sponges, worms, jellyfish and seaweeds. It is thought that this fauna is typical of the middle Cambrian, although the mode of preservation is untypical.

3 In Ordovician times living things had diversified even more and their fossils are abundant. A deep sea trough covered the area of the British Isles separating land masses of rising mountains in the north from low relief in the south.

This restoration has been based on an upper Ordovician fauna from Scotland. The seas towards the edge of the trough contained trilobites that grew up to 30cm (12in) long such as *Proteus* [1], *Tetraspis* [2], *Phillipsinella* [3], *Paracybeloides*

[4], *Sphaerocoryphe* [5] and *Remopleurides* [6]; brachiopods such as *Sampo* [7] and *Raphinesquina* [8]; snails such as *Cyclo-* nema [9] and *Sinuites* [10]; bivalves such as *Byssonychia* [11]; shelled cephalopods such as *Orthoceras* [12]; and echinoderms such as *Aulechinus* [13] and the sea-lilies [14]. This area persisted into Silurian times and fauna remained very similar.

chaos of the ocean beaches. Most fish had jaws which meant that, unlike their ancestors, they were able to bite and eat their neighbours. Many fish evolved bony armour as a means of self-protection against these and other predators. It was at that time, too, that the ancestors of the sharks appeared.

The conquest of the land by vertebrates was basically a matter of survival. The lungfish and lobed-finned fish lived in pools that periodically dried out. When this happened they were forced to move to the next water. As time passed the fish became better adapted and were able to spend much longer periods out of water. From these came the amphibians, which lived in water for only part of their life cycle. The process of mountain-building, begun in the Silurian period, reached its climax in the Devonian.

From forests to deserts

By the time the Carboniferous period [5] was under way, 345 million years ago, land plants had firmly established themselves and vast forests of ferns and giant club-mosses had spread over well-watered, fertile lowlands.

These were the swamps that were eventually buried and compressed to form our present coalfields. Amphibians of many shapes and sizes slithered in the green gloom of the undergrowth, while above them, between the patterned trunks, droned enormous insects, some of them with 60cm (24in) wing spans [5]. On higher ground, above the steaming valleys and deltas, a more open vegetation of ferns and primitive conifers flourished. Simple reptiles evolved from amphibian stock during the Carboniferous period, but did not become dominant until later, during the Permian period.

The transition from the Carboniferous into the Permian period [6, 7], 280 million years ago, was accompanied by a change, for reasons that are not clear, from forest to desert conditions, together with an ice age in the Southern Hemisphere. During that time the reptiles successfully spread over the land. They had an advantage over their amphibian ancestors in that they could live entirely out of water and had better constructed skeletons. Towards the end of the period some reptiles developed mammal-like features.

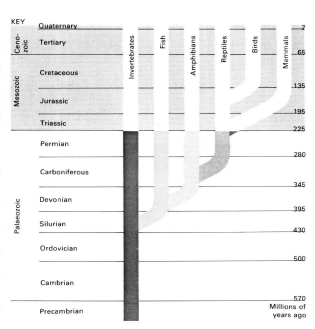

KEY		
Cenozoic	Quaternary	2
	Tertiary	65
Mesozoic	Cretaceous	135
	Jurassic	195
	Triassic	225
Palaeozoic	Permian	280
	Carboniferous	345
	Devonian	395
	Silurian	430
	Ordovician	500
	Cambrian	570
	Precambrian	Millions of years ago

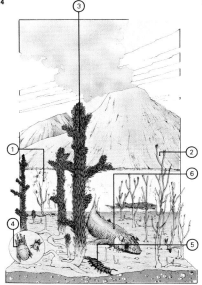

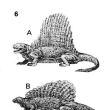

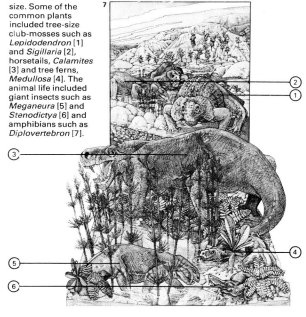

4 Northern Scotland had inland seas in Devonian times. On the shores of these grew psilophytes, the earliest land plants to possess a water-conducting system. They included *Psilophyton* [1], *Rhynia* [2] and *Asteroxylon* [3], 20cm (8in) tall. The earliest known land animals also lived there.

These were mites, *Protocarus* [4] and the millipede *Archedesmus* [5]. When the waters dried up, lungfish such as *Dipterus* [6] also came ashore. The remains of these organisms were preserved in great detail when a nearby volcano erupted and buried them in silica-rich deposits.

6 During the Permian period the humid forests and swamps gave way to mountains and deserts in the Northern Hemisphere. The reptiles had evolved by that time and these were far better adapted than the amphibians to withstand the rigours of desert life. They became diversified to fit the new conditions. One group, the pelycosaurs, developed strange backfins supported on elongations of the

vertebrae. These fins may have been temperature-regulating structures, absorbing heat when turned towards the sun and radiating heat when turned away. *Dimetrodon* [A] was a flesh-eating pelycosaur that attained a length of about 3m (10ft). *Edaphosaurus* [B] was another 3m long pelycosaur but one with gentle herbivorous habits.

5 During the Carboniferous period there was considerable erosion of newly-raised mountain chains and the debris that was washed off them formed vast deltas in the Northern Hemisphere. These deltas supported the great coal forests consisting of primitive plants grown to tree size. Some of the common plants included tree-size club-mosses such as *Lepidodendron* [1] and *Sigillaria* [2], horsetails, *Calamites* [3] and tree ferns, *Medullosa* [4]. The animal life included giant insects such as *Meganeura* [5] and *Stenodictya* [6] and amphibians such as *Diplovertebron* [7].

7 The Permian deserts of the southern continents were at times subjected to ice ages. In between these glacial periods, they supported a sparse vegetation of horsetails, conifers, ferns and seed-ferns. Among these lived a variety of reptiles that ranged from the extremely primitive to others that displayed advanced mammal-like characteristics, showing them to be the ancestors of the mammals to come. Fauna found in the Permian beds of South Africa includes primitive reptiles such as *Pareiasaurus* [1] and mammal-like types such as *Endothyodon* [2], *Lycaenops* [3], *Hofmeyria* [4], *Dicynodon* [5] and *Choerosaurus* [6].

Reptiles of the past

Once the lungfish and primitive amphibians had finally emerged from the water in the Devonian period (345–395 million years ago), animal life on land had begun. By the Carboniferous (280–345 million years ago) the first reptiles had appeared, turning their backs completely on their watery ancestry.

To do this, the aquatic amphibian larval stage – the tadpole – had to be dispensed with and the hard-shelled egg, protected from drying out by membranes, and hatching into a fully developed but tiny adult, was evolved.

Study of the fossil record
The skeletons of the earliest reptiles and amphibians were so alike that it is difficult to distinguish between them. The only accurate identification lies in the way the eggs were produced, which is almost impossible to tell from the fossil record.

The oldest fossil reptile known is *Hylonomus*, a lizard-like creature about 1m (39in) long, found preserved among the stumps of coal-forest trees of the upper Carboniferous. It was one of the cotylosaurs or "stem reptiles", a very simple group, from which all others developed. The cotylosaurs appeared in the Carboniferous, flourished in the Permian and vanished in the Triassic period.

The skeletons of these creatures differed from those of the amphibians in having more joints in the toes – a feature that was to give rise to the more specialized limbs of their successors – and a taller, narrower skull, hinting at a greater brain development. Reptiles also possess scales, which were lacking in their amphibian forebears. These scales are not analogous to those of the ancestral fish but were evolved afresh after the amphibians had lost them.

The cotylosaurs expanded rapidly into their newly conquered environment and gave rise to a large number of new and experimental forms. Many of these were short-lived, but one of the most successful groups was the thecodonts [2].

Thecodonts and dinosaurs
The thecodonts – the name means "socket-toothed" and refers to their dental arrangement – stood on their hind limbs and adopted a semi-upright mode of movement balanced by a long tail. This freed the forelimbs for other purposes such as grasping, although the animal could have resumed a four-footed stance when resting. The thecodonts were small creatures but they were to evolve into some of the biggest animals that ever lived – the dinosaurs.

The word *dinosaur*, meaning "terrible lizard", was coined by the pioneer palaeontologist Richard Owen, in 1842, to describe a number of reptilian finds that were coming to light in England at that time. "Dinosaur" is no longer a precise term scientifically, but is used as a convenient blanket category that includes the two orders Saurischia and Ornithischia.

The Saurischia – meaning "lizard hips" – are so named because the bones of the hip resemble those of a lizard [3]. The group is subdivided according to the arrangement of the bones of the feet – theropods had feet like mammals and sauropods feet like lizards. The theropods were the two-legged, meat-eating dinosaurs and they appeared in a great variety of sizes: *Podokesaurus* was chicken-

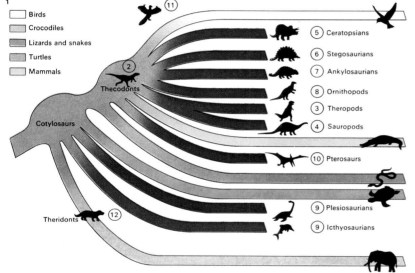

1 Birds
Crocodiles
Lizards and snakes
Turtles
Mammals

Thecodonts

Cotylosaurs

Theridonts

5 Ceratopsians
6 Stegosaurians
7 Ankylosaurians
8 Ornithopods
3 Theropods
4 Sauropods
10 Pterosaurs
9 Plesiosaurians
9 Icthyosaurians

1 The diversification of reptiles from amphibian-like cotylosaurs took place in the upper Palaeozoic and gave rise to the thecodonts and other groups. (Numbers refer to illustrations below.)

2 *Scleromochlus*

2 The thecodonts, or "socket-toothed" reptiles, were fairly small lizard-like animals that gave rise to some of the strangest beasts that ever lived. They adopted a two-legged stance, leaving the forelimbs free for grasping. This is evident in many of their descendants and although most of the dinosaurs reverted to a four-footed stance their forelimbs remained short. The thecodont *Scleromochlus* had a length of 1m (39in).

3 *Tarbosaurus*

4

4 The sauropods, herbivorous dinosaurs of the end of the Jurassic, were the largest beasts ever to live on land. *Atlantosaurus* (*Apatosaurus* or *Brontosaurus*) was 20m (66ft) long.

Atlantosaurus

5 *Styracosaurus*

5 The bird-hipped dinosaurs were herbivorous and most of them were equipped with armour of some sort that served as protection against the theropods.

In the ceratopsians the armour took the form of head shield and horns, leaving the rest of the animal unarmoured. The limbs helped in defence as well, the

hind limbs carrying the weight of the body and the forelimbs turning it quickly to face an attacker. The ceratopsians differed from each other in the number of horns. *Styracosaurus* lived in the Cretaceous.

3 The theropods were great carnivorous, lizard-hipped dinosaurs. The terrifying *Tarbosaurus* was 12m (39ft) long.

sized, whereas at the other end of the scale the *Tyrannosaurus* was 12m (40ft) long. The sauropods were the huge, long-necked vegetarian dinosaurs that flourished in the upper Jurassic and lower Cretaceous [4]. *Diplodocus* was 25m (82ft) long, *Brachiosaurus* weighed more than 50 tonnes.

The herbivores and early birds

The Ornithischia – meaning "bird hips" – had hip bones arranged like those of a bird and were herbivores. They are classified into four groups. Ceratopsia [5], rhinoceros-like with an arrangement of armour on the head, lived in the upper Cretaceous. *Triceratops* was typical of this group, with three forward-pointing horns and a bony neck frill. Stegosauria were late Jurassic and early Cretaceous forms that had armour mounted vertically down the back. *Kentrurosaurus* was similar to *Stegosaurus* [6] but had an armour of spines rather than plates. The Ankylosauria, squat creatures from the upper Cretaceous, had a tight mosaic of armour plates upon their backs and *Scolosaurus* [7] had a bony club on the end of its tail. The last of the four Ornithischia groups, the Ornithopoda [8], differed from the other three in having a bipedal stance and a lack of armour. One of these – *Anatosaurus* – possessed features that suggest it may have been semi-aquatic.

Some experts believe that the dinosaurs and the other large fossil reptiles, including the flying pterosaurs [10], were warm-blooded and more like mammals than reptiles in their behaviour. If this had been the case, then the animals shown in the illustrations would look much more active, and some would possibly have had fur.

The first birds [11] may have evolved directly from the bird-hipped dinosaurs and – if the warm blood theory is correct – it seems likely that feathers were first developed from scales for insulation and that the power of flight came later.

Some offshoots of the cotylosaurs returned to the sea [9]. But the most significant offshoot was that which first produced mammal-like reptiles with mammal-like teeth and limbs, and finally in the Triassic the first primitive mammals.

KEY

Cenozoic	Quaternary	2
	Tertiary	65
Mesozoic	Cretaceous	135
	Jurassic	195
	Triassic	225
Palaeozoic	Permian	280
	Carboniferous	345
	Devonian	395
	Silurian	430
	Ordovician	500
	Cambrian	570
	Precambrian	

Millions of years ago

(Invertebrates, Fish, Amphibians, Reptiles, Birds, Mammals)

6 The armour of the stegosaurians, as in *Stegosaurus*, consisted of a double row of bony plates mounted vertically along the spine. These beasts lived in the upper Jurassic.

6

Stegosaurus

7 The ankylosaurians were also a heavily armoured group but they were squat animals and their armour was arranged as a compact mosaic over their backs. Some form of weapon was usually present on the tail. The 6m (20ft) *Scolosaurus* lived in the upper Cretaceous.

7 *Scolosaurus*

8 *Iguanodon*

8 The ornithopods were bipedal, herbivorous, bird-hipped dinosaurs that usually lacked armour. *Iguanodon* of the lower Cretaceous was 10m (33ft) long.

9 *Plesiosaurus*　　*Ichthyosaurus*

9 The sea-living reptiles, such as *Plesiosaurus* and *Ichthyosaurus*, evolved from the cotylosaurian stock quite early.

10 The pterosaur group was another offshoot of the thecodonts. These were flying reptiles that had wings of thin membrane stretched between the hind legs and body and extended fourth digit in the forelimb. This enormous form recently found, from the upper Cretaceous was named only in 1975. Its wing span was 15m (49ft).

10 *Quetzalcoatlus*

11 *Archaeopteryx* was the earliest known recognizable bird and dates from the upper Jurassic. The presence of wings and feathers define it as a bird, but the skeleton is quite reptilian. The wings, instead of being the specialized flying limbs of modern birds, were really elongated thecodont forelimbs, complete with claws. The tail was a lizard's tail and the skull had teeth. The small breastbone shows it was a poor flyer. It was descended from the thecodonts, possibly via the bird-hipped dinosaurs.

11 *Archaeopteryx*

12 A

12 The line of the theriodonts, mammal-like reptiles, formed an early and perhaps the most important offshoot from the primitive cotylosaurs. They resembled mammals in their teeth, limbs and stance. Many types have been found ranging from reptiles with some mammal traits, such as *Titanophoneus* [A], to virtual mammals, for example *Oligokyphus* [B].

B

The age of reptiles

The age of reptiles began about 225 million years ago and ended, for no known reason, about 65 million years ago. The Mesozoic era, as it is called, was made up of the Triassic, Jurassic and Cretaceous periods. During that time reptiles dominated land, sea and sky, and evolved into what were some of the largest and most terrible land animals that ever lived – the mighty dinosaurs.

Reptiles return to the sea

The oldest period of the Mesozoic, the Triassic [1], saw a perpetuation of the desert environment that had been established during the preceding Permian period. Mountain ranges that had arisen during the Permian in the Northern Hemisphere were worn down to arid hills and plains with scattered salt lakes. Plant life differed little from that of the previous era, consisting mostly of ferns and horsetails in the moist areas and conifers such as *Voltzia* and *Araucarites*, which resembled modern monkey-puzzles, on the drier hill slopes. While the desert environment had changed little, this was certainly not true of its inhabitants. The complex interac-

tion of living things that prevailed in the oceans during the Palaeozoic was swept away and replaced by a new order of existence. The cephalopods of the time – early relatives of modern squids and octopuses – developed into ammonites, an advanced group with bodies encased in chambered shells that was to dominate the seas of the Mesozoic.

In the seas and estuaries aquatic reptiles were becoming prominent. Some groups, that had evolved from amphibians to lead a totally dry-land existence, found it advantageous to return to the sea. As a result, they adopted aquatic features such as webbed feet and streamlined bodies that enabled them to exploit their ancestral environment as expertly as their forebears had done.

On land, the reptiles were extremely versatile. *Kuehneosaurus* even took to the air on primitive gliding wings like those of the so-called flying lizards found in Malaysia today. The mammals had evolved, but they were tiny and left few remains. Among the land-dwelling reptiles was a lizard-like creature, a thecodont, no bigger than a turkey, that began to walk on its hind legs, leaving its

forelimbs free for grasping or balancing. This small creature was the ancestor of the great dinosaurs that were destined for future domination of the earth.

During the Jurassic period [2, 3] about 190 million years ago the single large continent that had been in existence during most of the upper Palaeozoic was splitting up as continental drift continued. The main crack ran from north to south, eventually widening to become the Atlantic Ocean, with Europe first separating from North America, followed by Africa from South America. Humid climates returned to the continents and lush vegetation grew over most of the world.

The coming of the dinosaurs

The jungles of the Jurassic consisted mostly of ferns and tree ferns, with stands of conifers. The Bennettitales – gymnosperms that resembled cycads – with their thick, stumpy trunks and crowns of fern-like leaves, provided the closest thing to flowers – rosettes of seed-bearing cones. Reed beds of horsetails grew by the shallow waters. These were the jungles in which the great dinosaurs of the

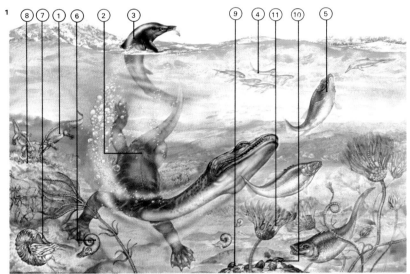

1 Central Europe in Triassic times was covered by a shallow, limy sea surrounded by low-lying desert plains. The animal life in this sea was completely different from earlier kinds. The sudden diversification of the reptiles that was taking place on land was evident in the sea as well and several groups adopted features that equipped them for a sea-going existence. *Placodus* [1] retained a familiar reptilian shape but had webbed feet and a flattened tail to help it to swim, and rounded, knob-like teeth for crushing shellfish. It attained a length of 2m (6.5ft). *Nothosaurus* [2] belonged to the group ancestral to the sea-living plesiosaurs of the Jurassic and Cretaceous – and showed an elongation of the neck and a shortening of the tail, both plesiosaur features. It ate fish and had a length of 3m (10ft). Most fish-like was the 2m (6.5ft) *Mixosaurus* [3], an early ichthyosaur. It had limbs modified into paddles and the start of a fish-like tail and dorsal fin. These creatures fed on fish such as *Thoracopterus*, [4] a flying fish, and *Semionotus* [5]. The invertebrate fauna was typical of the Mesozoic with the ammonites, cephalopods with coiled shells, very much in evidence. They differed from one another by the ornamentation on the shells, *Cladiscites* [6] having longitudinal ribs and *Trachyceras* [7] transverse ribs. Shellfish were a mixture of brachiopods like *Coenothyris* [8] and *Tetractinella* [9], and bivalves such as *Miophoria* [10]. Sea lilies like *Encrinites* [11] flourished. The bodies of these creatures were buried in the limy mud of the seabed, which was later solidified into limestone, then raised to form the Alps. These organisms were found fossilized in north Italy.

2 In the region of Bavaria during the Jurassic period there was a series of islands and quiet, shallow lagoons. Fine-grained limestone was slowly precipitated in them preserving the remains of animals and plants that lived in the area. The most significant fossil from these deposits is *Archaeopteryx* [1]. This is regarded as the first bird because it had wings and feathers, but some experts claim it was a reptile because of its lizard-like skull, tail and fingers. Also flying above the lagoons were pterosaurs, the gliding reptiles such as *Pterodactylus* [2] and the long-tailed *Rhamphorhynchus* [3]. They ranged in size between that of a sparrow and of an eagle. *Archaeopteryx* was crow-sized.

The sea abounded in swimming reptiles that were even better adapted to life in the sea than those of the Triassic. These included *Ichthyosaurus* [4], which was fish-shaped, *Plesiosaurus* [5] with its long neck, and *Steneosaurus* [6], a sea crocodile. Up to 7m (23ft) long, they ate fish such as the early banjo fish [7], *Macrosemius* [8] and *Mesodon* [9]. Also in the lagoons lived squid-like belemnites [10] and lobsters, prawns and king-crabs. In the undergrowth of the islands and along the mud banks of the shoreline ran the smallest dinosaur *Compsognathus*, [11] which was the size of a chicken, and *Homoeosaurus* [12], a lizard-like creature related to the modern tuatara of New Zealand.

Jurassic period lived. There was ample vegetation to support the herbivorous sauropods and bird-hipped dinosaurs, the prey of the carnivorous theropods. Also present were the gliding pterosaurs, crocodiles and the early mammals.

Most of the remains of those ancient land creatures that can be seen in museums today were preserved because they fell into water and were buried by sediments when they died. Such a rare and fortunate combination of circumstances occurred in Bavaria where, in upper Jurassic times, fine-grained limestone was gently precipitated into shallow lagoons. The animal and plant life preserved by this process has provided us with a full picture of the extreme specialization of reptiles into an abundance of flying and swimming types, from the appearance of the smallest dinosaurs to the first true bird.

Last of the dinosaurs

The Cretaceous period [4, 5], which began about 135 million years ago, was a time of spreading, shallow seas. There the swimming reptiles reached the peak of their dominance,

as did the ammonites. On land, ferns and cycads gave way to the more familiar willow, sycamore and oak. These lands supported the last and greatest of the dinosaurs. Flying reptiles also grew to enormous sizes at the end of the period and one pterosaur had a wing span of 15.5m (50ft).

At the end of the Cretaceous period a number of sweeping changes took place in the earth's animal life. The dinosaurs died out, as did the specialized swimming and flying reptiles, together with the ammonites and belemnites. Nobody knows for certain why this happened although many theories have been advanced to account for it. The most plausible explanation is that the environments, after tens of millions of years of stability, began to change, draining swamps, altering the climate and modifying vegetation. The dinosaurs found themselves unable to adapt to a radically new life-style. Whatever the reason, the distinctive, reptilian life form of the Mesozoic became completely extinct more than 65 million years ago, leaving the mammals with the opportunity to emerge into dominance.

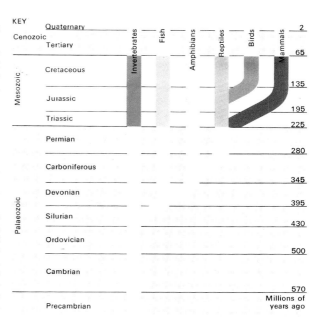

KEY

		Invertebrates	Fish	Amphibians	Reptiles	Birds	Mammals	
Cenozoic	Quaternary							2
	Tertiary							65
Mesozoic	Cretaceous							135
	Jurassic							195
	Triassic							225
Palaeozoic	Permian							280
	Carboniferous							345
	Devonian							395
	Silurian							430
	Ordovician							500
	Cambrian							570
	Precambrian							Millions of years ago

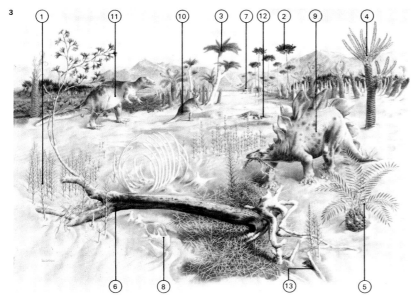

3 In the upper Jurassic of North America there is a sequence of rocks called the Morrison Formation which is famous for the bones of giant dinosaurs found in it. During the upper Jurassic the Rocky Mountains were just beginning to rise and material washed off them was spread by sluggish, winding streams to form a low fertile plain to the east. The sand-spits and mudbanks of these rivers, destined to form the rocks of the Morrison Formation, gave rise to a lush vegetation of horsetails [1], ferns, cycad-like plants and conifers. Among these browsed the herbivorous dinosaurs, which were preyed on by the theropods. The most common plants were conifers [2], tree ferns [3], *Williamsonia* [4] and other Bennettitales [5], and the ginkgo tree [6] which was widespread at that time. The herbivorous dinosaurs that fed on them included the huge sauropods *Atlantosaurus* (*Apatosaurus* or *Brontosaurus*) [7] and *Camarasaurus* [8], the armoured *Stegosaurus* [9] and the ornithopod *Camptosaurus* [10]. These were preyed on by the theropod *Antrodemus* (or *Allosaurus*) [11] and crocodiles [12] that were very like modern types. Primitive mammals [13] were present, but were still small and insignificant.

4 The Cretaceous landscape of North America differed from the Jurassic in that modern trees began to replace the forests of ferns and Bennettitales. The animal life, however, did not undergo any major change and the dinosaurs were still the dominant life form. Although many of these became truly enormous, some of them were of quite modest size. The last of the armoured dinosaurs were the ceratopsians ranging in size from *Triceratops* [1] 7m (23ft) to *Brachyceratops* [2] 2m (6.5ft). The theropods that fed on them were huge like *Tyrannosaurus* [3] at 12m (39ft), but there were also small theropods. *Ornithomimus* [4] was 4m (13ft) long and built like an ostrich. It probably fed on small lizards and eggs. They lived at the end of the Cretaceous just before the dinosaurs died out.

5 The seas of the Cretaceous period were rich in microscopic limy algae, the shells of which accumulated on the sea floor to form great deposits of chalk. The greatest sea creatures of the time were the reptiles: the 15m (50ft) plesiosaur *Elasmosaurus* [1], 66% of which was neck, the great sea lizard *Tylosaurus* [2] and the turtle with a 3m (10ft) shell (*Archelon*) [3]. Fish include the 3m (10ft) long herring *Portheus* [4] and the living "ghost shark" *Scapanorhynchus* [5]. Ammonites such as *Acanthoceras* [6], *Baculites* [7] and *Scaphites* [8] still existed. The sea-lily *Uintacrinus* [9] was a free-swimming organism. Above the water soared the pterosaurs *Pteranodon* [10] and *Nyctosaurus* [11]. Specialized sea birds like *Ichthyornis* [12] and flightless *Hesperornis* [13] existed at the time.

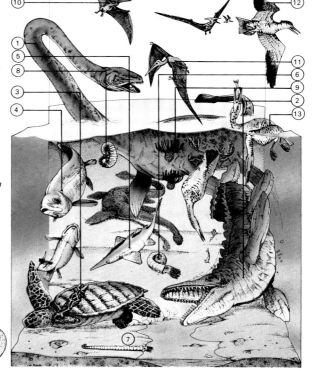

187

Mammals of the past

In the early days of reptile evolution during the Permian and Triassic periods, the line was established that was to lead to the mammals and eventually to man himself. The line was established by certain reptiles with features that later came to characterize the mammals.

Mammalian characteristics

These mammalian characteristics include warm-bloodedness and hair (unfortunately impossible to determine from fossil evidence), which are both temperature-regulating systems. A reptile can operate only with short bursts of activity followed by long periods of rest. A mammal, on the other hand, can live a continuously active existence. To support this it needs more food and a more efficient digestive system, thus different cutting, tearing and chewing teeth evolved. The palate appeared, enabling an animal to eat and breathe at the same time. The lower jaw became a more simple structure and an advanced hinge enabled it to be brought up to the upper jaw with great accuracy, allowing a more precise chewing action. With this simplification of the jaw the bones "left over" were incorporated into the ear and made hearing more efficient. But the most significant difference between mammals and reptiles is in reproduction. A mammal bears its young alive in a more or less fit state to survive. In contrast reptile development usually takes place in an egg at the mercy of the elements and any predators.

Ancient mammalian groups

Despite the advantages conferred by mammalian anatomy and physiology the new line did not catch on for some 160 million years. During the Mesozoic the mammals existed in several orders all of which consisted of small creatures resembling shrews. These differed from one another in their teeth, but very little is known about them for their remains are scattered and fragmentary. They suffered a fate similar to that of the early reptiles – extinction – and only one line really came to anything. These were the pantotheres that gave rise to practically all modern mammals. One of the ancient groups is represented today by the monotremes – the echidna and duck-billed platypus. These are true mammals, although they lay eggs, and were it not for their specialized life-styles they could provide a valuable insight into mammalian evolution.

The extinction of most orders of reptiles at the end of the Mesozoic left the environments empty and the small mammals expanded and diversified to fill the void. They produced hoofed animals to browse the forests and graze the plains, carnivores to feed on the herbivores, rodents to live in the undergrowth, primates to climb trees, bats to fly and whales to swim in the seas while the original mode of life was continued by the insectivores. All this evolutionary advance was established within a few million years of the extinction of the great reptiles.

Tracing the evolutionary lines

This spectacular evolution of the mammals began in the Tertiary period a mere 65 million years ago, so the remains of the extinct mammals tend to be better preserved than those of the great reptiles that preceded them and evolutionary lines of the horses are well known, from the tiny *Hyracotherium*

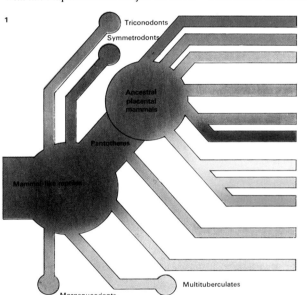

1 Mammals evolved in the Triassic from reptiles with mammal-like characteristics. They formed several groups, each possibly descended from the reptiles separately. Of these groups all have died out except two, and one of those – the monotremes – has only two living examples, the platypus and the echidna. The other group, the pantotheres, has diversified into the wide range of mammal life known today. Little is known about the other early groups because fossils are rare; they remain enigmatic creatures none larger than a cat, differing from each other in the shape and the arrangement of their teeth.

2 Megazostrodon

2 Megazostrodon was one of the earliest pantotheres and its remains have been found in Triassic rocks of Africa. It was about 10cm (4in) in length and was built rather like a reptile. It can be regarded as being very close to the ancestor of modern mammals, if not the actual ancestor, and its appearance was typical of that of all primitive mammals during the time of reptile dominance.

3 The marsupials, a group of mammals that carry their young in pouches, developed early and some grew to a large size. *Diprotodon* was a Pleistocene wombat as large as a grizzly bear.

4 The rhinoceros group is a division of the perissodactyls (odd-toed ungulates) that has existed for a long time. The earliest known is *Hyrachyus* [A], a dog-sized creature from the Eocene. This developed into *Baluchitherium* [B], the largest land mammal ever, 5m (16ft) high. Later members, such as the woolly rhinoceros [C], achieved the more modest size of today's rhinoceros.

3 Diprotodon

(eohippus) to the present-day *Equus*. The evolution of the titanotheres – elephant-sized rhinoceros-like herbivores – can be traced from the dog-sized *Eotitanops* of the Eocene to the massive *Brontotherium* that lived just before the extinction of the line some 40 million years later. But most evolutionary lines cannot be followed directly and only the remains of creatures that were offshoots from the main branch have been found. Many of the offshoots, such as the rhinoceroses and elephants, specialized in widely differing modes of life.

Certain mammals, however, showed what is known as convergent evolution in which unrelated creatures adopt similar life-styles and evolve to look like one another. *Coryphodon* of the Eocene behaved and looked like the hippopotamus but was not related to it. *Stenomylus* was a camel that looked more like a gazelle and lived on grassy plains rather than in deserts.

Carnivores as we know them developed in Oligocene times, quite late in the history of mammals. They were preceded by the archaic carnivores – the creodonts – which were quite generalized meat-eating creatures. These probably gave rise to the whales as well as to the advanced carnivores such as the great wolves and cave bears of the Quaternary. Most of the cats throughout the Tertiary and Quaternary (such as *Hoplophoneus* of the Oligocene and *Smilodon* of the Pleistocene) had long sabre-like teeth as their weapons. With these they made powerful slashing attacks, wounding their victims so deeply that they bled to death. This was an efficient way of killing the large, thick-skinned animals that abounded at that time. Modern cats, on the other hand, have smaller teeth and can run down fast grazing animals and kill them by breaking their necks.

Extinct side branches from the ungulate and elephant lines gave rise to the uintatheres – rhinoceros-like animals with six horns and powerful tusks – and chalicotheres, horse-like creatures with claws, not hoofs.

The primates developed rapidly throughout the Tertiary from tree shrews and tarsier-like creatures in the Eocene, through lemurs to monkeys and apes in the late Tertiary. One of these lines developed into man.

KEY

		Invertebrates	Fish	Amphibians	Reptiles	Birds	Mammals	
Cenozoic	Quaternary							2
	Tertiary							65
Mesozoic	Cretaceous							135
	Jurassic							195
	Triassic							225
Palaeozoic	Permian							280
	Carboniferous							345
	Devonian							395
	Silurian							430
	Ordovician							500
	Cambrian							570
	Precambrian							Millions of years ago

5 Extinct elephants differed from each other in the shape of the head. After the tapir-like Eocene ancestral *Moeritherium* [A], the body size soon reached that of today's elephants. *Trilophodon* [B] from the Miocene had four tusks, two in the upper jaw and two in the lower. Miocene *Deinotherium* [C] had a pair of downward-curving tusks in the lower jaw probably used as picks. *Platybelodon* [D], also from the Miocene, had its lower tusks expanded and flattened into a kind of shovel for scooping water-weed. *Mammuthus* [E], the Pleistocene woolly mammoth, was adapted to withstand cold.

5

6 The armadillo family was represented in the Pleistocene by glyptodonts such as 3m (10ft) long *Daedicurus*, which was heavily armoured and carried a bony club on the end of its tail.

6 Glyptodont *Doedicurus*

7 Brontotherium resembled a rhinoceros but was not closely related, being a member of the extinct titanotheres. It stood almost 3m (10ft) high and lived in the Oligocene.

7 *Brontotherium platyceras*

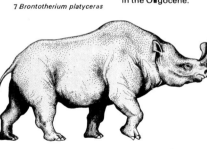

8 Sivatherium was a giraffe although it looked more like a moose than today's giraffes. It was heavily built and carried two pairs of horns on its head, a small pair above the eyes and a large pair farther back.

8

Sivatherium

9 Carnivorous mammals evolved to feed on the newly developed herbivorous mammals. The early carnivores, called creodonts, were quite unlike any we know today. They included animals such as *Oxyaena*.

Oxyaena

It was not until late in the Eocene or the Oligocene that their descendants, the more modern carnivores, replaced them. These carnivores

Cave bear
Ursus spelaeus

were, even then, divided into the groups that are now familiar – cats, dogs, bears, weasels and also seals. *Ursus spelaeus*, the cave bear, was one of the larger Pleistocene carnivores but modified its feeding habits and lived on roots and nuts. A very different descendant of the creodonts is the whale. It adopted its fish-like form rapidly, as in the Eocene *Zeuglodon*, but has since evolved both socially and mentally.

Zeuglodon

10 Primates began to evolve from Eocene insectivores that resembled tree-shrews. *Notharctus* was a lemur-like creature on the road to man.

10 *Notharctus*

189

The age of mammals

Earth's most recent geological history, the last 65 million years or so, is known as the Cenozoic era and is made up of the Tertiary and Quaternary periods. During that time landforms and climates changed slowly to the ones found on the earth today.

Mammals make their bow

The first 27 million years of the Tertiary is divided into the Palaeocene ("ancient period") and the Eocene [1] ("dawn of modern life") during which mammals first began to play a dominant role in the life of the planet. The complete extinction of the great reptiles of the Mesozoic left the forests and plains deserted; as a result, the then humble mammal evolved within a few million years into a vast number of different forms that could make full use of the empty environments. The first mammals appeared in the Mesozoic: tiny animals known principally from their teeth – their fragile bones are rarely preserved. After these came the marsupials (pouched animals) and the placentals (animals that are nourished inside the mother's body). The placentals became

dominant during the Eocene. So rapid was the evolution that all modern groups of mammals were represented, including maritime kinds such as whales and flying kinds such as bats. Birds also began to resemble their modern counterparts, although some became highly specialized and developed into creatures such as the ferocious, flightless, 2m (6ft) high *Diatryma*, which must have behaved rather like the smaller carnivorous dinosaurs. Bivalves and gastropods such as we have today appeared in the seas together with the kind of fish that we are familiar with and a species of large, coin-shaped foraminiferan.

The Oligocene, which embraced the next 12 million years of the Tertiary, provided a continuation of the Eocene environment during which basic stocks of animal life became specialized. Horses, which appeared early in the Eocene in the form of *Hyracotherium* (*Eohippus*), remained as forest-dwelling browsers but became a little larger. Rhinoceros-like creatures appeared such as *Arsinoitherium*, which had two small horns above the eyes and an enormous pair

pointing forwards above the nose, and the titanotheres, which were the size of elephants and carried great Y-shaped horns on their noses. These creatures only superficially resembled the true rhinoceroses that roamed the plains at this time.

Runners of the plains

During the next 19 million years – the Miocene [2] – the climate became more temperate. Grasslands flourished as the forests dwindled, and more animals became adapted to life in the open. Horses became longer-legged and their toes, which originally had numbered four in front and three behind, became fewer. In this way the animals became adapted to running on the plains, while their teeth became more suitable for eating grass than leaves. An early camel, *Protylopus*, was about the same size as the first horse. But in the Miocene, along with deer and other running creatures, it also grew leggier and better adapted to the plains. This was a time of upheaval in the earth. The Alps and Himalayas were being pushed up, the Andes were still rising and there was volcanic

1 Southern England in Eocene times was a subtropical lowland supporting swamps of modern plants such as swamp cypress [2], nipa palm [5], sabal palm [10] and magnolia [3]. Mammals were by then common in such environments.

In detail they might have looked odd to us, but in general their shapes and life-styles would not have differed much from today's animals. The hippopotamus-like *Coryphodon* [6] lived on river banks and fed on the roots torn up

by its powerful tusks. In the forests lived the tiny early horse *Hyracotherium* [8] that had a height of about 30 cm (12in) at the shoulder. It was adapted for life in the undergrowth and its teeth were suitable for cropping

leaves rather than grass. The birds, too, had spread and evolved into types that would be recognized today. *Halcyornis* [1] was a kingfisher that lived above the waters of the swamps and streams. *Lithornis* [11] was a vulture that

was very similar to the vultures flying today. The gannet-like *Odontopteryx* [9] was a coastal bird that fed on sea fish. Its beak was serrated with rows of barbs that resembled sharp teeth. Nobody knows for certain what colour these

birds were. The colours shown here are purely guesswork and based on the plumage of related modern birds. After the great reptiles of the Mesozoic had disappeared, only the more humble forms, such as lizards and snakes,

lived on. In this particular area the remains of crocodiles [4] and the river-turtle *Podocnemis* [7] have been found. Both these animals were virtually identical with types that are found in present-day rivers and swamps.

activity in North America accompanying the continuing upheaval of the Rockies.

After the Miocene came the comparatively short Pliocene, lasting about five million years. The huge, unwieldy mammals of earlier ages had died out and the animals that replaced them looked very much like the kinds we know today. The most important innovation was the appearance of an ape, *Australopithecus*, which walked upright on the plains of Africa and was advanced enough to use simple tools.

The coming of the ice ages

The lowering of temperatures at the end of the Pliocene heralded the coming of the Pleistocene [3] ice age about two million years ago. This was also the beginning of the Quaternary period – the age of man. The glaciers came and went four times in response to a changing climate, and the animal life of the Northern Hemisphere had to adapt to it. The animals living in what are now temperate latitudes were Arctic forms with thick fur which were able to subsist on conifers, birches and lichens. The woolly mammoth, mas-

todon and the woolly rhinoceros lived in this manner and fell prey to early man with his hunting skills. South of the ice sheets, *Smilodon*, a sabre-toothed "tiger", wrought havoc among the heavy, slow-moving ground sloths and the mastodons. Glacial conditions did not affect the whole of the world, nor were they continuous even in the glaciated areas. Warm periods occurred between the advances of the ice and during such times even elephants lived in latitudes as far north as the British Isles.

The Holocene, or Recent, which dates from the end of the Pleistocene, is such a short epoch that it cannot really be regarded as a distinct geological division. The modern pattern of climatic zones with different weather conditions seems to be exceptional and more stable worldwide climatic conditions seem to have existed throughout geological time – for example during the times when the deserts of the Permian and the forests of the Eocene covered the land. The last ice sheet has barely receded and it is possible that today's environment is another Pleistocene interglacial stage.

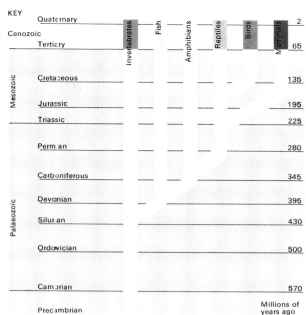

KEY

		Invertebrates	Fish	Amphibians	Reptiles	Birds	Mammals	Millions of years ago
Cenozoic	Quaternary							2
	Tertiary							65
Mesozoic	Cretaceous							135
	Jurassic							195
	Triassic							225
Palaeozoic	Permian							280
	Carboniferous							345
	Devonian							395
	Silurian							430
	Ordovician							500
	Cambrian							570
	Precambrian							

2 During the Miocene the forests that were typical of the lower Tertiary thinned out and were replaced by open grasslands. Running animals roamed these plains in herds. The plains of the lower Miocene of North America looked very much like modern savannas. The Rocky Mountains were still being pushed up and this process was accompanied by great volcanic activity. Primitive horses had evolved into *Parahippus* [1], which showed a reduction of toes to three on each foot (from the original four on the forefeet), and an adaptation of teeth to grazing. *Dinohyus* [2] was a pig as large as an ox. A corkscrew-shaped burrow [3], given the name *Daemonelix*, may have been occupied by some kind of rodent. *Diceratherium* [4] was a small, two-horned early rhinoceros.

Stenomylus [5] was a long-necked camel that lived in herds as do today's antelopes. *Moropus* [6] was related to the horse but had claws instead of hooves.

3 A good selection of Pleistocene organisms representative of an almost modern fauna was found in the tar pits of California, at Rancho La Brea.

These pits trapped unwary creatures and their struggles attracted scavengers and predators that were themselves trapped. As well as familiar-looking

animals such as rabbits, stoats, wolves, horses, bison and storks, there were creatures that are now extinct, such as mammoths and ground sloths.

1 Rabbit *Lepus* sp
2 Stoat *Mustela* sp
3 Giant ground sloth *Paramylodon* sp
4 Lion *Panthera atrox*
5 Mammoth *Mammuthus imperator*
6 Mastodon *Mammut americanus*
7 Stork *Ciconia maltha*
8 Camel *Camelops* sp
9 Dire wolf *Canis dirus*
10 Horse *Equus* sp
11 Bison *Bison antiquus*
12 Sabre-toothed "tiger" *Smilodon* sp
13 Giant condor *Teratornis* sp

Regions of the earth: zoogeography

Zoogeography is the study of the geographical distribution of animals which attempts to explain why, for example, kangaroos are found only in Australia and ostriches only in Africa. It is broad in its concepts and demands knowledge not only of present-day geography but also of the patterns and timings of changes in the location of land masses during the earth's development.

The drifting continents

The present-day distribution of animal species can largely be explained by the theory of continental drift. According to this theory the continents were once a huge land mass called Pangaea which – over millions of years – split into parts that drifted to their present positions, each taking with them their resident animals.

The theory of continental drift has firm backing in the fossil record. The fossil skull of *Lystrosaurus*, a reptile from the Triassic of over 200 million years ago, has been found in Antarctica. Similar fossils have been found in South Africa and the rock structures in which they were embedded are so closely linked that continental drift is the only plausible explanation currently available.

For convenience of study the earth is divided into six main zoogeographical regions [Key]. The ecological conditions of the regions vary, and even within the same region there may be huge variation. The Palaearctic region, which stretches from the Arctic north of Siberia to the tropical Far East, provides an extreme example of this.

Most animals have precise ecological requirements and for this reason they occupy specific ecological niches. Those with wide tolerances, such as rats, locusts or starlings, are regarded as pests while those with narrow tolerance may be rare species. Because many kinds of vegetation are common to all regions a certain number of common ecological niches are found in all of them. Each region has, for example, some animals that graze on grass, some that browse on foliage, and some that suck nectar from flowers. The species of animals filling similar niches vary from region to region, leading to characteristic interregional differences between fuana.

The continents became separate so long ago that in each one evolution has taken place "in parallel". Similar adaptations to similar ecological niches has produced so-called ecological equivalents [3]. Equivalent species are, of course, not identical and usually differ widely in their behaviour, physiology and anatomy.

Some species are without equivalents because of unsuitable physical and climatic conditions; for example, there are no truly arboreal leaf-eating mammals – the equivalents of the Ethiopian and Neotropical monkeys – in the Palaearctic and Nearctic.

The Australian example

Adaptation and the process of natural selection are the keys to an understanding of zoogeography. Their operation can best be described by the example of the continent of Australia. It broke off from a southern continental land mass, called Gondwanaland, at the end of the Mesozoic era just when the pouch-bearing mammals, the marsupials, were beginning to evolve from a primitive mammalian stock. It drifted eastwards and the marsupials were isolated from the

Key

Tundra
Coniferous forest
Deciduous forest
Temperate grassland
Marshland swamp
Desert
Pastoral and arable land
Tropical forest
Savanna

North America

Caribou
Brown bear
Wolf
Porcupine
Moose
Puma
Racoon
Lynx
Rocky Mountain goat
Bison
Coyote
Pronghorn
Whitetail deer
Peccary
Prairie dog
Skunk
Cottontail rabbit

South America

Nine-banded armadillo
Vampire bat
Spider monkey
Yapok
Sloth
Pudu
Capybara
Tamandua
Opossum
Peccary
Jaguar
Guinea pig
Puma
Vicuña
Chinchilla
Caenolestes
Giant anteater
Fairy armadillo

1 **The Nearctic and Neotropical regions** comprise the New World, but are quite distinct from one another. The fauna of North America resembles that of Eurasia – possibly a result of the existence (more than a million years ago) of a land bridge where the Bering Sea is today. Even the most "American" animals betray this ancient tie. The moose (*Alces alces*), an inhabitant of the coniferous forest belt, is the same species as the European elk. Brown bears (*Ursus arctos*), of which the grizzly is a particularly large race, are distributed throughout the Northern Hemisphere, and wolves (*Canis lupus*) still roam the forests of North America, Asia, and Europe. South of the Mississippi, the fauna is basically Neotropical because the area was re-colonized from South America during the last Ice Age.

South America was isolated for much of the Tertiary period. So in the absence of competition with successive new colonizations, its fauna evolved into countless species found nowhere else. Among the most curious and well known are the armadillos, different species of which are to be found both in open grassland and thick cover. All seven species of sloth come from the dense Neotropical forests.

2 **Alfred Wallace** (1823–1913) gave his name to the Wallace Line, his version of the sea barrier (between the islands of Indonesia) that prevented Australasian and Indo-Chinese species from intermingling.

3 **Ecological equivalents of the mole** have evolved in Australia and Africa. All have a similar shape of body and foot, but are not closely related. The Australian mole is a marsupial.

Marsupial mole (Australian)
Notoryctes typhlops

Large golden mole (Africa)
Chrysospalax sp

Mole (Palaearctic)
Talpa sp

later evolution of the placental mammals that took place farther north, although land bridges may have existed for many millennia between Australia and both South America and Antarctica, which the placentals failed to use. Faced with a welter of available niches and no mammalian competitors, the Australian marsupials slowly began to evolve into a mass of forms each suited to a particular niche, a process called adaptive radiation.

The Australian fauna can be described as relict – a fauna left behind by the march of evolution because of severe physical barriers which prevented the spread of animals. Today, many islands support relict faunas and the best known of these is Madagascar. Its fauna, rich in primitive mammals, birds and reptiles suggests what the fauna of Africa was like many millions of years ago.

Barriers to invasion
Faunal characteristics of regions are maintained by the presence of barriers preventing new invasions. If the barriers remained inviolable from their origins, faunal differences would today be greater than they actually are. But continental shifting of the earth's crust creates new barriers and destroys old ones. Each destruction allows a new invasion and mixing of species. Since the Pleistocene, which began about two million years ago, the fauna of Britain has received many elements of European fauna because of the temporary existence of land bridges linking it with northwest Europe. The Bering Sea is now a formidable barrier to animal dispersion but in earlier times it was a highway between the Nearctic and Palaearctic regions.

Animals with good powers of dispersal may not be deterred by even the most formidable barriers. Flying animals and ocean-dwelling species are frequently dispersed over several regions. Whether or not they can survive in another region depends on the existence of suitable food and climate. Sometimes animals are introduced accidentally into new regions.

The occurrence of species in inexplicable places serves to emphasize the fact that zoogeographical regions are arbitrarily drawn and not strict divisions.

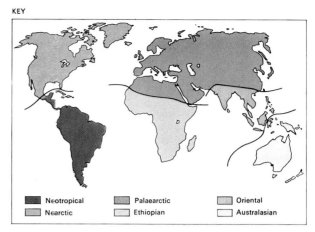

KEY

Neotropical
Nearctic
Palaearctic
Ethiopian
Oriental
Australasian

The six main zoogeographical regions of the world are all, apart from the Oriental region, continental land masses. For some purposes the Palaearctic and Nearctic are grouped together as the Holarctic. The divisions between regions are arbitrary and animals with good powers of dispersal may be found in more than one region. Each region has a characteristic fauna typified by certain species – the result of evolution at different places and times and of the different injections of species that the drifting land masses received.

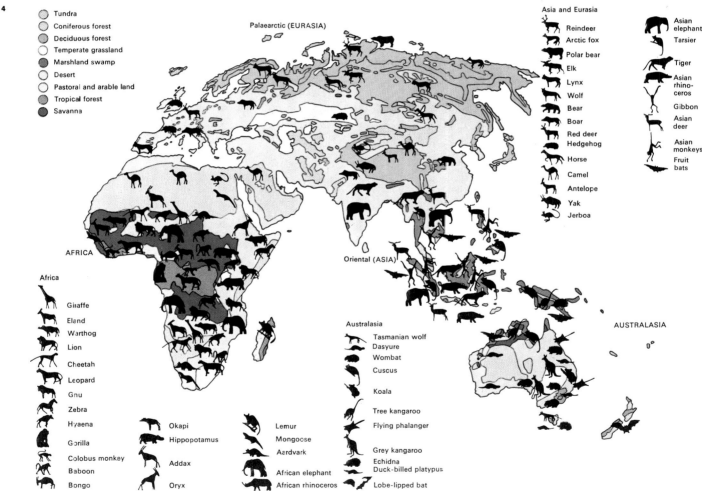

4

Tundra
Coniferous forest
Deciduous forest
Temperate grassland
Marshland swamp
Desert
Pastoral and arable land
Tropical forest
Savanna

Palaearctic (EURASIA)

AFRICA

Oriental (ASIA)

Australasia

AUSTRALASIA

Asia and Eurasia

Reindeer
Arctic fox
Polar bear
Elk
Lynx
Wolf
Bear
Boar
Red deer
Hedgehog
Horse
Camel
Antelope
Yak
Jerboa

Asian elephant
Tarsier
Tiger
Asian rhinoceros
Gibbon
Asian deer
Asian monkeys
Fruit bats

Africa

Giraffe
Eland
Warthog
Lion
Cheetah
Leopard
Gnu
Zebra
Hyaena
Gorilla
Colobus monkey
Baboon
Bongo

Okapi
Hippopotamus
Addax
Oryx

Lemur
Mongoose
Aardvark
African elephant
African rhinoceros

Tasmanian wolf
Dasyure
Wombat
Cuscus
Koala
Tree kangaroo
Flying phalanger
Grey kangaroo
Echidna
Duck-billed platypus
Lobe-lipped bat

4 The Palaearctic region stretches from western North Africa and the Persian Gulf to the north of Siberia. Its eastern limit is the Bering Sea. A large area – Arctic tundra and permafrost – is inhabited by polar bear, Arctic fox and reindeer. To the south lie coniferous forest and temperate deciduous woodlands. Both are rich in species; the Ice Ages and the consequent climatic effects caused all but the most tolerant species to move south. The Ethiopian region encompasses Africa from the Sahara southwards and includes southern Arabia. Straddling the Equator, the region is mostly tropical and typified by dry, highly seasonal grassland. Much of west and central Africa is covered by tropical rain forest and is inhabited by essentially arboreal species such as gorillas, monkeys and lorises. The grasslands support huge herds of grazing and browsing herbivores and their predators – the "big game" of Africa. Many of these species roamed far into the Palaearctic before the formation of the Mediterranean Sea barrier (evidenced by the discovery of fossil hippopotamuses under the streets of London). The Australasian region, separated from Gondwanaland during the Cretaceous, lacks placental mammals that evolved after that period. Its many primitive species, such as the duck-billed platypus, lungfish and flapfooted lizard, have never been in competition with the more successful forms that evolved later, and so have survived. Much of the region is desert and arid scrub and supports an impoverished fauna. The fertile coastal strip is occupied by marsupials, which have evolved into many specialized forms.

193

The basis of ecology

No organism alive on earth is an isolated individual. Every plant and animal is a member of a dynamic community known as an ecosystem [8]. An ecosystem is a complex that consists not only of living creatures but also of non-living matter and of radiant energy from the sun. The limits of any ecosystem – such as a forest or a stretch of sea-shore – are always arbitrary because they are drawn by man; this whole planet is one huge ecosystem. The concept remains relevant, however, to the study of ecology, a branch of biological science. The aim of ecology is to analyse ecosystems in detail and to account not only for the effects of each organism on the whole ecosystem but also for the influence of the system on the individual organisms of which it is composed.

Producers, consumers and decomposers
Energy is the basic essential of all ecosystems – without a ready supply of usable energy life cannot continue. Green plants make their own foods by the process of photosynthesis – using the plentiful atmospheric components carbon dioxide and water, and energy from

sunlight, they are able to synthesize sugars. These sugars retain, in their chemical bonding, some of the sun's energy and this can be released and used to build the more complex chemical compounds needed by the plant for building its structural and reproductive elements such as its conducting channels, flowers and seeds.

Unlike plants animals cannot make their own foods. Instead they "steal" energy from plants and from other animals by eating them – they can live only at the expense of other living organisms. Within any ecosystem green plants are thus known as producers or fixers of energy, while animals are consumers. The organisms that feed on the dead bodies of plants and animals, causing their decay, are decomposers. In any particular ecosystem producers, consumers and decomposers live together, rely on each other and adapt to each other. The more well defined their boundaries of influence become the more the ecosystem is said to become closed.

Complex ecosystems may include many thousands of species, all interrelating with each other to some degree. Their interactions

include feeding and the provision of food, but plants and animals can also provide various kinds of shelter or protection, nesting material and homes for each other.

Links in the food chains
The fundamental food-providing relationships between the organisms in an ecosystem are often expressed in terms of diagrams known as food chains [7]. Most commonly these consist of a sequence of species that are related to each other as prey and predators and most chains are tied to each other by cross linkages to form food webs that quickly become too complex to be mapped.

The energy flow within each food chain is, worldwide, remarkably constant and nearly always conforms to the "ten per cent rule". According to this rule ten per cent of energy is transferred at every link in the chain. Thus a herbivore obtains ten per cent of the calories in the herbs it eats and so on. This 90 per cent reduction in available energy determines the length of the food chain: the animals at the end of a chain, such as foxes, yield too few calories to make preying on them a

2 The tropical rain forest is the richest of all the biomes. Although large animals are few, small ones proliferate in an environment offering a huge supply of available energy.

1 The montane biome is found in polar, temperate and tropical regions. At the greatest altitudes organisms are few and their interrelations finely balanced to overcome energy shortages.

4 In the harsh conditions of the desert, plants and animals are restricted in numbers to those able to cope with the extremes of temperature and with the regular conditions of drought.

3 All the world's grasslands are classified by ecologists as the grassland biome. Although each one contains different selections of species they all share certain common characteristics such as a large variety of herbivorous animals. The savanna illustrated here also supports many predatory animals and carrion-feeding scavengers. The trees and bushes provide food and shelter for animals.

5 The temperate forest biome supports many fewer plant and animal species than its tropical counterpart. This is, however, one of the most complex natural biomes. Its trees are typically deciduous.

6 The life forms of the ocean are far greater in numbers than those that exist on land. They range from the largest of living creatures, the blue whale, to microscopic plankton.

7 All plant and animal communities are organized along essentially similar economic lines. Ecologists represent these in terms of diagrams known as food webs. In them, plants are the primary producers because they alone are able to fix solar energy and use it to synthesize complex food materials from the simple ingredients of carbon dioxide and water. The sun's energy is locked in chemical bonds within the plant and is used by herbivores. These animals are, in turn, preyed on by carnivores. In most ecosystems the food webs are complex and some degree of simplification is essential to make them easy to understand.

The desert food web [A] is essentially simple. In this North American example the cacti are the dominant fixers and producers, the coyotes and hawks the chief species of carnivores.

In temperate woodland [B] the complex food web reflects the fact that the biome has remained stable for millions of years. Insects are a significant food source for many woodland birds.

The plant pastures of the huge ocean biome [C] are the algae, the majority of which are minute diatoms. Every sea creature depends ultimately on the algae for its essential food supply.

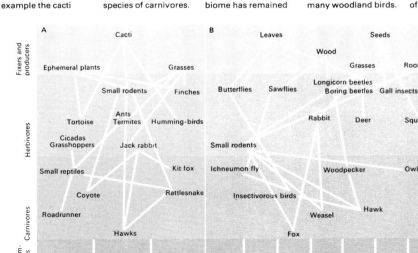

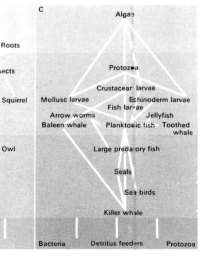

worthwhile method of obtaining energy.

Ecosystems do not arise suddenly but rather mature over many years [9], gradually becoming more complex. Generally speaking, the older the ecosystem the more species it is likely to contain. At its height, and in its final and most long-lived form, the ecosystem is known as a climax community.

The formation of an ecosystem

New environments (such as a newly-formed pond, a field recently devastated by fire or a glacier bed from which the ice has retreated) develop by succession. At first they attract only the hardy species of plants that can survive without shelter and a few animals capable of living among them. These pioneers – opportunist or fugitive species – are usually capable of rapid reproduction and the plants among them include lichens and mosses and many weed species. These early settlers in the chain of succession modify the environment by adding humus and nutrients to the soil, provide shelter from the sun and wind and make it more hospitable to other creatures. As more organisms move in, more habitats (known as ecological niches) become available. In a maturing ecosystem the early colonizers may die as stable species exert their superiority.

The surface of the globe can be divided ecologically into ten broad regions determined by their natural vegetation [Key]. These regions, along with such areas as the sea [6] and the polar ice caps, are often referred to as biomes. Some of the world's most complex ecosystems are to be found in the tropical rain forest [2] where productivity is high and conditions have been stable for many millions of years. Some of the most simple ecosystems occur in polar regions where there is less energy available and where few organisms have had time to adapt to the new environment exposed after the retreat of the ice sheets.

Just as ecosystems are built up so they may be destroyed, either naturally or, more probably, through the interference of man. The ecosystem is disturbed because one or more species is either wiped out or drastically reduced in numbers, thus changing the whole energy balance of the community.

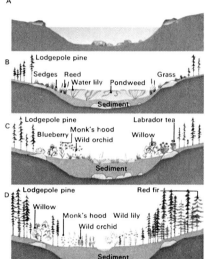

Ecologists divide the earth into regions known as biomes, each characterized by a pattern of natural vegetation or, as in the sea and in lakes and rivers, by the nature of its water. The land biomes vary from the tropical forest, teeming with plant and animal life, to the cold, treeless tundra carpeted with lichens, grasses and low-growing shrubs.

Evergreen trees and shrubs
Coniferous forest
Deciduous forest
Temperate rain forest
Monsoon forest
Tropical rain forest
Thorn forest
Grassland and savanna
Arid scrub and desert
Alpine tundra and ice

8 The basic unit of ecology is the individual plant or animal [A]. Many individuals, like this Arctic hare, are genetically distinct from others of their kind. Others, like the grass plants on which the hare feeds, reproduce vegetatively and form part of a "clone" of individuals, all of which are identical. The population [B] to which any individual belongs is isolated, either wholly or partly, from other populations of the same species. The Arctic hare of northern Greenland may thus differ genetically from that of Siberia. The ecosystem [C] is an assembly of animals and plants interacting with each other and the environment. Ecosystems can be simple or complex. A simple system could consist of unicellular algae on a tree trunk, insects and insect-eaters. A complex ecosystem such as that of the tropical rain forest may consist of thousands of interacting plants and animals.

9 Ecosystems do not arise overnight but evolve gradually. A shallow glacial tarn [A] is transformed to a forest [D] in a timespan of a few thousand years. The first living organisms to colonize it are algae whose spores are carried there by the wind. Larvae of flying insects feed on the algae and their debris accumulates on the bottom. Bacteria and unicells recycle the nutrient salts and small animals and plants enter the system. Surrounding sediments now fill the tarn [B], marsh plants spread inwards and land plants [C] replace them as the ground consolidates. The final, stable ecosystem is the climax community.

Montane life [D] tends to form a simple food web. In the Himalayas, whose web is depicted here, the herbivores are the predominant form of animal life and are preyed on by only a few carnivores.

The tropical rain forest [E] is the home of many highly specialized plants and animals, all of which take their place in the complex food web. Epiphytes, for example, are plants adapted to live on other plants. On the forest floor fallen leaves and dead wood are very rapidly decomposed.

The rich variety of savanna life [F] reflects a mature ecosystem. As in the woodland biomes, stable conditions persisting for many millennia have allowed many species to become established.

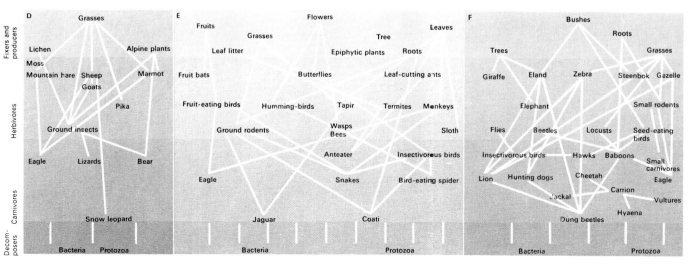

195

Isolation and evolution

According to the theory of evolution, all living plants and animals have derived from one or two simple ancestral stocks. As a convenient "labelling" system, scientists have contrived to classify more than one million animals and half a million plants into separate species. This classification is first made on the basis of physical appearance and, in organisms that appear to be similar, is confirmed by the fact that they do not interbreed freely in the wild. So, although it is possible in captivity to breed a cross between a male lion and a female tiger – a so-called "liger" – the lion and the tiger are nevertheless still classified as separate species.

The development of species

How then did different species develop from a single uniform population? If we imagine such a population inhabiting a fairly large area and having a large gene pool, then the forces of natural selection will, in time, produce groups of individuals that are adapted to live in certain habitats. If new habitats become available groups of animals or plants will colonize them and adapt to exploit them

fully. The original population will have undergone adaptive radiation [2].

The new groups may become distinct enough to be regarded as separate subspecies or races. Occasionally they may diverge farther from the parental stock in such a way as to prevent or decrease the ability to exchange genetic material. They then can be described as different species. By developing groups of specialist organisms, nature does not put all its "genes in one basket" and the presence of many specialist organisms ensures survival.

Geographical isolation

The most obvious barriers to reproduction – and often the first step in the process of specialization – is that of geographical isolation. This depends primarily on the habits and dispersal power of the species concerned. Birds have a large dispersal power but other creatures are limited in their range. Land snails on the Pacific islands, for example [Key], inhabit wooded valleys that are separated from each other by rocky ridges. In the valleys the frequency of different species of

snail varies and in some valleys particular species are wholly absent.

It seems that the original founders of the populations wandered quite by accident into one or more of the valleys and that the ridges between the valleys have kept them separate. Large physical barriers [1] such as oceans and mountains not only separate populations of the same species but also keep whole families and orders apart, as is shown by the marsupials of both Australia and South America.

Reproductive isolation also occurs when two or more species inhabit the same geographical area. The barriers to gene exchange can take two forms, acting either by deterring fertilization or by lowering the survival potential or fertility of any hybrids.

Isolation by habitat [3] is most common in plants but also evident in some animals. Individual plant species are often adapted to a specific soil condition and their hybrids are unable to survive in either of the two habitats.

Reproductive isolation

In animals one of the most important isolating mechanisms is mating preference,

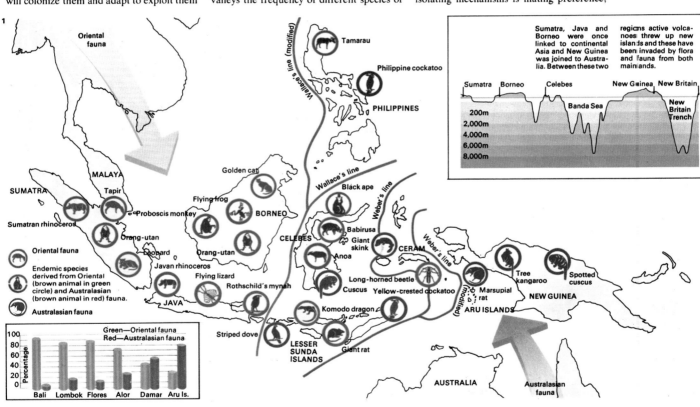

1 Natural barriers such as deep ocean channels separate the animals of the world into distinctive groups. These barriers have served as isolating mechanisms, keeping apart not only species but also whole families of animals. One striking example is the juxtaposition of the flora and fauna of South-East Asia and Australasia. The 19th-century biologist Alfred Russel Wallace (1823–1913) drew a dividing line where the fauna and flora of South-East Asia was distinct from that of Australasia. This was later modified by the naturalist Thomas Henry Huxley (1825–95). Even later, the German biologist Max Weber drew a new line marking the point of balance between the two groups. Later he modified this to mark the sea channel that forms a barrier to the spread of land mammals. Between Weber's and Wallace's lines is an area of faunal mixing where animals and plants have managed to cross the channels separating the islands.

2 Adaptive radiation can be seen in the cattle of South-East Asia which, although related, occupy different environments. The gaur is found in the hilly forests, while the wild buffalo, ancestor of the domestic variety, roams the upland swampy areas. The banteng lives in Java, where it is sometimes domesticated. Yaks, also domesticated, are adapted to withstand the cold of Tibet and central Asia. The small wild anoa lives in wet, hilly regions of the Celebes.

2 Gaur
Bos gaurus

Wild buffalo
Bubalus bubalis

Banteng
Bos banteng

Yak Bos grunniens

Anoa
Anoa depressicornis

which is widespread in animals inhabiting the same geographical areas. One of the most striking differences between closely related species is the ability of individuals to recognize breeding partners of the same species. Elaborate courtship displays and differences in the colour of an animal's external covering ensure that only individuals of the same species mate. Other cues to breeding partners may include smell, the distinctive calls of most species of birds and even the croaking of frogs and the singing of crickets. Without these stimuli many animals will not mate.

It was long believed that in animals the differences in the genitalia of different species led to reproductive isolation where, for example, the male genitalia could not physically fit into the female genitalia. Recent observations have shown that this kind of mechanical isolation does not operate in animals although in plants similar mechanical isolation ranks in importance with ethological (behavioural) isolation in animals. Differences in flower structure prevent pollination and fertilization between unrelated species. Plants exploit the activities of

animals to spread their genes, and differences in flower structure, colour, and scent attract various kinds of pollinators to different species. This reduces the chances of cross-pollination between two unrelated species. In California, for example, there are related species of a flower (*Penstemon*) that have different sized flowerheads. The different species are pollinated respectively by large bees, smaller bees and solitary bees.

Another possible isolating feature is found in the seasonal differences in reproductive activity, but this alone would be unlikely to keep species distinct.

If fertilization does occur between species it is quite likely that one of three mechanisms may have an effect. Often the first generation hybrid is rather weak because of abnormal genetic matching and rarely survives to reach reproductive maturity. The hybrids although healthy are often sterile, as is seen in the mule – the offspring of a horse and a donkey. When the hybrids are healthy and reproductively active, the final mechanism comes into play by ensuring a gradual breakdown of genetic harmony so that the hybrid line dies out.

KEY

1 *A. fulgens* (Miu)
2 *A. cestus* (Wailupe)
3 *A. fuscobasis* (Palolo)
4 *A. stewarttii* (Manoa)
5 *A. turgida* (Alea)

OAHU

Many species of agate snail (*Achatinella*) live on the Hawaiian island of Oahu, most of them in a small area of adjacent valleys. Individuals remain close to their point of hatching. The rocky ridges that separate the valleys are an effective barrier to migration for these relatively immobile animals. Over thousands of years the snails have been geographically isolated and, as a result, the species have remained quite distinct.

3 Zones that vary with altitude can lead to the formation of different habitats that are best exploited by different species of plants and animals. This can be clearly observed on Mount Kenya, which rises 5,199m (17,057ft) in equatorial Africa. High up on this mountain there is a specialized flora and mammals include the groove-toothed rat, the hyrax and the duiker.

Lower down the mountain, in the sub-alpine zone, tree heaths dominate the vegetation. Mammals are similar to those found higher up, but leopards and hunting dogs appear as predators.

The lower bamboo montane forest supports a wide variety of animals, including omnivores such as monkeys and pigs and browsing animals such as antelopes and elephants. Below the forest is the savanna, supporting animals that cannot live in the higher zones.

4 Vegetation zones in mountainous regions may not always be uniform because the high land forces winds laden with moisture to rise. As they do so, the water they carry is dropped and they blow down the other side of the range as waterless, desiccating airflows. A zonal plan of Mount Kenya shows that, where the rainfall is less heavy, areas of scrub and dry land replace the montane forest that covers the north of the massif. This affects many animals that depend on particular kinds of vegetation for their food and shelter.

A Dry scrub
B Afro-alpine
C Sub-alpine
D Moorland
E Bamboo
F Montane forest
G Savanna

Vertical scale in metres
4,900
4,300
3,600
3,000

B
C
E D
G F

0 Miles 10
Km 10 Scale

3

Giant *Senecio*
Giant lobelia
Mountain sedge
Everlastings
Afro-alpine zone
Groove-toothed rat
Rock hyrax
Duiker
Sub-alpine
Tree heaths
Moorland zone
Bamboo
Bamboo montane forest zone
Leopard
Bushbuck
Mole-rat
Monkey
Forest hog
Bush baby
Elephant
Bongo
Buffalo
Tree hyrax
Rhinoceros
Tropical forest trees
Savanna

17,000ft 5,200m
14,000ft 4,300m
11,000ft 3,300m
10,000ft 3,000m
8,500ft 2,700m
6,800ft 2,100m

5
A
B
C

B *G.g. brandtii*
A *G.g. glandarius*
F *G.g. atricapillus*

E *G.g. krynicki*
D *G.g. cervicalis*
C *G.g. japponicus*

D
E
F

5 More than 20 recognized races of jay are found in Eurasia, ranging from western Europe to the islands of Japan. The jay is primarily a bird of deciduous woodlands that feeds mainly on acorns wherever it can find them. But it is a versatile feeder and can live on a variety of seeds, fruits, buds, insects and even eggs and small birds. In some areas it has forsaken its usual habitat and has taken to coniferous forests, where it feeds on the seeds of certain pines. Many species of birds are widely distributed and show marked local variation in the colour of their plumage in different parts of their range. Where bird populations overlap, there is a gradual transition of features from one race to another, as the different populations freely interbreed. These populations are regarded only as sub-species or races. The birds at the extremes of their ranges are unlikely to interbreed and they may in time become separate species.

197

Northern grasslands

Grasses were among the last families of flowering plants to evolve. They have more than made up for lost time, however, and are now the dominant plants over much of the world. In the Northern Hemisphere grasslands occur in the prairies of North America, the steppes of Europe and central Asia and in areas extending along the valleys of some of the great rivers of eastern Asia [Key]. In general they are found where rainfall is too low to support tree growth, yet above the level at which semi-desert conditions prevail. Paradoxically, grazing animals and fire help to promote the growth of grasses.

Grasses and grass-eaters
Winds sweep unimpeded across the steppes and the prairies and the grasses make use of them for pollinating their flowers. Unlike those of many other plants, grass flowers [1] are not large, gaudy, sweet-scented or provided with nectar. Instead they possess the bare minimum of stamens with abundant lightweight, non-sticky pollen grains and stigmas to be pollinated by them. During their development the reproductive parts are

surrounded by protective scales, which later enfold the seed. The stamens and stigmas are ripe for only a comparatively short time, but the amount of pollen produced is so large that when the plant is shaken by the wind pollination is almost inevitable.

One reason why the grasses are successful lies in their compact with the grazing animals. The growing point in grasses lies close to the ground and when the foliage is cropped at a higher level it does little harm because the cut leaves simply continue to grow upwards with little or no interruption in the growth pattern. As a result the grazing animals can feed without damaging their food supply.

Large grazing animals are abundant in grasslands within the temperate areas. But individuals are relatively few where they were once present in huge numbers. Experts estimate that in 1700, before the coming of the white man to the prairies, more than 60 million bison [2] wandered over the plains, in addition to large numbers of pronghorn. In Asia, herds of wild horses, asses, saiga antelopes [3] and camels abounded. In both continents these large animals had been

almost eliminated by the end of the last century – in 1899 fewer than 550 plains bison survived. Careful conservation has gone some way to restoring the position, particularly with regard to the saiga and the pronghorn over part of their respective ranges. All these large mammals are herd animals that migrate with their food supply, so the grass is never overgrazed or badly trampled. They were once accompanied by predators, principally wolves, but as the great wild herds were destroyed the predators were also doomed.

A variety of animals
Small mammals were also once abundant on the plains but, unlike the large creatures, they were sedentary [4]. Some, such as the prairie dogs and susliks, lived in great colonies that at one time often included several million individuals in a single "township". These animals feed mainly on the grasses, eating the roots as well as the leaves, but often also renewing the plant growth because of their habit of gathering and storing seeds. The burrows of these animals are not very deep, but in digging them they turn the soil over, some-

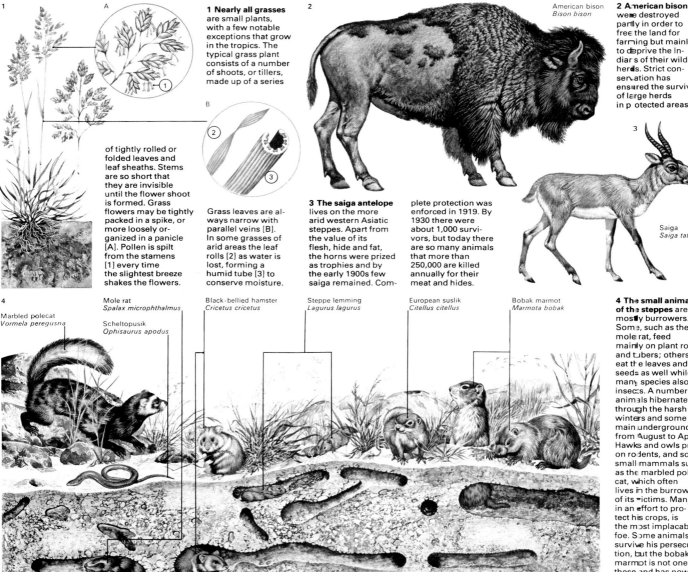

1 Nearly all grasses are small plants, with a few notable exceptions that grow in the tropics. The typical grass plant consists of a number of shoots, or tillers, made up of a series of tightly rolled or folded leaves and leaf sheaths. Stems are so short that they are invisible until the flower shoot is formed. Grass flowers may be tightly packed in a spike, or more loosely organized in a panicle [A]. Pollen is spilt from the stamens [1] every time the slightest breeze shakes the flowers.

Grass leaves are always narrow with parallel veins [B]. In some grasses of arid areas the leaf rolls [2] as water is lost, forming a humid tube [3] to conserve moisture.

American bison
Bison bison

2 American bison were destroyed partly in order to free the land for farming but mainly to deprive the Indians of their wild herds. Strict conservation has ensured the survival of large herds in protected areas.

3 The saiga antelope lives on the more arid western Asiatic steppes. Apart from the value of its flesh, hide and fat, the horns were prized as trophies and by the early 1900s few saiga remained. Complete protection was enforced in 1919. By 1930 there were about 1,000 survivors, but today there are so many animals that more than 250,000 are killed annually for their meat and hides.

Saiga
Saiga tatarica

Marbled polecat
Vormela peregusna

Mole rat
Spalax microphthalmus

Scheltopusik
Ophisaurus apodus

Black-bellied hamster
Cricetus cricetus

Steppe lemming
Lagurus lagurus

European suslik
Citellus citellus

Bobak marmot
Marmota bobak

4 The small animals of the steppes are mostly burrowers. Some, such as the mole rat, feed mainly on plant roots and tubers; others eat the leaves and seeds as well while many species also eat insects. A number of animals hibernate through the harsh winters and some remain underground from August to April. Hawks and owls prey on rodents, and so do small mammals such as the marbled polecat, which often lives in the burrows of its victims. Man, in an effort to protect his crops, is the most implacable foe. Some animals survive his persecution, but the bobak marmot is not one of these and has now disappeared from areas where it was once numerous.

times bringing up material of a different mineral type from below. As a result, the site of such a colony often displays a different type of soil from that of the surrounding area.

Birds of the grasslands include plant- and seed-eaters that parallel the mammals in their activities. Some, such as the bustards, are large and reluctant to fly, a propensity that has regrettably increased their rate of destruction. There are also many smaller species of birds, which feed both on seeds and insects. They in turn may fall prey to several species of falcons [8], hawks and eagles, which patrol the plains.

Other predators include large numbers of reptiles [9, 10], which may feed on eggs or young birds, as well as small mammals whose underground homes they can enter. The insects of the grasslands [11] are an important part of the fauna and repeat in miniature the pattern of primary herbivores and include hunters and scavengers, which are not well represented among the bigger creatures. Insects and other invertebrates may lack universal appeal, but their importance in the economy of the grasslands can scarcely be

overestimated because they stimulate soil fertility and help to hold erosion in check.

The most important of the grassland animals is man, who probably underwent important stages in his evolution in tropical grasslands and who has subsequently spread to other environments. He is, however, still mainly dependent on grasslands for his food – all cereals are cultivated grasses and most of man's domestic animals are species that still need grass for their survival.

The threat to the grasslands
It is ironical that although man has greatly increased the area of the world's grasslands, mainly by the destruction of forests which he has replaced with short-lived farm crops, he has in many areas of the world destroyed the grasslands too by overgrazing and returning too little to the soil. Sometimes, in his attempts to improve productivity with heavy-yield farm crops, he has ploughed up the grassland and harmed the delicate balance between plants and animals. This has led all too often to erosion and as a result deserts have encroached in many areas.

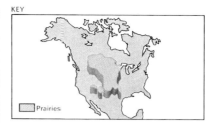

KEY

Prairies

The prairies of North America occupy much of the central part of that continent. They are almost entirely surrounded by forests and only in the southwest do grasslands yield to desert regions in the shadow of the Rockies.

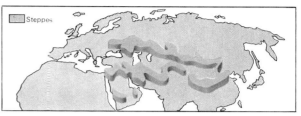

Steppes

In the Old World the natural grasslands or steppes start in eastern Europe and stretch across the Eurasian land mass in a great belt that is bounded on the south by semi-desert and scrub, and on the north by forests. North of Mongolia the steppe becomes discontinuous as it is broken by trees. But the steppe reasserts itself on the plains of Inner Mongolia and in the northeast of the Soviet Union.

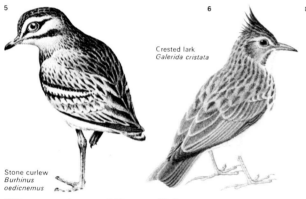

5

Crested lark
Galerida cristata

Stone curlew
Burhinus oedicnemus

5 The stone curlew was once common on grasslands from Britain to eastern Europe. As a result of its ground-nesting habit and reluctance to fly it has disappeared from many of its former haunts.

6 The crested lark, fearless of man, is common in dry Old World grasslands.

7 The burrowing owl, from the grasslands of the USA, lives in the disused burrows of prairie rodents.

6

7

Rattlesnake *Crotalus* spp

Burrowing owl
Speotyto cunicularia

9

8

Prairie falcon
Falco mexicanus

8 The prairie falcon is a bird of the arid southern and western parts of the prairies. It scans the ground from heights up to 30m (100ft) for the small animals and birds that make up its diet.

9 Rattlesnakes are common over much of the grassland and desert regions of North America. The amount of venom they produce far exceeds that needed to kill the rodents that form their usual prey. They use it to protect themselves against large predators, but they are not normally aggressive and try to avoid a confrontation by warning of their presence. Their unmistakable and menacing rattle is produced by "bells" of hard skin on the end of the tail.

10

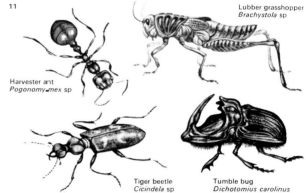

Bull snake
Pituophis sp

10 The bull snake, which grows up to 1.2m (4ft) long, is one of the many snakes found in the prairie region, where it preys on the abundant rodents. It is non-poisonous but suffocates its prey with its strong, constricting coils.

11

Harvester ant
Pogonomyrmex sp

Lubber grasshopper
Brachystola sp

Tiger beetle
Cicindela sp

Tumble bug
Dichotomius carolinus

11 Grassland insects include primary feeders on the herbage, such as grasshoppers, and more general feeders such as ants. Ants are smaller, but the huge colonies in which they live may alter the soil where they burrow. They also often scatter grass seeds, which they gather but do not always use. Predators such as ground beetles abound. The scavenging tumble bug fills a niche by breaking down the dung of other animals and returning it to the soil.

Grasslands of Africa

Savanna grassland covers much of Africa from the Sahara to the Cape. Only in the west does it give way to the dense tropical forest of the Zaïre River basin and the West African coast and to the Namib Desert. The grassland is not a uniform area, but ranges from dry steppe and sub-desert in the north and south-west (on the fringes of the Sahara and Kalahari deserts) through thorn scrub to open savanna woodland in the equatorial region. The landscape is bisected by water-courses and granite hills, and much of the southern area is modified by farming.

Distinct types of savanna

The transitions between the various types of savanna are distinct and each type has its own characteristic fauna and flora. The open woodlands are characterized by fire-resistant trees such as *Brachystegia* and *Isoberlinia* which are replaced in drier areas by acacias (thorn-scrub) and baobabs. The vegetation of the open areas is a mixture of grasses and herbs with occasional acacias.

Ecologically, the savanna is a highly complex system of interdependent components.

The food of the large herds of grazing animals – the grasses and herbs – is resistant to drought, fire and grazing and quickly recovers from their effects. Many savanna grasses reproduce by underground runners and can spread rapidly. The trees and bushes are protected from excessive browsing by long thorns on the trunks and branches.

The bewildering number of savanna animals exploit their environment in a variety of different ways. Among the browsers there is a vertical zoning of feeding habits [Key]. Giraffes and elephants feed on the trees – giraffes on the topmost shoots and elephants both on the upper shoots and the bark. Black rhinoceros, eland, kudu and gerenuk browse on lower shrubs and trees, with the lowest branches often only centimetres above the ground, providing food for steenbok and dik-dik. The grazing animals utilize the grasses in different ways. Zebras feed on the coarse tops, the leafy centre is eaten by wildebeeste and topi and gazelles crop the shoots at ground level. Underground roots and bulbs are eaten by warthogs.

Feeding on the herds of grazing animals

are a number of large predators. One of the best known is the lion [5], which lives in family groups and catches its prey in a short rush from cover. Two other cats are also to be found on the savanna: the leopard, which often kills from cover at waterholes, and the cheetah, which is capable of out-running gazelles over a short distance. Small packs of hunting dogs [6] roam the savanna, working as teams and pursuing their prey until it is exhausted and falls victim to their snapping jaws. The hyaenas are both predators and scavengers. No young animal is safe from them and no bone is too large for their powerful jaws to crack.

Rodents and hares

As well as the large and spectacular animals, many small animals, not often seen, live in the savanna. Like the larger animals they occupy a particular niche. The numerous rodents and hares feed on seeds and herbs and in turn provide food for foxes, snakes, small cats and birds of prey. Many kinds of mongooses, civets and weasels prey on the large numbers of small savanna animals. Two

1 Baobab tree
  Adansonia digitata
2 Umbrella thorn
  Acacia sp
3 Candelabra tree
  Euphorbia sp
4 Whistling thorn
  Acacia sp
5 Weaver bird nests
  Family Ploceidae
6 Red oat grass
  Themeda triandra
7 Giraffe
  Giraffa camelopardalis
8 Topi
  Damaliscalus korrigum
9 Impala
  Aepyceros melampus
10 Waterbuck
  Kobus defassa
11 Cape buffalo
  Syncerus caffer
12 Oxpecker
  Buphagus sp
13 Marabou stork *Lepto-ptilus crumeniferous*
14 Crested guinea fowl
  Guttera edouardi
15 Tawny eagle
  Aquila rapax
16 Common agama
  Agama agama
17 Banded mongoose
  Mungos mungo
18 Puff adder
  Bitis arietans

1 The African sav-anna, with its flat-topped acacias, supports large herds of herbivorous mammals. Their predators, being mainly nocturnal, are less often seen. But birds, from the busy weaver to the scavenging marabou stork and vegetarian guinea fowl, are as conspicuous as the game.

2 The masked weaver builds a nest of woven grass. Its foundation is a ring of knotted strands

to which the walls and then the roof are added. An entrance hole is left underneath the nest.

animals of especial interest are the meerkat and the ratel. The meerkat is a sociable burrowing animal living in large warrens and feeding on insects, spiders and millipedes. Also living in burrows, the ratel or honey badger is best known for its association with the honey guide – a small brown bird. When this bird finds a bees' nest it attracts the ratel with a particular call and leads it to the nest. The ratel digs out the nest with its powerful claws and both animals share in the feast.

Variety of bird life

Great numbers of birds inhabit the savanna, many more than live in the tropical rain forest. Ostriches, the world's largest birds, often accompany the grazing herds; so do storks and egrets, which feed on disturbed insects. The majestic kori bustard and the snake-eating secretary bird can also be seen striding through the grass. The nests of weaver birds [2], like huge fruit, adorn many acacias and overhead the circling vultures and soaring eagles scan the ground for carrion and small animals.

In summer vast numbers of birds migrate from Europe and Asia to the savanna regions to escape the severe northern winter. This enormous variety of birds feeds on seeds, berries and insects and is in turn preyed upon by hawks, owls and falcons.

Because of the seasonal rains, migration is a major feature of life on the savanna. Because the grass stops growing during the dry season the wildebeeste and other animals, such as the zebra, move to areas where rain has fallen and the grass is growing. During the migrations many thousands of animals, particularly wildebeeste, die through drowning and starvation. Scavengers and predators, notably vultures and hyaenas, take full advantage of this superabundance of easily obtained food. However, such is the richness of their environment that the numbers of wildebeeste soon recover from this apparent catastrophe.

The delicate balance of the savanna ecosystem is easily upset, especially by man. Fortunately, enlightened governments aware of the unique nature of the wildlife heritage have set aside large areas of the savanna as inviolable sanctuaries.

KEY

Vertical feeding patterns of savanna herbivores reduces competition between species. The giraffe [4] takes leaves 6m (18ft) up, leaving the lowest ones for the tiny dik-dik [11].

1 Springbok
2 Eland
3 Kudu
4 Giraffe
5 Warthog
6 Black rhinoceros
7 Elephant
8 Vervet monkey
9 Gerenuk
10 Steenbok
11 Kirk's dik-dik

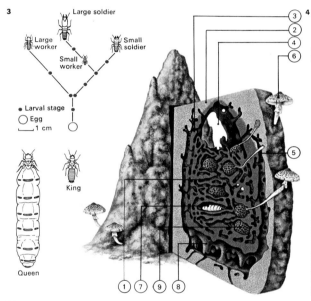

3

Large soldier
Large worker
Small soldier
Small worker
Larval stage
Egg
1 cm
King
Queen

1 Nest
2 Outer wall of termitarium
3 Wall of nest
4 Air space within nest
5 Chimneys to regulate the temperature of the nest
6 Fungus growing from mound
7 Queen's chamber
8 Base pillars supporting nest
9 Fungus gardens

3 A termite mound, with its blank exterior, gives no hint of the complex system of chambers and tunnels within. Thick, hard walls deter most predators and chimneys can be opened or closed to regulate the temperature. The mound is developed from a hole occupied by a mated male and female. The female (queen) only produces eggs, which become either workers or soldiers. Termites feed on wood and leaves brought in by foraging workers, and on fungus grown on faeces in special fungus gardens.

4 A typical savanna scene might have a herd of Grant's gazelle (*Gazella granti*) feeding on grass. The open plains are dotted with wide-topped acacias and clumps of small bushes.

5 Prides of lions, often with several adult males, are commonly seen lazing in the shade of a tree. Lionesses, which lack manes, do most of the killing, often working as a team to stalk prey. Lions themselves have no natural enemies except man.

5 Lion
Panthera leo

6 African hunting dog
Lycaon pictus

6 African hunting dogs are gregarious animals that live in packs of from 6 to 20 individuals. They hunt in an organized manner, taking it in turns to chase their animal prey until it is exhausted.

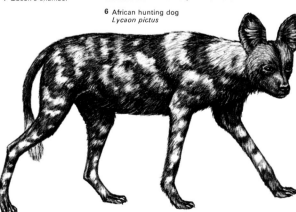

South American grasslands

From the Brazilian highlands and dense forests of Amazonia southwards to the barren lands of Patagonia, and from the eastern slopes of the Andes to the Atlantic Ocean, stretches a vast ocean of grassland. This is a region where droughts are frequent and where torrential downpours penetrate only the top layers of soil. As a result trees are unable to compete with shallow-rooted grass plants which take the available water.

Grassland and climate

The type of grassland, and therefore the kind of animals it supports, depends largely on rainfall and temperature. Hot, dusty summers, cold, windy winters and alternate periods of drought and heavy rainfall have produced the huge expanse of temperate grassland called the Argentine pampas [1]. On this plain, formed by layers of soil weathered from the Andes and carried eastwards by rivers towards the sea, the controlling influence of soil moisture on vegetation is clearly apparent. On the wetter eastern side the grass grows in tall, coarse tufts and eventually merges into forest. The fertility of

this part of the pampas has been made use of by man for wide-scale ranching and agriculture and, as a result, the landscape has been considerably modified. On the drier western and southern margins bare soil patches are found between clumps of prairie grass, and drought-resistant bushes and small scrub trees replace the grassland. Farther to the east the scrub gradually gives way to desert.

Animals large and small

Most of the large mammals that inhabited these grasslands when South America was an isolated continent disappeared about a million years ago when the isthmus of Panama was formed and carnivores travelled southwards from North America. The carnivores were unable to cope with conditions successfully, could not colonize and eventually died out. As a result the only animals found in the South American grasslands today are those that have managed to adapt to the climate and food supplies of the area. The only mammal of any size on the pampas is the pampas deer (*Blastoceros campestris*) [2] and the largest animal inhabitant is a bird,

the rhea (*Rhea americana*) [10]. This flightless bird, the South American counterpart of the ostrich, is well suited to an open habitat. Its long legs give it an elevated view of its surroundings and enable it to run at more than 50km/h (30mph). In spite of the absence of large mammals at ground level the grasslands teem with life. There may be as many as 1,000 surface insects per square metre (11 square feet), particularly grasshoppers and butterflies. They and the numerous lizards, snakes and spiders provide food for the predatory birds and mammals.

It is the small mammals, however, that have made best use of the ground cover and are most characteristic of the grasslands. Many of them solve the problems of climatic extremes and a lack of suitable hiding places by living underground for at least part of their lives and, as a result, burrowing rodents such as the viscacha (*Lagostomus maximus*) [4] and the noisy tucu-tucu (*Ctenomys talarum*) are among the most successful grassland inhabitants. For much of the day they remain hidden in their underground tunnels, coming out to feed only in the safety of darkness. The

1 **The flat Argentine pampas** covers more than 500,000 sq km (350,000 sq miles). In the east moist Atlantic winds promote the growth of rich, tall grass; but near the Andes a hot, dry climate produces an arid steppe land with bare soil patches between tufts of prairie grass and drought-resistant bushes. Many pampas animals in these harsh regions live underground for at least part of their lives.

Pampas deer
Blastoceros campestris

2 **The pampas deer** is one of the few herbivorous mammals of any size to be found on the South American grasslands. Its numbers have now been seriously reduced as a result of over-hunting and the destruction of its habitat by ranching and cultivation.

3 Giant armadillo
Priodontes giganteus

Six-banded armadillo
Euphractus sexcinctus

Pink fairy armadillo
Chlamyphorus truncatus

3 **The 21 species of armadillo** are among the most abundant and widespread of South American

mammals. Uniquely protected by plates of bone joined with skin, a few of them, such as the three-banded armadillo, can

roll themselves into defensive balls when threatened. But most escape predators by rapid burrowing.

Armadillos vary in size from the giant armadillo, 1.5m (5ft) in length and weighing up to 50kg (110lb),

to the fairy armadillo which is only 13cm (5in) long and spends most of its life underground. Arma-

dillos are regarded as pests in some areas but they do help man by ridding his crops of harmful insects and other small animals.

Three-banded armadillo
Tolypeutes matacus

Five-toed armadillo
Cabassous centralis

Hairy armadillo
Chaetophractus villosus

wild guinea pig (*Cavia aperea*) [4], which lives above ground, seeks protection by forming large colonies of up to several hundred and only emerges from tufts of grass to feed at dawn and dusk.

Rodents form a large part of the diet of grassland predators such as the pampas fox (*Dusicyon gymnocercus*) and Azara's oppossum (*Didelphis azarae*). But extremes of climate and erratic rainfall make food supplies unreliable, so many predators are omnivorous and will eat almost anything they can find. Numerous other predators, such as armadillos [3], skunks and anteaters, forage in the grass for the abundant insect life. Birds such as the smooth-billed anis (*Grotophaga ani*) and the aptly named cattle tyrant (*Machetornis rixosus*) [7] find their food by following grazing animals. They then catch the thousands of insects disturbed from the ground vegetation.

Many of the avian inhabitants of the pampas are migrant visitors. In autumn the golden plover (*Pluvialis dominica*) leaves the harsh conditions in the tundra and flies nearly 16,000km (7,500 miles) to winter in the grasslands before returning to its nesting grounds in spring. Many northern shore birds, such as redshanks and sandpipers, also migrate there in winter to feed on the fish and insects in marshy areas of the pampas.

The Chaco and its wildlife

Large expanses of water are also typical of an area to the north of the pampas called the Chaco – a transitional zone between the grassland and the tropical forests of the Amazon that lacks a specialized fauna but attracts animals from neighbouring zones. Sluggish rivers spread out over a large plain to form swamps. These become shallow lakes during the heavy summer rains and form a paradise for water-birds such as limpkins, ducks and noisy screamers [9]. The swamps, often inhabited by the giant anteater (*Myrmecophaga tridactyla*) [6], are interspersed with patches of grassland and deciduous forest. The forest is the home of the giant armadillo (*Priodontes giganteus*) [3] and the elusive maned wolf (*Chrysocyon brachyurus*) [5] which, although well adapted to plains life, seeks the shelter of trees.

KEY

☐ Grassland

☐ Savanna

☐ Deciduous forest and scrub

The grasslands of South America cover much of the continent east of the Andes. The vast, treeless Argentine pampas merges into forest in the wetter north-east and into scrub and desert to the west and south. In the north it is bounded by the Chaco – an area of deciduous woodland and scrub. The Venezuelan *llanos* and Brazilian *campos* are areas of tall-grassed savanna interspersed with forest.

4 Mara
Dolichotis patagona

Viscacha
Lagostomus maximus

Burrowing owl
Speotyto cunicularia

Cavy
Cavia aperea

5 Maned wolf *Chrysocyon brachyurus*

4 The viscacha digs a system of tunnels that it shares with the burrowing owl and the mara or Patagonian hare. By stripping the surrounding area of vegetation the viscacha can detect approaching predators. The pampas cavy nests at the base of grass tufts.

5 The shy, nocturnal maned wolf, like other plains predators, will eat almost anything from small animals to fruit.

6 Giant anteater
Myrmecophaga tridactyla

6 The giant anteater rips open termite mounds and scoops the termites with its long, sticky tongue. Its shaggy coat protects it from bites.

8 The crested caracara, a ground-dwelling falcon that is both hunter and scavenger, searches the grassy plains for food both living and dead with a characteristic head-bobbing walk.

9 The southern screamer haunts marshy areas of the pampas with harsh cries. Its slightly webbed feet enable it to walk on floating vegetation.

9

Southern screamer
Chauna torquata

10 Common rhea
Rhea americana

7 The cattle tyrant is a flycatcher that perches on the backs of hoofed mammals waiting for insects disturbed by the grazing animals. It has an erectile crest of feathers.

7

Cattle tyrant
Machetornis rixosus

10 The flightless rhea or South American ostrich, a bird that lives in open country, roams the pampas in flocks of up to 30, feeding on vegetation and insects. It stands up to 1.5m (4.6ft) tall and can detect approaching danger even in high grass. When threatened it can run faster than a horse.

8 Crested caracara
Polyborus plancus

Australian desert grasslands

Two-thirds of Australia's eight million square kilometres (three million square miles) is semi-arid land supporting only tough grasses [5] and scattered acacia and gum trees. Less than 50cm (20in) of rain falls during the year, daytime temperatures average 32°C (89°F) and a dry wind blows. Inhospitable to man, these grasslands are the home of many of the symbolic animals of Australia [1, 3] such as the kangaroo, emu, koala and kookaburra, as well as flocks of parrots, cockatoos and budgerigars – all animals that cope with the dry conditions.

Making the most of water

The "outback" of the interior has largely been created by drying winds from the southeast and west. These are forced to release their moisture prematurely by the Great Dividing Range in the east and cold currents off the west coast. The "desert" lies at the limit of penetration of rain from the north and south and consequently may one year receive rain in winter, another in summer and another not at all. The animals and plants are opportunists – they make the most of the

water when it comes. The dull porcupine grass (*Triodia* sp) [4] bursts into a yellow sea of blossom on the red desert sand. The mulla mulla (*Trichinium manglesii*) springs up and produces fluffy pink flowers. Aromatic gum trees such as the river gum (*Eucalyptus camaldulensis*), which line the banks of dry rivers and watercourses, bear unique flowers whose stamens provide vivid colours.

Many plants are adapted to prevent excessive loss of precious water. Acacia leaves are reduced to flattened leaf stalks and those of the desert oak (*Casuarina decaisneana*) have developed into needles which hang in fringes. Porcupine and cane grasses have thick, waxy cuticles and saltbush leaves (*Atriplex* spp) are coated in salts.

Australia's unique marsupials

The grasses and trees provide food, shelter and nesting materials for the animal inhabitants. These animals entered Australia when it was still part of the land mass known as Gondwanaland (which included present-day Africa, South America, South-East Asia and Antarctica before continental drift) and were

able to evolve unhindered by competition from placental animals when it became an island at the end of the Mesozoic era. The result is a unique set of animals with a trend towards non-aggressiveness, herbivorous diet and tolerance for arid conditions.

The name "koala" means "the animal that does not drink". These timid marsupials, or pouched mammals, live on an extremely restricted diet of eucalypt leaves, which provide all the moisture they need. The fat-tailed sminthopsis (*Sminthopsis crassicaudata*) stores fat "for a rainy day" in its tail. Kangaroos, which feed on triodia grass and saltbush, can survive long periods of drought. The wombat (*Vombatus* sp), like several of its neighbours, digs a long underground burrow, often more than 3m (10ft) long, where it can escape the excessive heat and dryness of the day and the cold of the night. Many smaller marsupials, such as the insectivorous bandicoots and jerboa marsupial mice, come out only at night. The primitive egg-laying echidna or spiny anteater, one of Australia's two monotreme genera, is heavily armed with spines and when threatened this

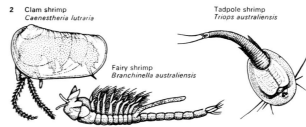

1 Koalas and kangaroos, which symbolize Australia, are only two of a wide variety of unique animals (see illustration 3) that have evolved since Australia became an island about 50 million years ago. In the vast, flat, desert-like interior covered with hardy grasses and fire-resistant eucalypt and acacia trees the marsupials – "pouched" animals – and even stranger monotremes, like the echidna, were able to adapt unhindered by competition from placental mammals. The number and diversity of birds are equally impressive. The emu is the

world's second largest bird and the parrot family has its stronghold here. The lizards, many of them nocturnal, have made the most successful adjustment to desert life. Amphibians, notably frogs, make use of temporary desert pools and there are numerous species of insects.

2   Clam shrimp
Caenestheria lutraria

Tadpole shrimp
Triops australiensis

Fairy shrimp
Branchinella australiensis

2 Tiny crustaceans like these shrimps take advantage of temporary freshwater desert pools, which form during rain storms. When the pools evaporate the crustaceans die, leaving behind covered eggs which survive periods of drought to hatch when conditions are favourable.

mammal can either roll into a ball or rapidly dig itself into the desert soil. It scoops up desert insects with its long, whip-like tongue.

Birds, reptiles and insects

Birds of the desert grasslands are usually seen more often than mammals. Colourful flocks of galahs, the most common cockatoos, nectar-feeding lorikeets and green and yellow budgerigars congregate round waterholes, often dug by ranchers for their cattle. These birds are constantly on the move in search of water and breed only after the rains. The uncertain water supply means that the kookaburra, Australia's largest kingfisher, no longer finds most of its food in pools but searches for insects and snakes on the ground and in the trees. The emu, Australia's largest bird, changes from a diet of seeds to new-grown grass when the rains arrive. The mallee fowl (*Leipoa ocellata*) lays eggs in mounds of vegetation and sand. The vegetation ferments with the aid of dampening rain and the sun's heat and incubates the eggs. By altering the amount of nest material the male mallee fowl is able to regulate the nest temperature

to the correct one to facilitate incubation.

Reptiles include dragon-lizards such as the frilled lizard, which erects a frill when threatened, and skinks, which scuttle over tree trunks and between clumps of porcupine grass in search of insects. Some of the snakes, such as the carpet python, are harmless to man, but most, like the death adder and king brown snake, are deadly. Male frogs call females to ponds only in wet conditions so that the tadpoles will develop quickly before the next dry season. As the water dries up the water-holding frog (*Cyclorana platycephalus*) digs deeply into the mud with its body full of water and lives in a cell of mucus until the next rains. In the period following the rains the grassland is full of insects – flying termites, bees, mantids, grasshoppers and beautiful butterflies.

The dingo is one of Australia's few placental mammals but it is not indigenous. Now considered by many to be the same species as the domestic dog (*Canis familiaris*), it probably returned to the wild after being introduced by the Aborigines when they arrived in Australia about 30,000 years ago.

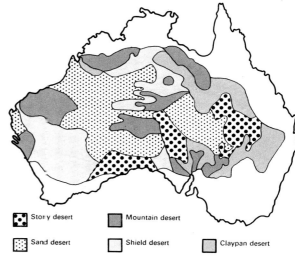

Stony desert   Mountain desert

Sand desert   Shield desert   Claypan desert

3 A selection of the birds and animals of Australia (shown in illustration 1), many of them unique to that continent and its offshore islands, is keyed here as follows:

1 Koala
Phascolarctos cinereus
2 Sulphur-crested cockatoo
Cacatua galerita
3 Galah
Kakatoe roseicapilla
4 Cockatiel
Nymphicus hollandicus
5 Budgerigar
Melopsittacus undulatus
6 Kookaburra
Dacelo gigas

7 Crimson winged parrot
Aprosmictus erythropterus
8 Lace monitor
Varanus varius
9 King brown snake
Pseudechis australis
10 Tawny frogmouth
Podargus strigoides
11 Sugar glider
Petaurus breviceps
12 Black kite
Milvus migrans
13 Dingo *Canis dingo*
14 Wallaroo
Macropus robustus
15 Great red kangaroo
Macropus rufus
16 Great grey kangaroo
Macropus giganteus
17 Emu
Dromiceius novaehollandiae
18 Northern native cat
Satanellus hallucatus
19 Brolga crane
Grus rubicunda

20 Bearded dragon
Amphibolurus barbatus
21 Red-tailed cockatoo
Calyptorhynchus magnificus
22 Wedge-tailed eagle
Aquila audax
23 Termite mound
24 Sand monitor
Varanus sp
25 Echidna
Tachyglossus aculeatus
26 Rainbow bee-eater
Merops ornatus
27 Hairy-nosed wombat
Lasiorhinus latifrons
28 Rufous rat kangaroo
Aepyprymnus rufescens
29 Turquoise grass parrot
Neophema pulchella
30 Long-nosed bandicoot
Perameles nasuta
31 Rainbow lorikeet
Trichoglossus haematodus
32 Frilled lizard
Chlamydosaurus kingi

33 Giant stick insect
Didymuria violescens
34 Little quail
Turnix velox
35 Agile wallaby
Wallabia agilis
36 Cicada
Platylomia sp
37 Common marsupial mouse
Sminthopsis murina
38 Carpet python
Morelia argus
39 Checkered swallowtail
Graphium sp
40 Jerboa marsupial mouse
Antechinomys spenceri
41 Death adder
Acanthophis antarcticus
42 Holly cross toad
Motaden bennetti
43 Common Australian crow butterfly
Euploea core

Two-thirds of Australia is flat desert formed by dry winds blowing across the interior. More than 2,000 species of plants and hundreds of animals have adapted to its aridity.

4 The characteristic grass of the Australian interior is the prickly porcupine grass (*Trioda*), which grows in huge pin-cushions in the red sand between salt-bush and blue-bush scrub. With an extensive deep root system and leaves with hard cuticles assisting moisture-retention, the grass is adapted to withstand long droughts and to exploit brief downpours.

5 Savanna woodland forms a continuous strip along the north coast and continues south inland from the Great Dividing Range with isolated areas in the south-west. The vegetation depends on over 50cm (20in) of rain a year. The trees (mainly eucalypts) grow fairly close together and the undergrowth is a mixture of grass and shrubs.

Northern pine forests

The northern pine forests stretch like a broad ribbon across the Northern Hemisphere, from the Pacific coast of Canada eastwards to the Kamchatka Peninsula [Key]. In North America and Europe these vast tracts of land are known as the boreal forest and in Asia as the taiga – a Russian word meaning a dark and mysterious woodland. The southern boundary is marked by the blending of the pine trees and the deciduous tree varieties, although much land has now been cleared for agricultural purposes. In the northern extreme the trees are hemmed in by the vast open wastes of the frozen tundra. This wide area of colonization contains very few tree species, but scattered at frequent intervals throughout the forest are glacier-scoured lakes and slow-flowing rivers that provide a varied habitat for many animals.

Food from the pine trees
The beaver (*Castor fiber*) [2], widespread throughout North America and now recovering its numbers in Europe and Asia after being almost hunted to extinction, creates its own particular niche in the habitat. Both the

male and female of the species, which pair for life, engage in felling trees with their sharp front incisors and damming a stream or river to form a pond with a high water level. In the middle of the pond they build a lodge (house) where they live and raise a family of up to 12 individuals. A supply of young twigs is stored under water as a food supply for use during the long, harsh winter.

The main supply of food for most of the resident animals is furnished by the coniferous trees, which provide an ample but unvaried diet of seeds, buds, bark and young needles. The yield of these coniferous products directly controls the numbers of animals that the forest supports. Many have become specialist feeders, particularly the birds and smaller mammals. The pine grosbeak (*Pinicola enucleator*) feeds on the young buds, seeds and needles, while the crossbills (*Loxia* spp) specialize in taking the buds, seeds and needles of pine and spruce. The habit of the nutcracker (*Nucifraga caryocatactes*) and the Siberian jay (*Perisoreus infaustus*) is to open the cones with their strong bills. The capercaillie

(*Tetrao urogallus*) and hazel grouse collect seeds that have dropped to the ground. The seed-eating mammals include the northern redbacked vole (*Clethrionomys rutilus*) and the wood lemming (*Myopus schisticolor*).

Many animals store up food for the winter. The red squirrel (*Sciurus vulgaris*) collects and hides cedar seeds and nuts in the hollows of trees and, for a special delicacy in the long, cold, dark months, impales mushrooms on spiked ends of branches.

From winter into summer
The brown bear (*Ursus arctos*) and the badger (*Meles meles*) hibernate and let the cold months pass them by as they relax in a deep sleep, using up a reserve of fat which they accumulate in the autumn when food is more plentiful. The spring and summer bring relief to the beleaguered forest and, as the snows melt, young tender shoots provide sustenance for the herbivores; and multitudes of insects, a plague to man but a boon to insectivorous birds and mammals, swarm through the forest. The moose (*Alces alces*) [4] (known as the elk in Europe and Asia),

1 **The northern edge of the pine forest** abuts on to the stark tundra terrain. The winters are long and severe, but in the forest the snow is loosely packed and the soil surface remains unfrozen for most of the year. This allows animals to burrow for the winter in comfort,

and ensures that insects are available for the insectivores. The evergreen conifers guarantee a continuous food supply of seeds, twigs and buds throughout the year and most of the animals spend their time in only a restricted area of the forest, rather than migrating.

Carpeting the forest floor is a layer of fallen pine needles which provides a habitat for carpenter ants, carabid beetles and spiders. Other spiders construct their webs under loose bark on tree trunks or decaying timber. Rotten wood is a home for insects such as ants.

1 Siberian jay
  Perisoreus infaustus
2 Pine marten
  Martes martes
3 Great grey owl
  Strix nebulosa
4 Lynx
  Felis lynx
5 Pine grosbeak
  Pinicola enucleator
6 Capercaillie
  Tetrao urogallus
7 Eurasian flying squirrel
  Pteromys volans
8 Northern bat
  Eptesicus nilssoni

9 Black woodpecker
  Dryocopus martius
10 Wood-boring beetle
11 Willow tit
  Parus montanus
12 Siberian ruby throat
  Luscinia calliope
13 Elk *Alces alces*
14 Wolverine *Gulo gulo*
15 Boar *Sus scrofa*
16 Willow grouse
  Lagopus lagopus
17 Brown bear *Ursus arctos*
18 Reindeer
  Rangifer tarandus
19 Wolves *Canis lupus*

20 Eurasian ground squirrel
  Eutamias sibiricus
21 Pupa of longhorn beetle
22 Pygmy shrew
  Microsorex hoyi
23 Siberian weasel
  Mustela sibirica
24 Ichneumon wasp
  Rhyssa sp
25 Siberian tit
  Parus cinctus
26 Crossbill *Loxia* sp
27 Pine weevil *Hylobius* sp
28 Nutcracker
  Nucifraga caryocatactes
29 Snowy owl

  Nyctea scandiaca
30 Blue hare *Lepus timidus*
31 Raven *Corvus corax*
32 Greenshank
  Tringa nebularia
33 Pika
  Ochotona hyperborea
34 Brambling
  Fringilla montifringilla
35 Stoat *Mustela erminea*
36 Arctic fox
  Alopex lagopus
37 Masked shrew
  Sorex cinereus
38 Root voles
  Microtus oeconomos

which is the largest member of the deer family, can then stop stripping the bark of trees or grubbing for mosses and lichens in the snow and resume browsing on small shrubs and trees or aquatic plants.

The moose may be the tallest animal in the forest but the heaviest in North America is a woodland race of bison (*Bison bison*); and in Eurasia it is the European woodland wisent (*Bison bonasus*). Once widespread throughout both continents, these animals have been hunted to near extinction and now survive only in protected herds of a few hundred individuals. Another animal rescued from extinction is the sable of Eurasia (*Martes zibellina*), which has for many years been hunted for its valuable fur. Active conservation management has succeeded in restoring the sable population to the numbers living more than 200 years ago.

Common predators
The sable is carnivorous and drives its cousins the ermine (*Mustela erminea*) and the Siberian weasel (*Mustela sibirica*) from its home range. But these two animals are numerous in other regions along with other members of the mustelid family, such as the common weasel (*Mustela nivalis*). Another predator which is widespread throughout all of the northern forests of the world is the lynx (*Felis [Lynx] lynx*), a member of the cat family. A highly efficient predator it specializes in hunting hares and its numbers fluctuate according to the available food supply.

The brown bear is perhaps the largest predator of the forest, but the most ferocious is the wolverine (*Gulo gulo*), which is amazingly strong for its size and takes prey as large as a reindeer or caribou. The wolf (*Canis lupus*) lives on the edge of the forest but it makes forays into the forest, hunting in well-disciplined packs to bring down animals such as reindeer and moose.

Smaller mammals, birds and fish fall prey to the marauders of the sky, such as the golden eagle (*Aquila chrysaëtos*), the goshawk (*Accipiter gentilis*), the snowy owl (*Nyctea scandiaca*) and in North America the bald-headed eagle (*Heliaeetus leucocephalus*), the esteemed national symbol of the United States.

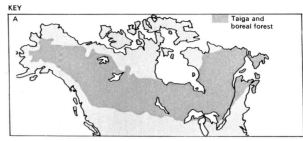

Taiga and boreal forest

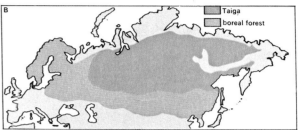

Taiga
boreal forest

Pine forests stretch across the Northern Hemisphere of North America [A], Europe and Asia [B]. To the south are mixed deciduous forests, to the north, sparse tundra vegetation.

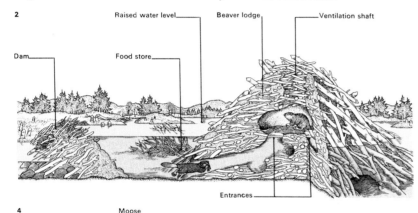

2   Raised water level   Beaver lodge   Ventilation shaft

Dam   Food store

Entrances

2 Beavers dam streams, creating ponds and flood pastures which change the face of the landscape. In the centre of the pond, both male and female of the species help in the construction of the lodge (house). Lodges vary in shape and size and usually consist of a "dining room", "living room" and "bedroom". The beavers, who mate for life, establish a family that is made up of the two most recent litters.

3 The beaver fells young trees with its powerful incisors and takes them to the dam or lodge in the middle of the pond.

3 Beaver
Castor fiber

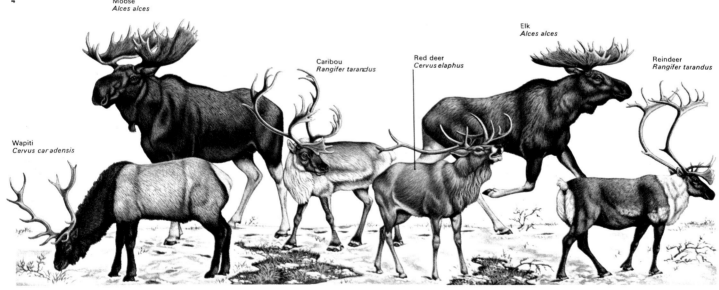

4   Moose
Alces alces

Elk
Alces alces

Caribou
Rangifer tarandus

Red deer
Cervus elaphus

Reindeer
Rangifer tarandus

Wapiti
Cervus canadensis

4 The tallest animal living in the forests of North America is the moose. It is identical with the elk that roams the European and Asian woodlands. The wapiti, which is confusingly known as the elk in North America, is second only to the moose in size. It is the American equivalent of the red deer, which is widespread in Eurasia, feeding on mountain pastures in summer and going down to the forest valleys in winter. Caribou are found in large herds throughout the tundra region of North America. They feed on low-growing vegetation, especially lichens or reindeer moss (*Cladonia* sp). In Europe and Asia caribou are called reindeer. There they lead a semi-domesticated life in herds under the care of nomadic herdsmen.

5

American mink
Mustela vison

5 The American mink is found in many parts of North America. It has webbed feet and thick fur and it is equally at home in water and on land. Mink are mainly active at night and feed on a wide variety of small animals. They are farmed for their valuable pelts, but many have escaped and returned to the wild. The mink is a member of the family Mustelidae, which includes weasels, stoats, polecats, badgers, wolverines, skunks and martens.

Northern temperate woodlands

Broadleafed forests, consisting of a mixture of deciduous tree species, are characteristic of much of the temperate zone of the Northern Hemisphere [Key]. In the past, forests of this sort were probably greater in extent, but they have been reduced in area by climatic changes and man's activities in creating more agricultural land.

Remnants of the former grandeur of the temperate woodlands can still be seen in China and North America. In forests there several plant species survive that are otherwise known only as fossils or cultivated species in Europe and western Asia. They include the dawn redwood (*Metasequoia glyptostroboides*), a deciduous conifer, and the maidenhair tree (*Ginkgo biloba*), both from China, and the tulip tree (*Liriodendron tulipifera*) from North America.

Man's activities in the forests have been almost entirely destructive. Over most of western Europe, for example, little of the original woodland remains. Many of the trees that do exist there have been planted or managed by man, who has subdued forests to his needs, often destroying them completely to make farmland. Whenever once forested land is abandoned in western Europe it is quickly covered by scrub. If this were left a succession of trees would gradually become dominant until, after two or three centuries, a completely new forest would arise.

The woodland community

The trees in temperate forests are generally smaller than those of tropical forests, but all the major plant groups are represented among the many species found in the northern woodlands. The trees themselves are dominant but there are other woody species, such as the honeysuckle and ivy, supported by the trees; and where there is enough light, there may be an undergrowth of hazel and hawthorn or other smaller trees. Below this is a ground layer of herbaceous plants that must complete their life cycle in the spring months due to the lack of light once the trees and underwood are in full leaf. When light and warmth increase in the early part of the year these plants, including snowdrops, primroses and bluebells, flower in rapid succession.

Most trees also flower in springtime, bearing catkins whose pollen is spread by the wind and would thus be impeded by leaves. The relatively few exceptions, particularly some of the more southerly woodland species such as limes and chestnuts, are insect-pollinated and flower later. Many kinds of lower plants, particularly the ferns and mosses, like the shade and dampness of the woodlands and are often abundant in forests. Fungi come into their own as the trees die for their function is largely that of recycling agents; they break down the woody tissues and return them to the soil.

Animals of the woodlands

Animals in temperate woodlands are mostly small. The largest European animals, the aurochs (*Bos primigenius*), wild oxen, became extinct in the seventeenth century and only a small number of European bison (*Bison bonasus*) survive in the eastern forests of the continent. In North America the wood bison (*Bison bison*) has been reduced in numbers by man and deer are now the largest creatures of most temperate woodlands. The

CONNECTIONS

See also
42 Woody flowering plants
64 Trees, shrubs and climbers
194 The basis of ecology
210 Woodlands of Australasia
192 Regions of the earth: zoogeography
106 The classification of insects
140 The classification of birds

In other volumes
192 The Physical Earth

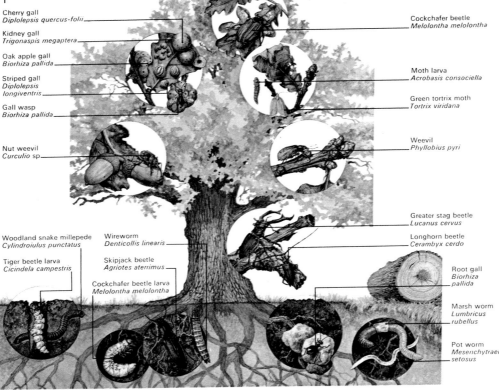

1 An oak tree supports a wealth of small animal life. In the spring swarms of caterpillars, bugs and beetles feed on the new leaves but do little lasting harm because the oak offsets the loss by putting out a late growth known as "lammas leaves". For some beetles the oak offers safety to their young. Larvae of the weevil *Phyllobius* live in twigs, those of the stag beetle in dead wood, and the nut weevil in acorns. Other species of larvae find safety in galls, the abnormal growths on a tree, which form where the minute grubs feed. In the soil beneath the tree a variety of worms break down fallen leaves and litter. Millipedes, worms and larvae feed on soft, rotting plants and roots. The voracious larva of the tiger beetle digs a deep hole in which to hide and catch unwary insects, while the colourful adult beetle hunts small forest invertebrates.

Cherry gall
Diplolepis quercus-folii

Kidney gall
Trigonaspis megaptera

Oak apple gall
Biorhiza pallida

Striped gall
Diplolepis longiventris

Gall wasp
Biorhiza pallida

Nut weevil
Curculio sp

Woodland snake millepede
Cylindroiulus punctatus

Tiger beetle larva
Cicindela campestris

Wireworm
Denticollis linearis

Skipjack beetle
Agriotes aterrimus

Cockchafer beetle larva
Melolontha melolontha

Cockchafer beetle
Melolontha melolontha

Moth larva
Acrobasis consociella

Green tortrix moth
Tortrix viridana

Weevil
Phyllobius pyri

Greater stag beetle
Lucanus cervus

Longhorn beetle
Cerambyx cerdo

Root gall
Biorhiza pallida

Marsh worm
Lumbricus rubellus

Pot worm
Mesenchytraeus setosus

2 The forest floor, with its fallen leaves and debris, supports a different microcommunity from that of growing trees. Hosts of small, often microscopic creatures are constantly active in breaking down the dead tissues of the forest litter and returning the components to the soil for use by the plants. Tiny insects bore into the tissues, preceding the many fungi, bacteria and other invaders that continue the process until the breakdown is completed. These organisms can be extremely specialized. In most cases they require a particular species of plant or tree, which will be tackled only when it is in a particular state of dryness or humidity. Many tiny creatures are the prey of hunters such as beetles and the slimy salamanders; the bigger kinds may be taken by birds and forest mammals. Foxes and badgers, for instance, will often eat worms and grubs when hungry.

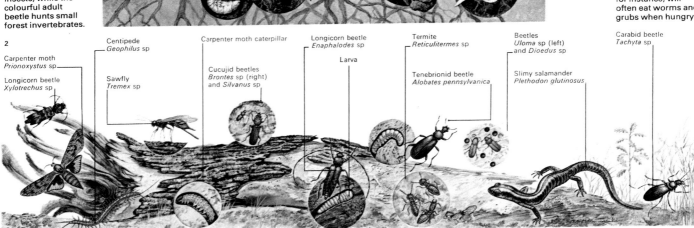

Carpenter moth
Prionoxystus sp

Longicorn beetle
Xylotrechus sp

Centipede
Geophilus sp

Sawfly
Tremex sp

Carpenter moth caterpillar

Cucujid beetles
Brontes sp (right) and *Silvanus* sp

Longicorn beetle
Enaphalodes sp

Larva

Termite
Reticulitermes sp

Tenebrionid beetle
Alobates pennsylvanica

Beetles
Uloma sp (left) and *Dioedus* sp

Slimy salamander
Plethodon glutinosus

Carabid beetle
Tachyta sp

biggest of these, the moose (*Alces alces*), is found in the more northern forest areas in Europe although it is more common in North America, where it is known as the elk.

The paucity of large animals in the northern forests does not mean that the forests lack animal life, but rather that its richness depends on specialization, which is better achieved by small creatures. The common oak (*Quercus robur*) is said to support more than 300 animal species [1]. Some of these, such as squirrels, depend partly on other plants. But many small animals are tied both to one kind of tree as well as to a particular part of it – a leaf, twig or root – for their specialized existence. At all levels, from the canopy to the ground, there is a web of interdependent species [3].

The yearly food cycle

With the springtime growth of the plants there is a resurgence among the animals. Many small creatures emerge from hibernation to feed on the new leaves and to become themselves the food of numerous predators. These predators include invertebrates, small birds and various insectivorous mammals.

Later in the year the first flush of animal and plant life disappears, but at no time are the woods empty of animal life. Even in winter grubs continue to bore their tunnels through the trunks of the trees. When a tree dies it still supports a host of organisms, such as fungi, ants and beetles, whose task is to assist in the speedy breakdown of the wood so that the soil is not impoverished by the loss of a tree. There are always small creatures, especially worms, at work in leaf litter on the forest floor, converting the autumn leaf fall into humus in the soil.

The trees influence life far beyond their own boundaries. Their need for water draws it from as far as their roots will reach. It is taken through the woody tissues and finally much of it is lost to the atmosphere via the leaves. A single large tree will, during spring and summer, pass hundreds of litres of water to the atmosphere in a day. The forest thus bypasses the normal slow route of water through the soil to watercourses and then to the sea, from where it is evaporated to form clouds from which rain falls again.

Forest is the natural ground cover wherever there is sufficient moisture in the temperate regions. In North America, only remnants are left of the great temperate deciduous forests that once covered much of the eastern half of the continent.

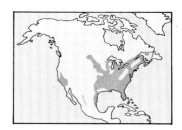

The temperate mixed woodlands of Europe extend from the British Isles across central Europe into the USSR. Beyond the Tibetan mountains are further areas of deciduous forest in eastern China. Oak, ash, beech and chestnut are typical trees of Europe's forests. Vast areas are being cleared for urban and agricultural development and in places the natural growth is being replaced with faster-growing conifers.

3

25 Nightingale *Luscinia* sp
26 Ash *Fraxinus* sp
27 Tree creeper *Certhia* sp
28 Woodpecker *Dendrocopos* sp
29 Tawny owl *Strix* sp
30 Grey squirrel *Sciurus* sp
31 Nut weevil *Curculio* sp
32 Galls
33 Purple emperor *Apatura* sp
34 Green woodpecker *Picus* sp
35 Elm *Ulmus* sp
36 Humble bee *Bombus* sp
37 Great tit *Parus* sp
38 Roe deer *Capreolus* sp
39 Badger *Meles* sp
40 Hart's tongue fern *Phyllitis* sp
41 Fly agaric *Amanita* sp
42 Wasp beetle *Clytus* sp
43 Primrose *Primula* sp
44 Ground beetle *Carabus* sp

1 Woodcock *Scolopax* sp
2 Fox *Vulpes* sp
3 Wood sorrel *Oxalis* sp
4 Horn of plenty *Craterellus* sp
5 Dormouse *Muscardinus* sp
6 Woodmouse *Apodemus* sp
7 Sparrowhawk *Accipiter* sp
8 Wood anemone *Anemone* sp
9 Bluebell *Endymion* sp
10 Violet *Viola* sp
11 Slow-worm *Anguis* sp
12 Oak *Quercus* sp
13 Honeysuckle *Lonicera* sp
14 Wood ant *Formica* sp
15 Wood warbler *Phylloscopus* sp
16 Birch *Betula* sp
17 Butterfly *Pararge* sp

18 Parasite wasp *Ichneumon* sp
19 Hornet *Vespa* sp
20 Bush cricket *Tettigonia* sp
21 Glow-worm *Lampyris* sp
22 Earwig *Forficula* sp
23 Red underwing moth *Catocala* sp

24 Hover fly *Myiatropa* sp

3 The trees of an English oak wood provide nourishment and shelter for many species of wildlife. Some species are primarily dependent on the trees for their food, while others are predators, finding their food among the herbivores. In its natural state, a broad-leaved woodland usually contains mixed plant species. If the soil is acid, birches are often present; where it is alkaline, ash trees and ferns can often be found. Elm usually grows in hedgerows or at the edge of a wood for it spreads by suckering, not seeds. Among the smaller plants the woody honeysuckle is supported by other trees. The spring-flowering herbs include primroses, violets, wood anemones, wood sorrel and bluebells. Two common large fungi present are horn of plenty and fly agaric. Of the woodland mammals only the dormouse and woodmouse are confined to the forest. Foxes, badgers, deer and squirrels use it as a source of food and shelter. The grey squirrel, an introduced species from America, has largely replaced the native red squirrel in Britain, but it is not present in the rest of Europe. Woodland birds include the ground-dwelling woodcock, rarely seen because of its camouflaged colouring and shy ways. The great tit is a resident, while the wood warbler and nightingale are migrants, taking advantage of the summer wealth of insects. The tree creeper searches the bark for small insects but the woodpeckers feed on grubs extracted with their long bills. Birds of prey are the day-hunting sparrowhawk and the nocturnal tawny owl, which feed on small mammals and other birds. Many of the insects have short adult lives and can be seen for only a few weeks in the breeding season.

Woodlands of Australasia

Despite its large area, only a small proportion of the Australian land mass is covered with true forest. Temperate forest covers the southern portion of the Great Divide, the mountain ridge separating the eastern coastal plain from the interior, extending to the south-eastern corner. In contrast, the forest of the northern coastal areas is typically tropical. Of the large islands that form the remainder of the Australasian region, New Guinea is clothed with tropical rain forest and New Zealand and Tasmania largely with temperate trees, although the last two have suffered extensive clearance by settlers.

The effects of isolation
The animals and plants of the region have close equivalents in other parts of the world but relatively few representatives. This is because of the long period of isolation from other major continents. The less advanced forms of life, including insects, reptiles and amphibians, are unique to the Australasian forests. Among them are many primitive or "relict" species – the hallmarks of millennia of isolation and of island living – including primitive New Zealand frogs (*Leiopelma* spp) which live near mountain tops, and two families of egg-laying mammal, the platypus (*Ornithorhynchus anatinus*) of temperate Australian streams and the spiny anteaters of eastern Australia and New Guinea.

Temperate and tropical trees
The trees of the temperate forests comprise many different genera unique to the world's southern continents. They include many species of aromatic gum (*Eucalyptus* spp) and southern beeches (*Nothofagus* spp). In the damper areas, giant tree ferns (*Dicksonia* spp) abound [1, 2].

Australasia's tropical forests, very similar to those of the Oriental region, include large-leaved tropical trees such as stilt-rooted mangroves (*Sonneratia* spp) and trees with edible fruits – coconut palms (*Cocos* spp), bananas (*Musa* spp) and breadfruit (*Artocarpus* spp). In the humid forest of New Guinea there are some peculiar "living fossils" found only in the southern Pacific region. They include the winter's barks of the family Winteraceae, the spiky monkey puzzles (*Araucaria* spp), and members of the primitive podocarp family, Podocarpaceae, of which the "plum yew" (*Podocarpus* spp) bears succulent "cones".

Like the trees among which they live, many of the mammals of the Australasian forests are unique and almost all (apart from bats and rodents) are marsupials. The canopy, inhabited by primates in other forests, is here the home of tree kangaroos (*Dendrolagus* spp) and the many species of arboreal phalangers, among them the recently rediscovered Leadbeater's possum [5] and the koala bear [4]. The single koala species is confined to the eucalyptus forests of eastern Australia. There, koalas are fully protected from the depredations of man, who once slaughtered them for their fur.

The predators within the forest are the varied species of native "cat", but no longer are the forests of the southeast roamed by the pouched "wolf" or its attendant "devil". The forest and scrub of Tasmania are now the only homes of the Tasmanian devil (*Sarcophilus harrisii*), which feeds on a wide variety of animal food, including carrion. When the pouched wolf or thylacine (*Thylacinus*

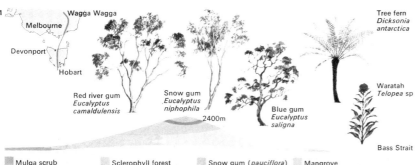

Red river gum *Eucalyptus camaldulensis* — Snow gum *Eucalyptus niphophila* — 2400m — Blue gum *Eucalyptus saligna* — Tree fern *Dicksonia antarctica* — Waratah *Telopea* sp — Bass Strait — Wagga Wagga — Melbourne — Devonport — Hobart

Black beech *Nothofagus* sp — Snow gum *Eucalyptus pauciflora* — 1600m — Mountain ash *Eucalyptus regnans*

Mulga scrub — Red river gum — Sclerophyll forest — Snow gum (*niphophila*) — Snow gum (*pauciflora*) — Mixed forest — Mangrove

Mountain ash — Snow gum (*pauciflora*) — Black beech — Beech forest

1 The effects of local climatic differences can be seen in the trees found along a line transect of a southern portion of the Great Divide and Tasmania. The westward-facing mountain slopes are much drier than those facing east and on them grow the dry, "hard-leaved" or sclerophyll forests of eucalyptus. Only in the moist areas can the species of tree ferns survive. The prevailing winds often warp the growth of beech and snow gum.

2 Tree ferns and eucalyptus abound at the edge of a typical "wet" forest area of New South Wales. In the depths of this type of forest the lyrebird builds its domed nest of sticks and moss on a tree stump or ledge.

3 The corroboree frog is at home in cold, damp moss at high altitudes in the Australian alps. Any accumulation of water in the moss is sufficient for it to lay its eggs in and for tadpole development.

3 Corroboree frog *Pseudophryne corroboree*

4 The koala (*Phascolarctos cinereus*) was once slaughtered for its fur, but the animal is now protected. Confined to the southeastern states of Australia, koalas feed on about 12 species of eucalyptus or gum tree. They rarely come to the ground and spend most of their lives in the tree tops eating leaves and young bark. They are mostly solitary, but an adult male usually has a small harem which he guards jealously. The maximum body length is about 85cm (33in).

5 Leadbeater's possum *Gymnobelideus leadbeateri*

5 Leadbeater's possum was at one time thought to be extinct as no specimens had been seen since 1909. However in 1961 the species was rediscovered in a dense mountain ash (*Eucalyptus regnans*) forest in the highlands near Marysville, Victoria. It is completely arboreal and its paws are adapted to its life-style, being very wide at the tips with strong, short claws. This possum displays great agility in pursuit of its insect prey which it catches by night.

cynocephalus) was more common, accompanying Tasmanian devils fed on the remains of the wolves' kill. The last specimen of thylacine to be seen was shot in 1930.

Fliers of the forests

New Zealand forests contain only two native species of land mammals, both of them bats, the wattled bat (*Chalinolobus tuberculatus*) and the New Zealand short-tailed bat (*Mystacina tuberculata*). Unusually, the chief herbivores are birds. The most spectacular of these, the moas, were hunted to extinction by the original migrants, the predecessors of the Maoris, for food and feathers. Other peculiar birds of New Zealand still living are the three species of flightless kiwis (*Apteryx* spp) and a rare ground parrot, the kakapo (*Strigops hapbroptilus*).

Some of the most strikingly coloured and vocal inhabitants of the Australasian forests are the many varieties of birds, a great number of which are found nowhere else in the world. The most magnificent of these are inhabitants of the tropical forests of New Guinea and include the birds of paradise

(family Paradisaeidae) the bowerbirds (family Ptilonorhynchidae) [6], and several unique species of pigeons, among them the slate blue and crested crowned pigeons (*Goura* spp), the largest of which is almost the size of a turkey – about 1m (39in) long.

Parrots abound among the branches of both tropical and temperate trees. One of the most familiar is the sulphur-crested cockatoo (*Cacatua galerita*), a popular cage bird the world over. Harsh and strident parrot calls resound through the forest canopy while in the densely vegetated undergrowth of the temperate southeast the male lyrebird [8] indulges in his ventriloquist song, throwing his voice in a remarkable assortment of calls borrowed from others.

From crown to floor, the forests are occupied by nectar-eating honeyeaters, insect-eating thornbills, fantails and Australian robins. Although some birds, such as the rock warbler of the Hawkesbury Sandstone area of Sydney, have a restricted range, others, such as the golden whistler [7], are widespread and show many local variations of bill shape or plumage.

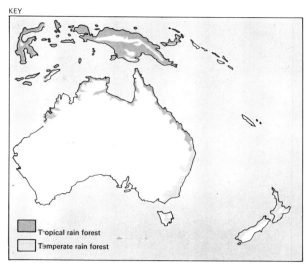

Tropical rain forest

Temperate rain forest

The forests of Australasia are of two distinct types: temperate and tropical, each supporting a unique selection of plants and animals. The temperate forest of southeast Australia is the one on which man has had the most impact, but even here high rainfall and fertile soil may produce "jungle".

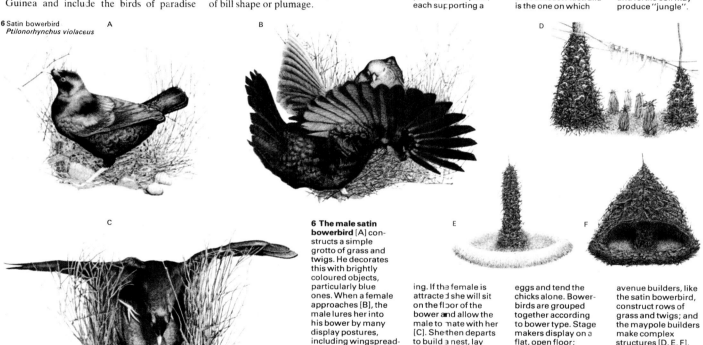

6 Satin bowerbird
Ptilonorhynchus violaceus

A   B   D   C   E   F

6 The male satin bowerbird [A] constructs a simple grotto of grass and twigs. He decorates this with brightly coloured objects, particularly blue ones. When a female approaches [B], the male lures her into his bower by many display postures, including wingspreading. If the female is attracted she will sit on the floor of the bower and allow the male to mate with her [C]. She then departs to build a nest, lay eggs and tend the chicks alone. Bowerbirds are grouped together according to bower type. Stage makers display on a flat, open floor; avenue builders, like the satin bowerbird, construct rows of grass and twigs; and the maypole builders make complex structures [D, E, F].

7

Louisiade Archipelago

Tasmania

Tanimbar Islands

7 The golden whistler is a bird found in many different forms on the islands round the Australian coast. Bill shape, as shown here, varies considerably between island races, and reflects the way in which each has become adapted to feed in different kinds of forest.

Golden whistler
Pachycephala pectoralis

8 Superb lyrebird
Menura superba

8 The secretive and shy lyrebird is rarely seen by man because it hides in densely vegetated gullies in the most inaccessible parts of the temperate forest. The female lyrebird is remarkably fastidious, for she carries all the droppings from the nest and puts them in a stream.

9 Tawny frogmouth
Podargus strigoides

9 The tawny frogmouth swoops by night on any insect, reptile or small mammal that it can find, mangling it with the bill for easy swallowing. This common bird species is 48cm (19in) long.

African rain forest

A tropical forest is a deceptive place. No other form of vegetation is so complex or supports such a teeming variety of animals. Yet a visit there may reveal little more than hosts of butterflies and an occasional bird. Since many of the animals are nocturnal, they stay hidden by day in their leafy sanctuary. Some remain in burrows in the ground litter and others seldom, or never, venture down from their arboreal homes. They are betrayed only by forest sounds; a distant crashing of monkeys through the foliage, for example, or the resounding shrieks of hornbills as they feed in the canopy.

The forest vegetation

Rain forests such as those of Africa are found in tropical regions where rain falls in heavy storms throughout the year. In such places temperatures average 27°C (81°F), varying little between night and day, and the air is moist and humid.

The vegetation of the rain forest [3], forever pushing upwards towards the sun, forms a series of distinct layers. On top, scattered tall trees or "emergents", many with flanged buttresses formed from root and trunk for support, reach above the thick-spread canopy layer. In the middle smaller trees with long, oval crowns compete for narrow shafts of sunlight penetrating the foliage. Below, the smaller trees, palms and shrubs exist in a twilight microclimate and merge into a relatively sparse ground layer of shrubs and tree seedlings. Matting the tree trunks are creepers and woody vines (lianas). Wherever the sunlight reaches the ground, herbs, shrubs, creepers and tree seedlings are woven into a dense undergrowth.

Everywhere in the forest is an ever-growing, evergreen tangle of plant life. Species that grow on other plants, the epiphytic ferns, lichens and orchids, lodge in sunny branches in the canopy or shady nooks farther down. Some rely on litter accumulating round their roots for food and moisture while others absorb water through hanging roots. Strangler figs bring death to their host trees; germinating in the tree bark, they send down roots that eventually surround the trunk and cause it to rot away.

The African rain forests contain more than 7,000 species of evergreen and deciduous flowering plants and the virtual non-seasonal climate means that at any time of the year there is a supply of leaves, flowers and fruits for the animals.

Life in the trees

The way in which the forest is formed into well-defined layers means that animals can live and feed in different habitats at different heights from the ground. Herbivores such as the colobus monkeys stay fairly rigidly within a particular layer, while predators, such as genets, are forced to make journeys from the ground to the canopy in search of bird, small mammal and insect prey.

The whole forest is a finely balanced ecosystem where the animals feed not only in different layers but also at different times. Monkeys and birds tend to be diurnal and most forest mammals nocturnal. At night the tiny bush babies such as the dwarf galago (*Galagoides demidovii*) emerge from their nests to search for flying insects [7] and fruit.

In the dense canopy the nocturnal and omnivorous Beecroft's tree hyrax

1 Okapi 1·6m
Okapia johnstoni

Water chevrotain 30cm
Hyemoschus aquaticus

Chequered elephant shrew 30cm
Rhynchocyon sp

Forest genet 40-50cm
Genetta tigrina

Banded duiker 50cm
Cephalophus zebra

Four striped squirrel 20cm
Funisciurus sp

Bongo 1·3m
Taurotragus eurycerus

1 Stripes and spots work equally well as disruptive camouflage for many shy forest mammals. Blending in with patches of sunlight and dark vegetation, nocturnal animals such as the banded duiker and genet remain motionless and hidden by day and emerge to feed at night. The okapi, a relative of the giraffe, is a daytime feeder. The horizontal stripes imitate the sun's rays as they penetrate the forest canopy.

The hoofed mammals are smaller than their relatives in open habitats. A compact body and small or backward-pointing horns facilitate movement through tangled bushes. Bongo and duiker pass under low bushes with odd, crouching runs. Dimensions given for the chevrotain, okapi, duiker and bongo are heights at the shoulder; for the genet, shrew and squirrel, they are head and body length.

2 Rarely seen but continuously heard, forest birds tend to leap among the trees rather than fly around the forest. High in the canopy the Gold Coast turaco and heavily built yellow-casqued hornbill crash noisily from branch to branch in search of their staple diet of fruit, while the grey parrot is often seen in small flocks flying over the tops of the trees. Many of the smaller birds feed on insects and other small invertebrates. The blue fairy flycatcher finds them in the trees, while the forest robin and Angolan pitta, aptly called the "jewel-thrush" because of its brilliant coloration, grub among the humus of the forest floor. The pitta is a particularly hard bird to detect as it has the ability to throw its voice like a ventriloquist. The Congo peafowl, a ground-dwelling bird related to the Asiatic peacock, was not discovered by Europeans until 1936.

2

Gold Coast turaco
Tauraco persa

Congo peafowl
Afropavo congensis

Blue fairy flycatcher
Erannornis longicauda

Yellow-casqued hornbill
Ceratogymna elata

Grey parrot
Psittacus erithacus

Angola pitta
Pitta angolensis

Forest robin
Stiphrornis erythrothorax

3

Dwarf galago
Red colobus

100ft
30m

Diana monkey
Python
Great blue turaco

50ft
15m

Chimpanzee
Grey-cheeked mangabey

25ft
7·5m

Gaboon viper
African civet
Tullberg's rat
Okapi

3 A variety of tree heights provides different levels at which forest animals may live. The tallest trees emerge through the canopy and may reach to 60m (200ft) from the ground. Below, a canopy of interlocking tree crowns 15 to 30m (50–100ft) high forms the roof of the forest. In the middle layer (7.5–15m [25–50ft]) trees with long, narrow crowns are less intermeshed. Trees up to 7.5m (25ft) high form a lower layer that merges with shrubs and herbs on the forest floor.

(*Dendrohyrax dorsalis*) screams loudly as it scurries up and down the tree trunks. The red-backed flying squirrel (*Anomalurus erythronotus*) [5] glides between trees on bat-like wings of skin and the African linsang (*Poiana richardsoni*) emerges fro ts nest in thick, tangled vines to hunt for insects and small vertebrates.

On the ground, night-browsing mammals such as the bushbuck (*Tragelaphus scriptus*) start to forage, while in the forest litter the African civet (*Civettictis civetta*) hunts for small rats, mice and shrews.

At dawn the monkeys begin to feed. Distinguished by their long black-and-white coats, the colobus (*Colobus polykomos*) sit in groups of up to 20 animals, eating leaves and grooming each other. All the permanently arboreal monkeys are lightweight animals able to reach the leaves and fruit at the very end of slender branches.

In the forest canopy the sound of crashing branches may be caused by black-cheeked, white-nosed (*Cercopithecus ascanius*) or diana monkeys (*C. diana*) leaping between trees, or it may be noisy hornbills, turacos [2]

and wood-hoopoes, jumping from branch to branch in search of fruit.

In the middle layer of trees, troops of grey-cheeked mangabeys (*Cercocebus albigena*) feed on fruit. The ranges of the monkeys may overlap but each species uses a different kind of leaf or fruit as its main diet. The monkeys share the trees with arboreal snakes, chameleons and frogs as well as invertebrates living in the soil around the epiphytic plants that grow on the trees.

The ground-dwellers

Not all forest primates are arboreal. The vegetarian gorillas (*Gorilla gorilla*), although they can and do climb trees, prefer to pick leaves from ground vegetation. These animals, together with the forest elephants, buffalo and the okapi (*Okapia johnstoni*), are among the few large terrestrial mammals of the rain forest. More numerous are the squirrels, mice, rats and tiny elephant shrews [1] that live in the ground vegetation. Beneath them a host of invertebrates carries on the job of breaking down the forest litter, thus releasing nutrients for the growing plants.

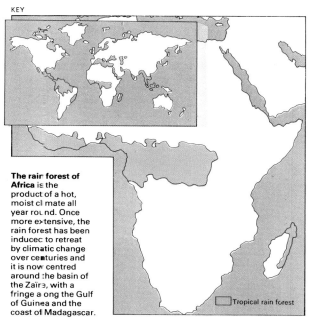

The rain forest of Africa is the product of a hot, moist climate all year round. Once more extensive, the rain forest has been induced to retreat by climatic change over centuries and it is now centred around the basin of the Zaïre, with a fringe along the Gulf of Guinea and the coast of Madagascar.

☐ Tropical rain forest

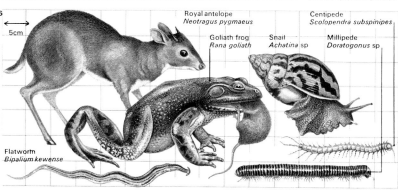

4 Along the banks of muddy rivers below the canopy the rain forest is an interwoven complex of different plant species. The crowns of small trees overlap and smooth, un-branched tree trunks are festooned with woody lianas and creepers. The undergrowth is a network of ferns, herbs and tree seedlings, while the forest floor is littered with rotting leaves, fruits and tree trunks, which provide nutrients for the growing plants. Very dense ground vegetation is usually characteristic of secondary forest. In virgin forest the scarcity of light at ground level confines undergrowth to scattered clumps of shade-tolerant shrubs.

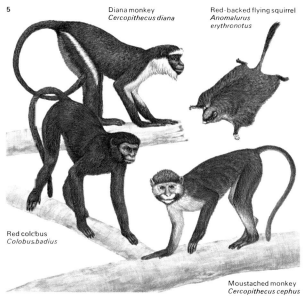

5

Diana monkey
Cercopithecus diana

Red-backed flying squirrel
Anomalurus erythronotus

Red colobus
Colobus badius

Moustached monkey
Cercopithecus cephus

Royal antelope
Neotragus pygmaeus

Centipede
Scolopendra subspinipes

Goliath frog
Rana goliath

Snail
Achatina sp

Millipede
Doratogonus sp

5cm

Flatworm
Bipalium kewense

5 The monkeys and squirrels are some of the most agile of the tree-top inhabitants, moving cautiously among the branches or taking spectacular leaps from tree to tree. The red-backed or scaly-tailed flying squirrel runs up tree trunks, steadied by pointed scales on the underside of its tail, and then glides between trees on wings of skin between elbows and feet. The diana and moustached monkeys use opposable thumbs to grip the branches and seem to crash their way haphazardly among the tree tops, unlike the red colobus which uses its hind legs to make precision jumps between branches.

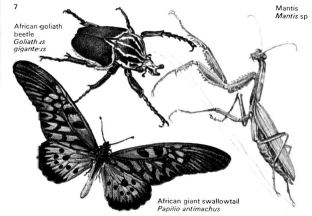

7

African goliath beetle
Goliathus giganteus

Mantis
Mantis sp

African giant swallowtail
Papilio antimachus

6 Extremes of size are common in tropical rain forests. A warm, moist climate has allowed some invertebrates and cold-blooded vertebrates to become giants, while life in dense undergrowth has led to the evolution of dwarf, herbivorous mammals.

The royal antelope stands only 30cm (12in) at the shoulder and is the smallest antelope in the world. In contrast, the goliath frog, of deep forest pools, can grow up to 80cm (32in) from nose to toe and weigh 15kg (7lb). It feeds on many small animals that include mice, rats and lizards. On the forest floor, feeding on dead plant material, live the giant millipedes that grow to 30cm (12in) in length. Also on the forest floor are the active predatory centipedes and the giant snails and flatworms.

7 Tropical forest insects are the biggest in the world. The goliath beetle may grow up to 10cm (4in) long and the unique wings of the poisonous swallowtail span 25cm (10in). Some predatory kinds of mantis are large enough to feed on flying lizards.

Forests of South-East Asia

Rain forests are the natural vegetation of major parts of South-East Asia and the Indian subcontinent [Key], but as a result of man's activities they have virtually disappeared in large areas, especially in India, Bangladesh and Indochina. Where they still exist, in climates of high constant temperatures and high rainfall, there grows the greatest diversity of plants found anywhere in the world. With animals ultimately dependent on plants for food, the forests have also provided for the development of one of the richest animal communities in the world.

Forest communities

Trees growing in the South-East Asian rain forests are taller than their related species in African and South American rain forests, with the "emergents" often reaching a height of more than 70m (225ft). These giant trees begin branching only at about 45m (147ft) above the ground, but produce wide, spreading crowns, equipped with thick, leathery leaves adapted in shape to allow rapid drainage of rain from their surface. Their most striking feature is a thick buttressed trunk: because the thin but nutritious soil supports only a shallow rooting system, the buttresses – often 11m (35ft) high – are needed to provide stability against buffeting winds and rain.

The animal community of the emergent zone is made up almost exclusively of birds, insects and bats. Because the climate lacks marked seasonal changes, trees can flower and fruit throughout the year, thus providing animals with a continuous supply of food. Hornbills (*Buceros* and *Aceros* spp) feed on the fruits of the "emergents" using their strong, curved bills to push through the foliage; butterflies, bees and wasps feed on the nectar of flowers, and fruit-eating bats, Megachiroptera, strip trees of their produce. Swifts (*Hemiprocne* and *Chaetura* spp) fly high above trees, feeding on insects, while eagles (*Ictinaetus* and *Spilornis* spp) and other birds of prey hover above, waiting to pounce on small vertebrates.

The forest canopy

Below the emergent zone is the forest canopy where the crowns of buttressed trees, fig trees and trees producing such exotic fruits as durians, mangosteens and rambutans merge to form a continuous layer of leafy vegetation about 45m (147ft) above the ground.

The canopy is a home and major food source for many animals. There, gibbons (*Hylobates* spp) and the Malayan and Sumatran siamangs (*Symphalangus syndactylus*) swing from branch to branch in search of fruit, leaves and flowers. Orang-utans, found only in Borneo and Sumatra, forage for fleshy fruits. The giant squirrel (*Ratufa indica*) leaps between trees in search of nuts, fruits and even birds' eggs. Minivets (*Pericrocotus* spp) flit between branches, feeding on insects, while the fruit-eating barbets (*Megalaima* spp) may be found nesting in holes in trees. Insects – butterflies, beetles, ants, wasps and bees – abound and provide food for insectivorous bats.

The middle zone of the rain forests is composed of trees adapted to shade and high humidity, 30–33m (98–110ft) high bearing deep, narrow crowns and trunks and branches covered in flowers. A tangle of ferns that live on trees, climbing lianas and vines and

1 South-East Asian forests contain perhaps the richest populations of animals and plants in the world. A single hectare of forest may contain 60 different species of trees and an island may support more than 150,000 species of animals. A few are shown in this picture, with a key on the opposite page.

Within forests most animals forage for food beside watercourses or lakes where vegetation is less dense and where light penetrates. Many animals in the forest are adapted for specialized feeding, for colonizing a particular niche or for moving about within the layers of the forest.

2 Ants form a bridge

Two leaves are drawn together

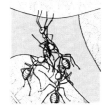

Leaf edges are held together before gluing

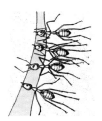

2 Weaving ants (*Oecophylla smaragdina*) are commonly found in all rain forests of South-East Asia. They build nests in trees and bushes by binding living leaves together with silken threads. Although the ants are unable to produce silk their larvae bear silk glands. Construction of a new nest commences when a team of worker ants forms a bridge between two leaves. While one set of workers draws the leaves together a second set commandeers the silk-secreting larvae (not shown) to spin a web of silk across the gap, thus glueing and sealing the leaves together.

parasitic "strangler" figs, wrapped and twisted around the branches and trunks of canopy trees, also add to the foliage.

Many of the animals of the middle layer are highly adapted for movement between trees. Several species of lizards (*Draco* spp), frogs (*Rhacophorus* spp), squirrels (*Aeromys* spp) and flying lemurs (*Cynocephalus* spp) bear extensible flaps of skin that allow them to glide large distances from tree to tree.

The binturong (*Arctictis binturong*) has claws and a prehensile tail for climbing, while the long tails of langurs (*Presbytis* spp) and proboscis monkeys (*Nasalis larvatus*) which live mainly in swamp regions are used only for balancing as they leap between the branches. The more terrestrial macaques (*Macaca* spp) possess much smaller tails. The calm, humid climate of the middle layer has also allowed flying insects to evolve enormous wings [3], which they use in their search for specific flowers and fruits.

The shrub layer
In primary rain forests the shrub layer is rather sparse and consists mainly of woody plants and young growing trees, which reach a height of about 5.5m (18ft). However, where light penetrates to the inner forest, or in secondary forests, this layer becomes much denser and bears numerous broad-leaved herbaceous plants. On the margins of rivers and lakes palm trees and mangroves thrive. Such areas are the home of the nocturnal tarsier (*Tarsius* spp) which leaps from branch to branch as it searches for lizards, spiders and insects. The leopard (*Panthera pardus*) sits on an overhanging bough awaiting its prey. Vipers (*Vipera* spp), pythons (*Python* spp) and tree-snakes (*Boiga* spp), which feed on small mammals, reptiles and birds, wrap themselves around the thick branches. Broadbills (*Calyptomena* spp), shamas (*Copsychus* spp) and flycatchers (*Terpsiphone* spp) fly swiftly through the foliage. But by far the most numerous animals are the insects – camouflaged stick and leaf insects, grasshoppers, crickets and cicadas, beetles, butterflies and moths and ants, bees and wasps – of which more than 100,000 species have been described. It is possible that a comparable number remain undiscovered.

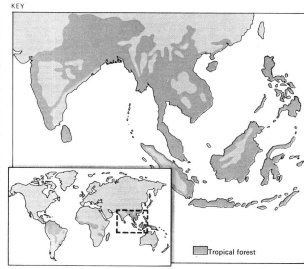

KEY

Tropical forest

Rain forests would grow in all the areas shown in blue on the map were it not for man. They are now found only in the Western Ghats of India, Sri Lanka, Burma, Thailand, Laos, Malaysia, the Philippines, Borneo, Sumatra and the Celebes.

3 The Atlas moth and the gliding frog are relatively common inhabitants of rain forests in South-East Asia. With a wing span of up to 30cm (12in) the Atlas is the largest Asian moth and has the greatest wing span of any lepidopteran. The gliding frog of Malaysia is equipped with webbed feet, which act as flat planing surfaces and allow it to glide more than 12–15m (40–50ft) between trees.

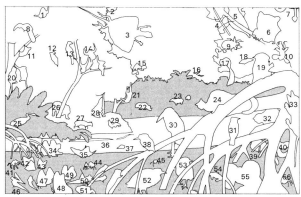

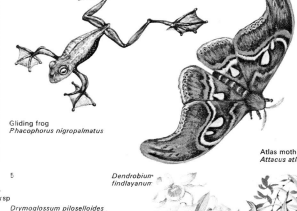

Gliding frog
Phacophorus nigropalmatus

Atlas moth
Attacus atlas

1, 14 Gibbons *Hylobates* spp
2, 11 Colugo *Cynocephalus* sp
3 Flying lizard *Draco volans*
4 Stick insect *Carausius* sp
5 Lesser-green broadbill *Calyptomena viridis*
6 Tarsier *Tarsius* sp
7 Green pit viper *Trimeresurus* sp
8, 13 Hornbills *Buceros* spp
9 Leaf insect *Pulchriphyllium* sp
10 Orchid mantis *Hymenopus coronatus*
12, 16 Flying foxes *Pteropus* spp
15 Brahminy kite *Haliastur indus*
17 Prevost's squirrel *Sciurus prevosti*
18 Tree shrew *Tupaia* sp
19 Yellow-crowned bulbul

Pycnonotus sp
20 Langurs *Presbytis* sp
21 Orang-utan *Pongo pygmaeus*
22 Sumatran rhinoceros *Didermoceros sumatrensis*
23 Malayan tapir *Tapirus indicus*
24 Banded linsang *Prionodon linsang*
25 Indian darter *Anhinga melanogaster*
26 Tiger *Panthera tigris*
27 Muntjac *Muntiacus* sp
28 Sun bear *Helarctos malayanus*
29 Jungle fowl *Gallus* sp
30 Adjutant stork *Leptoptilus dubius*
31 Black-naped blue monarch *Hypothymis azurea*
32 Fishing cat *Felis viverrina*
33 Moon rat *Echinosorex* sp
34 Crab-eating macaques

Macaca irus
35 *Graphium* butterfly
36 Crocodile *Crocodylus* sp
37 Small-clawed otter *Aonyx* sp
38 Indian three-toed kingfisher *Ceyx erithacus*
39 Binturong *Arctitis binturong*
40 Mangrove snake *Boiga dendrophila*
41, 45, 56 Soldier crabs *Dotilla mictyroides*
42 *Appias* sp
43 Butterfly *Prothoe* sp
44, 51, 54 Fiddler crab *Uca* sp
46, 50 Mudskippers *Periophthalamus* spp
47, 48 Swallowtails *Papilio* spp
49 *Atrophaneura* sp
52 White-breasted water hen *Amaurornis phoenicurus*
53 Blue-winged pitta *Pitta brachyura*
55 *Python reticulatus*

Dendrobium findlayanum

Drymoglossum piloselloides

Phalaenopsis heideperle

Vanda tricolor

Averrhoa bilimbi

Coelogyne massangeana

4 *Rafflesia arnoldi* [A] is a rare parasitic plant, bearing neither stem nor leaves, that grows on the branch of one species of vine in Malaysia. Measuring up to 1m (39in) in diameter, it is the world's biggest flower. [B] Germinating *Rafflesia* seeds produce strands of cells [1] that penetrate the water- and food-conducting channels [2] of the host plant to absorb water and nutrients. The bloom lives for less than a week.

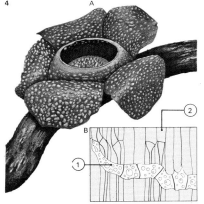

5 Mosses, lichens, ferns and orchids grow on branches of trees such as *Averrhoa*, where they receive sunlight and are nourished by humus in the bark. Accumulations of these epiphytes trap water and provide a suitable home for a varied host of invertebrate animals.

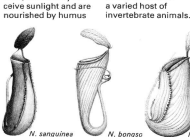

N. rajah

N. gracilis

N. sanguinea

N. bongso

N. ampullaria

6 The pitchers of the *Nepenthes* genus of plants are expanded leaves modified for catching insects. Ants, flies and beetles are lured to the pitchers by the scent of nectar and become trapped. Insects are probably eaten to obtain nitrogen, which is scarce in the heavily leached soil. There are about 60 different *Nepenthes* species which are found only on the Malayan archipelago.

New World tropics

Enveloping the basin of the great River Amazon in a shroud of dense vegetation is the greatest continuous mass of rain forest in the world. The abundance of water and the warm, humid climate support dense stands of trees that are festooned with creeping and climbing lianas. The forest floor receives little light and is the home of dark-coloured animals that prefer a cool, damp environment. High above is the forest canopy, a continuous collection of tree tops broken only by tall, scattered emergent trees. It is here that most of the light is filtered out and where an abundance of animal life resides.

The exuberant vegetation provides a wide range of habitats for a remarkable variety of animals, although they are not usually densely distributed and the visitor may see few of them. The lack of seasonal variation in temperature guarantees a supply of food for these animals throughout the year.

Fish, insects and birds
The number of different species of fish in the rivers and insects in this tropical rain forest is so vast that they have never been fully clas-sified. Small predators, especially the birds, grow fat at their expense, living in colonies along the banks of the rivers. The insects are remarkable not only for their numbers but also for their size and way of life. The Hercules beetle (*Dynastes hercules*) measures up to 15cm (6in) in length and is rivalled in size only by the bird-eating spiders (*Theraphosa* spp) which grow up to 25cm (10in) across, feeding on small animals of all kinds. Butterflies are remarkable for their beautiful colouring but some of the most fascinating insects are the ants. Leaf-cutter ants (*Atta* spp) dissect leaves and carry them off to build compost heaps where they grow and "farm" the fungi on which they feed.

It is the birds of the forest that represent the most diverse adaptations to the numerous habitats and which exploit the various food sources without undue competition. They include macaws, pigeons, tanagers and the delicate humming-birds, all of which feed almost entirely on plants and their products. The insectivores are specialized in their feeding habits: swifts take insects in flight, woodpeckers feed on bark insects, cuckoos eat venomous wasps and a whole family, the Formicariidae, feed entirely on the abundant supply of ants. High in the canopy hawks and owls patrol the forest, preying on reptiles, birds and small mammals.

Mammals and other land animals
In comparison to the birds, mammals are poorly represented. Land mammals tend to be small, like the shy brocket deer (*Mazama* sp) which is only 91cm (3ft) long. The habitat usually occupied by deer is here exploited by rodents such as the agouti (*Dasyprocta* sp) and the paca (*Cuniculus paca*). But the region can also boast the largest rodent in the world, the peaceful capybara (*Hydrochoerus hydrochaeris*), which lives in large family groups on the banks of the rivers. Two of the strangest creatures in this land of contrasts are the anteater and the tapir. The great anteaters or ant bears (*Myrmecophaga tridactyla*) are solitary animals that roam the swampy areas of the forest, while another anteater, the tamandua (*Tamandua tetradactyla*), has become arboreal. The tapir (*Tapirus terrestris*), a relation of the horse and

1 Epiphytic orchid *Oncidium* sp
2 Tree *Vochysia* sp
3 White-headed capuchins *Cebus apella*
4 Scarlet macaw *Ara macao*
5 Three-toed sloth *Bradypus tridactylus*
6 Howler monkeys *Alouatta* sp
7 Common opossum *Didelphis marsupialis*
8 Epiphytic orchid *Cattleya* sp
9 Bromeliad
10 Tiger butterfly *Heliconius ethillus*
11 Tree *Cecropia* sp
12 Brocket deer *Mazama* sp
13 Tamandua *Tamandua tetradactyla*
14 Termite nest
15 Scarlet ibis *Eudocimus ruber*

16 Brown coati *Nasua narica*
17 Epiphytes
18 Great anteater *Myrmecophaga tridactyla*
19 Roseate spoonbill *Ajaia ajaja*
20 Keel-billed toucan *Ramphastos sulfuratus*
21 Liana flowers *Cephaelis* spp
22 Ruby and topaz humming bird *Chrysolampis mosquitus*
23 Capybara *Hydrochoerus hydrochaeris*
24 South American river turtle *Podocnemis expansa*
25 Arrow-poison frog *Dendrobates* sp
26 Paca *Cuniculus paca*
27 Bird-eating spider *Theraphosa leblondi*
28 Tapir *Tapirus terrestris*

29 Jaguar *Panthera onca*
30 Giant water lily *Victoria amazonica*
31 Red and blue leaf-hopper
32 Leafcutter ants *Atta* spp

1 The dominant features of the forest are the trees and the water; as a result, most of the animals living there are either arboreal or aquatic and sometimes both. Every single niche in this enormous habitat is fully occupied with a bewildering collection of animals and plants. The huge trees are supported

by buttress roots in the often unstable soil and are festooned with long lianas. Epiphytes, such as the startlingly beautiful orchids and bromeliads, abound on tree trunks and branches. The numerous rivers provide homes for millions of fish. There are a number of aquatic or semi-aquatic creatures that use the rivers as their base. These include reptiles such as terrapins, alligators and anacondas. Wading birds such as the roseate spoonbill and scarlet ibis are also found on or by the river. Many animals live at the water's edge, browsing or aquatic plants. They include the tapir and the world's largest rodent – the capybara. The shy paca roams the forest floor amid millions of insects and many species of frogs and reptiles. But most animals live high up in the forest canopy. There troops of monkeys show off their acrobatic skills and the sloth lives up to its name with its funereal progress through the branches. Most of the birds also live in the canopy. They range in size from the big-billed toucan to tiny humming-birds that flit from flower to flower like iridescent jewels. The solitary jaguar is fond of water and often chases tapirs and capybaras right into the river. It also feeds on fish and alligators which it catches in shallow parts of the river.

rhinoceros, belongs to a family of animals that was formerly widespread but is now confined to Central and South America and similar habitats of southeastern Asia. It is nocturnal and feeds on vegetable matter. Other ancient residents include marsupials, made up of a group of opossums, of which the common opossum (*Didelphis marsupialis*) is found as far north as the southern states of the USA. The only aquatic marsupial present, the yapok (*Chironectes minimus*), hunts by night for frogs, fish and crustaceans.

In the higher levels of the tree canopy the inhabitants vary from extremely agile, acrobatic monkeys to slow-moving sloths – both the three-toed (*Bradypus tridactylus*) and the two-toed (*Choloepus hoffmani*) kinds. The monkeys differ from their counterparts in the Old World, having evolved quite separately. They are divided into two groups, the marmosets and the cebid monkeys. Marmosets are the smallest primates in the world – the pygmy marmoset (*Cebuella* sp) grows only 7–10cm (3–4in) in length. The cebids are more diverse in their appearance. They include the agile spider monkey (*Ateles* sp); the howler monkey (*Alouatta* sp), which has the most powerful vocal call of all the primates, and whose roars can be heard for miles; and the only truly nocturnal monkey, the large-eyed douroucoulis (*Aotus* sp). Bats also abound in the forest. Some, such as the fruit bats, are herbivorous, others are insectivorous and yet others entirely carnivorous. One of the most notorious is the vampire bat (*Desmodus rotundus*) which pierces the skin of warmblooded animals with its sharp teeth and drinks the fresh blood.

Numerous predators

The vast array of animal life supports many predators including larger reptiles such as the giant anaconda (*Eunectes murinus*), the smaller boa constrictor (*Constrictor constrictor*) and alligators and caimans. The largest mammalian predator is the jaguar (*Panthera onca*), one of the resident cat family that includes the puma (*Felis concolor*), ocelot (*Felis pardalis*), margay (*Felis weidii*) and the jaguarundi (*Felis yajouaroundi*).

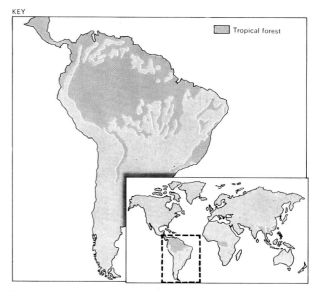

KEY

Tropical forest

The great South American rain forest covers the low-lying Amazon basin and extends eastwards to the Guianas and westwards to the foothills of the Andes mountains.

most abundant foods are insects and the fruits of plants, and many forest birds feed on these. Humming-birds feed on both insects and nectar. Many daytime birds of prey feed only on other birds but some like the eagles also eat mammals. Owls normally hunt at night and feed on any prey small enough to catch.

1 Crested eagle
Morphnus guianensis
2 White-collared swift
Streptoprocne zonaris
3 Hyacinth macaw
Anodorhynchus hyacinthinus
4 Short-billed pigeon
Columba nigrirostris
5 Blue-throated piping guan
Pipile pipile
6 Silver-throated tanager
Tangara icterocephala
7 Squirrel cuckoo
Piaya cayana
8 Rufous-browed pepper-shrike
Cyclarhis gujanensis
9 Ornate umbrella-bird
Cephalopterus ornatus
10 Cayanne jay
Cyanocorax affinis

11 Black-eared fairy
Heliothryx aurita
12 Fork-tailed woodnymph
Thalurania furcata
13 Bronze-tailed plumeleteer
Chalybura urochrysia
14 Band-tailed barb-throat
Threnetes ruckeri
15 Long-tailed hermit
Phaethornis superciliosus
16 Semi-collared hawk
Accipter collaris
17 Collared forest falcon
Micrastur semitorquatus
18 Spectacled owl
Pulsatrix perspicillata
19 Great jacamar
Jacamerops aurea
20 Rufous-capped spinetail
Synallaxis ruficapilla
21 Chestnut woodpecker
Celeus elegans
22 Red-billed scythebill
Campyloramphus trochilirostris
23 Black-throated trogon
Trogon rufus
24 Cock-of-the-rock
Rupicola rupicola
25 Red-capped cardinal
Paroaria gularis
26 Brazilian tanager
Ramphocelus bresilius
27 Grey-winged trumpeter
Psophia crepitans
28 Marbled wood-quail
Odontophorus gujanensis
29 Black-bellied gnat-eater
Conopophaga melanogaster
30 Grey-throated leaf-scraper
Sclerurus albigularis

2 The birds of South American forests are more numerous and vividly coloured than any others. The complex ecosystem provides an abundance of habitats and a very wide variety of food to sustain them. The unvarying climate ensures a year-round food supply and, as a result, birds do not have to migrate to escape harsh conditions. These factors also mean that birds need move only within a restricted area. Most of them have short, wide wings for manoeuvring between the trees. The swifts rarely enter the forest but hunt for insects over the canopy. Their wings are long and narrow for fast flying. The

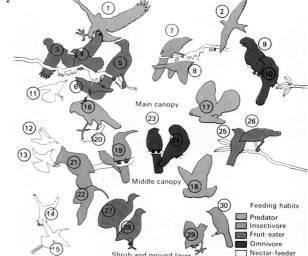

2

Main canopy

Middle canopy

Shrub and ground layer

Feeding habits

Predator
Insectivore
Fruit-eater
Omnivore
Nectar-feeder

217

Life in the desert

A desert is defined as an area with less than 25.5cm (10in) rainfall a year, supporting only a limited animal population. No two deserts are the same. They are of two types, known as cold or hot deserts [Key] depending on whether they are cold or warm in winter.

The distribution of deserts

Cold deserts are found in North America (the Great Basin), northern Asia (Gobi) and South America (Andean plateau-puna). The hot deserts are found in southern North America and Mexico (Mojave, Colorado, Arizona-Sonora, Baja California and Chihuahua), in South America (Chile and Peru coast), northern and southern Africa (Sahara, Kalahari-Namib) and Australia. The plant and animal life supported by the desert is as varied as the people who share the same environment. But there is one factor common to them all: to a greater or lesser degree they have had to adapt to the harshness of desert life caused by the lack of water and shelter and the extreme changes in temperature. Temperatures as high as 56.7°C (134°F) have been recorded in the hot desert

of Death Valley, California, and as low as −40°C (−40°F) in the Gobi and there are great changes in temperature from day to night. All living organisms in the desert – plants, insects, reptiles, birds and mammals – depend upon water in some measure if they are to live, but in various ways they have evolved a capacity for surviving on the barest minimum supplies during the frequent periods of drought.

Desert plants and insects

Desert plants (or "xerophytes", which is the name for plants living in an area where water is scarce) have adapted in a variety of ways [3]. Some plants can become dehydrated without permanent injury. Older leaves may dry up and die but young leaves, though drying and turning brown, will continue to grow when watered again.

Plants may also evade drought. The life cycle of a plant such as the Californian poppy is completed within a short space of time so that during the most arid months of the year it is below the surface in seed form where it can lie dormant for several years without rain [4].

An interesting phenomenon of some desert plants is that only part of the seed germinates at the first watering; other parts have to be subjected to several falls of rain before growth of the seedling is possible. This helps to ensure the survival of the species.

Drought resistance is the most usual method of survival. Some plants shed leaves, so that they need less water to stay alive. Others grow extraordinarily deep roots to tap the reservoirs of water deep underground. Mesquite and acacia trees, for example, grow roots to a depth of 15m (50ft), while those of shrubs such as *Atriplex halimus* grow to 18 or 21m (60–70ft). Another way is for plants to develop root systems over a wide but shallow area to take maximum advantage of a brief fall of rain. Such root systems can run for more than a kilometre. Some plants die back to ground level surviving the drought as tubers and bulbs, while others store excess moisture within their stems to be used up as required. Many succulent plants (*Aloe* spp, *Crassula* spp) have leaves that are adapted for water storage. The long, silky hairs covering such cacti as *Cephalocereus* spp also help

1 The Joshua tree (*Yucca brevifolia*) is one of the few plants able to survive in the New Mexican desert. It plays host to a large number of animals. Dependent insects include weevils (*Yuccaborus* spp) [3] and termites (*Paraneotermes* spp) [6] while the yucca moth (*Tegeticula*

alba) [5] is the tree's major pollinator. Small mammals, among them the pack rat (*Neotoma magister*) [7], build their homes at the tree base. Birds, including the small cactus wren (*Campylorhynchus brunneicapillus*) [1], Scott's oriole (*Icterus parisorum*)

[2] and the gila woodpecker (*Centurus uropygialis*) [4], attracted by insects, nest among the leaves. Reptiles such as the spotted night snake (*Hypsiglena torquata deserticola*) [8] and the night lizard (*Xantusia vigilis*) [9] soon arrive to consume some of the abundant food.

Key to species
1 *Eriocactus* sp
2 *Astrophytum* sp
3 *Mammillaria* sp
4 *Lobivia* sp
5 *Gasteria* sp
6 *Crassula* sp
7 *Stetsonia* sp
8 *Lithops* sp
9 *Haworthia* sp
10 *Cereus* sp
11 *Euphorbia* sp
12 *Stapelia* sp

2 The desert, often thought of as a barren, inhospitable place, can support a wealth of plant life. The hot deserts of Mexico and North America are the home of most of the world's 1,500 species of cacti. These plants, which range in height from less than 2.5cm (1in) to over 15.2m (50ft), are the ones most often associated with the desert landscape. The prickly pear (*Opuntia* sp) and the tall upright giant saguaro (*Cereus giganteus*) seen here are two of the most common and spectacular.

3 The cactus form is generally cylindrical or spherical to reduce surface evaporation. Few cacti (family Cactaceae) [1, 2, 3, 4, 7, 10] have leaves but many have sharp thorns or spines to discourage animals from eating them. The cactus form may be seen in other desert plants [11, 12] but most have succulent [5, 8, 9] or reduced [6] leaves. These plants belong to several families including the Crassulaceae [6], Stapeliaceae [12], Euphorbiaceae [11] Liliaceae [5, 9] and the Aizoaceae [8].

4 When rain falls in the desert after a dry period the land may be rapidly transformed into a colourful and lush meadow. The seeds have lain dormant, awaiting water to complete their germination before bursting into life.

to prevent water loss. Desert plants provide shelter [1] for many animals, including varieties of birds, reptiles and moths.

Insects exist in great numbers in the desert and they play an important part in the survival of the other desert-dwellers, some of which are exclusively insect-eating. The adaptation of these small desert inhabitants to the aridity and heat varies enormously [5, 8]. Some, such as the harvester ant, are physically unsuited to the terrain and so make their nests deep underground where they are only slightly affected by temperature. The harvester ant makes brief foraging expeditions above ground to collect seeds, most of which are stored against times of drought.

Many insects are nocturnal, sheltering during the day under rocks or just below the ground's surface. Diurnal insects (those that are active only by day) have either long legs to lift their bodies above the hot sands, or the ability to fly, or climb on to plants. Most insects avoid dehydration by secreting on their cuticles a thin layer of wax that is relatively impermeable to water vapour. This layer is also impermeable to oxygen and

carbon dioxide so a mechanism has evolved whereby the gases can cross the shell while water loss is kept to a minimum. Breathing is controlled via special apertures, known as spiracles, which are opened to facilitate respiration only when a sufficient level of carbon dioxide has accumulated in the body.

Reptiles – the best survivors
Reptiles form another group of desert creatures, and it is probably the group that feels most at one with its habitat [12, 13]. They are cold-blooded animals – unlike birds and mammals – which makes it impossible for them to maintain internal heat, so they have to rely on the sun to warm the surrounding air and ground in order to raise their body temperature. In extreme heat, however, their bodies become so hot that it is often necessary even for them to take shelter. They also have to find less exposed places at night to protect themselves from the cold. Snakes are more numerous in the desert than any other reptiles but because they are often nocturnal they are less obvious than the lizards that scurry about during the day.

KEY

Deserts are found in all the continents of the world and make up about 14 per cent of its total land mass. The Sahara, which is the largest desert, has an area of roughly 900 million hectares (3.5 million square miles).

Although deserts are easily recognizable, they are often difficult to define, but any area receiving less than 25.5cm (10in) of rain a year can be thought of as one. The boundaries of deserts are also hard to define as few deserts consist

of totally barren sand. Instead there is a fringe zone where desert vegetation merges imperceptibly with that of the surrounding area. Sand is not a regular desert feature. Deserts can consist of stones, rocks or even salt.

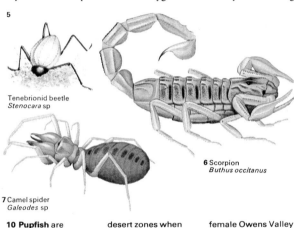

5

Tenebrionid beetle
Stenocara sp

7 Camel spider
Galeodes sp

6 Scorpion
Buthus occitanus

5 Tenebrionid beetles are typical of the many desert beetles; they live in sand dunes into whose safety they plunge at the first hint of danger.

6 Scorpions can withstand great heat. They live on spiders and insects but will also attack other scorpions.

7 Camel spider is a name for a solifugid, meaning "fleeing from the sun"; there are over 800 species, mostly nocturnal.

8 Some honey ants are living food stores accepting honey from the rest of the colony. Their abdomens distend to form food-filled globes from which others feed in drought periods.

8

Honey ant
Myrmecocystus melliger

9 Locusts are destructive desert grasshoppers: an average swarm is made up of 1,000 million individuals which, collectively, need to eat 3,000 tonnes of food each day to survive. Although there is a naturally high mortality rate, they present a serious problem to farmers wherever they are found as they are unpredictable and difficult to control. The young congregate *en masse* and march across country.

9

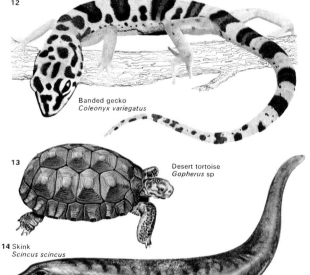

10 Pupfish are relics of an aquatic fauna that once populated certain desert zones when there were extensive lakes rather than arid sands. Male and female Owens Valley pupfish, together with several other species of fish, may still be found in the water pools of the American deserts, which are in fact vestiges of former great lakes.

10

Male

Owens Valley pupfish
Cyprinodon radiosus

Female

11 Spadefoot toad
Scaphiopus couchi

11 The spadefoot toad, so-called because of the digging appendages on the hind feet, is one of several species of desert toads. It lives in a burrow for most of the year, emerging at night to feed on insects. If it rains the toad goes

to the nearest pool, where males and females mate and the eggs are laid. The tadpoles then develop

rapidly and by the time the waters have dried up have become mature adults, ready for burrowing.

12

Banded gecko
Coleonyx variegatus

13

Desert tortoise
Gopherus sp

14 Skink
Scincus scincus

12 The banded gecko is one of a great many species of gecko inhabiting desert regions. It is nocturnal, hiding under rocks during the day and foraging for insects at night.

13 Desert tortoises are perfectly adapted to their environment. The heavy shell, a protection from extreme heat (and cold) and from predators, houses two storage sacs for excess moisture.

14 The skink is sometimes known as "sand fish". It seems to swim through the sand, pushing itself along with its flattened toes which are fringed with scales. It spends much of its time just beneath the surface looking for insects.

Desert birds and mammals

There are fewer birds and mammals, especially large ones, in deserts than in other regions, but a surprising number do manage to survive in this harsh environment. They can only do so by adaptations of their behaviour, life processes and external coverings.

The adaptable camel

The camel (*Camelus dromedarius*) [Key] is, because of its usefulness to man, one of the commonest and best known of the larger desert mammals. Unlike other large, indigenous desert mammals such as the addax (*Addax nasomaculatus*) [2], which is threatened with extinction, camels are increasing in numbers.

The adaptations that enable the camel to live in desert conditions have been closely researched and the findings probably apply also to other large desert mammals. The camel has a remarkable ability to withstand heat, and its body temperature can fluctuate by as much as 6–7 Centigrade degrees (about 12 Fahrenheit degrees) without causing it distress. During the day the camel's body may heat up slowly to 40°C (105°F), and

during the night may drop to 33.9°C (93°F). The camel's hump, which is a fat store, contributes indirectly to the control of its body temperature. The lack of insulating fat beneath the skin allows heat to escape easily and the animal's thick, woolly fur serves as insulation against the entry of heat.

The camel is also unusually adapted to reducing water loss from breathing, sweating and excretion. If man lost water amounting to 12 per cent of his body weight he would die of "explosive heat" because, as water is lost, blood decreases in volume and becomes thicker, slowing down the circulation. Internal heat can no longer be dissipated from the skin and death quickly follows. The camel, however, can lose up to 25 per cent of its body weight in water and still maintain its blood volume by removing water from the body tissues. The camel's sweating reflex is not activated until its body temperature reaches about 40°C (105°F).

Antelopes and cats

The increasingly rare addax is a member of the large cattle family (Bovidae) which

includes the antelopes and sheep. It is even more resilient than the camel in that it does not need to drink water at all, because it takes in the moisture it needs by eating plants. Simply by sniffing the air, the addax can locate from great distances an area where rain has fallen. As a result, it can live in arid regions that no other mammal could tolerate, not even the well-adapted scimitar-horned oryx (*Oryx dammah*) – another animal related to the addax – which is at home in the Libya and Sahara deserts.

A well-known desert antelope is the graceful and fleet-footed gazelle (*Gazella* sp) of which there are several species, but the few remaining animals are a sad indictment of the "sportsmen" who shoot them with high-powered rifles and even machine-guns from motorized vehicles – proving that even swift species such as these can eventually be wiped out by such methods.

Relatively few large carnivores live in deserts: the cat family is well represented, but the only big cat is the cheetah (*Acinonyx jubatus*), the fastest land animal in the world. It is exceptional in its habits because, unlike

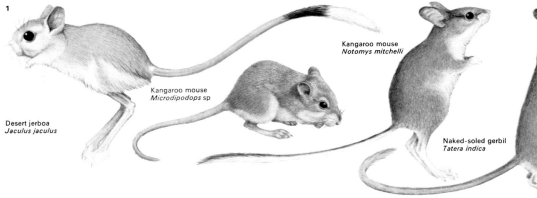

1 Desert rodents exist in large numbers, having successfully adapted to their environment by living in burrows. They include the desert jerboa, kangaroo mouse and naked-soled gerbil. They can make long, swift leaps with their powerful hind legs to escape from predators. Using their tails as rudders, they can alter course in mid-air. They are also protected by their sensitive hearing.

Kangaroo mouse *Microdipodops* sp

Desert jerboa *Jaculus jaculus*

Kangaroo mouse *Notomys mitchelli*

Naked-soled gerbil *Tatera indica*

Addax *Addax nasomaculatus*

2 One of the larger desert mammals is the addax. This antelope often falls victim to the hunter, who prizes its meat and fine skin. As a result, it is now a very much endangered species. Like many other desert dwellers, it does not need liquid water to survive, but obtains the necessary fluids from its food. It has a well-developed sense of hearing but lacks the speed to flee successfully from danger.

3 Pallas' cat lives in burrows or among the rocks of the Asian deserts. It is an agile hunter, whose body length does not normally exceed 50cm [20in]. It feeds on a variety of small birds and rodents.

3 Pallas' cat *Felis manul*

4 A hardy desert animal from Australia is the red kangaroo. Although widely distributed across the continent, its numbers have been reduced by hunting and the fencing of sheep ranges. Its urinary system, like that of the ground squirrel and kangaroo rat, has evolved to cope with desert conditions and its urine is highly concentrated. This desert mammal drinks only in periods of extreme heat and drought. Kangaroos such as this species are famed for their jumping ability. Recent studies have shown this to be a more efficient way of moving than the normal four-legged gait of other mammals.

4 Red kangaroo *Macropus rufus*

5

Kit fox *Vulpes velox*

Fennec fox *Fennecus zerda*

5 The kit fox and the fennec are similar but unrelated desert mammals. The kit fox is from the New World, the fennec from the Old, and both have adapted independently but in the same way to desert life. This is a case of convergent evolution – the development of similar characteristics by animals or plants of unrelated species. Both are nocturnal and live in burrows. They feed on insects, lizards, rodents and birds which they detect with their large and sensitive ears.

the other members of its family, it does not stalk and pounce on an unsuspecting prey, but runs it down. The small desert cats [3] tend to be nocturnal, lying in wait for other creatures of the night – their prey.

Very few members of the dog family inhabit deserts, but those that do are all similar in colour and appearance [5]. They are generally small with large ears and usually hunt by night. Such animals are omnivorous.

Rodents are the commonest of the small desert mammals. Most are nocturnal and pass the day in burrows. There, where the air is more humid than it is outside, they remain cool. Many do not drink, obtaining all their water from plants. Some are known to obtain water by the oxidation of the carbohydrates in their food, which is mainly dry seeds. They also have highly efficient kidneys which extract most of the water from their urine. Jumping as a means of movement has been developed in most of these desert rodents [1]. Although they all look similar, they are not closely related and come from different continents. Jerboas (*Jaculus* spp) and gerbils (*Tatera* spp) are from Africa and Asia. *Mic-*

rodipodops are from North America and *Notomys* from Australia.

Birds of the desert

Many birds live in the desert, from the tiny elf owl (*Micrathene whitneyi*) [6] to the flightless ostrich (*Struthio camelus*), the largest bird in the world. Desert birds may feed on seeds or water-rich plants such as cacti but many, such as the roadrunner (*Geococcyx californianus*) and the elf owl are predators, getting all the liquid they need from other animals. Birds that are mixed feeders, such as Pallas' sandgrouse (*Syrrhaptes paradoxus*), are often migrants or nomads.

Ostriches are well adapted to desert life. They obtain the water they need from the plants that they eat and merge well into the pale tones of the desert landscape when squatting on the ground. If threatened while on the nest, which is just a shallow scoop in the earth, the hen bird presses her long neck flat down on to the sand, making herself difficult to distinguish. This attitude has perhaps given rise to the mistaken belief that ostriches bury their heads as a defence.

KEY

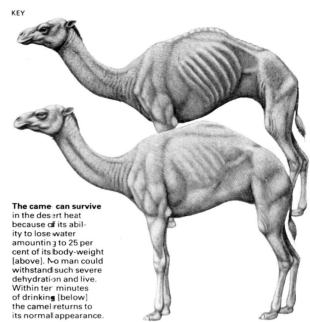

The camel can survive in the desert heat because of its ability to lose water amounting to 25 per cent of its body-weight [above]. No man could withstand such severe dehydration and live. Within ten minutes of drinking [below] the camel returns to its normal appearance.

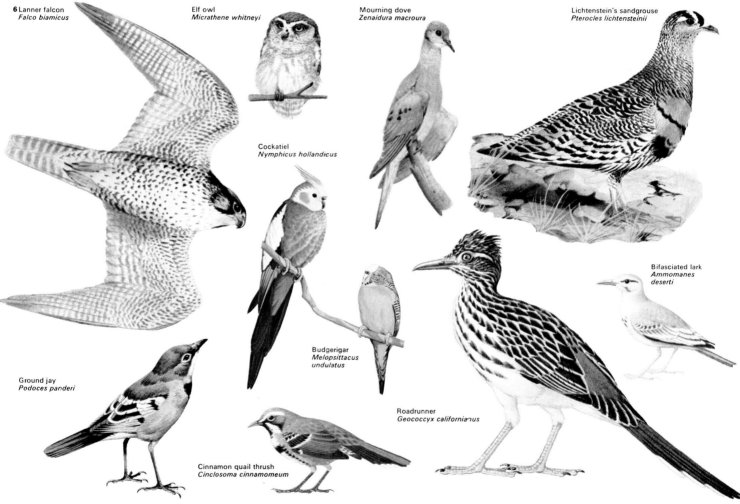

6 Lanner falcon
Falco biamicus

Elf owl
Micrathene whitneyi

Mourning dove
Zenaidura macroura

Lichtenstein's sandgrouse
Pterocles lichtensteinii

Cockatiel
Nymphicus hollandicus

Ground jay
Podoces panderi

Budgerigar
Melopsittacus undulatus

Bifasciated lark
Ammomanes deserti

Roadrunner
Geococcyx californianus

Cinnamon quail thrush
Cinclosoma cinnamomeum

6 Typical desert birds include the lanner falcon, which feeds mainly on other birds. It has been a popular bird with falconers for thousands of years in Africa and Asia. The elf owl is one of the smallest of the owls, about 15cm

[6in] long. It often nests in cacti using the abandoned nests of the gila woodpecker. The elf owl feeds on insects and occasionally on small birds. The mourning dove can withstand a variety of climatic conditions and so is at

home in the desert. It breeds almost throughout the year. Lichtenstein's sandgrouse is unusual in that the male carries water to its offspring by soaking its breast feathers; the young drink from these. The water also cools unhatched

eggs. The budgerigar is the native name for a little Australian parakeet well known as a cage bird in many parts of the world. In the wild it lives in large flocks in the semi-deserts of Australia, as does the widespread cockatiel.

The roadrunner is noted among desert birds for its expertise in killing snakes. It does this by a series of quick stabs with its long, pointed beak. It rarely flies, but is a very fast, agile runner. The bifasciated lark has an unusually

strong bill for digging out grubs and ant-lions from the sand to supplement its monotonous diet of seeds. The ground jay, so called because it has become almost totally ground-based, uses its wings only to increase its running speed. Its

grey plumage effectively camouflages it in the desert. It nests in burrows or beneath bushes. The cinnamon quail thrush can survive the most severe drought. It seldom drinks and gets its water content from eating insects.

Mountain life

Mountains make up about five per cent of the earth's land mass. The world's great mountain ranges are not confined to any one latitude but are found in tropical and temperate regions and in all continents. Specifically, the world's major mountain ranges are the mountain ridge from Alaska to the southern tip of South America which includes the Rockies and the Andes; the Appalachians of eastern North America; the various ranges of the Eurasian land mass, including the Alps, Pyrenees, Caucasus and Atlas mountains; and the Himalaya range [2, 3] between the Palaearctic and Oriental regions. In Africa there are several isolated mountainous regions in Ethiopia, Kenya and Rwanda. Australia has one major range – the Great Divide of the eastern coast.

Mountain zones

The plants and animals of mountains comprise different species according to locality but all share adaptations essential to survival in this habitat and some of these adaptations are very similar to those of plants and animals found in polar regions. The range of life on a

mountainside is very wide, for in terms of climate a difference in altitude of 70m (200ft) is equivalent to 1° of latitude – roughly 110km (70 miles). This is due to sharp falls in temperature and rises in precipitation up the mountainside.

Mountain plants reflect changes of climate by showing a definite vertical zonation. This is modified by the distance from the Equator. Mount Kilimanjaro (5,595m [19,340ft]), in Africa, shows all the possible mountain zones – tropical forest, deciduous forest and coniferous forest which merges into the alpine zone at the tree line. Topping the alpine zone are snow and ice. Mountains in temperate regions such as the Sierra Nevada [1] have no tropical zone while those at even higher latitudes, as in the Northern Rockies, have no temperate zone. Herbivorous animals, from insects to mountain goats, show a similar zonation because their distribution is limited by food sources, but mountain carnivores tend to be more mobile.

As in all other habitats, the plants and animals of mountains are affected by such factors as soil, aspect, cover and competition.

And because of the harsh environment of the higher zones the number of species that are able to survive is inevitably limited.

Plant adaptations

Plants, as in all ecosystems, are the basis of the mountain food chains. Plants of the alpine zone [5] are either short, compact and virtually stemless or have supple shoots and stems. These adaptations enable them to survive cold, strong winds. Many mountain plants have extensive root systems to provide anchorage against the wind and to increase their chances of obtaining water on dry slopes. The common alpine cushion plants stand only a few centimetres high yet their roots may penetrate almost a metre (39in) down into the soil. Water conservation within alpine plants is enhanced by such adaptations as waxy coverings on the leaves and the alpine edelweiss (*Leontopodium alpinum*) has leaves covered with hairs whose function is to trap heat and reduce loss of water. To combat freezing, plants such as the mountain crowfoot (*Ranunculus glacialis*) have a rich cell fluid that acts as an antifreeze. The alpine

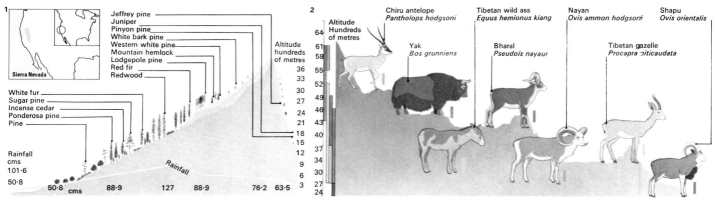

1 The Sierra Nevada of California is a classical example of the botanical zonation typical of most mountains. Shown here in cross-section it illustrates the influence of rainfall, altitude, slope and aspect in determining the distribution of mountain vegetation. Although the climate is temperate, deciduous trees are replaced by semi-desert scrub because of low rainfall. The western slope has a higher rainfall and thus a richer and more varied selection of plants than the dry easterly slopes of the mountain.

2 Several species of large, hoofed animals (ungulates) live in herds with wide-ranging territories on the steep slopes and high grazing areas of the Himalayas and the Hindu Kush. Herds of chiru antelope roam over large areas of scant vegetation. Yaks, protected by dense, matted coats, also survive in high bleak areas where they often feed only on mosses and lichens. The Tibetan wild ass has the greatest vertical range of the animals shown here. It can withstand long periods without food and water. The Tibetan gazelle and the bharal, nayan and shapu sheep are all less well equipped to endure the cold and so live in lower regions, feeding on the grassland plains.

3 The Himalayan alpine zone extends near to the summit of Mt Everest 8,848m (29,030ft) high. Animals are limited in numbers and range by availability of food. At 5,000m (16,500ft), Alpine choughs (*Pyrrhocorax graculus*) [1] eat insects and worms. The golden eagle (*Aquila chrysaëtos*) [2] flies far in search of carrion or small animals. The Tibetan pika (*Ochotona ladacensis*) [3], feeds on small green plants, storing some for winter. Stalking, rather than waiting in ambush, the snow leopard (*Uncia uncia*) [4] preys on both the Himalayan ibex *Capra ibex sakeen*) [5] and bharal sheep (*Pseudois nayaur*) [6], which roam the slopes in search of small plants.

soldanella (*Soldanella alpina*) can even melt surrounding snow by releasing heat from the breakdown of carbohydrate within its cells.

In the alpine zone the summer growing season lasts only a few weeks. As a result most alpine plants are perennials. The cushion pink (*Silene acaulis*) may take ten years to flower, but when it does bloom several hundred flowers may appear. This low but concentrated reproductive rate probably occurs because it takes a very long time for the plant to build up sufficient food reserves to enable it to flower. Many mountain plants are self-pollinating, a mechanism that ensures fertilization in a situation where there are few flying insect pollinators. For the same reason wind-pollinated species outnumber insect-pollinated ones.

Animal life on mountains

Many mountain animals tend to stay on high slopes and peaks throughout the year but can avoid extreme conditions in a number of ways. Larger herbivorous animals of the alpine zone, such as sheep, goats and antelopes – and their attendant carnivores –

migrate to lower, snow-free slopes in winter, returning to the higher regions in spring. Marmots (*Marmota* spp), and to a lesser extent ground squirrels (*Citellus* spp), hibernate. The marmots gorge themselves with food in the summer and lie dormant in burrows during the winter, slowly using up their stored energy supplies. The European snow vole (*Microtus nivalis*) remains active in winter in its underground home, using hoarded food. Pikas (*Ochotona* spp) often remain above ground in winter, living on dried plant material they have previously hidden beneath rocks [8].

Birds [7] are well adapted to altitude. Their dense covering of feathers provides excellent insulation. Many insects can tolerate cold conditions above the tree line because their body fluids have a low freezing-point. The glacier flea (*Isotoma saltans*) can withstand being frozen solid for short periods, while springtails (Collembola) have even survived being trapped in ice for several years. Most insects of high altitudes, however, remain dormant over the winter in hollows beneath snow-covered rocks [6].

KEY
Mountains close to each other such as the Tatra and more southerly Bucegi mountains of central Europe, have similar vegetation, but the altitudes at which individual zones occur may vary. This variation is a reflection of differences in climatic conditions.

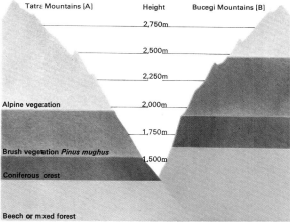

Tatra Mountains [A]   Height   Bucegi Mountains [B]

2,750m

2,500m

2,250m

Alpine vegetation   2,000m

1,750m

Brush vegetation *Pinus mughus*

1,500m

Coniferous forest

Beech or mixed forest

4 *Parnassius charltonius charltonius*

Baltia shawii

Orchid
Dendrobium nobile

Bhutan glory
Armandia lidderdalei

4 Butterflies are found at nearly all mountain levels. *Parnassius charltonius charltonius* lives at heights of 5,000m (17,000ft) in the Himalayas. Most mountain butterflies are darker than related lowland species, which allows them to absorb the maximum of heat and protects them from the sun's ultraviolet rays. Among common species of the mountain regions of Turkestan, Tibet, Mongolia and the Himalayas are *Baltia shawii* and *Armandia lidderdalei*. Like most mountain insects they fly close to the ground to avoid high winds. They feed on nectar obtained from orchids and other flowering alpine plants.

5 Alpine lily
Lilium bulbiferum

Alpine saxifrage
Saxifraga aizoon

Trumpet gentian
Gentiana kochiana

Mountain avens
Dryas octopetala

Edelweiss
Leontopodium alpinum

Stemless thistle
Carlina acaulis

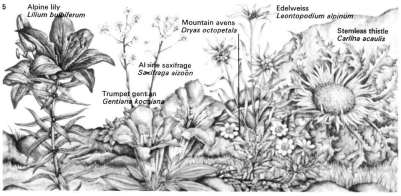

5 Mountain plants are most often short, supple and compact with leaves and petals that show a variety of adaptations to cold and windy conditions. The alpine lily is a perennial plant and is among the tallest of all alpine species. But it has an extremely supple stem and thus can bend in the wind. The well-known edelweiss shows another adaptation to mountain life. The whole plant is covered with a thick layer of white hairs which act as a heat trap. The trumpet gentian has large, thin and flat leaves to absorb a maximum amount of radiant energy. The mountain avens and alpine saxifrage are both very small, low-lying perennial plants, and are particularly common in northern Europe. The stemless thistle is found in rocky places of mountain regions in western Europe.

6

7 Snow finch
Montifringilla nivalis

7 The snow finch, (*Montifringilla nivalis*), is commonly found in mountainous parts of Eurasia. It lives at heights of up to 4,500m (15,000ft) and feeds on insects and the seeds of various alpine plants.

8 The Asian pika (*Ochotona ladacensis*) occupies a habitat 5,400m (18,000ft) up in the Himalayas. Its small, round body and short, stubby ears are adaptations to reduce heat loss.

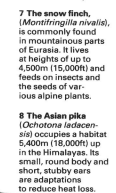

8 Asian pika
Ochotona ladacensis

6 The insects and other arthropods of the high Himalayas are usually confined to three locations. Hollows under rocks shelter attid spiders [1], psuedoscorpions [2], carabid beetles [3] and millipedes (Diplopoda) [4]. Springtails (Collembola) [5] and mites (Acarina) [6] live on rock surfaces. Beside meltwater streams stoneflies (Plecoptera) [7] and mayflies (Ephemeroptera) [3] are found. Many live on seeds and pollen deposited on the mountain by wind.

Polar regions

The northern and southern polar regions of the earth [Key] share the same hostile characteristic of intense cold. In almost every other way they are different. The Arctic is almost entirely frozen ocean surrounded, for the most part, by land. The Antarctic, on the other hand, is a land mass pressed down beneath a great raft of ice, a lost continent that at one time, before the continents drifted apart, was much warmer.

The polar climate

The common feature of both regions – their coldness – stems from their positions where the earth's axis passes through its surface. When one pole is tilted briefly for its summer towards the sun, the other is tilted away into an endless night. They are cold because they receive their sunlight more obliquely than any other part of the world, for received heat is proportional to the angle at which the sun's rays strike the surface. Finally the cold is intensified because the ice acts as a mirror that reflects solar energy back into space.

The contrast between the polar regions is most marked in the forms of life that each

supports. The Arctic numbers among its inhabitants one of the largest and most powerful of predators, the polar bear (*Thalarctos maritimus*) [1]. The Antarctic boasts as its only true terrestrial animals a handful (63 species) of insects, the largest of which is an unusual species of wingless mosquito.

Ruthless trackers and ocean giants

The polar bear has adapted perfectly to its life in the Arctic, the frozen ocean. It is able to spend most of its life in water, insulated from the cold by thick, greasy fur. A solitary animal, the polar bear patrols the ice in search of its prey, particularly the ringed seal (*Pusa hispida*) [1]. When the seal's head appears through a breathing hole in the ice the bear stuns it with a single blow from its powerful front leg. During the harshest of the winter months pregnant female bears hole up in ice caves beneath the surface [2]. The young are born in March or April, usually two at a time. The family leaves its den within the next two months – once the sun has become warm enough – and for the next two years the cubs learn how to hunt.

The polar bear is hunted for its meat by the Eskimo. The only other mammal that tracks it closely is the Arctic fox (*Alopex lagopus*) [1]. In lean times the fox will even live off the bear's droppings.

The sub-polar waters round the edges of the permanent Arctic ice are covered only by broken pack ice during the summer and these nutrient-rich waters support hosts of seals, including the ribbon seal (*Histriophoca fasciata*), the hooded seal (*Cystophora cristata*) [6] and the walrus (*Odobenus rosmarus*) [1], hunted by the Eskimo for its meat and for its valuable ivory tusks.

The marine mammals of both north and south include the remnants of once huge populations of whales, reduced to a fraction of their former numbers by over-hunting. Among them are the blue whale (*Sibbaldus musculus*) – at a weight of around 135 tonnes, the largest living mammal – which feeds by straining the abundant pastures of plankton and small crustaceans through a horny sieve that hangs at the back of its mouth.

One of the most fascinating creatures of the Arctic waters is the narwhale (*Monodon*

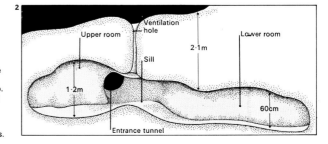

1 Two carnivorous predators, the polar bear (*Thalarctos maritimus*) [1] and the Arctic fox (*Alopex lagopus*) [2], head the food chain of the ice cap surrounding the North Pole. The polar bear feeds chiefly on the seals of the species *Pusa hispida* [3] and the young walruses (*Odo-* *benus rosmarus*) [4], although it may itself fall prey to the killer whale (*Orcinus orca*) [5]. The Arctic fox feeds mainly on the carrion left behind by the polar bear. Seals feed on abundant fish and squid while the walrus grubs on the ocean bottom for molluscs and crustaceans. In the early spring young seals and polar bears are born in caves shaped out beneath the snow. The polar bear is protected from the cold by a thick fur coat that traps a layer of insulating air. In contrast the seal is insulated by a layer of subcutaneous fat or blubber.

2 As the winter approaches all pregnant female polar bears, most other females and some adult males seek out a rock or ice den and spend the cold months in sleep. Here, too, the young are born – usually two cubs which remain with their mother for two years.

monoceros), the unicorn of the sea, which has a long single tusk. The narwhale feeds mainly on squids and is an amiable animal, often using its tusk as a sword for "fencing" matches with its fellows.

Considerable areas of North America and Eurasia fall within the Arctic Circle. These are the bleak, treeless tundra lands that support only grasses, mosses and lichens. During the summer they are visited by wandering herds of caribou (the reindeer of Europe) and swarms of insects breed in the meltwater pools. The insects and the fresh, transient vegetation attract millions of migrant birds to feed, away from the crowded temperate regions. Of the 100 or more birds that breed in the tundra all but five or six are migrants.

The Antarctic seals
Summer and winter, the great mass of the Antarctic is lifeless, containing as it does 90 per cent of the world's ice. Only on its fringes is life to be found in the form of huge colonies of penguins and seals.

The four species of seal for whom Antarctic waters are home are the southern-most mammals in the world: the Weddell seal (*Leptonychotes weddelli*) [9], the crabeater seal (*Lobodon carcinophagus*), the Ross seal (*Ommatophoca rossi*) [8] and the leopard seal (*Hydrurga leptonyx*) [10].

Each has its own particular life-style; the Weddell seal always remains inshore while the others range far out to sea. The crabeater seal earns its name from its feeding habits; it gulps in large quantities of water and squirts this out through fan-like teeth that form a fine mesh at the gum margins, to sieve out the small crustaceans. The leopard seal is a vicious predator, cruising the edges of the ice pack in pursuit of its favourite dish, the Adélie penguin (*Pygoscelis adeliae*). The leopard seal preys also on the pups of other seal species, on squids and on fish.

The Adélie is the most common of the Antarctic penguins, of which there are five species; the largest is the emperor (*Aptenodytes forsteri*) [5] which stands about 1m (39in) high. Other birds are well represented, including the skua (*Catharacta skua*), which preys on penguin eggs and chicks. The skuas range in size from 44–56cm (18–23in).

KEY

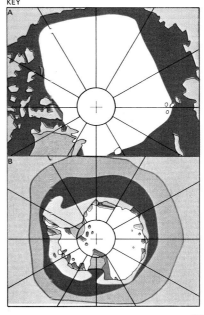

3 The gyrfalcon lives throughout the Arctic regions taking birds after a chase on the wing instead of striking from above like the peregrine falcon.

3 Gyrfalcon
Falco rusticolus

4 Glaucous gull
Larus hyperboreus

4 A greedy predator – like some other gulls – the glaucous gull nests on Arctic and North Atlantic cliffs and islands. It takes a wide variety of prey ranging from small mammals to the eggs of other birds.

5 Largest of the penguins, the emperor breeds on the ice packs and small islands around the coastline of the Antarctic continent. The female lays a single egg which is then passed to the care of the male, who carries it on his feet for the two-month incubation period. The female spends this time at sea but returns for hatching and brooding. The emaciated male now returns to the sea to feed greedily.

6 Hooded seal
Cystophora cristata

6 The grotesque nose of the hooded seal inflates during the breeding season to intimidate his rivals and – paradoxically – to attract females.

7 Banded seal
Histriophoca fasciata

7 The banded seal, found exclusively in the Arctic waters of the Pacific south and west of the Bering Strait, is one of the smallest of seal species.

10 The hunter of the southern oceans is the leopard seal. This solitary animal hunts along the edge of the pack ice in search of its favourite prey –

the Adélie penguin. This continual persecution produces one of the sights of Antarctica – the reluctance of individual Adélies to enter the

water (the time when they are most vulnerable to attack). The leopard seal varies its diet with squid, fish and the pups of other seals.

Emperor penguin
Aptenodytes forsteri

8 The Ross seal is one of the rarest of seal species and is located on the fringes of the Antarctic ice pack.

8 Ross seal *Ommatophoca rossi*

10 Leopard seal
Hydrurga leptonyx

9 Weddell seal *Leptonychotes weddelli*

9 The southernmost mammal, the Weddell seal, chews breathing holes in the ice when the sea is frozen over.

Adélie penguin
Pygoscelis adeliae

225

Life on the tundra

The great ice sheets that covered much of North America and Eurasia in geologically recent times have now retreated to the far north and the high mountains. Round their fringes, however, there remains a vast cold region known as "the barren grounds" in North America and the tundra (from the Finnish for a treeless plain) in the Old World. Despite their different names, the two areas are similar and have about 75 per cent of plant and animal species in common. The area occupied by the tundra lies between 60° and 70°N latitude and begins roughly where the average temperature of the warmest month reaches 10°C (50°F). Temperature is one of the limiting factors for tree growth.

Harsh climatic conditions

Within the tundra, darkness and cold are the two factors that act as the principal controls on life. At the Arctic Circle the sun does not rise above the horizon at the winter solstice, and north of the circle the darkness is more prolonged. Over much of the tundra there is no sunlight for several months of the year. This is compensated for, to some extent, by the continuous daylight of summer, but the plants and animals cram much of their activity into this short period of light.

The cold is equally important and part of the geological inheritance. The subsoil is totally frozen, a condition known as "permafrost" [Key]. It is at least 610m (2,000ft) deep in Greenland and may be deeper in other places. Only the surface soil thaws in the summer. This is known as the active zone and may be as deep as 3m (10ft), or as shallow as 7.6cm (3in), but in either case it supports all life in the area, both plant and animal. Yet the soil in it is likely to be waterlogged because water cannot drain through the frozen zone. The frequent freezing and thawing causes the soil to rise in small "blisters", and stones collect in the hollows between them. The resulting formations are known as polygons and stone rings. Such precipitation as there is almost all falls as snow, equivalent to at most 50.8cm (20in) of rainfall a year.

There are some high mountains in the tundra, but mostly the land is low-lying, dotted with little lakes and pools and criss-crossed by small meandering rivers. In the "low arctic" of the southern tundra, plant cover may be complete; farther north in the "high arctic" strong winds may sweep away the scanty soil and rooted plants become restricted to crevices and sheltered places. Lichens may be the only plants visible.

Tundra food chains

The larger plants of the tundra are mostly woody, forming knee-high forests of birch and willow, with mature trees often only a few inches high [1]. Many members of the heather family are present. Most of these are berry-bearers and form an important part of the diet of many tundra animals – even the polar bear will gorge itself on cranberries in the autumn. The woody plants are augmented and often overgrown by herbaceous plants, which in the summer provide a brief blaze of colour, dying as winter nears.

The animal life of the tundra is surprisingly rich for an environment that seems so inhospitable. Large animals include the musk ox [3] and in the summer the reindeer (caribou in North America). Among smaller

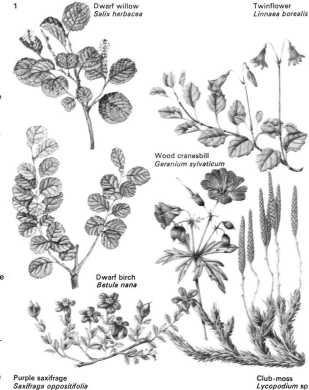

1 The vegetation of the tundra is almost all prostrate. A dwarf birch that grows up to 1m (39in) high is exceptional. Usually this plant is much smaller, as are dwarf willows, even when they are mature "trees". Beneath birch and willow is a rich growth of mosses and lichens which form the equivalent of a shrub layer in a temperate forest. Here *Lycopodium*, a club-moss, is found. This small plant may spread over the ground for many hectares, but is a poor reminder of the great scale trees of the past to which it is closely related. The wood cranesbill is one of the brilliant summer flowers of the tundra. Purple saxifrage is also found on many southern mountains, forming part of a flora referred to as "arctic alpine". *Linnaea* commemorates the great Swedish botanist Carl Linnaeus, who made a journey to the tundra of Finland.

Dwarf willow
Salix herbacea

Twinflower
Linnaea borealis

Wood cranesbill
Geranium sylvaticum

Dwarf birch
Betula nana

Purple saxifrage
Saxifraga oppositifolia

Club-moss
Lycopodium sp

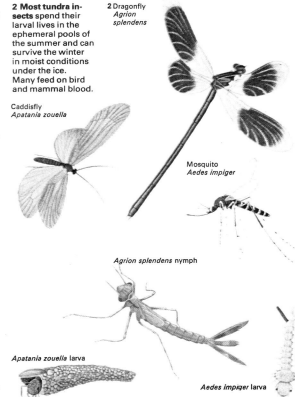

2 Most tundra insects spend their larval lives in the ephemeral pools of the summer and can survive the winter in moist conditions under the ice. Many feed on bird and mammal blood.

2 Dragonfly
Agrion splendens

Caddisfly
Apatania zouella

Mosquito
Aedes impiger

Agrion splendens nymph

Apatania zouella larva

Aedes impiger larva

3 Musk oxen (*Ovibos moschatus*) are the largest animals found in northern Canada and Greenland. They have disappeared from the Old World, which they inhabited in the Ice Age. Their extremely thick coats protect them from cold, but they become exhausted if they have to cross thick snow in search of food. If threatened, the herd bunches together, presenting lowered heads and heavy horns to the attacker – providing a good defence against wolves.

4 Arctic fox
Alopex lagopus

4 The arctic fox is a circumpolar tundra resident. It exists in two forms: the blue form which is smokey grey all the time, and the brown form which changes to white in the winter. The population fluctuates with the availability of food, following the rise and fall of lemming numbers. Man is the fox's chief enemy: he kills it for its thick coat.

predators are wolves, arctic foxes [4] and wolverines, which feed mainly on the arctic ground squirrel, voles and the lemmings. These mammals are adapted to the cold with thick, warm fur and, under the skin, a heavy layer of fat. The danger of frostbite to their extremities is reduced by their compact shape and their short faces, ears and tails.

Migratory birds

Few birds are resident in the tundra, but every year, drawn by the migratory urge, millions of ducks, geese and waders [5], among others, travel to the far north to nest and feed. This is the time of plenty for the arctic fox, which haunts the bird colonies, taking what it can of weak or injured chicks. Those that it cannot eat it hides and some of these caches remain, frozen, as supplies for the winter. The basis for much of the bird life on land is the abundance of insects, including mosquitoes and other biting flies [2].

The sea is as rich as the land is poor and the summer flush of plankton supports a huge web of life, including seals, whales and sea birds [6] most of which migrate to the far

north to feed on the abundance of the waters. For a short time the temperature rises and activity and life abound. By August, however, the hours of daylight have become noticeably shorter, the temperature drops and the first sleet heralds the winter. Insects die, leaving their eggs and larvae to winter beneath the ice covering the pools or in the shelter of snow-covered vegetation. The birds, their reproductive mission complete, fly southwards in great flocks to spend the winter months in temperate and subtropical areas. Reindeer and caribou, followed by the wolves, migrate south, leaving only the musk oxen and the hardy arctic fox.

Yet even in summer the tundra, for all its apparent abundance, is an area where species are few but populations are large, and where the delicate balance of predators and prey is easily upset. Wildlife populations, often made up of prey such as small mammals, may fluctuate wildly between superabundance and paucity. The lemmings [8] are the best known of the animals that undergo this population swing, but they are by no means the only species to do so.

KEY

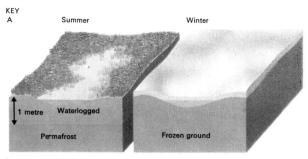

The ground surface of the tundra thaws in summer but stays waterlogged as the permafrost impedes drainage [A]. In winter it freezes, but small mammals [B] survive in burrows.

5 Small waders, such as dunlin, turnstone and knot, migrate north each spring to nest in the tundra.

Like them, long-tailed duck and brent geese return south to feed on mudflats and estuaries once their

families are reared. The snowy owl is a resident but moves slightly south in the hardest of weather.

5 Dunlin *Calidris alpina*

Turnstone *Arenaria interpres*

Knot *Calidris canutus*

Long-tailed duck *Clangula hyemalis*

Brent goose *Branta bernicla*

Snowy owl *Nyctea scandiaca*

6 Arctic loon *Gavia arctica*

6 Both skuas and loons breed on the tundra but they migrate to the oceans in winter. The arctic loon or black-throated diver feeds on fish. The skua eats voles and lemmings or fish it steals from gulls.

Arctic skua or parasitic jaeger *Stercorarius parasiticus*

7 The herbivorous ptarmigan [1], lemming [2] and arctic hare [3] are the principal prey of a range of predators, such as the gyr and peregrine falcons [4, 5], short-eared owl [6], rough-legged buzzard [7],

pomarine skua [8], stoat [9], arctic fox [10], wolverine [11] and wolf [12]. Because the predators tend to be nomadic, wandering from poor areas to richer ones, there is not much danger of their food becoming scarce.

8 Norway lemmings have a 4-year cycle of abundance [A]. A small population can, under favourable conditions, build up to great numbers. These attract predators, but lemmings continue to multiply and deplete their food source. Popularly supposed to commit mass suicide, many lemmings die through attempting to migrate. They may drown while crossing rivers or when they reach the sea. A few remain to restart the cycle. Yearly lemming activity [B] is above and below the snow.

8 A

1st year

2nd year

3rd year

4th year

B

January   February   March   April   May   June   July   August   September   October   November   December

Life on islands

Oceanic islands produced by the eruption of undersea volcanoes are, for some time afterwards, unsuitable for any kind of life. Eventually, however, the lava cools and the area begins to become habitable. Chance is an important factor in colonization, but some organisms will manage to reach the land [Key] and in time even the most remote islands will become populated.

Certain kinds of vegetation and animals are adapted to travel to islands by air or sea [1]. Some plants produce numerous spores or very light airborne seeds, such as those of the daisy family (Compositae) that stand a better chance of making a sea crossing than do the large, heavy fruits of forest trees. Some plants have seeds that may be carried by birds, either hooked to their plumage or in mud on their feet. In a few cases seeds are carried in a bird's digestive system but the speed at which birds digest their food limits the chances of the seeds surviving long journeys. Some plants, notably the "coco de mer", the rare Seychelles double coconut, have well-protected floating seeds adapted to be spread by the sea. But it is doubtful whether most seeds can survive immersion for more than a few weeks.

Among the animals, many tiny creatures are carried in the upper air currents. Aphids and minute spiders form a large part of this "aerial plankton". Most die before they are dropped but a few survive to fall on to dry land and colonize new areas. At lower altitudes large insects and birds whirled along by a storm may make landfalls thousands of miles from their starting-points.

The successful colonies

Water travel is more difficult for land-dwelling animals. Gigantic rafts of vegetation [3], which are sometimes swept down to the sea by tropical rivers in flood, can become floating islands able to carry a number of small, tree-living animals to a distant landfall. The most successful colonists, however, are the reptiles, and the fauna of most tropical islands includes lizards and tortoises. Inside the timbers of a vegetation raft the grubs of beetles and other insects continue to gnaw their way, unaware of their changed circumstances. They may hatch out far from the place where the eggs were laid. The beetle fauna of many islands consists largely of wood-boring forms.

Once ashore on an oceanic island a degree of luck must be present to ensure survival. An animal arriving on a barren island will soon die from lack of food or water. But should the basic needs of life be available the new arrivals may find themselves in an open environment in which the ecological niches remain unfilled. The adaptability inherent in every living organism enables the newcomers to occupy some part of the available living space. In many instances [8] the descendants of a single species have diverged widely to form several new and different species adapted to the conditions on an island.

Certain trends can be seen among animals and plants on islands. One is increase in size for, in general, the lack of competition allows larger organisms to take what food is available. If predators are rare the need to be small or swift for protection will disappear. Giant forms, particularly of reptiles, can be found on tropical islands. Birds and insects are often markedly larger than their ances-

CONNECTIONS

See also
196 Isolation and evolution
152 Islands birds
32 Evolution: classical theories
244 Endangered birds

In other volumes
124 History and Culture 2

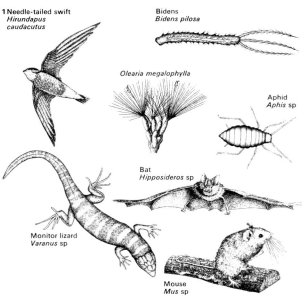

1 Needle-tailed swift
Hirundapus caudacutus

Bidens
Bidens pilosa

Olearia megalophylla

Aphid
Aphis sp

Bat
Hipposideros sp

Monitor lizard
Varanus sp

Mouse
Mus sp

1 Many kinds of organisms are likely to colonize islands. They include a number of medium-sized birds such as the swift, since smaller ones are unlikely to survive the journey and big ones are not so easily blown off their course. The birds may carry fruits like those of Bidens hooked in their plumage. Tiny insects, such as aphids, might be carried by air currents, along with plumed fruits of plants such as Olearia. Lizards can survive long-distance sea travel on vegetation rafts, as also can rats, mice and other small animals. Bats, too, will make short sea journeys and even survive long ones.

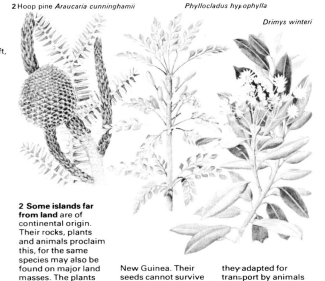

2 Hoop pine Araucaria cunninghamii

Phyllocladus hypophylla

Drimys winteri

2 Some islands far from land are of continental origin. Their rocks, plants and animals proclaim this, for the same species may also be found on major land masses. The plants illustrated grow in New Guinea. Their seeds cannot survive in seawater nor are they adapted for transport by animals or the wind.

3 Tropical rivers in flood will sometimes sweep downstream huge masses of timber and creepers tangled together into rafts of vegetation. As these floating islands reach the sea they are still the home of many animals. But each day farther out from land these creatures become fewer. First to go are the bigger mammals, such as the monkeys, which may have been clinging to the branches and cannot survive the reduced size of their habitat and the cooler sea temperature. The reptiles and some insects, which have lower metabolic rates and are relatively inactive, can last longer and perhaps make a successful landfall in a place where they can survive.

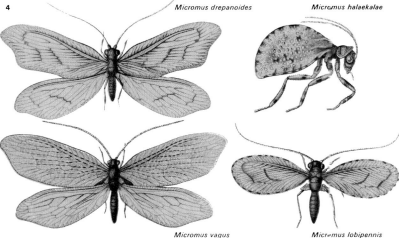

Micromus drepanoides

Micromus halaekalae

Micromus vagus

Micromus lobipennis

4 The lacewings that have reached the Hawaiian islands have undergone great changes to adapt them to life where flight can mean destruction by being blown out to sea. In some species, such as Micromus drepanoides and M. vagus, wings have become large and heavy, with angular corners, thus making flight almost impossible. In other forms the hindwings are almost totally absent (M. lobipennis). Some species with reduced hindwings have forewings that are studded with thickened areas that increase their weight and hamper flight, as in M. halaekalae.

tors, and plants may evolve from herbaceous types into tree-like forms.

Islands are windy places and organisms that have been blown there may well be blown off again. Successful colonizers often lose their power of flight and many species of flightless birds [5] are known in remote places. Flightlessness and large size are often combined and giant flightless birds, such as the extinct elephant birds of Madagascar, and outsize flightless insects, such as the lacewings of Hawaii [4], inhabit remote islands. Among plants, the fruits may become heavier and less likely to drift out to sea again.

The chance to survive

Although large animals cannot reach distant islands they are sometimes cut off from the main population by changes in sea-level. When this happens the opposite kinds of changes occur to those found in birds on remote islands. Large mammals make heavy demands on their environment and because islands tend to be restricted in size the smaller creatures, whose needs are simpler, stand the best chance of survival. Examples of this can

be seen on many continental islands. The race of sika deer (*Cervus nippon*) found in Japan, for example, are smaller than the mainland forms. An extreme case is that of elephants and hippopotamuses isolated on Mediterranean islands some two million years ago, during the Pleistocene, by rising sea-levels. Fossil evidence shows that these creatures developed into miniature forms as small as Shetland ponies.

The arrival of man

Plants and animals of remote islands often survived in a very delicate ecological balance until the arrival of man. To humans, the large, slow creatures meant little more than a convenient food supply and in many cases [6] they were virtually wiped out soon after their discovery. The lack of natural enemies often led the animals to be so trusting that their destruction was the more certain. The introduction of domestic animals, especially pigs and goats, has in many cases also caused the destruction of the unique flora of some islands, as well as bringing the native animals to the point of extinction.

KEY

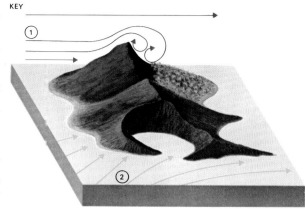

An island rising from the ocean bed, far from any land, can be colonized by plants and animals in only a limited number of ways. Those that do reach it arrive largely by accident. If the island stands in the path of regular strong winds, such as the trade winds [1], spores and lightweight seeds and tiny animals may be blown there. If it lies within a hurricane belt, storms may bring larger insects and birds that have been blown off course, while sea currents [2] may deposit vegetation carrying seeds and also small creatures.

5 Flightless cormorant
Nannopterum harrisi

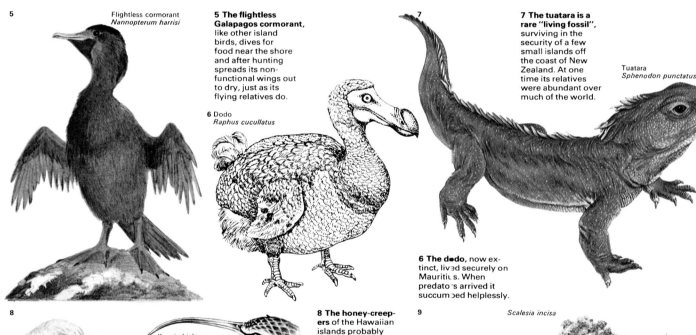

5 The flightless Galapagos cormorant, like other island birds, dives for food near the shore and after hunting spreads its non-functional wings out to dry, just as its flying relatives do.

6 Dodo
Raphus cucullatus

7

7 The tuatara is a rare "living fossil", surviving in the security of a few small islands off the coast of New Zealand. At one time its relatives were abundant over much of the world.

Tuatara
Sphenodon punctatus

6 The dodo, now extinct, lived securely on Mauritius. When predators arrived it succumbed helplessly.

8

Laysan finch
Psittirostra cantans

Seeds

Kauai akialoa
Hemignathus procerus

Insects, nectar

Amakihi
Loxops virens

KAUAI

OAHU

Insects

MOLOKAI

Maui parrotbill
Pseudonestor xanthophrys

Akepa
Loxops coccinea

LANAI

MAUI

Insects

HAWAII

Akiapolaau
Hemignathus wilsoni

Palila
Psittirostra bailleui

Seeds

Kona finch
Psittirostra kona

8 The honey-creepers of the Hawaiian islands probably have as ancestors a species of tanager (a widespread American bird family) that used, long ago, to migrate there. Their beak shapes show that some feed on nectar, others on insects or seeds. Among them they fill a variety of ecological niches.

9 The plumed fruits of dandelion-like plants (family Compositae) are carried long distances by the wind. On remote islands they often develop into tree-like plants adapted to extremes of climate or other conditions. *Dubautia* spp and *Argyroxiphium* spp of Hawaii and *Scalesia* spp of the Galapagos are all woody and tree-like in form.

9

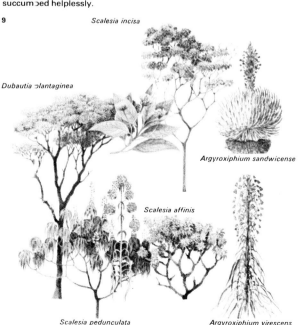

Scalesia incisa

Dubautia plantaginea

Argyroxiphium sandwicense

Scalesia affinis

Scalesia pedunculata

Argyroxiphium virescens

Lakes and rivers

The overwhelming mass of the world's water is oceanic, its freshwater lakes and rivers covering a minute area by comparison. However, fresh waters provide a wide variety of habitats and these are occupied by a diverse selection of specialized plants and animals. For the purposes of plant and animal study rivers are divided into "reaches", each characterized by a dominant fish species. The classification is based on European rivers but can be adapted to suit rivers anywhere.

The reaches of a river

The typical river begins high in the mountains as a small stream, with occasional sharp falls, trickling over rocks. This is referred to as the upper cascade reach, an area of cold, clear and pure water with few inhabitants although some insects, and the bullhead (*Cottus gobio*), may be present. Downstream, where the water is still clear and cold and the slope still steep, an amalgamation of small torrents forms the trout reach. There is a sharp increase in the number of fish species, many of them using the pebbles that accumulate in sheltered spots for their breeding places.

Salmon will penetrate upstream to the lower part of the trout reach, but they thrive much better in larger rivers with deeper, pebbly areas. These are found in the grayling reach, often called the minnow reach in Britain.

In the grayling reach the well-oxygenated water still flows rapidly but it is deeper and more suited to the salmon. The turbulence of the water prevents many plants from rooting at the edge of the stream but the high oxygen content encourages a wealth of small animals such as crustaceans, worms, insects and their larvae. The grayling reach is followed by the barbel reach. The slope of the river is easing, the flow less headlong, but it is still cool, clear and well oxygenated.

The last reach before the estuary is the bream reach, which winds slowly across a flat landscape. The water is warm and murky with suspended silt. The oxygen level is lower but it may be augmented in the summer by the many plants growing in what is often biologically the richest part of the river supporting a wide variety of fish and other forms of animal life.

The last stage of the river's life is the estuary where salt water from the ocean contaminates and changes its character. Estuarine waters are a unique environment in their own right and are not strictly part of the river, which shares them with the sea. Some riverine species – the pike and perch among them – may occur in several reaches but most are restricted to the section that is their native habitat.

Life in lakes and ponds

A lake may be part of a river, or its birthplace, and ponds may speckle the valleys and plains through which the river passes. The difference between a lake and a pond is based on the depth of the water. Area alone does not turn a pond into a lake although a large pond is sometimes given the nominal status of one. In a pond rooted plants may grow from any part of its silty bed because its water is uniformly warm and poorly oxygenated in summer. On the edge of the lake conditions may be pond-like but beyond, where the lake's bed may fall to great depths, no rooted plants can grow. The water is cold and dark even in midsummer and separated from the

1 No part of the living space provided by a pond is neglected. Plants root in the muddy bottom or float on the surface. Some animals, such as *Tubifex* worms or midge larvae, bury themselves in the silt but most are more active, gliding over the mud like the planarians, undulating through the water like the leech or jogging upwards like the water fleas. The larvae of pond insects are the food of newts and fish, which are in turn eaten by birds such as herons.

1 Dragonfly *Aeschna grandis*
2 Water starwort *Callitriche* sp
3 Great diving beetle *Dytiscus marginalis*
4 Great pond snail *Lymnaea stagnalis*
5 Pea mussel *Pisidium amnicum*
6 Tubifex worm
7 Moorhen *Gallinula chloropus*
8 Pond skater *Gerris* sp
9 Newt tadpole *Triturus vulgaris*
10 Hornwort *Ceratophyllum demersum*
11, 12 Water spider and nest *Argyroneta aquatica*
13 Great ramshorn snail *Planorbis corneus*
14 Toad tadpole *Bufo bufo*
15 Larva of great diving beetle *Dytiscus marginalis*
16 Blood worm *Chironomus* sp
17 Mayfly *Cloeon dipterum*

18 Arrowhead *Sagittaria sagittifolia*
19 Dragonfly nymph *Libellula quadrimaculata*
20 Water boatman *Notonecta glauca*
21 Wandering snail *Lymnaea pereger*
22 Three-spined stickleback *Gasterosteus aculeatus*
23 Hydra *Hydra oligactis*
24 Great crested newt *Triturus cristatus*
25 Damselfly nymph *Agrion virgo*
26 Aplecta *Aplecta hypnorum*
27 Cyclops *Cyclops* sp
28 Common frog *Rana temporaria*
29 Horse leech *Haemopsis sanguisuga*
30 Ivy-leafed duckweed *Lemna trisulca*
31 Minnow *Phoxinus phoxinus*
32 Frog-bit *Hydrocharis morsus-ranae*

33 Mallard *Anas platyrhynchos*
34 Kingfisher *Alcedo atthis*
35 Damselfly *Coenagrion puella*
36 Water crowfoot *Ranunculus aquatilis*
37 Water scorpion *Nepa cinerea*
38 Toad spawn *Bufo bufo*
39 Saucer bug *Ilyocoris cimicoides*
40 Water flea *Daphnia* sp
41 *Bithynia tentaculata*
42 Freshwater sponge *Spongilla lacustris*
43 Water mites *Hydrarachna globosa*
44 Caddisfly larva *Limnephilus flavicornis*
45 Painter's mussel *Unio pictorum*
46 Water louse *Asellus aquaticus*
47 *Planaria gonocephala*

upper water – warm and containing life forms – by a thermocline or region of abrupt temperature change. There is some mixing of the two water zones during winter storms but in summer the two water bodies exist separately in the same basin.

Plant life in lakes and rivers includes some planktonic floating organisms – richer in lakes and ponds than in rivers, where they would be swept away by the current. In the still waters of the pond larger plants such as frogbit and bladderwort are rootless, but most are attached to some kind of firm substrate. Occasionally they grow up through several feet of water – the water-lily is one such plant. A common feature of stream plants is the possession of dissected underwater leaves that reduce the drag of the flowing water, plus broad-surface leaves that catch the sunlight and thus effect photosynthesis. In their protected environment most water plants have tissues that are unable to support themselves in air yet have great power to resist the relentless tug of the water.

Ponds, lakes and rivers are stocked with a rich animal life wherever the worst effects of

man's presence have not polluted the water and reduced it to lifelessness. Lakes are often large enough to support a fish population of such numbers that a commercial fishing industry can also be maintained. And most ponds, however small, have enough fish to attract a crowd of young anglers.

Amphibians, birds and mammals
Apart from the species that visit fresh water to drink, lakes and ponds are also inhabited by creatures of two worlds, the amphibians who spawn in the waters and whose tadpoles lead an aquatic life until they metamorphose. Even afterwards they must remain in or close to a moist environment to prevent drying out. Many birds, some of them adapted for swimming by having webbed feet and dense feathers naturally waterproofed with oil, form part of the freshwater populations, as do many mammals. Some of these, such as the muskrat and the coypu, are herbivorous; others, such as the mink and the otter, are carnivores. Every major phylum except echinoderms (starfish and sea-urchins) is represented among freshwater invertebrates.

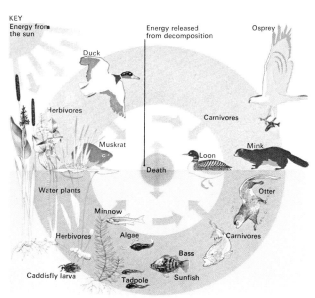

A river, pond or lake is a largely self- / contained microcosm taking energy from / the sun and minerals from the land.

2 Mayfly
Rhithrogena semicolorata

Caddisfly
Plectrocnemia conspersa

2 In the upper cascade reach the rush of a mountain torrent is usually unbearable for most animals. The few inhabitants spend only their immature phases here. Some caddis larvae attach themselves to boulders by silk webs / to avoid being swept away, while black fly larvae hang on with hooks at the rear end. Some mayfly larvae are streamlined swimmers; those with flattened forms cling firmly to large stones and let the water flow past.

Caddisfly larva net
Plectrocnemia conspersa

Black fly larva
Simulium sp

Mayfly nymph
Baetis rhodani

Mayfly nymph
Rhithrogena semicolorata

Caddisfly larva net
Hydropsyche fulvipes

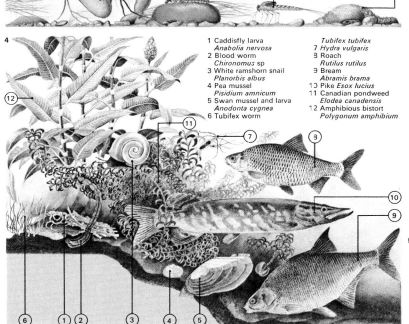

1 Caddisfly larva
Anabolia nervosa
2 Blood worm
Chironomus sp
3 White ramshorn snail
Planorbis albus
4 Pea mussel
Pisidium amnicum
5 Swan mussel and larva
Anodonta cygnea
6 Tubifex worm

Tubifex tubifex
7 Hydra *vulgaris*
8 Roach
Rutilus rutilus
9 Bream
Abramis brama
10 Pike *Esox lucius*
11 Canadian pondweed
Elodea canadensis
12 Amphibious bistort
Polygonum amphibium

4 In lower reaches, dense growth such as Canadian pondweed – introduced into Europe where it choked many watercourses like a form of natural pollution – tangles with *Potamogeton* and amphibious bistort. In the silt bed *Tubifex* worms and / some midge larvae are buried. Their red haemoglobin enables them to use what little oxygen the water contains at this level. A few freshwater dwellers are filter feeders, like the swan mussel and the small related pea mussel. / Many more are plant feeders like the ramshorn snail that glides over plants. Caddis larvae, of different species to those of fast streams, protect their bodies with cases made of sand grains or vegetable debris, some neatly, some in a / higgledy-piggledy manner. Deep-bodied, slow-moving fish such as roach and bream are found here, making a rich living off the wide range of plants and animals in the habitat. In turn, they are preyed on by the fierce, lurking pike.

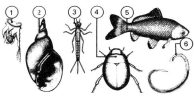

3

3 Some animals can survive even in the hot springs that bubble up from the earth in areas of volcanic activity. To the heat is added a high mineral content but the conditions do not seem to discomfit a New Zealand water snail (*Lymnaea tomentosa*) [2] and a damsel fly nymph of the genus *Ischnura* [3] which occur in water up to 35°C (95°F). A beetle (*Laccobius* sp) [4], some rotifers [1] and the carp (*Cyprinus carpio*) [5] can stand even higher temperatures and a nematode (*Tylocephalus* sp) [6] has been recorded in water up to 80°C (176°F). This is the limit of life: above it, protein coagulates and organisms no longer survive.

5 Pond dwellers endure greater changes of living conditions than those of any habitat other than the sea-shore. The contrast for them is most marked in winter when silt / accumulates and the surface freezes over. At this time frogs, fish and other animals retire to hibernate in the muddy bed of the pond, but those that remain active rarely / suffer from an oxygen deficiency. As the ice forms an air bubble is trapped underneath it and from this sufficient gas diffuses into the water to sustain the creatures living there.

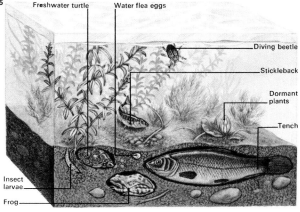

5 Freshwater turtle

Water flea eggs

Diving beetle

Stickleback

Dormant plants

Tench

Insect larvae

Frog

Wetlands: marshes and swamps

Wetlands are the most difficult of all the major environments to define and describe concisely. They may be regions of meandering rivers in the Arctic or of ephemeral lakes in the tropics. They may be acid upland bogs, with a flora and fauna quite different from that of alkaline lowland fens only a few miles away, or they may be areas of slow-flowing water such as the Sudd or the "River of Grass" of the Everglades of the USA. The one factor that all of these habitats share is water – usually poorly oxygenated – to which their inhabitants are adapted in an astonishing richness of life. The term "wetlands" has been coined to cover all of these varied habitats; as normally used it excludes lakes and rivers and saltwater estuaries.

Luxuriant plant life
Few woody plants can stand continuously waterlogged conditions; thus the luxuriant plant growth often found in wetlands is usually made up of herbaceous plants. Some wetland plants are adapted to slight waterlogging, others to life afloat in stagnant water and many to intermediate conditions.

The extreme acidity of many wetland areas means that nutrients, trapped in the peat at the bottom of the water, are not available to the plants. Some have overcome this by developing a carnivorous way of life. Using modified leaves, pitcher plants, Venus fly-traps, sundews and butterworts, for example, trap insects or other small animals which they consume by the use of enzymes. The minerals from the tissues of these animals compensate for the deficiency of minerals in the environment.

Animal life of wetlands
The animal life of wetlands is usually very rich; these regions are often regarded by man as areas of little use so that animals, hounded elsewhere, find in them a secure haven. Some large mammals are specialized for a wetland life, with adaptations of the feet, in particular. These often spread to take the creature's weight on sinking ground. Even reindeer, which spend much of their time in swampy tundra areas, have toes that spread to take their body weight at each step.

Other mammals and birds may have swamp-adapted feet [Key], but it is in some of the lower animals that the most complete adaptations to a wetland life take place. Crocodiles and many other reptiles are well adapted to life in tropical swamps where, because they are air-breathers, the deoxygenation of the water does not bother them. Many tropical amphibians have adaptations involving parental care of the young, a reflection of the ephemeral quality of many of the ponds and waterways that are their homelands.

Fish of swampy areas are usually deep-bodied animals, better able to weave between thick-growing vegetation than the streamlined fish of fast-flowing or open waters. Most can survive reduced oxygen levels and in the tropics where, even before total evaporation takes place, the warm water can hold very little oxygen, many have the ability to breathe dry air. This they do in a variety of ways – in some the gill chambers have become richly supplied with blood capillaries and act as lungs; in others, such as some of the loaches, part of the gut takes over the task of absorbing oxygen and in the

1 Matetite reed
  Phragmites sp
2 Hammerhead stork
  Scopus umbretta
3 Hippopotamus
  Hippopotamus amphibius
4 Saddle-billed stork
  Ephippiorhynchus senegalensis
5 Sitatunga
  Tragelaphus spekei

6 Black crake
  Limnocorax flavirostra
7 Water cabbage
  Pistia stratiotes
8 Swamp worm
  Alma emini
9 Bichir
  Polypterus sp
10 Catfish
  Malapterurus sp
11 Papyrus

Cyperus papyrus
12 Malachite kingfisher
  Corythornis cristata
13 Herald snake
  Crotaphopeltis hotamboeia
14 African spoonbill
  Platalea alba
15 Lily trotter
  Actophilornis africana
16 Water-lily

Nymphaea sp
17 Shoebill
  Balaeniceps rex
18 Squacco heron
  Ardeola ralloides
19 Marsh mongoose
  Atilax paludinosus
20 Snail
  Biomphalaria sudanica
21 Lungfish
  Protopterus aethiopicus

1 A swamp in the Upper Nile valley is dominated by papyrus, the paper reed, which grows to a height of 3.5m (12ft) or more. Other plants, which mask the open water, include water-lilies and water hyacinths. The largest of swamp animals is the hippopotamus, which inhabits shallow rivers and lakes over much of Africa. Hippos usually live in groups, leaving the water at night to feed. The sitatunga is far more secretive than the hippo and is rarely seen. If in danger it will submerge with only its nostrils showing, for long periods if need be. The many birds are almost all long-legged relatives of the herons. They feed on small animals, particularly insects, which swarm throughout the swamp.

mailed catfish even the stomach has become specialized to serve as a lung.

The true lungfish [2] are all to be found in the tropics of the southern continents, living in streams or pools that are liable to deoxygenation or even complete drying out. It is from relatives of ancestors of the lungfish, which may have been very similar to modern genera, that the first land-dwelling animals, the early amphibians, are thought to have evolved, for they were able to survive the desiccation of their swampland habitats.

Among the invertebrates most freshwater snails breathe by means of lungs rather than gills and so they are well placed to survive should their ponds dry up. Most of these creatures can stand adverse conditions – even the freezing of their swampy homes in wintertime – by growing a mucus shield over the mouth of the shell and entering a state of dormancy until more favourable conditions allow them to return to full activity.

All of the insects that inhabit swamps are also air-breathers, as are the larvae of midges and mosquitoes. These have snorkel-like tubes at the hind ends of their bodies, which they can push up to the surface of the water. A number of other larvae, such as those of dragonflies and water beetles, have gills, but in both these cases the adults breathe dry air because they leave the water for at least part of their lives, often in order to find a mate and thus complete the life cycle.

Man and the wetlands

Because wetlands cannot be farmed easily man has tended to ignore them until relatively recently. But today, with the need for more land, many swamp and marsh areas have been drained. This has often reduced the numbers of disease-carrying organisms, such as mosquitoes, adding a further impetus to the drainage programmes. Nevertheless, wetlands have an importance beyond their own boundaries because they often act as a reservoir of water for distant areas and are frequently the seasonal home of many species of migratory birds. Naturalists and conservationists have realized this for many years and some of the earliest conservation attempts, such as those in North America, have been on behalf of wetlands.

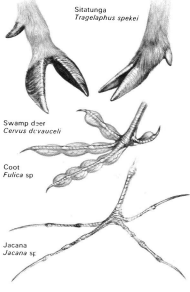

Sitatunga
Tragelaphus spekei

Swamp deer
Cervus duvauceli

Coot
Fulica sp

Jacana
Jacana sp

Animals of wetlands run the risk of becoming bogged down in the soft, quaking ground that lies at the water's edge, so they frequently develop large, weight-spreading feet. Among birds the long toes of herons, waders and moorhens are examples of this adaptation. It takes an extreme form in the jacana, which can even walk on floating vegetation. Some waterside birds have partly webbed feet. In the coots and finfoots the adaptation takes the form of lobes along the toes which, as well as spreading their weight, help them to swim. Both the swamp deer of India and the sitatunga, an African antelope, have long, loosely-jointed toes.

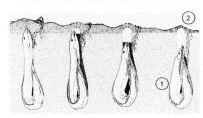

African lungfish
Protopterus sp

2 The lungfish is related to extinct creatures that were the first vertebrates to live on dry land. Today it is found in tropical regions of the three southern continents. The South American and Australian forms can survive complete deoxygenation of the water by breathing air. The African species *(Protopterus)* can live through total dehydration of its environment by burrowing into the mud while it is still damp and enveloping itself in a mucous cocoon. This dries to form a protective case [1]. Air from the disturbed mud filters through a porous "lid" [2] at the top of the case. During this dormancy it lives on reserves of fat.

3 The boat-billed heron is a bird of Central and South America that lives mainly in the freshwater parts of mangrove swamps. It is secretive and nocturnal, somewhat like the night herons of the Old World. Little is known of its habits, but it has been observed standing or walking slowly in shallow water, using its broad bill to scoop up shrimps and other invertebrate prey.

3 Boat-billed heron
Cochlearius cochlearius

4 The matamata, a South American turtle, has a bossed carapace that resembles a lump of dead, waterlogged wood. The flaps of tissue that dangle from its head and neck look like inviting scraps of food to the small denizens of the unruffled tropical waters where the matamata lives. The inquisitive amphibians or fish do not discover their mistake until it is too late. In spite of its grotesquely flattened head, the matamata has huge if feeble jaws and a greatly distensible throat. It sucks in a huge volume of water, bearing any nearby small animals irresistibly with it.

4 Matamata
Chelys fimbriata

5 Much of the vast Amazon basin is drained by meandering, swampy streams. Electric fish such as the electric eel *(Electrophorus electricus)* can stun their prey with charges of up to 600 volts. Here, too, lives the arapaima, the world's largest freshwater fish, weighing up to 90kg (200lb). Some of the nearby faster flowing waters abound with shoals of piranha, *(Serrasalmus natteri)* which are among the most voracious of all vertebrates. Hunting in "packs", they can reduce a human being to a skeleton within minutes.

1 *Symphysodon discus*
2 *Hyphessobrycon innesi*
3 *Boulengerella lateristriga*
4 *Anostomus anostomus*
5 *Carnegiella strigata*
6 *Leporinus fasciatus*
7 *Prochilodus insignis*
8 *Metynnis schreitmuelleri*
9 *Serrasalmus natteri*
10 *Gymnotus carapo*
11 *Gymnorhamphichthyes hypostomus*
12 *Electrophorus electricus*
13 *Oxydoras niger*
14 *Pseudoplatystoma fasciatum*
15 *Ancistrus cirrhosus*
16 *Auchenipterus nigripinnis*
17 *Corydoras myersi*
18 *Arapaima gigas*
19 Young *Arapaima gigas*
20 *Osteoglossum bicirrhosum*

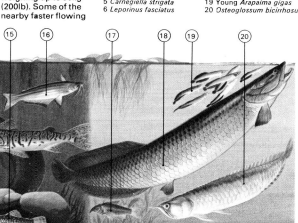

Salt marshes and coastal swamps

The boundaries between sea and land are not always exact and throughout the world complex transition zones exist. In the temperate regions of the world they are represented by salt marshes; while in the tropical regions mangrove swamps [2] flourish on many coasts. Both areas are highly productive and present a habitat that has a unique and populous fauna. The plant species that compose the marsh and swamps are specially adapted to withstand the presence of salt water and periodic immersion by tides [Key].

Salt marsh succulents

Many plants of the salt marsh are succulent, their stems and leaves swollen with water stored in special tissues. Salt marsh plants suffer from a lack of water in much the same way as desert plants because it is difficult for them to extract water from the sea. The reason for this so-called "physiological drought" is that the concentration of mineral salts in seawater is similar to that inside the plant cells and, as a result, little water is able to move into those cells. Thus storing water overcomes the problem.

Another problem faced by salt marsh plants is oxygen lack. To overcome this many plants develop aerial roots that grow like periscopes above the surface of the mud. The tolerance of salt concentrations, and of the length of time of inundation by the tide, varies from species to species and thus there is a characteristic division into zones [1].

In the mangrove swamps of South-East Asia growing near to the sea are *Sonneratia* spp [2] with a wild array of aerial roots sprouting from both branches and trunks. The stands behind them are of *Rhizophora* spp [4], which have roots that lie just above or below the ground. Growing behind these are *Bruguiera* spp with their roots buried in the mud, leaving only small spikes jutting above the surface. The mangrove roots trap sediment that accumulates to form muddy banks; these present a new habitat to be colonized by more trees. In this way the stilted forest takes over the shoreline in a slow but sure march into the sea.

Some of the most characteristic animals of the mangrove forest are mudskippers [5, 6], tiny fish that can live out of water and walk across the mud. Millions of small fiddler crabs (*Uca* spp), each with one outsize claw, scuttle along the mud seeking refuge beneath it when the tide rises or danger threatens. The vast number of molluscs, crustaceans and fish that live on the quantities of organic debris provide prey for monitor lizards (*Varanus salvator*), sea crocodiles (*Crocodylus porosus*) and various extremely venomous fast-swimming sea snakes such as the banded sea snake (*Laticauda colubrina*).

Mangrove birds and mammals

The mangrove snake (*Boiga dendrophila*) lives on birds that flock to the mangrove swamps to take fish and shellfish. These include the graceful fish eagles (*Haliaeetus leucogaster*) and the tall adjutant storks (*Leptoptilos javanicus*). Mammals include the long-tailed macaque (*Macaca crus*) which is also known as the crab-eating monkey. As their name suggests, the members of the clan spend their time on the mudflats watching for any crabs that disappear down their burrows. When the crab reappears it is skilfully grabbed, torn apart and devoured. Found

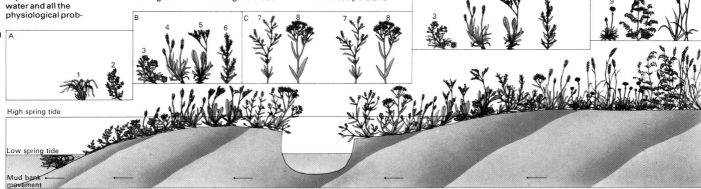

1 **On a typical temperate salt marsh** a distinct division of plants can be seen, which reflects the plants' abilities to withstand periodic immersions by the tide and thus their exposure to salt water and all the physiological problems this involves. The primary colonizers [A] of the bare mud are the eel grass (*Zostera* sp) [1] and saltwort (*Salicornia* sp) [2]. At the beginning of the 20th century the sea cord grass (*Spartina townsendii*) became a major colonizer of the bare mud zones of many European salt marshes. The general marsh community [B, C, D] contains a number of different plants, among them sea spurry (*Spergularia* sp) [3], sea plantain (*Plantago maritima*) [4], sea lavender (*Limonium* sp) [5] and sea blight (*Sueda maritima*) [6]. On the hummocks and edges of creeks [C] are found sea purslane (*Halimione portulacoides*) [7] and sea aster (*Aster tripolium*) [8]. These grow only in the better-drained areas. At the edge of the marsh, in the areas with the least likelihood of inundation, are found [E] thrift (*Armeria maritima*) [9], sea wormwood (*Artemesia maritima*) [10] and sea couch grass (*Agropyron* sp) [11].

High spring tide

Low spring tide

Mud bank movement

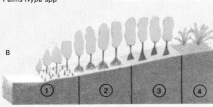

2 **Moving forever seawards,** the mangrove forests [C] claim new territory for the coastline. A section through a mangrove swamp [B] shows the distribution and zonation [A] of the various species of mangrove. The pioneer mangrove (*Sonneratia* sp) [D] has large numbers of upright aerial roots (pneumatophores) by means of which it is able to breathe. Lateral roots form from these [E].

1 *Sonneratia* zone
  S. griffithii, S. alba

2 *Rhizophora* zone
  R. mucronata, R. apiculata

3 *Bruguiera* zone
  B. parviflora,
  B. gymnorniza
  B. cylindrica, B. sexangula

4 Palms *Nypa* spp

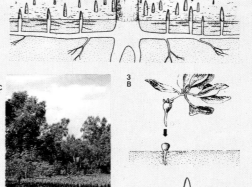

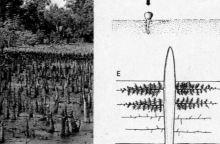

3 **The seeds** of the *Rhizophora* mangrove germinate before they leave the parent plant [A]. When they fall from the tree their roots stick in the mud [B] and the seedlings become established before the tide can wash them away.

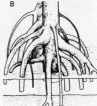

4 **The knee roots** of *Bruguiera* [A] and the stilt roots of the *Rhizophora* mangrove [B] differ from those of *Sonneratia*. They form a dense network around the stem.

exclusively in the mangroves of Borneo are the rare proboscis monkeys (*Nasalis larvatus*) [7], grotesque-looking creatures with large nasal appendages that hang over the mouths and chins of the male. Despite their appearance they are peaceful animals. They live in troops of 15 or 20 and feed on leaves of the *Sonneratia caseolaris* mangrove.

Marshland communities

The salt marshes of the world, although not as dramatic in appearance as the mangrove swamps, are no less productive. Many of the estuaries and coastal marshes serve as nurseries for a wide variety of animals. Many fish and invertebrates lay their eggs in these sheltered regions and the newly hatched young are less vulnerable in the protective, shallow waters. The division of plant species into zones is also affected by the tidal range and reflects individual tolerances to salt concentrations and periods of covering by the tide. The lower reaches, which may be submerged at all times, may be colonized by eel grass (*Zostera* spp), which provides food for the brent geese (*Branta bernicla*), or by

meadows of turtle grass (*Thalassia* spp), the food of the green turtle (*Chelonia mydas*). Adjoining this zone are expanses of saltwort (*Salicornia* spp), or marsh cord grass (*Spartina* spp), which are tolerant to high salt concentrations. These provide detritus vegetation that feeds a multitude of molluscs, crustaceans and birds.

Farther towards firm ground is a general salt marsh community that is accustomed to the salt concentrations but not to prolonged immersion in seawater. The richest feeding grounds for visiting birds are those that are exposed for long periods each day. The open marsh provides food for widgeon and brent geese and the maze of channels and pools criss-crossing the area yields food for opportunist feeders such as gulls and shelduck [10]. The best adapted of all birds are the waders, such as the redshank [11].

The area of the Camargue in southern France is a patchwork of fresh and salt-water marshes adjoining the Mediterranean. A multitude of birds and flocks of greater flamingos present a dramatic spectacle among the less colourful species.

KEY

Plants of the genus *Salicornia* are among the first colonizers of sandy and muddy shores. There are about 35 species world-wide, commonly called saltwort, glasswort or samphire, all so similar in appearance that they are almost impossible to tell apart. Small plants have an upright growth but larger plants tend to sprawl. They have minute leaves and a succulent form, which are adaptations to conserve water. Although they live in wet conditions the soil contains a lot of salt and this prevents water from being freely available to the plants. These and other plants adapted to living in a salty environment are known as halophytes.

5 The mudskippers of the mangrove swamps are fish that live as much out of the water as in it. Each species [B] feeds on a different diet and occupies a separate niche in the mangrove mud. *Boleophthalmus boddaerti* is found at the seaward edge. Mudskippers exert strong territorial rights at breeding time. *Periophthalmus chrysopilus* builds a circular burrow in the mud to which it attracts a female with a series of leaps. In

contrast *Periophthalmodon schlosseri* makes its home on the firmer mud within the fringes of *Avicennia* and is carnivorous; *B. boddaerti* feeds on algae at the seaward edge of the swamp. Courtship of the mudskippers involves a typical mouth-to-mouth display [A].

5

A   *B. boddaerti*

B

6 As the tide rises the mudskipper *Periophthalmus chrysopilus* climbs the mangrove trees [A] and clings on with a sucker [B].

B

Periophthalmodon schlosseri

Periophthalmus chrysopilus

Boleophthalmus boddaerti

7 A

B

7 The mangrove forests of Borneo and South-East Asia are the home of the grotesque and amiable proboscis monkey (*Nasalis larvatus*) [A]. The silvered langur (*Presbytis cristatus*) [B] and dusky langur (*P. obscuras*) [C], live in the same habitat.

C

10 The shelduck is a familiar inhabitant of salt marshes of Britain and western Europe.

8A

8 The seed-eating seaside sparrow (*Ammospiza maritima*) [A] and the long-billed marsh wren (*Telmatodytes palustris*) [B] an insect-eater, live in east coast North American salt marshes.

11 A typical wading bird of the salt marsh is the redshank. t probes in the mud for food with its long beak.

B

9

9 The red-headed honey eater (*Myzomela erythrocephala*) is one of 20 species virtually confined to mangrove swamps in Australasia.

12 Brown pelican *Pelecanus occidentalis*

10 Shelduck *Tadorna tadorna*

11 Redshank *Tringa totanus*

12 The brown pelican has a large bill with a distensible pouch that it uses to catch the fish on which it feeds. It lives on the coasts of tropical America.

235

The sea-shore: life between the tides

The shore lies between the land and the sea, allied to both yet belonging to neither. It is a zone defined by the daily rise and fall of the tides, washed by salt water, but exposed to the damaging effects of drying air.

Types of beaches

The type of beach is dependent on its hinterland – hard rocks give a cliffed and rocky shore while softer rocks give a sandy or muddy beach – and on the effectiveness of the waves in eroding the rocks.

On shingle or pebble beaches it is impossible for plants to grow because the action of the waves causes the stones to rub together and grind off any life form that attempts to gain a foothold. The only animals on a pebble beach exist on the strand line at the edge of the high tide mark.

As well as the composition of the beach other factors that have an important influence on its fauna and flora are its aspect and degree of slope. Shallow, sloping beaches offer a far greater area for the development of animals and plants than do steeper beaches. The effect of the tide on sea-shore life is also altered by the degree of slope, because the steeper the slope the fiercer are the waves. On a steeply sloping beach the waves have a greater scouring action, preventing all but the most tenacious animals and plants from securing a foothold.

Once seaweeds do obtain a hold on shallow, rocky shores they exert a great modifying influence on wave action. Beneath their sheltering fronds large numbers of less well adapted plants and animals are able to find a secure home.

Life on beach zones

The beach can be divided into a series of zones [7] according to how far it is influenced by the water. At the land edge is the splash zone, normally wetted only by sea spray but still affected by the maritime influence. Below this is the upper shore extending from the level of high spring tides down to the average high tide level. The middle shore runs from there to average low tide level and the lower shore from that point to extreme low spring level. A walk from the upper shore to the low tide line shows that the plants and animals change with their level on the beach. This zonation is one of the most characteristic features of the sea-shore.

The beach as a whole is the most variable of all environments. When the tide is out the drying effects of the wind and sun threaten the plants and animals of the upper shore. If they are not to become desiccated they have to be well protected. In summer the beach and the rock pools may warm up considerably, but they are cooled in a moment when the water comes splashing back. At low tide the salinity of rock pools may increase with evaporation or decrease if there is heavy rainfall. The acidity of rock pools is low during the daytime when the plants are photosynthesizing but may increase sevenfold at night when they are producing carbon dioxide. Most plants and animals are adapted to live in a narrow range of temperatures, salinity and acidity but the organisms of the shore can stand continuous large variations in their environment.

The plants of the shore are almost all seaweeds [7]. These are entirely different from land plants in that they have no roots, stems,

2 The sand hopper lives on the strand line near the top of the beach where the last energy of the waves has thrown the detritus of the sea. Huge numbers of these small animals are found living in decaying seaweed, sand and even fine gravel, which may be fairly dry. They are valuable scavengers of dead material and in turn create an abundant food supply for shore birds such as turnstones on the upper shore.

1 Barnacles, sea snails and seaweeds are typical of sea-shore life. Barnacles [A] feed by opening their shelly plates when submerged, catching minute, suspended food particles. Many sea snails, such as the dog whelk [B], eat flesh. A dog whelk rasps through the shells of barnacles and mussels with its strong, file-like tongue (radula) to reach the unprotected animal inside. Seaweeds do not have roots but attach themselves to rocks with strong, branched holdfasts [C].

Sand hopper *Orchestia gammarella*

3 Sand burrowers

Parchment worm
Chaetopterus variopedatus
1 Mouth
2 Funnel
3 Fan
4 Parchment tube

Peppery furrow shell
Scrobicularia plana

Sea potato
Echinocardium cordatum
5 Respiratory funnel
6 Feeding area
7 Oral tube feet
8 Sanitary tube

Structure of tentacle
9 Ciliated surface
10 Food groove
11 Muscle fibres

Sea mouse
Aphrodite aculeata

Sand gaper
Mya arenaria

Amphitrite johnstoni

Lug worm
Arenicola marina

Worm cast

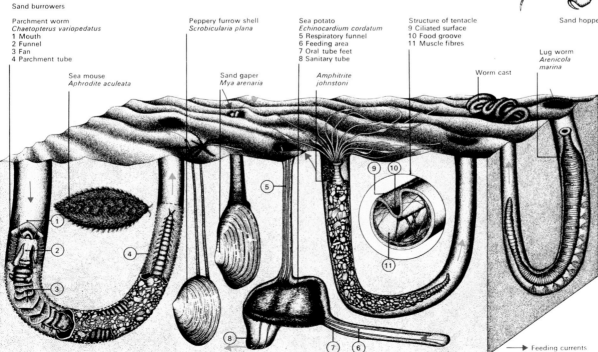

Feeding currents

3 The surface of a sandy beach at low tide gives few clues to the amount of life hidden beneath it. The principal inhabitants are worms and bivalves of many kinds, but other animals such as burrowing sea-urchins and crabs may also be present. Most bivalves depend on the planktonic richness of the sea for their food: when the sand is covered at high tide they push siphons up to the surface and pump a flow of water that circulates through their gill system. By this means oxygen is removed and food particles are trapped. Worms may use specialized tentacles or, like the lugworm, eat sand to swallow the tiny inhabitants of the water film round each grain.

leaves, flowers or fruit. Supported and bathed in seawater, they absorb all the nutrients they need directly from the sea. All photosynthesize, but in many the green of the chlorophyll is masked by other pigments that assist photosynthesis at low light levels or screen the chlorophyll from intense light. Defended against desiccation by sticky mucilage, they are often unattractively slimy objects when found at low tide or when cast up on the beach. Under water they are transformed, for their structure allows the fronds to be carried towards the light.

Animals of the shore
Among the animals of the shore, most of the major groups, or phyla, are represented. Frequently brightly coloured and of bizarre shapes, to many people they are one of the great attractions of the seaside. On the middle and upper shore in particular, many are protected against the battering force of the waves by heavy shells, although others creep into burrows or cracks in rocks either in times of storm or at low tide. Most are more or less sedentary but they produce planktonic

larvae that float off in the water and may colonize other beaches.

The feeding patterns of beach animals are complex. A few animals eat the seaweeds, but the big algae are largely inedible, although their smaller relatives of the open sea are the basis of all life in the oceans. Some animals are carnivores [1], some are scavengers [2] and many are filter feeders, finding food in what they can strain from the floating life of seawater. A variant is found in the sand eaters [3], which ingest vast amounts of sand in order to devour the tiny animals whose homes are the jackets of water that surround each sand grain.

Apart from recreation man has comparatively little use for the sea-shore. In some places minor industries are based on collecting molluscs or crustaceans or even the algae. Unfortunately most seaweeds cannot be digested by man or most other large animals unless processed to provide useful minerals (particularly iodine) and mucilage. Some seaweeds also yield alginates. These chemicals are the basis of "instant" desserts and other edible products.

KEY

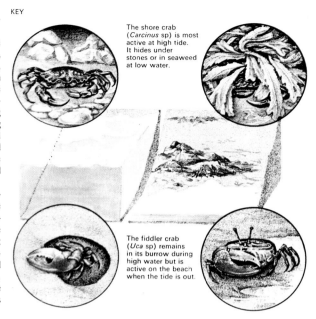

The shore crab (*Carcinus* sp) is most active at high tide. It hides under stones or in seaweed at low water.

The fiddler crab (*Uca* sp) remains in its burrow during high water but is active on the beach when the tide is out.

4 Starfish live mostly on the lower shore although they may be cast higher up the beach by the tide. *Linckia laevigata* comes from Pacific coral reefs which are a special type of shore.

5 Sea slugs are snails without shells but unlike land slugs they are frequently beautiful. This one is crawling over a red seaweed. They owe their protection, in part, to their unpalatable taste.

7 Most living things are closely zoned to a narrow part of the shore and no single species of plant or animal is found in all zones. Among algae, green species live on the higher levels of the shore although channelled wrack, a brown weed, is often found on rocks at the landward edge of the beach. Brown weeds generally belong to the middle of the shore although the oar weeds extend below the bottom of

the tide line. The fragile-looking red weeds grow on the lower shore. Animals are also zoned. Sometimes there is an association between a particular species and a seaweed, such as that between the flat periwinkle and bladder wrack. Those that can endure the greatest desiccation and varied conditions live at the top of the shore, while those that require greater stability are found closer to the sea.

6 Many species of birds feed on the small animals of the shore. Gulls will eat almost any kind of food but most of the others are specialists and will eat only a narrow range of organisms.

Waders such as the redshank probe the mud with long, thin bills. The oystercatcher can open cockle shells at a blow. The bills of eider ducks crush the shells of crabs and sea-urchins.

1 Redshank *Tringa totanus* 7
2 Oyster-catcher *Haematopus ostralegus*
3 Shelduck *Tadorna tadorna*
4 Black-headed gull *Larus ridibundus*
5 Eider *Somateria mollissima*
6 Little tern *Sterna albifrons*

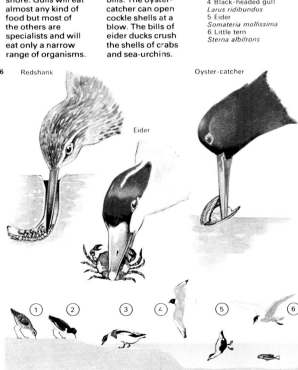

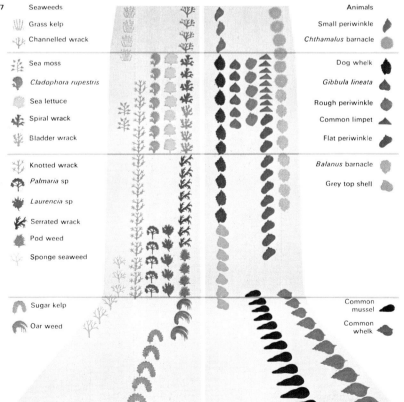

Seaweeds		Animals
Grass kelp		Small periwinkle
Channelled wrack		*Chthamalus* barnacle
Sea moss		Dog whelk
Cladophora rupestris		*Gibbula lineata*
Sea lettuce		Rough periwinkle
Spiral wrack		Common limpet
Bladder wrack		Flat periwinkle
Knotted wrack		*Balanus* barnacle
Palmaria sp		Grey top shell
Laurencia sp		
Serrated wrack		
Pod weed		
Sponge seaweed		
Sugar kelp		Common mussel
Oar weed		Common whelk

Life in the oceans

More than two-thirds of the earth's area is taken up by the oceans. These great bodies of water not only cover a huge area but also have immense depth – on average more than 3,650m (12,000ft). Thus the oceans offer a vast and immensely varied three-dimensional living space to countless plant and animal species of great diversity.

The richness or poverty of marine life at any one place is largely determined by the current systems of the ocean. Where two currents, or a current and a land mass, interact so as to draw deep water to the surface, they carry fresh nutrients into the upper, lighted zone and as a result animal and plant life thrive on a rich food source.

The pyramid of life

The oceans are, for the greater part, cold, dark and comparatively still. This would appear to give life little chance, yet it is thought that all life started in the sea and continues to flourish there. All the modern phyla of animals can be found in the sea and some have never left it. An important animal group virtually unrepresented in the oceans is the insects. Instead the crustaceans (the shrimps and their relations) are extremely abundant in the oceans in terms of numbers and species. The smallest of these crustaceans are dependent for food on small marine plants that are the basis of a vast pyramid of life [5] that includes the squids, fish, birds and mammals with which the seas abound. So wide is the base of the pyramid that the natural destiny of more than 90 per cent of all sea creatures is to be swallowed by other animals.

Although nearly all oceanic organisms are found in the upper, light regions, even the greatest depths support some life, dependent on a slow rain of organisms from above. Creatures living on the sea-bed are referred to as benthic, while those living nearer the surface are known as pelagic. These may be subdivided into those strong enough to swim against the currents when they wish – the nekton – and those too small or feeble to do anything but drift with the current – the plankton. Small plants are, in turn, known as phytoplankton [1], while planktonic animals are the zooplankton [Key]. Members of the zooplankton may be the young of large creatures, which during this phase of their lives disperse as widely as possible, or they may be – as are the arrow worms and minute crustacean copepods – a permanent planktonic component. In either case they are the basis of all larger life forms in the sea.

Pastures of the deep

If an attempt is made to assess the productivity of the sea in terms of the dry weight of the plants produced in a year, a given area of the sea appears to be as rich as an equal area of land and the richest estuaries are equivalent to land–growing forests. The big difference is that whereas on land plants have developed into multicellular forms large and strong enough to support themselves, in the sea this has never been necessary for the water supports the individual cells. Many of the millions of plant cells that go to make up a forest are for support or are employed as conduction channels and are not involved in photosynthesis and thus in food production. In the open sea, where plants are all single-celled, every one of them is capable of photosynthesis and is potentially productive.

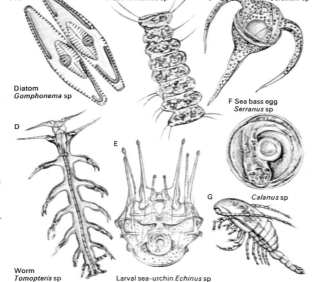

1 Mostly invisible to the naked eye, plankton lives all through the upper oceans. If collected in a fine-meshed net it can easily be observed through a low-powered microscope. The most important elements are the plants, including diatoms [A], green algae [B] and dinoflagellates [C] These single cells may cling together in chains. Using minerals from the water and the sun's energy, plants act as primary producers of food and are eaten, along with animal elements of plankton such as fish eggs [F] and small animals such as worms [D], copepods [G] and larvae [E]. These in turn are eaten by other members of the pyramid whose base is the plankton.

Diatom *Gomphonema* sp
B *Scenedesmus* sp
C *Ceratium* sp
F Sea bass egg *Serranus* sp
G *Calanus* sp
Worm *Tomopteris* sp
Larval sea-urchin *Echinus* sp

2 The wandering albatross, the largest of all sea birds, is an inhabitant of the southern oceans that only comes ashore, on remote islands, to breed. Its wings are ideally shaped for effortless gliding.

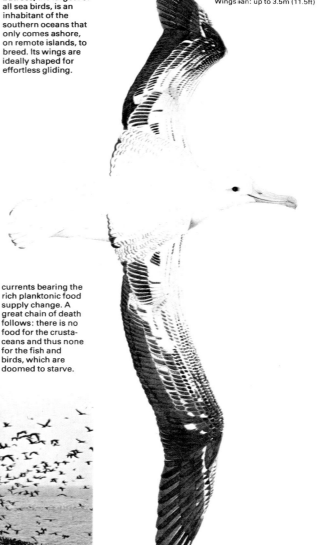

Great wandering albatross
Diomedea exulans
Wingspan: up to 3.5m (11.5ft)

3 Strange animals live far down in the cold, dark depths of the sea – all of them carnivores or scavengers dependent for food on the sunlit waters above. Many of these creatures bear luminous organs. In fish, crustaceans and squids the luminous organ can be a complex structure with a lens and a reflecting layer that increases the penetration of light through the water. The light shed may be used to help locate prey or discomfort predators, or as a pre-mating signal.

4 Huge numbers of cormorants, pelicans and other birds off the west coast of South America feed on the anchovies that thrive in the cold waters of the Humboldt Current. The droppings of these birds return minerals to the sea but certain islands are used for nesting and here the accumulated dung or guano makes a thick deposit which is removed and used as fertilizer. Occasionally and inexplicably the currents bearing the rich planktonic food supply change. A great chain of death follows: there is no food for the crustaceans and thus none for the fish and birds, which are doomed to starve.

The distribution of the phytoplankton is far from uniform over the oceans. As on land, there are apparently barren areas or deserts where plant life is sparse and in contrast, places where it is so abundant that the water is opaque and green or brown in colour. Because plant life requires both sunlight and nutrient salts, phytoplankton is produced only in the first few hundred metres or so below the surface, to a depth at which sufficient sunlight penetrates.

The need to strain food from water has given rise to an enormous diversity of filtering mechanisms in almost all groups of marine invertebrates, from the minute animals that drift in the plankton to those that live buried in or lie upon the floor of the ocean. By maintaining a constant current of water through their bodies these animals are able to extract the nutrients they need.

Creatures of the reefs

A very special ocean habitat is that of the coral reef. The reef itself is a complex of coral colonies whose architects are minute coral polyps, coelenterate animals related to the jellyfish and sea anemones. The polyps are able to extract calcium carbonate from seawater and use it to create solid support. They grow only in warm seas where the winter temperature never falls below 20°C (68°F), often in close association with algae. The two have a complex interdependence because the algae release substances useful to the coral polyps and in return are supplied with nutrients. The polyps by making important mineral nutrients – particularly phosphates and nitrates – available to the algae help promote their healthy growth.

The high rate of nutrient turnover in the reef results in a complex and highly organized community of living creatures; it has been estimated, for example, that there are more than 3,000 species of animals living in the Great Barrier Reef. These include an abundance of jellyfish and sea anemones and of brightly coloured fish, shrimps, crabs, starfish, sea slugs and sea cucumbers. In many animals the colour is used as a camouflage. Against its many-hued background, the boldly striped clownfish, for example, merges most effectively with its surroundings.

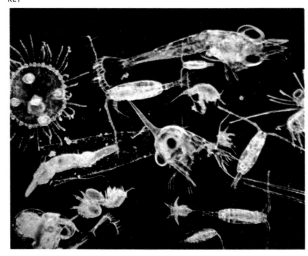

Plankton contains some adult animals but most of this mass of floating marine organisms consists of larval forms. Kept afloat in the water by means of weight-spreading spines or tiny beating cilia, they are swept well away from their parents. Some species become sedentary when adult.

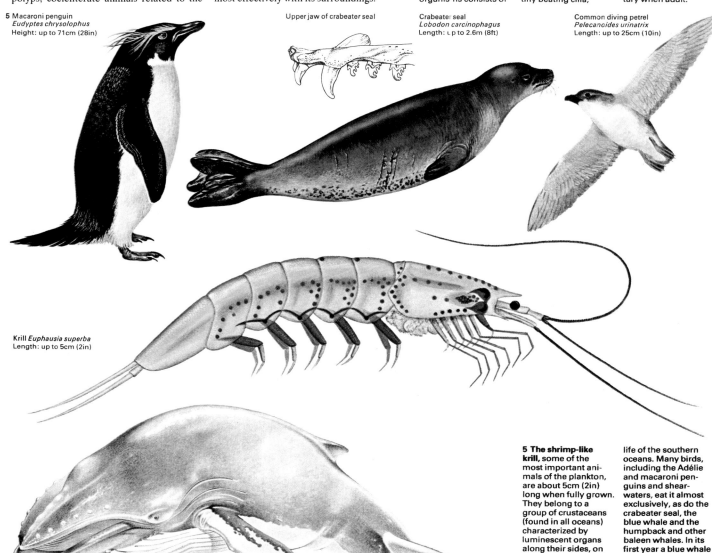

5 Macaroni penguin
Eudyptes chrysolophus
Height: up to 71cm (28in)

Upper jaw of crabeater seal

Crabeater seal
Lobodon carcinophagus
Length: up to 2.6m (8ft)

Common diving petrel
Pelecanoides urinatrix
Length: up to 25cm (10in)

Krill *Euphausia superba*
Length: up to 5cm (2in)

Humpback whale *Megaptera novaeangliae*
Length: up to 15.2m (50ft)

5 The shrimp-like krill, some of the most important animals of the plankton, are about 5cm (2in) long when fully grown. They belong to a group of crustaceans (found in all oceans) characterized by luminescent organs along their sides, on their undersides and heads. *Euphausia superba* is the most important species of the Antarctic seas for it supports much of the warm-blooded life of the southern oceans. Many birds, including the Adélie and macaroni penguins and shearwaters, eat it almost exclusively, as do the crabeater seal, the blue whale and the humpback and other baleen whales. In its first year a blue whale eats up to 450 tonnes of krill. Commercial harvesting of krill may further reduce the chances of survival of the already overhunted great whales.

Animals of the ocean

The oceans of the world are a continuous mass of some 5,000 million tonnes of water. But far from being uniform from surface to floor the ocean is divided into several regions [Key], each with its own typical forms of life. The most spectacular of these are animals and their distribution is determined by the interaction of such factors as light, temperature, pressure, salinity, currents and waves.

The sunlit zone
The smallest animals of the ocean are concentrated in the uppermost or euphotic zone [1] into which the most light penetrates and are known as zooplankton. Most of them are copepods, krill and other small crustaceans but the zooplankton also includes the eggs and larvae of many sea creatures, worms, comb jellies, sea snails and jellyfish.

The smallest zooplankton are the chief herbivores of the ocean, grazing on the microscopic plants or phytoplankton that form the basis of the ocean food chains. Larger zooplankton, including small fish and also invertebrates such as large jellyfish, live either as carnivores or as detritus or "carrion" feeders living on any dead matter.

All parts of the oceans are inhabited by fish, but those of the superficial euphotic zone are largely zooplankton feeders and many are small, immature fish that are themselves the prey of larger fish. The herrings (*Clupea* spp) and their allies, however, consume phytoplankton as a significant proportion of their diet. It is because they can use the ocean's plant resources that the herrings provide the bulk of man's ocean harvest.

While fish can swim in search of food, many large jellyfish are passive surface drifters. The Portuguese man-of-war (*Physalia physalis*) has a float with a "sail", which puts it at the mercy of the sea winds.

The number of small fish present in the euphotic zone tends to increase at night. These fish migrate upwards from the lower, pelagic zone and this migratory movement seems to be a protective mechanism because predators are generally less active during the hours of darkness. In addition, these migrations help the small fish to conserve their energy resources, for by moving into deeper, cooler waters in daytime they metabolize their food more slowly because their body temperatures are lower.

At night the upper layer of the ocean is illuminated by the light produced biologically by the constituents of the plankton. It has been suggested that rather than putting the plankton at a disadvantage it protects them, for the light makes plankton feeders such as small fish easily visible to the larger ocean carnivores; the very presence of fish seems to excite the plankton into light production.

Sea-dwelling mammals, such as whales, seals and dolphins can dive deep into the ocean, but regularly inhabit the uppermost layer of the open ocean because they must come to the surface to breathe. The baleen whales are further limited by the distribution of their food supply.

Predation and protection
The middle or mesopelagic zone of the ocean holds myriads of fish, accompanied by larger invertebrates such as squids, octopuses and prawns. This is the habitat of the sea's active predators, but it is one with no shelter. In order to survive the threat of large carnivores

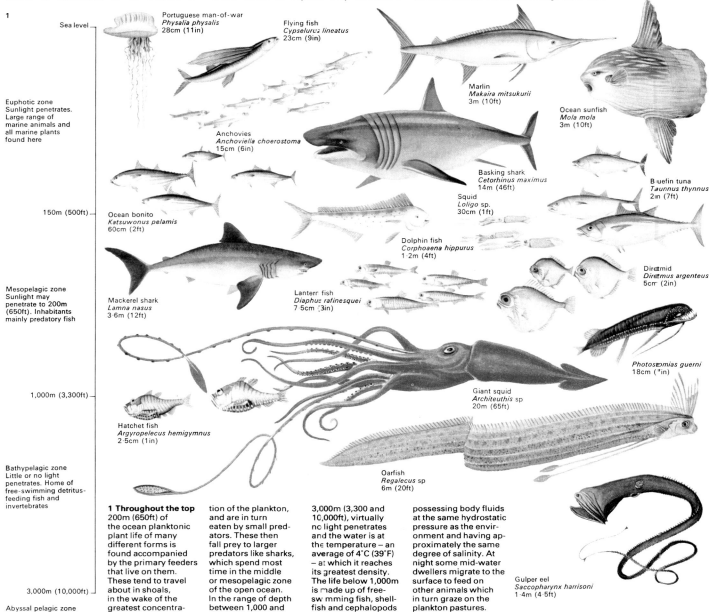

1

Sea level

Portuguese man-of-war
Physalia physalis
28cm (11in)

Flying fish
Cypselurus lineatus
23cm (9in)

Marlin
Makaira mitsukurii
3m (10ft)

Ocean sunfish
Mola mola
3m (10ft)

Euphotic zone
Sunlight penetrates. Large range of marine animals and all marine plants found here

Anchovies
Anchoviella choerostoma
15cm (6in)

Basking shark
Cetorhinus maximus
14m (46ft)

Bluefin tuna
Taunnus thynnus
2m (7ft)

150m (500ft)

Ocean bonito
Katsuwonus pelamis
60cm (2ft)

Squid
Loligo sp.
30cm (1ft)

Dolphin fish
Corphoaena hippurus
1·2m (4ft)

Diretmid
Diretmus argenteus
5cm (2in)

Mesopelagic zone
Sunlight may penetrate to 200m (650ft). Inhabitants mainly predatory fish

Mackerel shark
Lamna nasus
3·6m (12ft)

Lantern fish
Diaphus rafinesquei
7·5cm (3in)

Photostomias guerni
18cm (7in)

1,000m (3,300ft)

Hatchet fish
Argyropelecus hemigymnus
2·5cm (1in)

Giant squid
Architeuthis sp
20m (65ft)

Bathypelagic zone
Little or no light penetrates. Home of free-swimming detritus-feeding fish and invertebrates

Oarfish
Regalecus sp
6m (20ft)

Gulper eel
Saccopharynx harrisoni
1·4m (4·5ft)

3,000m (10,000ft)

Abyssal pelagic zone

1 Throughout the top 200m (650ft) of the ocean planktonic plant life of many different forms is found accompanied by the primary feeders that live on them. These tend to travel about in shoals, in the wake of the greatest concentration of the plankton, and are in turn eaten by small predators. These then fall prey to larger predators like sharks, which spend most time in the middle or mesopelagic zone of the open ocean. In the range of depth between 1,000 and 3,000m (3,300 and 10,000ft), virtually no light penetrates and the water is at the temperature – an average of 4°C (39°F) – at which it reaches its greatest density. The life below 1,000m is made up of free-swimming fish, shellfish and cephalopods possessing body fluids at the same hydrostatic pressure as the environment and having approximately the same degree of salinity. At night some mid-water dwellers migrate to the surface to feed on other animals which in turn graze on the plankton pastures.

the creatures of the mesopelagic must either be armed with powerful defensive apparatus, such as the stinging cells of the jellyfish *Rhizostoma* spp and *Cyanea* spp, or be adept swimmers equipped with sensory apparatus efficient enough to detect the approach of potential enemies. For this reason the most streamlined of all fish, both prey and predators, are found in the mesopelagic zone and include such species as the bonito (*Katsuwonus pelamis*) and the mackerel shark (*Lamna nasus*). Other protective mechanisms of pelagic fish include shoaling behaviour [2] and bioluminescence [3].

Deep-water dwellers

Animals of deep or bathypelagic waters are largely dependent for food on the rain of debris from the mesopelagic and euphotic zones above them. In this habitat, as yet largely unexplored, more than 2,000 species of fish and about the same number of large invertebrates have so far been discovered. Many have been located with the help of baited cameras; others have been found in the stomachs of whales and swordfish.

The problems of life at great depth – inky darkness, cold and crushing pressures – have resulted in the evolution of many curious but beautiful species [6]. The majority of deep-sea fish are 30cm (1ft) or less in length and most swim with their mouths permanently agape. Although often dark-coloured, more than 60 per cent of all deep-sea animals have light-producing organs. Their bioluminescence is used for recognizing neighbours and mates and confusing predators.

On the sea-bed the fauna varies according to the distance from the surface, but most bottom dwellers are detritus feeders. Most of these benthic species are found in the comparatively shallow waters of the continental shelves. There, mussels and other bivalves, fan worms, sea cucumbers, crabs, sea-urchins and their near relatives cohabit with flatfish.

The life of the abyssal ocean is much less well documented, but explorations to depths of more than 2,000m (6,500ft) off the coast of California have revealed a fauna that includes species known to be adapted to the near-freezing temperatures of Arctic waters.

KEY
Littoral zone | Shallow water zone | Deep water zone | (Pelagic zone)

Continental shelf
Continental slope
Ocean platform

Sea-level
Euphotic zone
Mesopelagic zone — 183m
— 610m
— 1,219m
Bathypelagic zone — 1,829m
— 2,438m
— 3,048m
— 3,658m
— 4,267m
Abyssal pelagic zone — 4,877m

The ocean layers from the translucent surface waters through the twilight zone to the depths of eternal gloom provide a range of habitats to which the ocean's wealth of species are adapted. The "conventional" fish shape of the tuna and the shark [illustration 1], which live in the euphotic and mesopelagic zones, is a marked contrast to the highly developed forms of the abyssal zone [illustration 6]. The herbivores of the ocean are concentrated in the euphotic zone, carnivorous predators in the pelagic zones and detritus feeders on the sea-bed. Materials are constantly carried back to the surface by the upwelling and mixing of seawater.

2 Many fish such as the sweetlips (*Gaterin* sp) live in shoals – enormous masses of fish that act as one. This habit may be a defence mechanism, individuals being protected by their large numbers. Shoaling or schooling may also be an aid to reproduction as most fish in a shoal are of a similar age and size. There appear to be no leaders and the direction of movement seems to be determined by the complete shoal.

3 The hatchet fish (*Sternoptyx diaphana*) is typical of the luminous fish of the bathypelagic. It possesses "cold light" luminous organs in its body that supply the only light to waters of this depth.

Its large mouth is typical of predatory deep-sea fish. In the sunless depths there are very few prey species and a large mouth is therefore essential if a fish is to obtain enough food to survive.

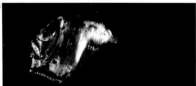

4 The catshark or skamoog (*Holohalaelurus regani*) is a member of the family (*Scyliorhinidae*) of small sharks or dogfish that are found worldwide. This species inhabits coastal waters of Africa. Like many of this family it feeds mainly on or near the sea-bed. Its sharp, needle-like teeth point towards its throat. This arrangement ensures that even if slippery prey struggles, it is securely held.

5 The stone fish (*Synanceja verrucosa*) is typical of the bottom-feeders in that it is superbly well camouflaged against the colours and textures of its background. As a further means of discouraging its enemies it is equipped with

sharp dorsal spines. When these come in contact with an enemy a poison is injected that is near-deadly to man. The stonefish normally lurks on the sea-bed to wait for its prey on which it "pounces" using its powerful pectoral fins.

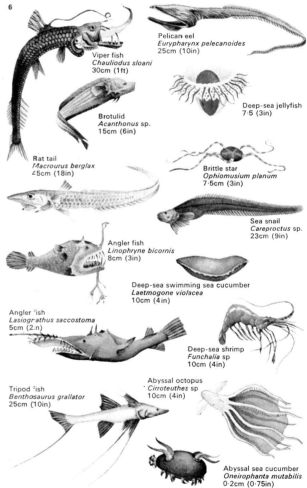

6

Viper fish *Chauliodus sloani* 30cm (1ft)

Pelican eel *Eurypharynx pelecanoides* 25cm (10in)

Brotulid *Acanthonus* sp. 15cm (6in)

Deep-sea jellyfish 7·5 (3in)

Rat tail *Macrourus berglax* 45cm (18in)

Brittle star *Ophiomusium planum* 7·5cm (3in)

Angler fish *Linophryne bicornis* 8cm (3in)

Sea snail *Careproctus* sp. 23cm (9in)

Deep-sea swimming sea cucumber *Laetmogone violacea* 10cm (4in)

Angler fish *Lasiognathus saccostoma* 5cm (2in)

Deep-sea shrimp *Funchalia* sp 10cm (4in)

Tripod fish *Benthosaurus grallator* 25cm (10in)

Abyssal octopus *Cirroteuthes* sp 10cm (4in)

Abyssal sea cucumber *Oneirophanta mutabilis* 0·2cm (0·75in)

6 Below 3,000m (10,000ft) life in the world's oceans includes a range of animals, many of them weirdly shaped, adapted to living in near-freezing water at extremely high pressures. The only glimmers of light in this pitch-dark region come from the bioluminescent organs that many of them possess. About 75% of the free-swimming species of fish in this region, representing 90% of the individuals, have light organs. These fish are small, few larger than 30cm

(1ft) in length. Many deep-sea, bottom-dwelling fish such as the rat tails grow a little longer, although much of this elongation is accounted for by the tail. Few of these benthic creatures have light organs, but manage to feed in darkness.

241

Endangered mammals

Extinction is a natural enough process: we know from fossils that there have been countless thousands of animal species on the earth that no longer exist today. Man had nothing to do with their disappearance. But in the last few hundred years the pace of extinction has quickened: since 1600 we have lost at least 36 species of mammals and another 120 are now in danger.

A few of these are species that have simply reached the end of their natural timespan. Evolution has passed them by and they are gradually declining under the relentless competition of animals better adapted to live and breed. But at least four out of every five endangered animals are rare because of man's deliberate or unthinking actions.

Man's responsibility

Man has always been a hunter. Indeed, some zoologists believe that ancient man played a part in the downfall of many American mammals, presenting these beasts with an enemy they were not evolved to cope with. However, the traditional hunting of wild animals for food and skins can rarely have caused any

extinctions. When an Eskimo hunted the polar bear [Key] with dog sled and spear, his prey had a good chance of fighting back or escaping. Today, the snowmobile and the repeating rifle make killing very much easier and external demand has pushed up the value of skins. If the polar nations – the United States, Canada, Denmark, Norway and the USSR – had not signed a convention in 1973 prohibiting all hunting except for scientific purposes or by traditional methods, the polar bear might well have become extinct.

The threats to mammals

Three factors have totally altered the effects of man the hunter on wildlife: modern technology, the world market and man's explosive increase in numbers. Guns, telescopic sights, aircraft and Jeeps with headlights have shifted the odds dramatically against the hunted and the prospect of selling a skin for a high price has increased the hunter's motivation even further.

Greed has sent many species to the edge of extinction, a process that started when seamen from sailing ships clubbed to death

thousands of puppy seals on the Arctic ice, boiling them down in huge vats to extract the blubber. The same process has continued into the 1970s as Soviet and Japanese whaling fleets have used explosive harpoons and giant factory ships to decimate the whales of the Antarctic Ocean, just as the British, the Norwegians and the Americans exterminated the Arctic whale herds a century ago.

Most dramatic of all has been the plight of the blue whale [5], the largest animal ever to have lived on our planet. Thirty metres (100ft) long and weighing up to 10 tonnes, three times heavier than the most massive dinosaur, the blue whale was once so plentiful that 200,000 of them lived in the Antarctic. By 1963, there were fewer than 1,000. Since then, under complete protection in all oceans, the blue whale seems to have slowly increased in numbers.

Most endangered mammals, however, are not threatened by any deliberate act of man. Some, especially those that live in water, suffer from the effects of pollution [6]. The Pyrenean desman, for instance, is an aquatic mole, an insectivorous animal not

1 The fashion for fur coats threatens the survival of spotted cats of all kinds. The cats most in danger include the leopard, from which this coat was made, the jaguar, ocelot and snow leopard.

2 The tiger was once common throughout Asia. An inhabitant of wooded areas and jungle, it has suffered much from the reduction of its habitat but its greatest threat comes directly from man due to the increased availability of firearms. Reserves have been established in India for the protection of the tiger.

3 The remaining 50 protected specimens of the Javan rhino are still threatened by poachers in Indonesia. The powder made from its horn is erroneously believed to be a sexual stimulant.

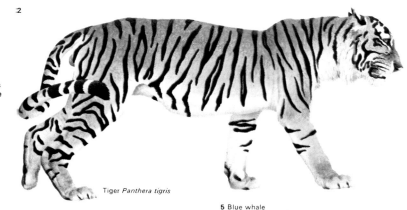

Tiger *Panthera tigris*

5 Blue whale *Sibbaldus musculus*

Javan rhinoceros *Rhinoceros sondaicus*

Kouprey *Bos sauveli*

4 The kouprey, a wild cow of the Cambodian forests, has probably already been hunted to extinction by soldiers. It could have been important in improving the strains of Asian domestic cattle.

5 The blue whale is the largest animal the earth has ever known. It has been almost exterminated for its meat and oil but it is now thought to be recovering, although fewer than 1,000 remain.

6 Wrecked supertankers, such as the *Torrey Canyon,* shown here burning violently after being bombed before it sank off the Scilly Islands in March 1967, threaten birds and mammals. Oil discharged by such wrecks impairs the insulating effect of fur, and as a result sea otters, for example, may die of exposure. Other threats at sea include dumped pesticides which contaminate food supplies.

7 The monk seal is extinct in the Caribbean, nearly gone in the Mediterranean, but survives in Hawaii. The sea otter still thrives off California, but the Pyrenean desman, an aquatic mole, is on the decline.

Pyrenean desman *Galemys pyrenaicus*

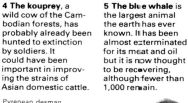

Monk seal *Monachus sp*

Sea otter *Enhydra lutris*

unlike a water shrew, which swims in the clear mountain rivers of southern France, Spain and Portugal. The development of those streams for irrigation or hydroelectric power, or their pollution by pesticides or other chemicals, deprives the desman of the pure, highly oxygenated water that is its only possible habitat.

Introduction and destruction

Another hazard is the introduction of alien forms of wildlife. The monotremes or egg-laying mammals of Australia have suffered from many such introductions, starting with the dingo or hunting dog which the Aborigines took with them 10,000 years ago.

Undoubtedly the greatest danger of all is the destruction of a natural habitat [8]. As man ploughs up the prairies, fells the rain forests, dams the rivers, drains the marshes and builds roads, towns and cities everywhere, the specialized habitats of animals with restricted distributions are squeezed until they disappear. Many mammals vanish with their habitats. The great prairies of the United States, over which the Sioux once

hunted bison, are today ploughed and turned over to grain and cattle. Bison survive only in a few reserves and smaller mammals such as the prairie dog and the black-footed ferret that preys on it have almost vanished [9].

The ethical argument for saving endangered mammals is a powerful one: do we want our children to know a rhinoceros, say, only from picture books? But the practical reasons are even more compelling: we need wild mammals as genetic resources for the future. Cattle, for example, are unsuited to many parts of Africa and Australia and natural populations of wild game, bucks and kangaroos are already being enclosed and shot instead. In Asia, the rare wild cow called the kouprey [4] could play a role by hybridization with the domesticated humped zebu cattle of India and the Far East.

Conservation is indivisible. To save a rare mammal such as the indri [9] we must preserve the Madagascan rain forest that is its home. And in conserving the rain forest, we safeguard the wild plant *Coffea bertrandii*, which may one day allow the plant breeders to develop caffeine-free coffee.

Polar bear
Thalarctos maritimus

Wildlife's main enemy is man: the polluter, the destroyer of wilderness, the introducer of alien species and the hunter. Today at least 120 mammal species are in danger of extermination. The polar bear (*Thalarctos maritimus*), however, is now on the increase under international protection.

8 The biggest threat to wildlife is the damage to habitats caused by developments like the Trans-Amazonian Highway. The destruction of the wilderness destroys the homes of the animals with it.

Pygmy hog *Sus salvanius*

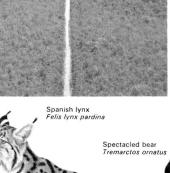

Spanish lynx
Felis lynx pardina

Spectacled bear
Tremarctos ornatus

Indri
Indri indri

Black-footed ferret
Mustela nigripes

9 The pygmy hog lives in the *terai*, a belt of grass and woodland along the Himalayan foothills that is being carved up by local cultivation, tea estates and forestry. The indri is a handsome tree-living lemur from the Madagascar rain forests which are increasingly being cleared by felling and burning. The spectacled bear lives in forests on the high slopes of the Andes; although human settlement has reduced its numbers, it is still widespread in Ecuador and Bolivia. The Spanish lynx needs wilderness to hunt its prey and is now confined to a few inaccessible mountain sierras, and the Doñana National Park in the Guadalquivir delta. The black-footed ferret lives in the prairies of the American West, fast disappearing under the plough. This cultivation destroys the burrows of the ferret's main prey – small rodents called prairie dogs.

10 When man allows domestic livestock such as these goats, to escape he upsets the natural ecological balance. Because they are such voracious feeders, goats threaten the food supplies of native mammals.

11 The onager has been pushed by domestic livestock on to the most barren pastures in northern Iran. The insect-eating Cuban solenodon is threatened by competition from the introduced mongoose, while the thylacine, very rare and confined

to Tasmania, was exterminated on the Australian mainland when the Aborigines took the dingo with them from Asia.

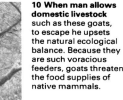

11 Onager
Equus hemionus onager

Cuban solenodon
Solenodon cubanus

Thylacine
Thylacinus cynocephalus

Endangered birds

There are about 350 species and sub-species of birds in danger of extinction, but they are not spread evenly among the continents. Most at risk are birds with a naturally restricted range: those that live in remote or small habitats or on oceanic islands.

The regional scene
Of the total, there are 30 endangered species in the Palaearctic (Eurasian) region; eight in Europe and North Africa and 22 in Asia north of the Himalayas. On the whole, Palaearctic birds are widely distributed over their huge land·mass, but local birds of prey such as the Spanish imperial eagle [4] are still at risk.

The Ethiopian region (Africa) with 16, the Oriental region (South-East Asia) with 38 and the Nearctic region (North America) with 39, have relatively few endangered birds. But species such as the ivory-billed woodpecker [2], which requires undisturbed forest swamps, and the whooping crane [1], which breeds on remote subarctic lakes, have suffered from the loss of their habitat and from overhunting respectively.

The Neotropical region (South and Central America), and even more so the Australasian region, have more primitive bird fauna than the great continents of the Northern Hemisphere. Their unique forms of bird life are probably – on an evolutionary time scale – on the way out. And man has greatly hastened the process by destroying habitats and by introducing predatory mammals such as dogs, cats, pigs, goats, stoats, rats and foxes. These a tack ground-nesting or flightless birds or damage the vegetation on which the birds feed. As a direct result, 69 Neotropical and 41 Australasian birds were on the endangered list in 1975.

The greatest number of species at risk, 117 in all, come from the oceanic islands where isolated environments house only a relatively few species that originally emigrated from the continents. Once on the islands, many of these birds evolved distinct races and even totally new species and the restricted size of their new homelands meant that they never built up large populations. Man has threatened these birds in three main ways: by destroying vegetation, by intro-

ducing competitors and by overhunting.

The Pacific Ocean islands have an alarming total of 84 endangered species. The Hawaii group is a striking example, with no fewer than 29 species at risk.

Overall, 32 per cent of all endangered bird species are believed to be rare primarily because of natural causes. Hunting threatens another 24 per cent, introduced predators 11 per cent and introduced competitors three per cent. For the remaining rarities, destruction of habitat is the greatest threat, endangering 30 per cent of rare birds.

Organizations to the rescue
The statistics of threatened birds, together with those of animals and plants, are compiled by the International Union for Conservation of Nature and Natural Resources (IUCN), an international scientific organization with headquarters in Switzerland. The IUCN has compiled detailed *Red Data Books* for all endangered mammals and birds and for some reptiles, amphibians, fish and plants.

An IUCN fact sheet describes the present

1 The whooping crane, a survivor from the age of the dinosaurs, was already rare and declining when the white man first colonized North America. No nests were found between 1922 and 1955 and then the only remaining colony was discovered in Wood Buffalo National Park, Canada. In 1975 there were about 50 whoopers. The birds migrate 3,700km (2,300 miles) to winter in Texas and risk being shot *en route* by sportsmen who mistake them for sandhill cranes. Whooping cranes lay two eggs a year but rear only one chick, so some eggs have been taken for captive breeding in the hope of eventually supplementing the wild flock.

1 Whooping crane
Grus americana

2 The ivory-billed woodpecker is one of the rarest birds in the world. It is possibly extinct now in southern parts of the United States. An inhabitant of primeval swamp forests, it is

Ivory-billed woodpecker
Campephilus principalis

extremely shy and may desert its nest even if it is only watched. Indian chiefs once adorned their belts with its bill and plumes; now tree felling has removed most of the big trees in which it breeds.

Puña grebe
Podiceps taczanowskii

3 The flightless puña grebe is confined to the shallow water of Lake Junín, 4,084m (13,400ft) up in the Peruvian Andes. The lake is becoming increasingly polluted

with mine effluent, sewage and run-off from eroded farmland. Hopefully, Peru may soon declare this potential tourist attraction a national park.

4 The Spanish imperial eagle once ranged over Morocco, Algeria, Spain and Portugal. By 1975 it bred in only a

few remote Spanish sierras and in Europe's last great wilderness, the Coto Doñana national park near Seville. This is probably Europe's most threatened bird, with fewer than 100 individuals surviving.

Although it has been well protected in its Doñana breeding grounds, it may soon die out completely unless farmers outside the protective limits of the park can be persuaded to stop shooting it.

Imperial eagle
Aquila heliaca

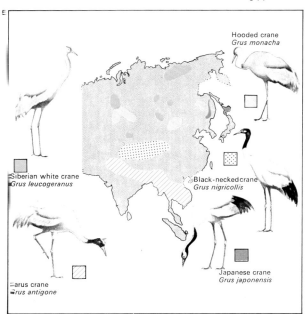

Hooded crane
Grus monacha

Siberian white crane
Grus leucogeranus

Black-necked crane
Grus nigricollis

Sarus crane
Grus antigone

Japanese crane
Grus japonensis

5 The Asian cranes are survivors from the warm, wet swamps of the Pleistocene. Never common, their numbers have been greatly reduced by hunting and the loss of the wide marshlands in which they nest and winter. The magnificent Japanese crane plays a major part in the nation's folklore and legend and was rigidly protected by the medieval nobility. It breeds rapidly in captivity and 33 per cent of its population is now in zoos. Carefully protected today, it is slowly on the increase. The hooded crane, which winters in Japan, is faring even better. Less is known of the black-necked crane, which nests on remote lakes deep in the highlands of central Asia.

Highway Patrol restore order after local policeman gunned down black teenager

Highway Patrol Captain Ronald Johnson hugs a woman in Ferguson as tensions eased yesterday

Picture: AP

e of sex assault

Medal sent back over Gaza deaths

ALAN HALL

A DUTCHMAN who was awarded Israel's highest civilian honour for saving a Jew in Nazi-occupied Europe has sent his medal back after six of his relatives were killed by an air strike in the Gaza strip.

In 2011, the Yad Vashem Holocaust museum declared 91-year-old lawyer Henk Zanoli and his late mother, Johana Zanoli-Smit, Righteous Among the Nations for having saved a Jewish child, Elhanan Pinto, during the Nazi occupation of the Netherlands.

Mr Zanoli's great-niece, Angelique Eijpe, is a Dutch diplomat who serves as deputy head of her country's mission in Oman. Her husband, Ismail Ziadah, was born in the al-Bureij camp in central Gaza.

On Sunday, 20 July, an Israeli jet dropped a bomb on the Ziadah family's home in al-Bureij, killing matriarch Muftiyah, 70, three of her sons, her daughter-in-law and her 12-year-old son.

Mr Zanoli returned his medal to the Israeli embassy in The Hague, where he was honoured just three years ago.

"For me to hold on to the honour granted by the state of Israel, under these circumstances, will be both an insult to... my courageous mother who risked her life," he said

Vultures 'more at risk than rhinos' after poison kills 39

JANE FIELDS
IN HARARE

CONSERVATIONISTS have warned that vultures are "more threatened than rhinos" after the bodies of 39 poisoned vultures were discovered in southern Zimbabwe.

The decomposing carcasses were discovered on a farm at Fort Rixon, about 370 kilometres from the capital Harare, according to Kerri Wolter, from vulture conservation group Vulpro in South Africa.

The birds – most of them endangered white-backed vultures – appear to have died after feeding on a cow carcass laced with cyanide or aldicarb, which are both fast-acting poisons.

First reports suggested the vultures were victims of a farmer trying to rid his land of predators such as jackals or hyenas rather than poachers, who have

been blamed for fuelling the decline of vulture populations in southern Africa over the past few years.

The sight of vultures congregating high above the African bush is a sure sign of a recent kill. But the scavengers are hated by poachers because they alert rangers to their presence. Poachers lace the carcasses of elephants and other game with cyanide or pesticides such as temik so the birds die before they can take to the skies again.

The practice gives the poachers time to saw off tusks and disappear into the bush. It has contributed to devastating declines in vulture populations in Zimbabwe, Namibia and South Africa, where the numbers of some species have fallen by 50 to 80 per cent.

Farmers who use poisons are also to blame. Ms Wolter said: "While there probably still are more vultures than rhinos, the

Vultures are in decline across southern Africa   *Picture: Getty*

soning from poachers and farmers and both are as detrimental"

Zimbabwean conservationist Clive Stockil, who has worked in the southern Save Valley Conservancy for decades, said: "While there probably still are more vultures than rhinos, the

hunting and killing of a rhino is very specialised and will kill one or two rhinos at a time. There have been cases of several hundred vultures being killed in one incident. This has the potential of bringing the vulture population to extinction."

The poison attack killed 39 white-backed vultures

Mr Stockil, who was presented with a Tusk Lifetime Achievement Award by Prince William last year, said vultures were now "more threatened than rhinos".

In Zimbabwe's Gonarezhou National Park 184 vultures were killed in 2012 when an elephant carcass was laced with temik, a pesticide sold as rat poison.

Last year researchers in the park found only five active lappet-faced vulture nests, down from 40 in the 1970s.

Figures for the total population of the eight vulture species that occur in the region are almost impossible to obtain. Vulpro said there are 3,800 breeding pairs of Cape Griffons, among the most vulnerable. There are about 25,000 rhinos left in Africa.

The dead vultures included two Cape Griffons, three lappet-faced vultures and 39 white-backed vultures. 16.8.2014

and former distribution of every threatened species, its estimated numbers, the presumed reasons for its decline, the numbers held in captivity and their breeding potential, and finally the protective measures already taken and those proposed. The sheets are colour-coded: green pages are for those species that have recovered so well that they are off the immediate danger list and red for those that are now on the verge of extinction.

Experts from IUCN decide, on a strictly scientific basis, what must be done to save each species and draw up action plans in consultation with the authorities in each country. But implementing these schemes is another matter and here the World Wildlife Fund (WWF) plays a key role. WWF has an international organization based, like IUCN, in the small Swiss town of Morges, near Geneva, but it also has national groups in many countries. It is a propagandist and fund-raising body, charged with cajoling and persuading national governments to take action to save the world's living heritage and with raising the money that alone makes IUCN's plans possible. IUCN and WWF work closely together, but they have quite distinct roles. WWF cannot lay claim to IUCN's scientific expertise, which goes far beyond the sphere of endangered species to include every aspect of the rational use of natural resources, whereas IUCN would not wish to involve itself in WWF's activities.

People must help

Together, IUCN and WWF have pulled a number of birds back from the brink and focused attention on many others. Without them, the Coto Doñana in southwestern Spain would not have provided a sanctuary for the imperial eagle [4] and the Galapagos hawk [10] might have disappeared entirely. But although money and advice are necessities, in the last resort conservation depends on the determination of the local people. Unless, for example, the Peruvians cease to pollute Lake Junín, the puña grebe [3] will become extinct, and the Filipinos themselves must decide whether they prefer to have their monkey-eating eagles [6] live and flying over their forests or dead, stuffed and mounted as specimens in glass cases.

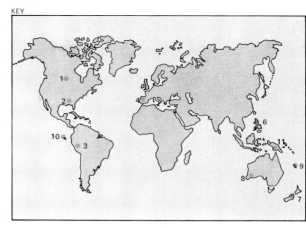

Birds in danger are found in all the zoo-geographical regions of the world. The species illustrated on these pages are keyed numerically in this map. Nearly a third of the world's total of endangered birds resides on oceanic islands. The chief reason for this is that they have tended to become highly specialized in habitats that are severely restricted in area. Man has posed a three-fold threat to bird species through destruction of their habitats, by the introduction of predators and competitors, and by hunting.

6 Monkey-eating eagle
Pithecophaga jefferyi

6 The immense monkey-eating eagles of the Philippines have been reduced to fewer than 100. This results from the destruction of their forest habitat and a demand for specimens dead or alive.

7 Kakapo
Strigops habroptilus

8 Noisy scrub bird
Atrichornis clamosus

7 The New Zealand ground owl-parrot or kakapo began to decline with the arrival of the Maoris. Predators introduced from Europe and deforestation have reduced its numbers to fewer than 100.

8 The noisy scrub bird of Western Australia lives in dense coastal scrub. Until rediscovered in 1961 at Two People Bay, near Albany, it had not been recorded since 1889 and was thought to be extinct.

10 Galapagos hawk
Buteo galapagoensis

10 The Galapagos hawk is a type of buzzard that lives only on the Galapagos Islands. Habitat destruction by goats and shooting by chicken farmers have reduced the population to about 200 birds. Protective measures, for example the removal of the introduced goats, are proving successful.

11 King of Saxony bird of paradise
Pteridophora alberti

9 Kagu
Rhynochetos jubatus

11 The King of Saxony bird of paradise is one of the world's rarest birds and now on the verge of extinction. Its range extends from the Snow Mountains to the Central Highlands of New Guinea. The export trade in plumes was banned in 1924, but illicit trading continues. Its survival is also threatened by the destruction of its forest habitat.

9 The mysterious kagu, a virtually flightless heron, is confined to the remote forests of New Caledonia. It is threatened by introduced predators such as cats, pigs, rats and especially dogs.

Endangered species

A great deal is known about some 300 mammals and birds thought to be in danger of extinction, but very little about the many other threatened species. One estimate suggests that 20,000 plant species may be threatened, but detailed information exists for only a few hundred of them. There is a little more data about amphibians, and about endangered reptiles such as snakes, turtles, lizards and crocodiles, as well as freshwater fish, at least in Europe and North America. But there is an almost total lack of knowledge about endangered butterflies [7-10] and other invertebrates, nor is it known how many marine inhabitants are threatened by man's pollution of the sea.

Value of wildlife
In all, 50,000 or even 100,000 species of animals and plants may be endangered, largely as a result of man's activities, and despite the Washington Convention (1973) banning trade in rare species and the various kinds of products made from them.

Some of the world's disappearing wildlife has no obvious value to man but because all species of plants and animals are ultimately interdependent the loss of one or two of them can critically affect others in such a way that species of economic importance may themselves become involved.

Many endangered species, however, have actual or potential economic value in themselves. The marine turtles [1], for example, are the basis of an industry in meat, eggs, shell and oil, and if they were carefully harvested or even farmed they could provide an income for many more years. Komodo dragons [2] may seem useless curiosities, but tourists visiting Indonesia pay to see these 3m [10ft] carnivorous monsters.

Some fish species may seem useless at present, but the need to produce food from the most unlikely sources may give them an unexpected value in future. The endangered Moapa dace [5], for example, which is found only in a few warm springs in Nevada, might one day be used to breed fish that could be grown commercially in the heated effluent from electrical power stations. Sport fishermen willingly pay substantial sums to fight unusual fish like the Gila trout [6].

Insects are often involved in biological control of pests as natural predators or parasites to reduce the numbers of a plant or animal harmful to man. The relentless spread of impenetrable thickets of a South American cactus called the prickly pear (*Opuntia* sp) across the sheep lands of Queensland was stopped only with the help of an unexciting moth from the Argentine, *Cactoblastis cactorum*. Its caterpillars feed naturally on prickly pear, and when the moth was introduced to Australia 60 years after the cactus, it soon gnawed the pest to the ground.

Endangered plants
Endangered plants have perhaps even more potential uses than threatened animals, although some flowers are worth saving for their beauty alone. The fragrant Calabrian primrose [11], for example, might prove a profitable line for seed and garden firms. But other plants have provided man with a remarkable variety of useful products. Drugs such as aspirin and the heart medicine digitalin; drinks such as tea and coffee; all of his vegetables and fruits; pepper, nutmeg and

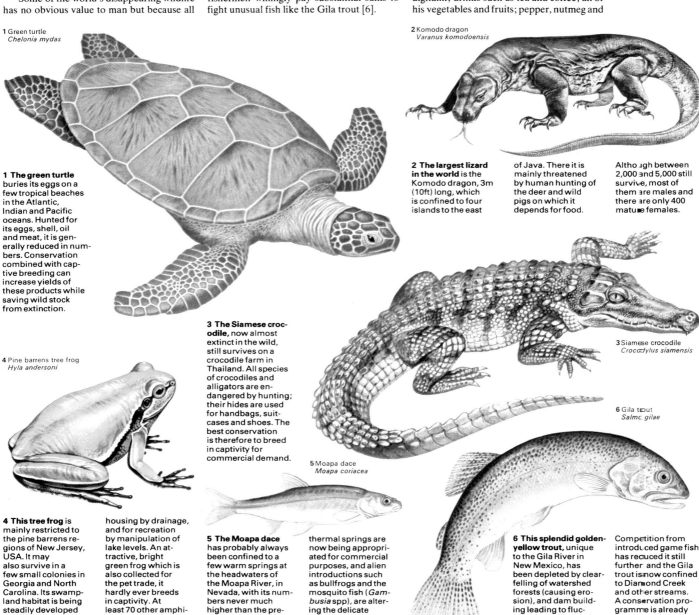

1 Green turtle
Chelonia mydas

1 The green turtle buries its eggs on a few tropical beaches in the Atlantic, Indian and Pacific oceans. Hunted for its eggs, shell, oil and meat, it is generally reduced in numbers. Conservation combined with captive breeding can increase yields of these products while saving wild stock from extinction.

2 Komodo dragon
Varanus komodoensis

2 The largest lizard in the world is the Komodo dragon, 3m (10ft) long, which is confined to four islands to the east of Java. There it is mainly threatened by human hunting of the deer and wild pigs on which it depends for food. Although between 2,000 and 5,000 still survive, most of them are males and there are only 400 mature females.

4 Pine barrens tree frog
Hyla andersoni

3 Siamese crocodile
Crocodylus siamensis

3 The Siamese crocodile, now almost extinct in the wild, still survives on a crocodile farm in Thailand. All species of crocodiles and alligators are endangered by hunting; their hides are used for handbags, suitcases and shoes. The best conservation is therefore to breed in captivity for commercial demand.

6 Gila trout
Salmo gilae

5 Moapa dace
Moapa coriacea

4 This tree frog is mainly restricted to the pine barrens regions of New Jersey, USA. It may also survive in a few small colonies in Georgia and North Carolina. Its swampland habitat is being steadily developed for industry and housing by drainage, and for recreation by manipulation of lake levels. An attractive, bright green frog which is also collected for the pet trade, it hardly ever breeds in captivity. At least 70 other amphibians are endangered.

5 The Moapa dace has probably always been confined to a few warm springs at the headwaters of the Moapa River, in Nevada, with its numbers never much higher than the present 500–1,000. The thermal springs are now being appropriated for commercial purposes, and alien introductions such as bullfrogs and the mosquito fish (*Gambusia* spp), are altering the delicate ecological balance.

6 This splendid golden-yellow trout, unique to the Gila River in New Mexico, has been depleted by clear-felling of watershed forests (causing erosion), and dam building leading to fluctuating water levels. Competition from introduced game fish has reduced it still further and the Gila trout is now confined to Diamond Creek and other streams. A conservation programme is already showing good results.

other seasonings; jute and other fibres; timbers, dyes and hundreds of other substances were all developed from what were originally wild plants. Man's warfare has endangered, if not exterminated, one of them. The source of the yellow pigment gamboge yellow, the tree *Garcinia hanburyi*, has been severely threatened by defoliation in its native habitat in Vietnam and Cambodia.

What to do?

For a few of these endangered animals and plants special reserves can be set up, but all must rely on man for the preservation of their habitats. The chain of reserves established by the Indian government to save the tiger from extinction is of equal value to hundreds of less dramatic but equally endangered species. In practice, there is no way of conserving the tiger in the wild without conserving the complex ecosystem of which it forms a part. Similarly, protection for the giant otter means protection for a whole area of the Amazonian rain forest.

There are some endangered species, of which the marine turtles and the crocodiles

are good examples, where the risk comes from over-exploitation by man. Here there are several ways in which conservation can be designed to ensure that a renewable resource remains to enrich future generations. One method is to limit the number of animals that may be killed or eggs that may be taken. Countries from Borneo to the West Indies are trying to do this for the green turtle [1] with the aim of taking no more than a suitable crop. Another technique is to collect turtle eggs and hatch them in captivity, releasing the young when they are a year old and past their most vulnerable stage. This method can replenish populations and sometimes even increase wild stocks.

A third system is to breed the animals entirely in captivity, as happens on crocodile farms in several countries [3], so that the best-quality hides can be collected from known sources and a complete ban placed on killing animals in the wild. A fourth possibility is semi-domestication. One day, herds of adult turtles may be pastured on fields of seaweed and cropped for eggs and meat, just as chickens are cropped by a poultry farmer.

The use of insecticide sprays, the fashion for reptile-skin handbags and wallets, and the incorporation of pressed flowers into goods such as bookmarks together threaten a huge number of species. As many as 20,000 plant species may be in danger of extinction and, unlike the threatened mammals and birds, the number of endangered plants, insects, fish, reptiles and amphibians is largely unknown.

7 Insecticides, collectors and the destruction of habitats are three main threats to butterflies, but little is known about the decline of many once common species such as the splendid Apollo, found over much of Europe and northern Asia.

7 Apollo butterfly
Parnassius apollo

8 Large copper butterfly
Lycaena dispar

8 The last individual of the British race of the large copper butterfly was recorded in 1848. Extinction resulted from over-collecting and local disappearance of the food plant of its caterpillars, the great water dock, with the draining of the fens in eastern England. The Dutch race, which has become rare for similar reasons, was introduced to Wood Walton Fen, England, from the Netherlands in 1927.

9 Victoria birdwing
Ornithoptera victoriae

10 Brown glasswing
Dircenna varina

11 Calabrian primrose
Primula palinuri

9 Collectors and forest clearance for agriculture threaten the Victoria birdwing in New Guinea. A single birdwing once fetched $1,875 at an auction in Paris.

10 The Trinidad subspecies of the southern brown glasswing of Ecuador is confined to the southern part of the island. It lives in scrub, much of which has been destroyed by agricultural fires.

11 The Calabrian primrose, a sweet-smelling species, is limited to two areas near Cape Palinuri, in southern Italy. Greatly reduced by grazing and picking, it needs total protection in the wild.

14 Cooktown orchid
Dendrobium bigibbum

12 Poor Knights brush lily
Xeronema callistemon

13 St Helena redwood
Trochetia erythroxylon

12 This New Zealand brush lily, with its brilliant, sword-shaped flowers, is already extinct on the mainland. The offshore Poor Knights and Hen and Chickens Islands were made reserves and protect the species from the feral pigs.

13 This unique redwood, now down to a single tree, stands on the isolated island of St Helena in the South Atlantic. The island was once thickly forested but has been eaten bare by goats introduced since its discovery by Europeans in 1502.

14 The Cooktown orchid is the state emblem of Queensland and, although protected, is avidly collected from the north Australian rain forests for sale and cultivation. Its flower hangs from the small pockets of humus in which it lives.

247

Destructive man

Of all the creatures alive on earth man is the most destructive. As Mark Twain expressed it, "Man is the only animal that can blush or needs to". For millions of years man has been destroying his environment, by activities usually attributed to his intelligence. What is meant, of course, is that man's superior intellectual ability, his power to reason, has enabled him to seek and find ways of exploiting his environment that have never been discovered by other animals. This intelligence has largely protected him from the adverse effects of this exploitation.

The overcrowded planet

It is now becoming increasingly clear that the earth's resources are not limitless. The speed of change and the mumber of new methods of extracting and exploiting natural resources continue to increase. The greater "wealth" produced by these new methods enables an ever-growing human population to survive on the planet, although paradoxically it does not raise the overall standard of life for these millions – there are more people today than ever before living at unacceptably low stan-

dards. And as the human population grows it becomes necessary to devise even more ingenious methods of supporting the increased numbers.

There are many signs that the earth's natural systems are fast approaching the limits of their ability to cope with growing human numbers, and advanced technology is rapidly using up natural resources. The innate ability of the environment to recover from attacks on its fabric is also in doubt.

Vanishing forests

Initially, destructive man was no more than a minor nuisance to nature. His tree-felling was of no great consequence because the forests soon returned to their former area and density after man had moved on. This system of agriculture [7] is still practised in some areas: unfortunately, due to expanding human populations and diminishing available land, the slash-and-burn cultivators return to the same patch of forest within two or three years, rather than after a generation or more. As a result, the soil has no time to recover between successive onslaughts.

Apart from the demands of agriculturists, both shifting and static, the world's great forests are under increasingly acute pressure from the timber industry. The tropical rain forests, thought to be the most ancient wild habitats on earth, are fast vanishing. It has been estimated that every minute some 5.6 hectares (14 acres) disappear. The results are manifold. Interesting, attractive and important animal species are exterminated and huge numbers of valuable and potentially valuable plants wiped out. The significance of this is that all man's cultivated vegetables were once wild plants and that many drugs he depends on are plant derivatives.

The rain forests often grow on poor, infertile soil and when the forest cover is removed all that is left is barren desert. Forest destruction has led to erosion, flooding and adverse climatic changes. But still the clearance goes on, partly because trees represent a good source of income but also for more primitive reasons. Man still regards wild forest – the "jungle" – as a challenge and something to be tamed.

Other habitat types are being destroyed

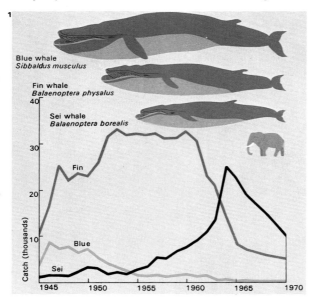

1 The whaling industry is probably the most efficient, ruthless and myopic of all man's industries based on the exploitation of the animal kingdom. So effective has whale-catching become, with its huge factory ships, sonar, radar and spotting aircraft, and its explosive harpoons, that the great whales are easy prey. The whales, properly harvested, could provide an enormous and continuing source of fat and protein. Instead the industry, until recently, greedily overkilled the whales, reducing species after species to near extinction. The graph shows the decline in numbers killed of three species.

Blue whale *Sibbaldus musculus*
Fin whale *Balaenoptera physalus*
Sei whale *Balaenoptera borealis*

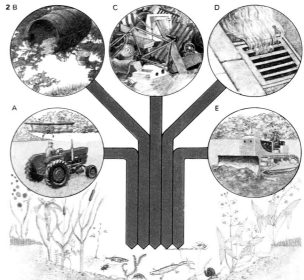

2 Modern man has little use for ponds as watering places for animals, but instead of preserving these thriving ecosystems he has destroyed many of them through pollution and land recla- mation. Pond life is threatened by the excessive use of fertilizers, pesticides and herbicides [A], by the rusty residue from dumped cans and other metal [B] and by rubbish-tipping on a larger scale [C]. Other pollutants reach ponds as run-off from road drains [D] and contain oil, rubber deposits, petroleum and tar. Bulldozers [E] of reclaimers obliterate ponds.

3 Grasslands can support a wide range of wild animals – grazers, browsers, predators and burrowers – all making use of different parts of the system [B, C, D]. Many are migrants, which allows the grass to regenerate. Some grasses are well adapted to withstand fire and one species sends its seedlings spiralling below ground away from the heat [A]. The coming of man and too many domestic cattle has spelt disaster [E]. The large herds compact the soil, overgraze the grass and are both fecund and protected.

4 The destruction of grassland by cattle is particularly severe in the Masai lands of East Africa. And the Masai do not cull their cattle for food but rather regard them as wealth.

around the world for much the same reasons. Other forms of destruction, however, are more insidious because their effects are less immediately obvious. Man has introduced alien animal and plant species into new environments, often with motives no more "evil" than to bring familiar creatures with him to a new land to remind him of home. The consequences have been disastrous when the native fauna and flora have been unable to compete successfully with these introductions.

Man the overhunter and polluter

Man has overhunted with ferocious prodigality, both for food and for sport. He has tried to graze his cattle on semi-arid lands where they do not do well but simply oust the better adapted native wild animals. He has introduced rabbits, donkeys and goats to oceanic islands as a future source of food with an unfailingly tragic effect on the simple and fragile ecosystems. In the sea he has overfished [1] to such an extent that yields have declined drastically.

Man's greatest threat to his environment lies in his technology. Rivers and inland waters, and even some seas, have been stripped of their wildlife by the mounting tide of noxious industrial waste and untreated sewage that is pumped into them. While the Rhine and Danube rivers are grossly polluted, and the Baltic and Mediterranean seas are fast becoming sterile, it is still barely recognized that it costs more to counteract the effects of pollution than to install non-polluting processes initially.

The construction of dams for irrigation or hydro-electric power can have serious ecological drawbacks. The Aswan High Dam in Egypt has blocked the nutritious flow of silt to Lower Egypt and has affected the inshore fisheries in the eastern Mediterranean by reducing the flow of Nile water to the sea; the silt itself, building up behind the dam, is already causing great concern.

The chemical revolution on the farm has increased yields but may ultimately prove self-defeating. An insect pest may become resistant to one pesticide, while some pesticides inadvertently kill the natural predators of the pests.

KEY

Grey squirrel
Sciurus carolinensis

Man's introduction of alien or exotic species has resulted in the destruction of the environment because the effects of the introductions were unforeseen and often disastrous. The grey squirrel, for example, was introduced to Britain from its native America, for no good reason, in the early 19th century. It has now spread [A] to a large part of the UK, ousting the native red squirrel (*Sciurus vulgaris*) [D] which is unable to compete. The grey squirrel is a pest because of its destruction of the bark of trees such as beech, sycamore, larch and oak [B]. It also feeds on cereals [C] as well as buds, shoots, birds' eggs and young birds [E]. In fact it finds Britain an ideal home.

Key to birds

1 Partridge
2 Pheasant
3 Lapwing
4 Skylark
5 Dunnock
6 Wren
7 Blackbird
8 Corn bunting
9 Yellowhammer
10 Chaffinch
11 Robin
12 Blue tit
13 Whitethroat
14 Great tit
15 Songthrush
16 Carrion crow
17 Long-tailed tit
18 Kestrel
19 Greenfinch
20 Moorhen
21 Reed bunting
22 Sedge warbler

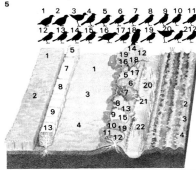

5 As man's numbers have increased, along with his growing technological expertise, so the demand for land has risen. All man's activities seem to need more and more land. And not only is the land itself vanishing but the acres that remain are being farmed ever more intensively. This diagram shows the effect on bird life of the modern trend towards one-crop agriculture and its ruthless removal of such traditional country features as hedges, streams, ponds and trees. These features are thought by some farmers to interfere with efficient agricultural practice in an age when farming is increasingly mechanized. Fortunately farmers are coming to realize that the retention of pockets of natural habitat, even on the most modern of farms, can repay them amply. Apart from aesthetics, the insect-eating birds they support, for example, can be more efficient than any chemical insecticide. Here, 22 species have been reduced to four by the removal of the stream and hedges, and the altered environment has not attracted any new species, either breeding or non-breeding.

Key to symbols

Breeding resident	Breeding resident lost from habitat

6 Early European settlers were lured to the prairies of the American West by a vast, fertile land with native wildlife

and vegetation well adapted to the region's extremes of climate. Man, however, ploughed the land, planted crops (mostly wheat) and introduced domestic cattle. No longer held together by protective grasses, the soil simply blew away during a prolonged drought in the 1930s, leaving a dust bowl. At the time, the farmers were driven away but some land has now been reclaimed.

7 The great forests of West Africa have long been felled by man to provide space for growing crops. The technique shown here is called slash and burn. Because most of the forests grow on rather infertile soils the cleared areas will support a crop for only a year or two. The people then move on to a new area of forest and repeat the process. Unfortunately, due to diminishing forests and increasing human populations, the cultivators are returning ever sooner.

Constructive man

"Nature is the oldest thing on earth," wrote Max Nicholson, formerly Director General of Britain's Nature Conservancy, "but nature reserves are among the youngest." It is only in the past century that active attempts have been made to preserve and nurture nature's treasures. The world's first national park, at Yellowstone, in what are now the states of Wyoming, Montana and Utah, was founded in 1872.

Connections with hunting

Conservation on a limited scale, practised often by lone saints, dreamers and enthusiasts, has a much longer history, however. Faeroe Islanders and Icelandic fishermen have been gathering sea birds and eggs for centuries on a controlled basis that gives them a good harvest while allowing the birds to maintain their population at a stable level, year after year. Henry VIII of England introduced laws governing the seasons during which certain animals could be hunted, while the Inca kings of South America reserved exclusive rights to wear robes made from the wool of the vicuña [6].

Nature conservation, and especially animal conservation, has long been closely connected with hunting. Indeed, the need for conservation arose as the result of man's hunting activities. The first marked impact of man on his environment, as measured by the extinction of animal species, took place some 10,000 to 15,000 years ago. Fossil records show how man in Paleolithic times began to gain ascendancy over his fellow inhabitants of the earth. Whole species disappeared in Britain, Europe and – spectacularly – in the Americas, as man spread down from the Bering Straits, exterminating giant sloths, bears, lions, wolves, bison, mastodons, mammoths, and some enormous birds of prey.

Early conservationists were invariably hunters. Their aim was to husband their quarry and to make sure there would always be an abundance of prey. The early nature reserves were often royal hunting preserves. Nowadays the conservation movement is much more broadly used, but hunters still play an important part.

Modern conservation began to gather momentum about the turn of the century.

Britain's largest nature conservation body, the Royal Society for the Protection of Birds, was set up in 1899, while in northern Europe the first national parks were established in 1909, in Sweden.

Internationally, there are four main bodies concerned with nature conservation. They are the Fauna Preservation Society (founded as the "Society for the Preservation of the Wild Fauna of the [British] Empire" in 1903); the International Council for Bird Preservation (founded in 1922); the International Union for Conservation of Nature and Natural Resources (IUCN – founded in 1948); and the World Wildlife Fund (founded in 1961). IUCN is in some sense the senior of these four bodies. The World Wildlife Fund has branches in nearly 30 countries and is a voluntary, fund-raising organization whose main aim is to produce money to finance the conservation programmes worked out by IUCN.

Safeguarding the habitat

On a national level there are scores of conservation organizations; in Britain, for

1 Wildlife conservation is locally achieved by the setting up of reserves and national parks. To live a natural life most animals need a fairly large area in which conditions are as close as possible to their wild habitat and in which human disturbance and development is minimal. This does not necessarily mean the exclusion of all humans; many reserves in North America, and this one in East Africa, are visited by many tourists. Also, it is often necessary to have a staff of game wardens to guard against poachers and encroachment by wood cutters, farmers or any other kind of developer.

3 Yellowstone National Park in the United States was the first such park in the world and is still one of the biggest and best. It is visited by millions of people each year who have the chance to see creatures such as this bull moose (*Alces alces*). At certain times of year, city-like traffic jams develop on the park's road system. The future prospect is that access to the world's nature reserves will have to be controlled ever more strictly if damage to the natural habitat is to be avoided. In the developing countries, increasing pressure is being felt on wildlife reserves as city people seek spiritual refreshment from the strain of urban life.

2 Nature reserves need not be large. To protect rare orchids like these, or a single plant threatened with extinction, such as the Indonesian *Rafflesia*, areas smaller than an acre can be set aside.

4 The eggs of the green turtle are a great delicacy in Asia. In Malaysia, conservationists run a hatchery to protect enough eggs to ensure that turtles will keep returning to the beaches.

example, there are local conservation trusts covering every county, besides several large bodies such as the National Trust which have nature conservation as one of their aims; there are also a host of large and small societies devoted to the study of birds or lichens or reptiles or some other branch of natural history, and each is actively involved in conservation.

Following the impetus given by the establishment of Yellowstone National Park in the United States [3], the method of saving animals and plants by safeguarding the habitat itself has spread widely. There are now some 1,200 areas listed in the United Nations List of National Parks and Equivalent Reserves; these parks, in a hundred countries, are paralleled by countless smaller areas with similar objectives around the world. They include game reserves [1], nature reserves, forest reserves, sites of special scientific interest, areas of outstanding natural beauty and marine reserves. The world's largest national park is at present the North-East Greenland National Park, of 8,000 square kilometres (2 million acres),

while at the other end of the scale are many nature reserves of less than an acre, often for the protection of a single plant [2]. In Indonesia, for instance, reserves have been created to protect *Rafflesia*, the largest flower in the world, but this method of conservation still makes any species vulnerable to chance destruction.

The vital task of conservation

People everywhere are coming to realize the importance and urgency of saving wild animals, wild plants and wild places for future generations [4, 5, 7, 8]. Nearly 1,000 different kinds of animals are threatened with extinction, and perhaps 20,000 plants (ten per cent of all the world's flowering plants).

The 1972 Conference on the Human Environment in Stockholm, held under UN auspices, and the subsequent establishment of the United Nations Environment Programme, are indications that the nature conservation movement has now reached maturity and that conservation is being accepted as one of the most important tasks of our time.

Polar bears that approached too close to human settlements in the Arctic regions as scavengers were once shot. Today they are drugged, lifted by helicopter and carried to places of safety.

5 Greater bird of paradise
Paradisaea apoda

6 Alpaca
Llama pacos

Llama
Llama peruana

Vicuna
Vicugna vicugna

5 The greater bird of paradise is one of a number of birds endangered by the demand for their brightly coloured feathers. Strong conservation laws are needed to control this trade.

6 The members of the camel family found in South America are the alpaca, llama and vicuña together with a fourth species – the guanaco. Only the vicuña and the guanaco are wild; the other two animals have long been domesticated. The vicuña once numbered millions in its home in the High Andes and was carefully conserved by the Inca kings who alone were allowed to wear cloth made from its wool. Since the arrival of the Spanish, however, the animal has been heavily exploited for its valuable wool, which is considered the finest in the world. By 1970, vicuña numbers had dwindled to only a few thousand. But since then, reserves have been established to protect the species in Peru and Bolivia, and conservation measures have produced a slow increase in numbers.

7

7 The orang-utan is endangered by the demands both of zoos and of private people who consider it a status-symbol pet. Conservation authorities in Borneo and Sumatra (the only places where it is found) now confiscate any animals illegally owned and send them to centres like this one in Sumatra, where they are retrained for life in the wild before being released.

8 Osprey
Pandion haliaetus

8 Ospreys at Loch Garten, Scotland, are protected by a 24-hour watch maintained on their nesting tree during the breeding season. They are one of the examples of protected birds of prey. Such birds used to be persecuted as a result of a widespread, but erroneous, view that they are pests that take domestic animals or compete with human hunters for the same prey. In fact, a healthy population of birds of prey is necessary to keep other species in balance.

251

Zoos and botanic gardens

Whether the animals ever actually went into the Ark two by two is a matter for conjecture. But without doubt mankind has been making collections of living animals for a very long time. The ancient Egyptians, the Chinese and the Romans all had their court menageries; so too did the kings of England – at the Tower of London – for some 600 years until the nineteenth century. The first zoo existed in Austria in the fifteenth century; others followed, until in 1752 Emperor Franz I founded the Garden at Schönbrunn outside Vienna, the oldest zoo still in existence. The Madrid Zoo opened in 1775, the Paris Zoo in 1793 and the London Zoo in 1826.

Zoo development
The disadvantage of older zoos is that they were designed more for the convenience of spectators than that of animals [Key]. As a result these once-proud establishments now have to struggle to find the increasingly large sums of money necessary to bring their enclosures up to the more enlightened standards of today [1, 2, 3].

Many German zoos were destroyed during Word War II and so had to be completely rebuilt. Fortunately, this process coincided with new developments in zoo design and architecture and these zoos are now among the best in the world.

A more open kind of zoo is the parklike establishment, typified by the Hamburg Zoo, which was laid out by the great Karl Hagenbeck in 1907; the Rome Zoo, designed by the same architect; the Zoological Society of London's Whipsnade Park; and the San Diego Wild Animal Park. There is another new kind of zoo, a commercially motivated development of the park type, whose advantages can best be appreciated by first examining what purposes zoos should serve and indeed whether they are justifiable at all.

While the early zoos existed purely to display exciting or exotic animals for the amusement of their royal owners, nowadays responsible zoologists and zoomen (the two are not necessarily synonymous) generally agree that it is only justifiable to keep animals in captivity for education, conservation, or scientific research. In other words, a zoo should have a serious scientific purpose, with the exhibition of animals purely as "entertainment" or for status very much a subsidiary factor in zoo planning.

Education and research
It is not difficult for a zoo to run a good educational programme. Much depends on adequate labelling of the exhibits, but the zoo can also publish and distribute informative guidebooks and organize school lecture tours. Education can often succeed by simply displaying the animals attractively; it is not educational for the public to see decaying food inside dirty cages or filthy water opaque with excrement.

The contribution made by zoos to conservation [5] is more uneven. It is said that zoos can help by breeding rare animals [7] and in due course reintroducing them to the wild. However, there are not many successful examples. The European bison – once reduced to a few captive specimens but now thriving in more or less natural conditions on the Polish-Russian border – is one. Another is the successful breeding of ne-ne or Hawaiian geese at the Wildfowl Trust in

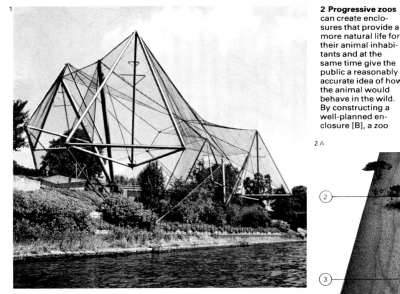

1 The modern aviary at London Zoo, designed by Lord Snowdon, manages to combine good architecture with satisfactory surroundings for the creatures that live inside. This is proved by the number of birds that regularly breed within its confines. But it is not only impressive to look at from the outside; it is a walk-through aviary which allows visitors to enter the birds' own environment, however briefly. This is an enjoyable – and educational – experience, even for the hundreds of families who keep birds as pets, and enables visitors to learn a great deal about the birds' habits.

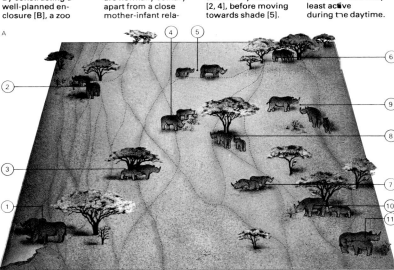

2 Progressive zoos can create enclosures that provide a more natural life for their animal inhabitants and at the same time give the public a reasonably accurate idea of how the animal would behave in the wild. By constructing a well-planned enclosure [B], a zoo could approximate the surroundings in which a female black African rhinoceros and her young offspring would go through a typical 24-hour activity pattern. In their savanna grasslands, [A], these animals are normally solitary apart from a close mother-infant relationship [11]. An intrusive adult male will be driven off [9]. A typical day might begin with a drink [1] about 2am. Then the mother and young would follow well-worn paths to areas where they feed on scrubby plants and woody herbs [2, 4], before moving towards shade [5]. From 9am until late afternoon, they rest and sleep [3, 6,], often cooling themselves by wallowing in dust-filled depressions [7]. Their activity increases towards nightfall [8, 10]. Unfortunately for zoo visitors, rhinos are naturally least active during the daytime.

2 A

3 Rhinos need hard ground to walk on (soft, muddy pasture is unnatural to them) and they need dusty wallows to lie in. They also need trees or man-made structures to rest beneath or rub against. Visitors to zoos sometimes get the impression that an animal's enclosure is in some way unsuitable merely because it does not correspond with the kind of surroundings they would like to live in themselves. Good zoos explain each animal's needs.

B

Gloucestershire, England, and their reintroduction to Hawaii [4].

Many good zoos, like good botanic gardens [8], have substantial programmes of scientific research. Often this research tells us more about the behaviour of captive animals than about their wild relatives. Nevertheless, this is important if it means that zoo animals will live longer, happier lives than they would otherwise. Some zoos, especially New York's Bronx Zoo, maintain an active programme of field research, often with great emphasis given to conservation.

Conventional city zoos are constantly improving their facilities, partly to make them more attractive to visitors [9] but even more with the animals' health and happiness in mind. Some years ago, London Zoo spent £250,000 on a rhino and elephant pavilion that was unsuitable for the animals inside because insufficient thought was given to their needs. The same mistake would not be made today. A new enclosure for big cats at the same zoo, completed to mark the Zoological Society of London's 150th anniversary in 1976, is a marvellous example of how to dis-play the optimum number of animals in a confined space so that the needs of both public and animals are best served.

Safari parks

One modern zoo development, that looks at first sight to be an excellent one, has some retrograde aspects. It is the mushrooming growth of the drive-through "safari park". Unfortunately, too many safari parks are owned and run by organizations that are commercially motivated and education, conservation, and indeed even animal husbandry are sometimes secondary to the profit motive. Many keepers are young, inexperienced and poorly paid and unless someone is interested or knowledgeable enough to train them they soon leave. The animals' welfare may be subordinated to the public's enjoyment. Where this happens, the animals can develop unnatural behaviour patterns.

The good zoo, therefore, will increasingly be supported out of public funds, or be a non-profit-making society. It will concentrate on education and science purposes and will specialize in small groups of animals.

The original raven cage at Regent's Park, London, now empty, symbol-izes the historical zoo concept – that of a menagerie where it was more im-portant to create an interesting enclosure than to consider the creature inside.

4 Ne-ne *Branta sandvicensis*

4 The ne-ne, or Hawaiian goose, illustrates how zoos can play a part in conserving wild creatures. This bird was declining in its native Hawaii and was in danger of becoming extinct when some were sent to Britain. There, with great expertise, they were bred in captivity until some could be sent back to their native habitat. Several hundred have so far been reintroduced in this way.

5 The Arabian oryx is a fleet-footed inhabitant of the Arabian peninsula; its misfortune is that it was a favourite prey of hunters, who, armed with modern firearms and mounted on Jeeps, not horses, have probably exterminated the animal in the wild. Luckily, a few were saved in time and now there exist flourishing captive herds at two American zoos; the first was in Phoenix, Arizona.

5

Arabian oryx
Oryx leucoryx

6 A lone gorilla or other animal was once a common sight in zoos that felt they had to have a "representative" collection with many different animals. Not only were they without the knowledge needed to cope with thousands of species but they also tended to collect solitary individual beasts. Now it is considered better to have breeding pairs and family groups of fewer creatures.

7 Breeding is usually regarded as the yardstick for judging a zoo – on the reasonable assumption that an animal is unlikely to breed if it is not happy. But some animals defy their keepers' best efforts. Cheetahs, for example, used never to breed in zoos. In several zoos the problem has now been solved. Part of the solution requires the separation of the males and females until the time is ripe for them to mate.

8

8 Botanic gardens are for plants what zoos are for animals. Both are collections of wild species. Like zoos, botanic gardens aim to "breed" their charges and to educate their visitors. They conduct much scientific research and can cultivate endangered species. Like zoos, they can act as gene banks (or seed banks) which could be important in producing new varieties of useful plants. This is the Victorian palm house at Kew Gardens, England.

9

9 Young polar bears are potent crowd-pullers. However richly endowed zoos may be, nearly all depend largely on income from the public. Many zoos have found that attendance increases dramatically when they have a notable animal birth. It is a happy coincidence that baby polar bears are so popular, for their species is endangered. If a zoo can breed them successfully, it can fulfil serious aims and benefit financially as well.

253

Table of animal taxonomy

Modern systems of animal (and plant) classification are based on the work of the Swedish botanist Carl Linnaeus (1707–1778). Because his knowledge of animals and plants was limited, his classification has been extensively changed, but its principles remain the same.

Linnaeus divided living things into groups of diminishing size—class, order, genus and species—on the basis of common characteristics. Thus members of the same species have most characteristics in common and can breed successfully, while members of the same class have the fewest common characteristics. Normally only members of the same species produce viable offspring and for this reason the species can be considered as the taxonomic unit of breeding. Taxonomists have since added the phylum—a group of one or more genera.

The system given here is used throughout *The Natural World* and is one of a number of widely accepted systems. The construction of animal classification systems is so complex, and so subject to individual judgement, that in 1898 an International Commission on Zoological Nomenclature was set up. The commission is the authority that decides what the thousands of new species discovered each year shall be called.

Plant classification is covered on pages 38 to 39.

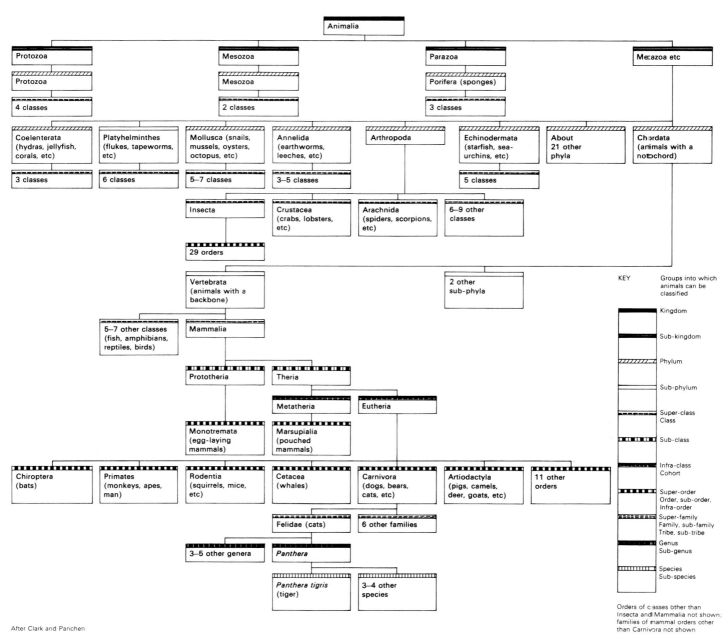

After Clark and Panchen

INDEX

269

Picture Credits

Every endeavour has been made to trace copyright holders of photographs appearing in *The Joy of Knowledge*. The publishers apologize to any photographers or agencies whose work has been used but has not been listed below.

Credits are listed in this manner: [1] page numbers appear first, in bold type; [2] illustration numbers appear next, in parentheses; [3] photographers' names appear next, followed where applicable by the names of the agencies representing them.

16–7 Eric Hosking. **18–9** O. S. F./ Bruce Coleman Ltd. **20–1** [2] Institute of Molecular Evolution. **22–3** [2] Jane Burton/Bruce Coleman Ltd. **26–7** [2] Gene Cox/Bruce Coleman Ltd; [3] Gene Cox. **28–9** [Key] M. H. F. Wilkins; [6] Gene Cox. **30–1** [5] Mansell Collection. **36–7** [6] C. James Webb; [7] C. James Webb; [8] C. James Webb. **40–1** [5] Heather Angel; [11] Heather Angel. **42–3** [1] Heather Angel; [3] Dr D. A. Reid; [5] Dr D. A. Reid; [6] Dr D. A. Reid; [8] Brian Hawkes. **44–5** [1] University of Leeds: Dr Eva Frei and Professor Preston. **48–9** [2] Heather Angel; [3] Heather Angel; [4A] Eric Hosking; [4B] R & C Foord/N.H.P.A. **50–1** [3] Heather Angel. **52–3** [5] A-Z Botanical Collection; [8] Heather Angel; [10] Heather Angel; [11] P. H. Ward/Natural Science Photos. **54–5** [4] Laboratory of Tree-Ring Research, University of Arizona; [9A] A-Z Botanical Collection; [9B] Botanical Collection. **56–7** [4] Bruce Coleman/Bruce Coleman Ltd; [5] W. F. Davidson; [6] Arne Schmitz/Bruce Coleman Ltd; [10] Claude Nardin/Jacana; [11] F. H. C. Birch/Sonia Halliday. **66–7** [9] Ron Boardman. **76–7** [5D] P. H. Ward/Natural Science Photos; [6] Francisco Erize/Bruce Coleman Ltd. **80–1** [7] Ronan Picture Library. **82–3** [1] O.S.F./Bruce Coleman Ltd; [2] Allan Power/Bruce Coleman Ltd; [4] Jane Burton/Bruce Coleman Ltd; [6] Allan

Power/Bruce Coleman Ltd. **86–7** [8] Gene Cox/Bruce Coleman Ltd; [10] K. S. Seymour. **88–9** [3] Oxford Scientific Films; [4] Heather Angel; [5] Heather Angel; [7] ZEFA; [8] Heather Angel; [9] Australian News & Information Bureau; [10] Dr J. D. George/British Museum [Natural History]. **90–1** [4] Oxford Scientific Films; [5] Jane Burton/Bruce Coleman Ltd; [7A] S. C. Bisserot/Bruce Coleman Ltd; [8] Heather Angel; [9] Heather Angel; [10] Bruce Coleman/Bruce Coleman Ltd; [11] Isobel Bennett/Natural Science Photos. **92–3** [6] Dr D. P. Wilson. **94–5** [3] J. L. Mason/Ardea Photographics; [7A] Dr D. P. Wilson; [7B] Dr D. P. Wilson. **100–1** [2] Heather Angel; [4] O.S.F./Bruce Coleman Ltd; [5] O.S.F./Bruce Coleman Ltd; [9A] Jane Burton/Bruce Coleman Ltd; [9B] Jane Burton/Bruce Coleman Ltd. **102–3** [3] P.H. Ward/Natural Science Photos; [4] A. Bannister/ N.H.P.A.; [6] P. H. Ward/Natural Science Photos; [8] P.H. Ward/Natural Science Photos. **118–9** [3] Dr D. P. Wilson; [4] Heather Angel. **122–3** [5] Dr D. P. Wilson; [6] Dr D. P. Wilson; [11] Heather Angel; [13] Heather Angel. **126–7** [2] Jane Burton/Bruce Coleman Ltd; [3] Dr D. P. Wilson. **136–7** [4] P. Kirkpatrick/Frank W. Lane; [8] N. Myers/Bruce Coleman Ltd. **150–1** [2] L. Lee Rue III/Bruce Coleman Ltd; [4] Nina Leen/Life © Time Inc.

1976/Colorific. **164–5** [1] J. Whitman/Ardea Photographics; [6] Jeff Foott/Bruce Coleman Ltd. **172–3** [4] Ron Boardman. **174–5** [6A] W. R. Hamilton/Imitor; [6B] W. R. Hamilton/Imitor; [6C] W. R. Hamilton/Imitor. **176–7** [Key] Institute of Geological Sciences; [1B] Oxford Scientific Films; [5] A. C. Waltham; [9A] W. R. Hamilton/Imitor; [9B] C2M/Natural Science Photos. **178–9** [5] Heather Angel; [6] Peder Aspen; [9] Heather Angel; [13] W. F. Davidson. **180–1** [5] James Allan. **192–3** [2] Mary Evans Picture Library. **194–5** [1] A. J. Sutcliffe/Natural Science Photos; [2] C. J. Pruett/Natural Science Photos; [3] M. Stanley Price/Natural Science Photos; [4] Dick Brown/Natural Science Photos; [5] A. Leutscher/Natural Science Photos; [6] P. H. Ward/Natural Science Photos. **200–1** [4] Lyn Cawley. **202–3** [1] Francisco Erize/Bruce Coleman Ltd. **204–5** [4] J. A. Grant/ Natural Science Photos; [5] N. McFarland/Natural Science Photos. **210–1** [2] Hans & Judy Beste/Ardea Photographics; [4] John Brownlie/Bruce Coleman Ltd. **212–3** [4] P. H. Ward/Natural Science Photos. **218–9** [Key] Picturepoint; [2] P. Morris/Ardea Photographics; [4] C. Banks/Natural Science Photos; [9] Jane Burton/Bruce Coleman Ltd. **226–7** [3]

Brian Hawkes. **[230–1** [3] Picturepoint. **234–5** [Key] R. Scott/Institute of Terrestrial Ecology; [2c] Eric Hosking; [3A] P. Morris/Ardea Photographics; [6A] Ivan Polunin/N.H.P.A. **236–7** [1A] Heather Angel; [1B] Dr D. P. Wilson; [1c] Joyce Pope; [4] Isobel Bennett/Natural Science Photos; [5] P. Scoones/Photo Aquatics. **238–9** [Key] Dr D. P. Wilson; [3] Peter David/Seaphot; [4] Hans Dossenbach/Natural Science Photos. **240–1** [2] Christian Petron/Seaphot; [3] Seaphot; [4] Christian Petron/Seaphot; [5] Allan Power/Bruce Coleman Ltd. **242–3** [1] David Strickland; [6] Bill Eppridge/Life © Time Inc. 1976/Colorific; [8] Douglas Botting; [10] M. Stanley Price/Natural Science Photos. **246–7** [Key] Kim Sayer. **248–9** [4] Hans Reinhard/Bruce Coleman Ltd. **250–1** [Key] Joe Rychetnik/Transworld; [1] Horst Munzig/Susan Griggs Picture Agency; [2] Rex Graham Reserve, Mildenhall; [3] L. Lee Rue IV/Bruce Coleman Ltd; [4] Robert Schroeder/Bruce Coleman Ltd; [7] Nigel Sitwell. **252–3** [Key] Zoological Society of London; [1] Zoological Society of London; [3] Spectrum Colour Library; [6] Zoological Society of London; [7 Spectrum Colour Library; [8] Spectrum Colour Library; [9] Spectrum Colour Library.

Artwork Credits

Art Editors
Angela Downing; George Glaze; James Marks; Mel Peterson; Ruth Prentice; Bob Scott

Visualizers
David Aston; Javed Badar; Allison Blythe; Angela Braithwaite; Alan Brown; Michael Burke; Alistair Campbell; Terry Collins; Mary Ellis; Judith Escreet; Albert Jackson; Barry Jackson; Ted Kindsey; Kevin Maddison; Erika Mathlow; Paul Mundon; Peter Nielson; Patrick O'Callaghan; John Ridgeway; Peter Saag; Malcolme Smythe; John Stanyon; John Stewart; Justin Todd; Linda Wheeler

Artists
Stephen Adams; Geoffrey Alger; Terry Allen; Jeremy Alsford; Frederick Andenson; John Arnold; Peter Arnold; David Ashby; Michael Badrock; William Baker; John Barber; Norman Barber; Arthur Barvoso; John Batchelor; John Bavosi; David Baxter; Stephen Bernette; John Blagovitch; Michael Blore; Christopher Blow; Roger Bourne; Alistair Bowtell; Robert Brett; Gordon Briggs; Linda Broad; Lee Brooks; Rupert Brown; Marilyn Bruce; Anthony Bryant; Paul Buckle; Sergio Burelli; Dino Bussetti; Patricia Casey; Giovanni Casselli; Nigel Chapman; Chensie Chen; David Chisholm; David Cockcroft; Michael Codd; Michael Cole; Terry Collins; Peter Connelly; Roy Coombs; David Cox; Patrick Cox; Brian Cracker; Gordon Cramp; Gino D'Achille; Terrence Daley; John Davies; Gordon C. Davis; David Day; Graham Dean; Brian Delf; Kevin Diaper; Madeleine Dinkel; Hugh Dixon; Paul Draper; David Dupe; Howard Dyke; Jennifer Eachus; Bill Easter; Peter Edwards; Michael Ellis; Jennifer Embleton; Ronald Embleton; Ian Evans; Ann Evens; Lyn Evens; Peter Fitzjohn; Eugene Flurey; Alexander Forbes; David Carl Forbes; Chris Fosey; John Francis; Linda Francis; Sally Frend; Brian Froud; Gay Galfworthy; Ian Garrard; Jean George; Victoria Goaman; David Godfrey; Miriam Golochoy; Anthea Gray; Harold Green; Penelope Greensmith; Vanna Haggerty; Nicholas Hall; Horgrave Hans; David Hardy; Douglas Harker; Richard Hartwell; Jill Havergale; Peter Hayman; Ron Haywood; Peter Henville; Trevor Hill; Garry Hinks; Peter Hutton; Faith Jacques; Robin Jacques; Lancelot Jones; Anthony Joyce; Pierre Junod; Patrick Kaley; Sarah Kensington; Don Kidman; Harold King; Martin Lambourne; Ivan Lapper; Gordon Lawson; Malcolm Lee-Andrews; Peter Levaffeur; Richard Lewington; Brian Lewis; Ken Lewis; Richard Lewis; Kenneth Lilly; Michael Little; David Lock; Garry Long; John Vernon Lord; Vanessa Luff; John Mac; Lesley MacIntyre; Thomas McArthur; Michael McGuinness; Ed McKenzie; Alan Male; Ben Manchipp; Neville Mardell; Olive Marony; Bob Martin; Gordon Miles; Sean Milne; Peter Mortar; Robert Morton; Trevor Muse; Anthony Nelthorpe; Michael Neugebauer; William Nickless; Eric Norman; Peter North; Michael O'Rourke; Richard Orr; Nigel Osborne; Patrick Oxenham; John Painter; David Palmer; Geoffrey Parr; Allan Penny; David Penny; Charles Pickard; John Pinder; Maurice Pledger; Judith Legh Pope; Michael Pope; Andrew Popkiewicz; Brian Price-Thomas; Josephine Rankin; Collin Rattray; Charles Raymond; Alan Rees; Ellsie Rigley; John Ringnall; Christine Robbins; Ellie Robertson; James Robins; John Ronayne; Collin Rose; Peter Sarson; Michael Saunders; Ann Savage; Dennis Scott; Edward Scott-Jones; Rodney Shackell; Chris Simmonds; Gwendolyn Simson; Cathleen Smith; Les Smith; Stanley Smith; Michael Soundels; Wolf Spoel; Ronald Steiner; Ralph Stobart; Celia Stothard; Peter Sumpter; Rod Sutterby; Allan Suttie; Tony Swift; Michael Terry; John Thirsk; Eric Thomas; George Thompson; Kenneth Thompson; David Thorpe; Harry Titcombe; Peter Town; Michael Trangenza; Joyce Tuhill; Glenn Tutssel; Carol Vaucher; Edward Wade; Geoffrey Wadsley; Mary Waldron; Michael Walker; Dick Ward; Brian Watson; David Watson; Peter Weavers; David Wilkinson; Ted Williams; John Wilson; Roy Wiltshire; Terrence Wingworth; Anne Winterbotham; Albany Wiseman; Vanessa Wiseman; John Wood; Michael Woods; Owen Woods; Sidney Woods; Raymond Woodward; Harold Wright; Julia Wright

Studios
Add Make-up; Alard Design; Anyart; Arka Graphics; Artec; Art Liaison; Art Workshop; Bateson Graphics; Broadway Artists; Dateline Graphics; David Cox Associates; David Levin Photographic; Eric Jewel Associates; George Miller Associates; Gilcrist Studios; Hatton Studio; Jackson Day; Lock Pettersen Ltd; Mitchell Beazley Studio; Negs Photographic; Paul Hemus Associates; Product Support Graphics; Q.E.D. [Campbell Kindsley]; Stobart and Sutterby; Studio Briggs; Technical Graphics; The Diagram Group; Tri Art; Typographics; Venner Artists

Agents
Artist Partners; Freelance Presentations; Garden Studio; Linden Artists; N.E. Middletons; Portman Artists; Saxon Artists; Thompson Artists

271